Soybean: Production, Breeding and Management

Soybean: Production, Breeding and Management

Edited by Albert Marinelli

SYRAWOOD
PUBLISHING HOUSE
New York

Published by Syrawood Publishing House,
750 Third Avenue, 9th Floor,
New York, NY 10017, USA
www.syrawoodpublishinghouse.com

Soybean: Production, Breeding and Management
Edited by Albert Marinelli

International Standard Book Number: 978-1-68286-381-7 (Hardback)

Cataloging-in-publication Data

Soybean : production, breeding and management / edited by Albert Marinelli.
p. cm.
Includes bibliographical references and index.
ISBN 978-1-68286-381-7
1. Soybean. 2. Soybean--Breeding. 3. Soybean products. 4. Soybean--Genetic aspects. I. Marinelli, Albert.
SB205.S7 S67 2017
633.34--dc23

Printed in China.

TABLE OF CONTENTS

PREFACE

Soybeans are part of the family of legumes. Soybean is commercially processed into various items such as OPI, soymilk, etc. This book elucidates the concepts and innovative models around prospective developments with respect to soybean management. Soybean farming is dependent on weather quality as well as variety of seed ranging from high yield varieties to herbicide resistant seeds. This book is compiled in such a manner, that it will provide in-depth knowledge about theory and practice of soybean harvesting. From theories to research to practical applications, case studies related to all contemporary topics of relevance have been included in this book. The extensive content of this book provides the readers with a thorough understanding of the cultivation of soybean, from production to storage, as well as soybean breeding and management.

This book unites the global concepts and researches in an organized manner for a comprehensive understanding of the subject. It is a ripe text for all researchers, students, scientists or anyone else who is interested in acquiring a better knowledge of this dynamic field.

I extend my sincere thanks to the contributors for such eloquent research chapters. Finally, I thank my family for being a source of support and help.

Editor

1

Reference Gene Selection for Quantitative Real-Time PCR Normalization in *Reaumuria soongorica*

Xia Yan[1,2], Xicun Dong[3], Wen Zhang[2], Hengxia Yin[2], Honglang Xiao[1], Peng Chen[4], Xiao-Fei Ma[2]*

1 Key Laboratory of Inland River Ecohydrology, Cold and Arid Regions Environmental and Engineering Research Institute, Chinese Academy of Sciences, Lanzhou, P. R. China, **2** Key Laboratory of Stress Physiology and Ecology in Cold and Arid Regions of Gansu Province, Cold and Arid Regions Environmental and Engineering Research Institute, Chinese Academy of Sciences, Lanzhou, P. R. China, **3** Department of Radiobiology, Institute of Modern Physics, Chinese Academy of Sciences, Lanzhou, P. R. China, **4** Faculty of Plant Science and Technology, Huazhong Agricultural University, Wuhan, P. R. China

Abstract

Despite its superiority for evaluating gene expression, real-time quantitative polymerase chain reaction (qPCR) results can be significantly biased by the use of inappropriate reference genes under different experimental conditions. *Reaumuria soongorica* is a dominant species of desert ecosystems in arid central Asia. Given the increasing interest in ecological engineering and potential genetic resources for arid agronomy, it is important to analyze gene function. However, systematic evaluation of stable reference genes should be performed prior to such analyses. In this study, the stabilities of 10 candidate reference genes were analyzed under 4 kinds of abiotic stresses (drought, salt, dark, and heat) within 4 accessions (HG010, HG020, XGG030, and XGG040) from 2 different habitats using 3 algorithms (geNorm, NormFinder, and BestKeeper). After validation of the ribulose-1,5-bisphosphate carboxylase/oxygenase large unite (*rbcL*) expression pattern, our data suggested that histone H2A (*H2A*) and eukaryotic initiation factor 4A-2 (*EIF4A2*) were the most stable reference genes, cyclophilin (*CYCL*) was moderate, and elongation factor 1α (*EF1α*) was the worst choice. This first systematic analysis for stably expressed genes will facilitate future functional analyses and deep mining of genetic resources in *R. soongorica* and other species of the *Reaumuria* genus.

Editor: Enrique Hernandez-Lemus, National Institute of Genomic Medicine, Mexico

Funding: This work was supported by the "One Hundred Talents" Project of the Chinese Academy of Sciences (Grant No. 29Y127E71) and National Natural Science Foundation of China (No. 91125025). The funders had no role in study design, data collection and analysis, decision to publish, or preparation of the manuscript.

Competing Interests: The authors have declared that no competing interests exist.

* Email: maxiaofei@lzb.ac.cn

These authors contributed equally to this work.

Introduction

Gene expression analysis is increasingly important in many fields of biological research. Understanding gene expression patterns is expected to provide insight into complex regulatory networks and will help to identify which genes are relevant to new biological processes. Recently developed methods to measure transcript abundance have been frequently applied in many model biological systems; however, these approaches require the same normalization procedures as traditional mRNA quantification methods. Real-time quantitative polymerase chain reaction (qPCR) is one of the most common technologies used for gene expression and transcriptome analysis and is characterized by high sensitivity and specificity, good reproducibility, a widely dynamic quantification range, and high-throughput capacity for a limited number of target genes [1–3]. Despite its superiority over other methods available for evaluating gene expression, qPCR remains underused due in part to conflicting results and weak repeatability. Traditional reference genes, such as actin (*ACT*), ubiquitin (*UBQ*), alpha-tubulin (*TUA*), eukaryotic initiation factor 4A-2 (*EIF4A2*), 60S ribosomal protein L2 (*L2*), TIP41-like protein (*TIP41*), elongation factor 1-alpha (*EF1α*), cyclophilin (*CYCL*), histone H2A (*H2A*) and DNAJ-like protein (*DNAJ*), are mostly used in model plants and crops [4–6] and do not always maintain stable expression levels among different tissues, experimental conditions [7–12], or species [7,11–16]. Systematic validations of reference genes have mainly focused on models and important crop species, such as Arabidopsis [17], rice [6], poplar [18], soybean [19], wheat [14], barley [20], tomato [21], grape [22], and potato [5]. However, with the increasing importance of agronomy and ecology, greater numbers of non-model plants are being assessed with molecular functional analyses, and it is crucial to identify proper reference genes for new species. To date, reference gene selection has been performed for some non-model plants like bamboo [23], peach [15], *Caragana intermedia* [12], and eggplant [24], but seldom in desert plants.

Reaumuria (Tamaricaceae) plants are perennial xeric shrubs widely distributed in arid regions in North Africa, Asia, and southern Europe. *R. soongorica* (2n = 22, 778 Mb genome size) [25] is a constructive and dominant species of desert ecosystems in central Asia [26]. It has evolved typical phenotypes of desert plants, such as an extremely thick cuticle; hollow stomata; specialized leaf shape; and a root system that reduces the transpiration rate, increases water use efficiency, and maintains stem vigor to survive desiccation [27,28]. Thus, *R. soongorica* is a good species to study the molecular mechanisms of drought adaptation. Many studies have attempted to elucidate the mechanisms of the adaptation and tolerance of *R. soongorica* under the extreme drought conditions [29–33], which suggested its more and more importance of ecology. It is in urgent to systematically identify appropriate references genes for *R. soongorica*, for better understanding of the processes and molecular functions of its potential gene resources since its transcriptome analysis had been released [34].

Three statistical algorithms packages have been developed to assess reference gene stability: geNorm [35], NormFinder [11], and BestKeeper [36]. In this study, the stability of 10 candidate reference genes used in many plants (*ACT, UBQ, TIP41, CYCL, EF1α, EIF4A, H2A, L2, DNAJ*, and *TUA*) were assessed based on their expression abundance in two contrasting populations under four stress conditions. To validate the best candidate reference genes, the expression levels of the ribulose-1,5-bisphosphate carboxylase/oxygenase (*rbcL*) homolog were analyzed in comparison to 10 reference genes. This work will facilitate future studies on *R. soongorica* gene expression, speed up functional analyses of special genetic resources, and improve our understanding of the molecular mechanisms of drought adaptation.

Materials and Methods

Ethics Statement

R. soongorica is a species widely distributed in Shashichang, Jingtai County, Gansu province and other arid regions, and it is not included on any list of endangered or protected species. Before collecting the samples, oral permission was obtained from the local management of forestry after sending introduction letters from the CAREERI (Cold and Arid Regions Environmental and Engineering Research Institute, Chinese Academy of Sciences).

Plant Materials and Treatments

Seeds of four accessions (HG010, HG020, XGG030, and XGG040) were collected from two different habitats (Honggu [HG] N36°15′20.16″, E103°1′29.82″, annual precipitation [AP] ~400 mm based on WorldClim database 1950–2000; Xiaogangou [XGG] N36°54′21.60″, E103°53′31.44″, AP ~150 mm) were sampled in the year 2011. After 2 years in frozen storage at −20°C to completely break physiological dormancy, the seeds were surface sterilized with 0.1% sodium hypochlorite and rinsed with distilled water. The sterilized seeds were then grown in MS-medium with photosynthetically active radiation under long day condition (16/8-hour light-dark cycles); the relative humidity was 75%, and the temperatures were 25°C in the light and 22°C in the dark. After 2 weeks, 10 seedlings from each accession were collected for RNA extraction. For drought stress, seedlings were subjected to 15% PEG (polyethylene glycol) for 1 hour. Heat stress treatment was performed at 38°C. For salt stress, plants were treated with 100 mM NaCl for 3 hours. For dark stress, plants were wrapped with foil paper for 2 hours. Control plants were kept at 25°C during light hours on MS-medium.

Total RNA Isolation and cDNA Synthesis

Total RNA was extracted using an E.Z.N.A. plant RNA kit (Omega Bio-tek), and the remaining DNA was removed by RNase-free DNase (Omega Bio-tek) according to the instruction manual. Total RNA concentration and purity were determined with ratio optical density (OD) 260/280 by NanoDrop1000 (Thermo Scientific). All the samples passed QC with the ratio OD260/280 between 1.9 and 2.2 and ratio OD260/230 <2.0. RNA integrity was verified by 1.5% agarose gel electrophoresis with two clear bands of 28S/18S ribosomal RNA. In each treatment and control, three biological replicates of RNA samples were extracted from at least 10 seedlings of each genotype first, and then mixed equally as genotype representative sample. For each representative sample, 1 µg total RNA was reverse transcribed in a 20-µl volume reaction with oligo dT primers by the RevertAidTM First Strand cDNA Synthesis Kit (Fermentas). In total, we got 20 cDNA samples (four genotypes with four treatments and one control) to be analyzed in this study. The cDNAs were diluted 1:10 with nuclease-free water prior to the qPCR analyses.

Selection and Validation of Candidate Genes

Candidate reference genes were screened based on the transcriptome *de novo* assembly sequences [34] with lengths > 1 kb and transcription abundance FPKM (fragments per kilobase of exon per million fragments mapped) values >5. Unigenes were annotated by National Center for Biotechnology Information (NCBI) non-redundant protein (Nr) database with E-values less than $1e^{-28}$. A reciprocal BLAST hits approach was then conducted against Arabidopsis with a permissive E-value cutoff of 10^{-3} to obtain hits to confirm the gene homologs in *R. soongorica*. The final list of 10 candidate reference genes included *ACT, EF1α, TUA, TIP41, DNAJ, UBQ, CYCL, EIF4A2, H2A*, and *L2*, which were assessed in later expression analyses (Table 1).

rbcL gene (ribulose-1,5-bisphosphate carboxylase/oxygenase large subunit, chloroplast) was chosen for the validation of reference genes. We tookUnigene19142_A, a *rbcL* homolog in the *R. soongorica de novo* transcriptome sequences (from Shi et.al. 2013) to predict its peptide sequence and gene structure with the GENSCAN Web Server at MIT (http://genes.mit.edu/GENSCAN.html). The protein sequence was then blasted against Nr database in NCBI and got the best hit ribulose-1,5-bisphosphate carboxylase/oxygenase large subunit (chloroplast) in *Eucalyptus umbra* (YP_008575886.1, with total score of 978, coverage of 100% and identity of 99%). The protein sequences of the *rbcL* homolog was well aligned with those from other species in File S3.pdf, which suggested the high homologous of *rbcL* genes to each other. The annotation information of the ten reference genes and one validation gene were listed in the Table S1.

qPCR Primer Design and Verification of Amplification

Primers of each reference gene were designed by the platform OligoPerfect Designer (Invitrogen), with optimal melting temperatures 60°C, optimal GC content 50%, optimal primer size 20 bp, and product size was between 100 and 300 bp (Table 1, and Figure S1). Each primer showed good specificity and efficiency (File S1 and S2). PCR efficiencies per amplicon were calculated by LinRegPCR [37–39], and the mean PCR efficiency between 1.8 and 2.0 from three technology replicates was regarded as the efficiency of one sample. The final primer efficiency was obtained from the average of all sample amplified with each pair of primer.

qPCR amplification products of the 10 candidate reference from one genotype (XGG030) were sequenced with ABi 3730 automated DNA Sequencer. Both the product sizes and sequences of amplificiaton product were consistent with the unigenes from *de novo* sequencing (File S4).

qPCR

qPCR experiments were conducted in the Stratagene Mx3000P QPCR system (Agilent Technologies), in which combined gene amplification, detection and data analysis steps with FAM, HEX, ROX, and TET filters. A 20-µl reaction system volume was used for amplification. Each reaction contained 10 µl DyNAmo Flash SYBR Green qPCR Kit Master Mix (Thermo Scientific), 0.5 µl each of forward and reverse primer, 0.2 µl F-402 buffer, 2 µl cDNA synthesized from total RNA. The PCR program contained

Table 1. Candidate reference gene descriptions.

Gene Symbol	Length (bp)	Tm (°C)	LinRegPCR Efficiency	5′-Sequences	3′-Sequences	GeneID	Gene Length	Nr-Annotation
ACT	145	80.57	1.834	AGCAAGGTCAAGACGAAGGA	TGTTGCTATTCAGGCTGTGC	CL448.Contig2_A	1784	Actin
CYCL	287	83.56	1.946	CCCTGAAGCATGAAGCTAGG	GCCTTGATCTGGTGTTTGGT	Unigene10547_A	1193	Cyclophilin
DNAJ	141	81.91	1.911	GATTATCTCAGCGCGCCTAC	GATTGGTTTCGGGTCTTTCA	Unigene46962_A	819	DnaJ-like protein
EF1α	128	81.88	1.929	CCCCAGGTTAATCCTTCCAT	GGAAAGAAGCCCAAGGAATC	Unigene18820_A	1127	elongation factor 1 alpha
EIF4A2	116	77.24	1.854	TCAGCAACATTTGCAGGAAG	TTTGGAAGAAAGGGTGTTGC	CL9396.Contig2_A	1717	Eukaryotic initiation factor 4A-2
H2A	164	81.54	1.875	GGCGCTGCAAAGAAGTCTAC	AGGACCTCAGCAGCAAGGTA	CL9675.Contig2_A	818	Histone H2A
L2	280	82.9	1.885	TCATTGCGAACGAAACTCTG	ACGGTGACAATTCAGGGAAG	CL8967.Contig2_A	1342	60S ribosomal protein L2
TIP41	110	79.16	1.899	GGTTTCTGCTCTTGCGTTTC	CCGTTTAATGATGGGGAATG	CL7670.Contig1_A	1134	TIP41-like protein
TUA	264	82.82	1.893	CAACAAGCCGCTAAGGAAAG	GCCTTCGTCCACTGGTATGT	CL12940.Contig1_A	1916	alpha-tubulin
UBQ	113	80.16	1.878	CCGGCCTTCTGGTAAACATA	TTGAGGAAGGAATGCCCTAA	CL5323.Contig1_A	471	Ubiquitin
*rbcL**	185	80.93	1.890	AGAGCACGTAGGGCTTTGAA	GACAACTGTGTGGACCGATG	Unigene19142_A	5856	Ribulose-1,5-bisphosphate carboxylase oxygenase

Notes:
* used for normalization validation under stress conditions in four accessions.

an initial denaturation step of 5 min at 95°C, followed by denaturation for 15 s at 95°C, annealing for 30 s at 60°C, and extension for 30 s at 72°C for 40 cycles. Three technical replicates were used in qPCR experiments, and the average Cq values were used for quantification. The amplification curves and dissociation curves were shown in File S1 and S2. No-template controls were used as negative control for each pair of primers. The PCR products were checked by 1.5% agarose gel electrophoresis to verify the specificity and expected sizes of amplicons. The amplification curve was analyzed by LinRegPCR for efficiency correction and baseline subtraction [37–39].

Statistical Analysis

Three different types of Microsoft Excel-based software – geNorm, NormFinder, and BestKeeper – were used to rank the expression stabilities of reference genes across all of the experimental sets. The candidate reference genes were ranked by geNorm based on the expression stability value M, which was calculated for all genes investigated (the lower the M value, the higher the gene's expression stability) [35]. NormFinder program is a Visual Basic application tool for Microsoft Excel used to determine the expression stabilities of reference genes that ranks all reference gene candidates based on intra- and inter-group variations and combines both results into a stability value for each candidate reference gene [11]. BestKeeper identifies the most stably expressed genes based on the coefficient of correlation to the BestKeeper Index, which is the geometric mean of the candidate reference gene Cq values. BestKeeper also calculates the standard deviation (SD) and the coefficient of variation (CV) based on the Cq values of all candidate reference genes [36]. Genes with SD >1 are considered unacceptable [10]. Thus, with BestKeeper package the best reference genes should be the most stable ones, which should exhibit the lowest CV ± SD [40]. This analysis was done using the Cq values after efficiency correction. Following qPCR data collection, Cq values corrected after efficiency correction by LinRegPCR were converted to relative quantities using the formula: $2^{-\Delta Cq}$, in which ΔCq equals each corresponding Cq value minus the minimum Cq value. The sample with the maximum expression level (the minimum Cq value) was used as a calibrator and was set to a value of 1. Relative quantities were used for geNorm and NormFinder, while BestKeeper analysis was based on untransformed Cq values.

Results

Expression levels of candidate reference genes across all samples

A total of 10 candidate reference genes of *R. soongorica* were chosen based on pBlast similarity against the nr database (Table 1), and their transcript levels were quantified under different stress conditions. The relative expression levels of the candidate reference genes were determined as Cq values, and the transcripts of these genes showed different levels of abundance in pooled samples (as Total), and each subset of treatments in all four genotypes (PEG, Heat, Dark, and Salt) (Figure 1). The mean Cq values of the genes ranged from 22–40, with most between 28 and 31 across all tested samples. Among all the candidate genes, *TUA* had the lowest Cq value (24.8), indicating its highest level of expression; *EIF4A2*, *H2A*, and *L2* were moderately expressed; and *ACT* was expressed at the lowest level (35.1 of Cq value). Stability was regarded as the most important characteristic for reference genes, and *ACT*, *EIF4A2 and H2A* showed the least gene expression variation (CV of 1.42%, 2.32% and 3.01%

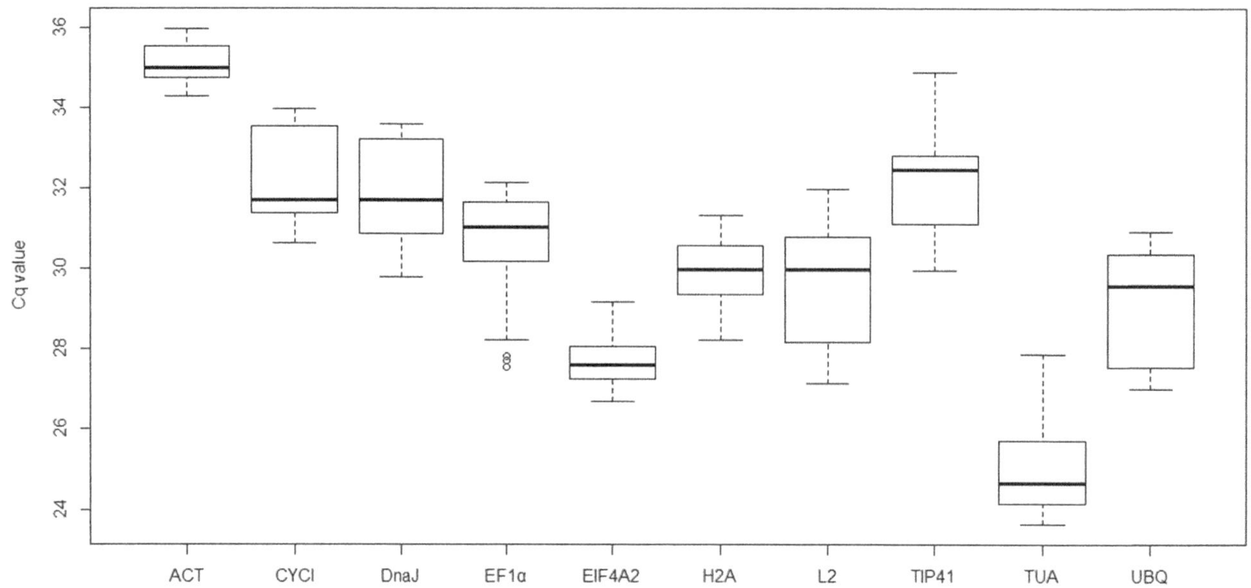

Figure 1. Expression levels of candidate reference genes across all samples. Lines across the boxes depict the medians, boxes indicate the interquartile range, whiskers represent 95% confidence intervals, and open circles represent outliers.

respectively), while *EF1α* (4.82%) and *TUA* (5.20%) were the most variable within the total group.

Expression Stability of Candidate Reference Genes under different treatments geNorm analysis

The expression stabilities of the 10 reference genes were assessed using the geNorm software [35] in the pooled and separate treatments (PEG, heat, dark, and salt) across four accessions from two contrasting habitats. Figure 2 shows the ranking of the tested genes on the basis of their expression stability (M). From the pooled data, the order from the most stable gene to the least was: *H2A/EIF4A2*, *DNAJ*, *TIP41*, *CYCL*, *ACT*, *UBQ*, *TUA*, *L2*, and *EF1α* (Figure 2A). Their calculated "M" values were 0.82, 0.94, 1.03, 1.14, 1.34, 1.48, 1.55, 1.65, and 1.71, respectively. Successive elimination of the least stable genes led to the identification of *H2A* and *EIF4A2* as the most stable genes. However, we found that *CYCL* and *DNAJ* were the most stable genes in PEG treatment (Figure 2B), *EF1α* and *DNAJ* in heat (Figure 2C), *H2A* and *EIF4A2* in dark (Figure 2D), and *TIP41* and *UBQ* in salt treatment (Figure 2E). *EIF4A2* was the moderate stable gene in individual PEG, heat, and salt treatments, which is inconsistent with the CV analysis of the pooled data.

geNorm also performs stepwise calculations of the pair-wise variation (Vn/Vn+1) between sequential normalization factors (NFn and NFn+1) to determine the optimal number of reference genes required for accurate normalization. A large variation means that the added gene had a significant effect and should preferably be included to calculate a reliable normalization factor [35]. The cut-off value of 0.15 was proposed by the geNorm program [35]; below this cut-off value there is no need of inclusion of an additional control gene. As shown in Figure 3, when all the samples were taken into account, the pairwise variation V2/3 was 0.311, higher than 0.15, indicating that the addition of the third reference gene was necessary to normalize gene expression. *H2A* and *EIF4A2* were recommended as the best reference genes based on their M value (Figure 2A). However, for individual treatments, V2/3 values were 0.092, 0.058, 0.057, and 0.104 in Heat, PEG, Salt, and Dark treatments, respectively. It is apparent that the inclusion of a third gene did not have a significant effect in separate treatments.

The expression stabilities of the 10 reference genes were also assessed among genotypes (Figure 4). In the HG genotypes (HG010 and HG020), the *CYCL*, *H2A* and *EIF4A2* were most stable; while in XGG genotypes (XGG030 and XGG040) were *EIF4A2*, *H2A* and *TUA*. Among individual genotypes, *H2A* and *EIF4A2* were frequently observed in the top of the most sable genes. However, compared to *H2A* and *EIF4A2*, *ACT* and *UBQ* which were usually used as reference gene for qPCR were not a good choice.

NormFinder analysis

NormFinder analysis results (shown in Table 2) revealed that the expression of three candidate reference genes, *EIF4A2*, *ACT*, and *H2A* had lower stability values (the lower, the better) across all the samples pooled together. However, the stability ranking of genes in individual treatments were significantly different; for example, the best reference genes were found as *UBQ* in PEG, *EF1α* and *DNAJ* in Heat, *DNAJ* in Dark, and *UBQ* and *CYCL* in Salt, and all the genes harbored the stable values <0.1. Interestingly, according to NormFinder analysis, *ACT* was suggested as a good reference gene among the four treatments, which is inconsistent with the conclusion from the geNorm analysis. However, the results from the NormFinder and geNorm analyses were consistent, suggesting that *EIF4A2* and *H2A* are stable reference genes.

BestKeeper analysis

The results of the BestKeeper analysis are shown in Table 3. Four candidate reference genes (*EIF4A2*, *ACT*, *H2A*, and *DNAJ*) in the Total group showed an SD value <1, which qualified as reference genes. When taking into account the separate treatments, all the candidate genes showed SD values <1 in individual groups. Here, only the top two genes were listed, as *UBQ* and *CYCL* in PEG treatment, *EF1α* and *DNAJ* in heat treatment, *DNAJ* and *TIP41* in dark treatment, and *UBQ*, *CYCL* in salt treatment. Further data processing using Pearson correlation

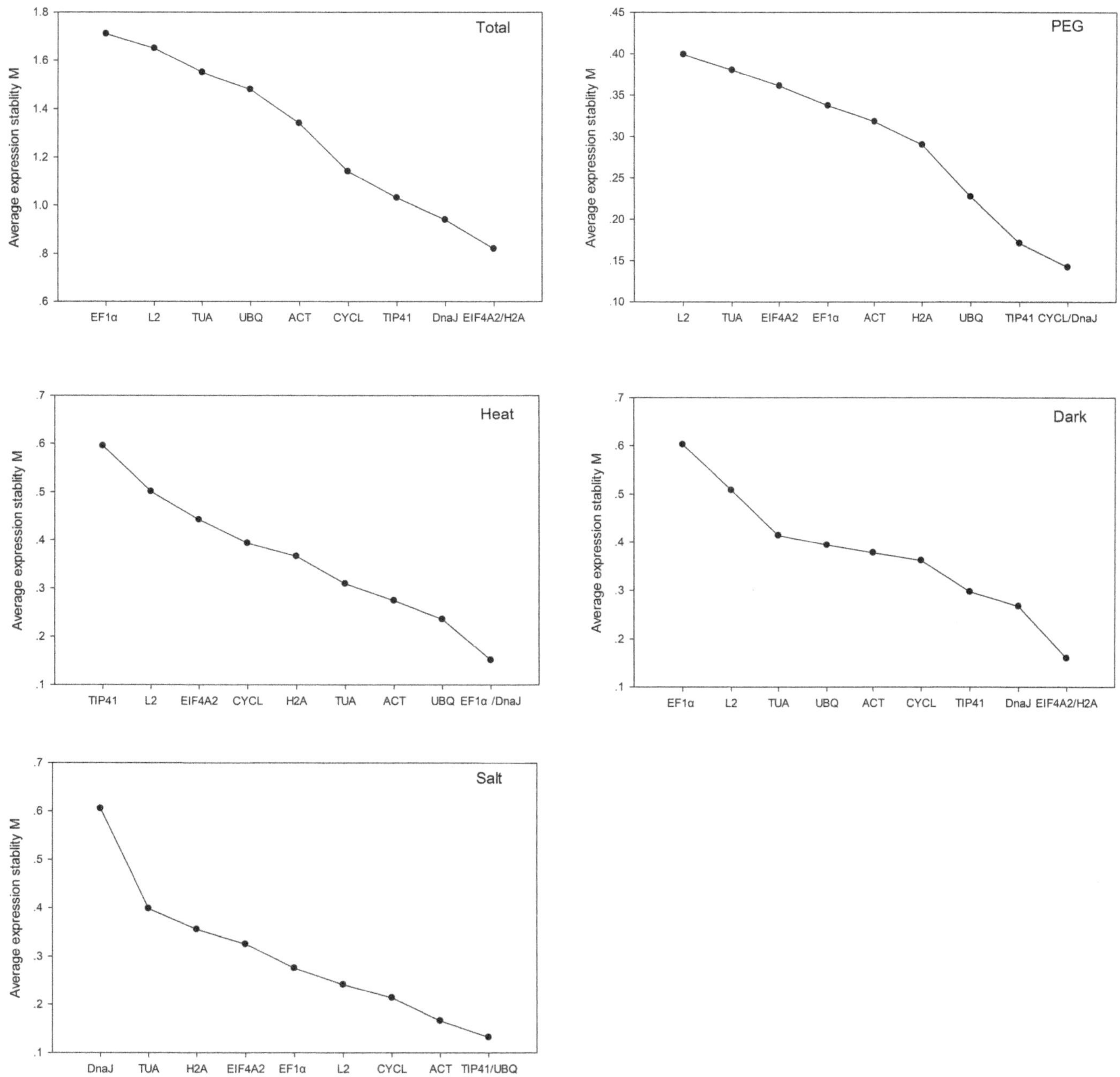

Figure 2. Gene expression stability and ranking of 10 reference genes calculated by geNorm. Mean expression stability (M) was calculated following stepwise exclusion of the least stable gene across all the treatment groups. The least stable genes are on the left, and the most stable ones are on the right.

coefficient and regression analysis (r) showed that *ACT* exhibited the lowest correlation with other candidate genes (data not shown, $r = -0.676$, p-value = 0.001), suggesting its higher independence. *DNAJ* was the most correlated to the other nine genes in pool ($r = 0.68$, p-value = 0.001). Pairwise correlation revealed that *TIP41* and *TUA* ($r = 0.804$, p-value = 0.001) were highly correlated. As these two genes were not stable in pooled data, we did not take these two genes as good reference. Although *EIF4A2* and *H2A* was the second most correlated gene pair ($r = 0.773$, p-value = 0.001), considering their higher scores in geNorm and NormFinder, *EIF4A2* and *H2A* could be selected as two likely reference genes with good correlation.

Reference Gene Validation

To validate candidate reference gene selection, the expression pattern of *rbcL* was analyzed against all 10 candidate reference genes (Figure 5). RuBisCO is an enzyme involved in the first major step of carbon fixation, a process by which atmospheric carbon dioxide is converted by plants to energy-rich molecules, such as glucose. In chemical terms, it catalyzes the carboxylation of ribulose-1,5-bisphosphate (also known as RuBP). It is probably the most abundant protein on Earth [41,42]. It was obviously down-regulated due to its light-dependent characteristic, and *rbcL* has also been shown to be down-regulated in soybean under stress [43,44]. In our results, *rbcL* was down-regulated against all genes

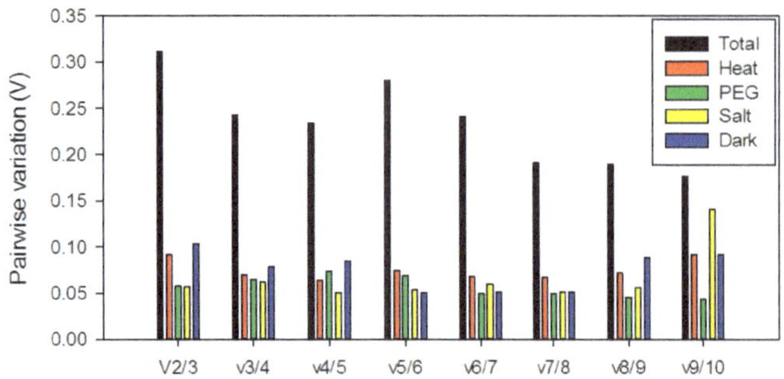

Figure 3. Determination of the optimal number of reference genes required for effective normalization. Pairwise variation (Vn/ Vn+1) analysis between the normalization factors (NFn and NFn+1) was performed by the geNorm program to determine the optimal number of reference genes and was carried out for qPCR data normalization in various sample pools.

except PEG treatments in three accessions (HG010, XGG030, and XGG040). However, in PEG treatments, *rbcL* was slightly down-regulated against *EF1α*, *CYCL*, *H2A*, *DNAJ*, and *UBQ*, which were the most stable genes in PEG. It was strongly up-regulated compared to *TUA*, *ACT*, *EIF4A2*, and *L2*, which were the most unstable genes in PEG treatment. In accession HG020, It was obvious that its expression level was down-regulated against most candidate reference genes, except in PEG and salt treatments. This could be due to the divergent genetic backgrounds between this genotype and the other three genotypes. Actually, the huge genetic differentiation within populations was supported by a phylogeography study based on chloroplast genes (unpublished data), and different genotypes may exhibit different physiological responses under drought conditions. Thus, the high expression level of *rbcL* in HG020 accession may be related to local environment adaptation. Except this, the validation results against the most stable genes were all consistent with the results from soybean [43,44], suggesting that the *DNAJ* and *H2A* genes could be used as practical reference genes in multiple treatments.

Discussion

qPCR has significantly improved the detection and quantification of expression profiles of genes in distinct biological samples. The main advantages of this technique are based on its high sensitivity, high specificity, and broad quantification range [45] [46]. For these reasons, qPCR is the first choice for detecting and quantifying gene expression profiles. An appropriate internal control gene is required for valid qPCR analyses. The ideal control gene should have relatively stable expression in distinct biological samples, which can be from different cell types, developmental stages, and samples exposed to different experimental conditions, even for different genotypes.

R. soongorica is regarded as a plant that is tolerant to extremely drought conditions [26–28]. Moreover, *R. soongorica* is a typical recretohalophyte, and its molecular mechanism of salt resistance remains unclear [47]. Only a few studies have focused on the molecular mechanisms of adaption to extreme environments [27]. Since his shrub species is prevalent in ecological engineering and restoration to mitigate further desertification of the arid land in northwest China, with its numerous novel genetic resources (around 40% of genes were newly found in its transcriptome sequences [34]),, it is convenient and feasible to focus on molecular functional analyses of the expression of unique genes. However, it is first necessary to identify stable reference genes.

In this study, the four stresses of drought, salt, dark, and heat were investigated in different *R. soongorica* genotypes to identify stable reference genes. Three different statistical approaches (geNorm, NormFinder, and BestKeeper) were used to evaluate the stability of the 10 most commonly used candidate reference genes. The most prominent observation was that each type of software produced a different set of top-ranked reference genes. This is not unexpected because the three programs employed different algorithms and analytical procedures. The top three

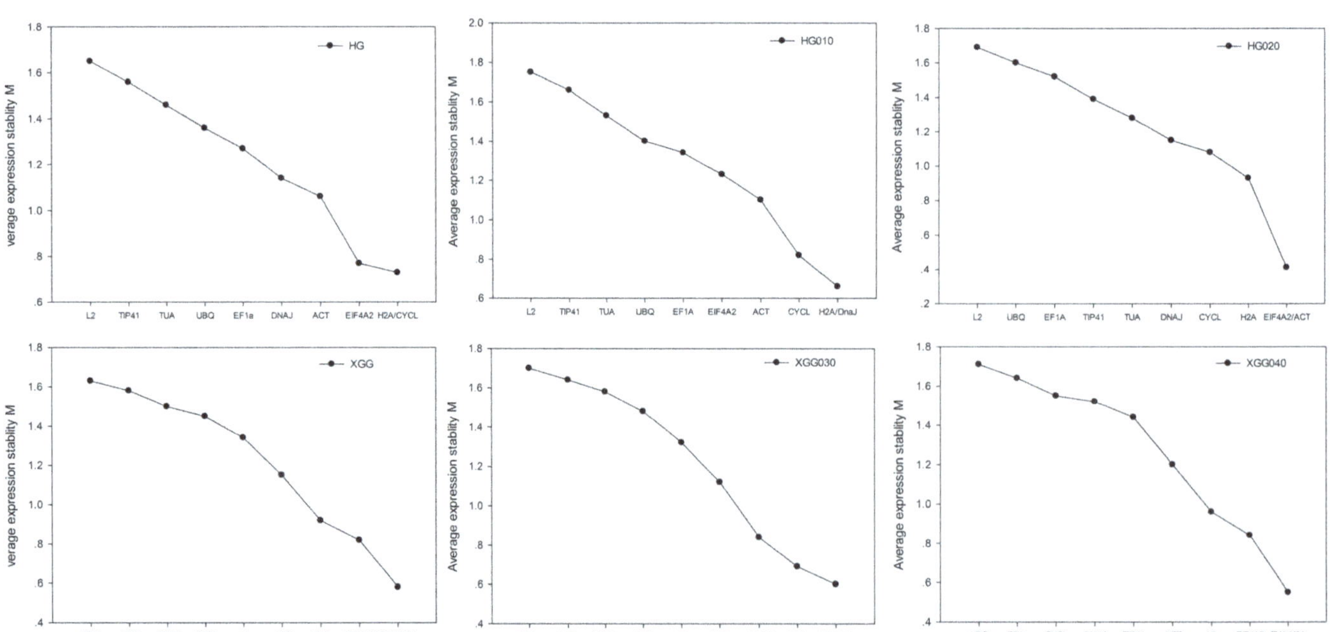

Figure 4. Expression stability and ranking of housekeeping genes by geNorm analysis within and among populations. Mean expression stability (M) was calculated following stepwise exclusion of the least stable gene.

Table 2. Expression stability of the reference genes calculated by NormFinder software.

Total	SV	PEG	SV	Heat	SV	Dark	SV	Salt	SV
EIF4A2	0.368	*UBQ*	0.073	*EF1α*	0.053	*DNAJ*	0.093	*UBQ*	0.043
ACT	0.4	*CYCL*	0.144	*DNAJ*	0.062	*TIP41*	0.147	*CYCL*	0.063
H2A	0.573	*DNAJ*	0.186	*ACT*	0.203	*ACT*	0.173	*TIP41*	0.11
DNAJ	0.677	*H2A*	0.19	*TUA*	0.216	*EIF4A2*	0.198	*TUA*	0.203
CYCL	0.874	*TIP41*	0.212	*UBQ*	0.238	*CYCL*	0.215	*L2*	0.209
TUA	1.017	*EF1α*	0.212	*CYCL*	0.244	*UBQ*	0.254	*EF1α*	0.21
UBQ	1.046	*ACT*	0.231	*H2A*	0.253	*TUA*	0.302	*ACT*	0.235
TIP41	1.119	*EIF4A2*	0.234	*EIF4A2*	0.382	*H2A*	0.311	*H2A*	0.32
L2	1.157	*TUA*	0.287	*L2*	0.451	*L2*	0.515	*EIF4A2*	0.329
EF1α	1.185	*L2*	0.291	*TIP41*	0.632	*EF1α*	0.628	*DNAJ*	0.978

Note: SV, stability value.

Table 3. Expression stability of the reference genes calculated by BestKeeper software.

Total	SD	CV	PEG	SD	CV	Heat	SD	CV	Dark	SD	CV	Salt	SD	CV
ACT	0.43	1.22	*ACT*	0.09	0.26	*EF1α*	0.12	0.39	*ACT*	0.19	0.54	*L2*	0.08	0.26
EIF4A2	0.49	1.75	*H2A*	0.12	0.4	*DNAJ*	0.12	0.39	*DNAJ*	0.2	0.63	*UBQ*	0.12	0.4
H2A	0.74	2.47	*UBQ*	0.19	0.69	*ACT*	0.18	0.51	*TIP41*	0.23	0.76	*EF1α*	0.14	0.45
DNAJ	0.98	3.07	*EF1α*	0.2	0.73	*H2A*	0.21	0.71	*TUA*	0.23	0.94	*TIP41*	0.16	0.48
TUA	1	4.01	*L2*	0.26	0.84	*CYCL*	0.22	0.71	*EIF4A2*	0.24	0.87	*CYCL*	0.16	0.52
TIP41	1.07	3.32	*TIP41*	0.28	0.8	*UBQ*	0.24	0.78	*CYCL*	0.25	0.73	*ACT*	0.22	0.62
CYCL	1.08	3.33	*CYCL*	0.33	1.05	*TUA*	0.25	1.01	*H2A*	0.34	1.11	*EIF4A2*	0.3	1.08
EF1α	1.16	3.78	*DNAJ*	0.37	1.22	*EIF4A2*	0.38	1.39	*UBQ*	0.35	1.27	*H2A*	0.39	1.28
UBQ	1.26	4.33	*EIF4A2*	0.51	1.79	*L2*	0.45	1.62	*EF1α*	0.56	1.79	*TUA*	0.4	1.61
L2	1.33	4.5	*TUA*	0.58	2.13	*TIP41*	0.74	2.32	*L2*	0.69	2.44	*DNAJ*	1.24	3.86

Note: Reference genes were identified as the most stable genes, i.e. those with the lowest coefficient of variance (CV) with standard deviation (SD).

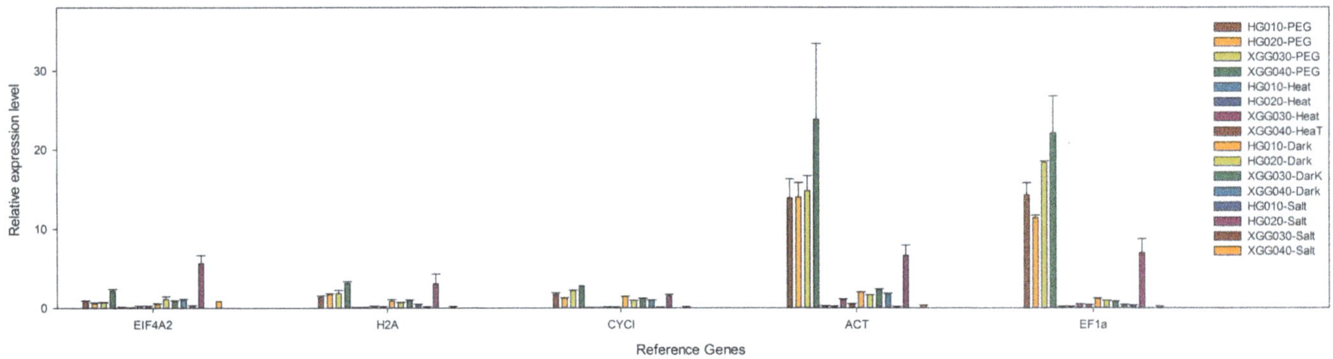

Figure 5. Relative quantification of *rbcL* expression using 10 reference genes for normalization in the four accessions. The results are represented as mean fold changes in relative expression compared to untreated samples.

stable reference genes were *EIF4A2*, *H2A*, and *DNAJ* using geNorm. *EIF4A2*, *ACT*, and *H2A* had lower stability values across all samples using NormFinder, which was consistent with the BestKeeper results. We used a previously described method [48] to determine the best reference gene based on all three algorithms (Table S2). According to the frequencies of the reference genes in the rank crossing the difference treatments (PEG, Salt, Dark, Heat and Total), *H2A* and *EIF4A2* were the best reference genes, *CYCL* was moderate stable, and *EF1α* was the fourth choice. This was supported in the subsequent validation analysis with *rbcL* (File S4). All the 10 genes are normally expressed under all the treatments in the 4 genotypes (Figure S2), indicating that normalization of total RNA was sufficient. In PEG treatments, *rbcL* was slightly or down-regulated against *EF1α*, *CYCL*, *H2A*, *DNAJ*, and *UBQ*, which were the most stable genes in accordance with results in Soybean [43,44]. Once we identify practical internal reference gene, it is easy to characterize the true expression level for a candidate gene.

ACT has been considered as one of the best reference genes for assessing gene expression in many plant tissues and under different experimental conditions [49–51]. In present study, GeNorm, NormFinder, and Bestkeeper analyses demonstrated that *ACT* was not that stable and exhibited a lower expression level compared to other good reference genes, which suggested that it is not a good reference gene for *R. soongorica* under such conditions. Actually, the instablity of *ACT* homolog genes was also confirmed in two other plant species [12,15]. The *ACT* homolog gene exhibited similar low expression levels in *Caragana intermedia* and *R. soongorica*, which are widely spread throughout the arid zone in China. Even though *C. intermedia* and *R. soongorica* belong to different families, it is likely that both of them harbor similar pathways involved in strong drought- and salt-resistance ability.

The other two most commonly used reference genes are *UBQ* and *TUA*. *UBQ* showed highly stable expression in Arabidopsis [52] and tomato [9], but the *UBQ* gene was ranked differently across different treatments in the present study. It was listed on the top under PEG with NormFinder; however, it was in a moderate stable position with geNorm and BestKeeper under other treatments. It was suggested to be an inappropriate internal control for qPCR studies at different developmental stages in rice [53] and soybean [54]. *TUA* has also been widely used as a reference gene in studies in water lily and soybean [16,19]. However, some studies revealed that *TUA* did not satisfy certain basic requirements for use as an internal control [18]. In our results in *R. soongorica*, *TUA* was once of the least stable genes according to geNorm, BestKeeper, and NormFinder analyses together with *rbcL* expression level. Consequently, *ACT*, *UBQ*, and *TUA* should be used with caution as an internal control in *R. soongorica*.

Conclusions

To our knowledge, this study is the first to systematically analyze reference genes for qPCR in *R. soongorica* under different abiotic stress conditions (drought, salt, dark, and heat) among four accessions. Analysis of expression stability using geNorm, NormFinder, and BestKeeper revealed that *EIF4A2* and *H2A* were appropriate reference genes under different abiotic stress conditions, whereas *EF1α* showed relatively low expression stability. In contrast, the most commonly used reference genes *ACT*, *UBQ*, and *TUA* were inappropriate. This work will benefit future gene expression studies assessing different abiotic stress conditions in *R. soongorica* and other *Reaumuria* species.

Supporting Information

Figure S1 Agarose gel electrophoresis for PCR products of 10 candidate reference genes and *rbcL*.

Figure S2 Expression levels of candidate reference genes across all the four accessions.

Table S1 Annotation information of 10 reference genes and the one validation gene from the *de novo* transcriptome sequences.

Table S2 Ranking of the best reference gene based on the all three algorithms.

File S1 Contains 11 figures from Figure S3 to S13 representing the qPCR amplification curves of 10 candidate reference genes (*ACT, CYCL, DNAJ, EF1, EIF4A2, H2A, L2, TIP41, TUA and UBQ*) and one validation gene *rbcL*, respectively.

File S2 Contains 11 figures from Figure S14 to S24 representing the melting curves of qPCR products of the 10 candidate reference genes (*ACT, CYCL, DNAJ, EF1, EIF4A2, H2A, L2, TIP41, TUA and UBQ*) and one validation gene *rbcL*, respectively.

File S3 Alignment of the protein sequences of *rbcL* homologs.

File S4 Contains 11 sequence alignment files from Fasta S1 to S11,for verification of the homologs of 10 reference genes and one validation gene *rbcL*.

Acknowledgments

We thank for the instructions on the data analysis from Jianfeng Zhu, and appreciate the constructive and helpful suggestions from Dr. Jinping Li and the two anonymous reviewers.

Author Contributions

Conceived and designed the experiments: XY XD X-FM. Performed the experiments: XY WZ HY. Analyzed the data: XY PC X-FM. Contributed reagents/materials/analysis tools: HX X-FM. Wrote the paper: XY XD X-FM.

References

1. Fink L, Seeger W, Ermert L, Hanze J, Stahl U, et al. (1998) Real-time quantitative RT-PCR after laser-assisted cell picking. Nat Med 4:1329–1333.
2. Heid CA, Stevens J, Livak KJ, Williams PM (1996) Real time quantitative PCR. Genome Res 6: 986–994.
3. Higuchi R, Fockler C, Dollinger G, Watson R (1993) Kinetic PCR analysis: real-time monitoring of DNA amplification reactions. Biotechnology (NY) 11:1026–1030.
4. Brunner AM, Yakovlev IA, Strauss SH (2004) Validating internal controls for quantitative plant gene expression studies. BMC Plant Biol 4: 14.
5. Nicot N, Hausman J, Hoffmann L, Evers D (2005) Housekeeping gene selection for real-time RT-PCR normalization in potato during biotic and abiotic stress. J Exp Bot 56:2907–2914.
6. Jain M, Nijhawan A, Tyagi AK, Khurana JP (2006) Validation of housekeeping genes as internal control for studying gene expression in rice by quantitative real-time PCR. Biochem Biophys Res Commun 345: 646–651.
7. Die J, Roman B, Nadal S, Gonzalez-Verdejo C (2010) Evaluation of candidate reference genes for expression studies in *Pisum sativum* under different experimental conditions. Planta 232:145–153.
8. Gimenez C, Mitchell VJ, Lawlor DW (1992) Regulation of photosynthetic rate of 2 sunflower hybrids under water-stress. Plant Physiol 98:516–524.
9. Lovdal T, Lillo C (2009) Reference gene selection for quantitative real-time PCR normalization in tomato subjected to nitrogen, cold, and light stress. Anal Biochem 387:238–242.
10. Migocka M, Papierniak A (2011) Identification of suitable reference genes for studying gene expression in cucumber plants subjected to abiotic stress and growth regulators. Mol Breeding 28:343–357.
11. Zhong HY, Chen JW, Li CQ, Chen L, Wu JY, et al. (2011) Selection of reliable reference genes for expression studies by reverse transcription quantitative real-time PCR in litchi under different experimental conditions. Plant Cell Rep 30:641–653.
12. Zhu J, Zhang L, Li W, Han S, Yang W, et al. (2013) Reference gene selection for quantitative real-time PCR normalization in *Caragana intermedia* under different abiotic stress conditions. PloS one 8:2.
13. Saito T, Matsukura C, Ban Y, Shoji K, Sugiyama M, et al. (2008) Salinity stress affects assimilate metabolism at the gene-expression level during fruit development and improves fruit quality in tomato (*Solanum lycopersicum L.*). J Jpn Soc Hort Sci 77:61–68.
14. Paolacci A, Tanzarella O, Porceddu E, Ciaffi M (2009) Identification and validation of reference genes for quantitative RT-PCR normalization in wheat. BMC Molecular Biology 10:11.
15. Tong Z, Gao Z, Wang F, Zhou J, Zhang Z (2009) Selection of reliable reference genes for gene expression studies in peach using real-time PCR. BMC Molecular Biology 10:71.
16. Luo H, Chen S, Wan H, Chen F, Gu C, et al. (2010) Candidate reference genes for gene expression studies in water lily. Anal Biochem 404:100–102.
17. Czechowski T, Stitt M, Altmann T, Udvardi M, Scheible W-R (2005) Genome-wide identification and testing of superior reference genes for transcript normalization in *Arabidopsis*. Plant Physiol 139:5–17.
18. Xu M, Zhang B, Su X, Zhang S, Huang M (2011) Reference gene selection for quantitative real-time polymerase chain reaction in *Populus*. Anal Biochem 408:337–339.
19. Hu R, Fan C, Li H, Zhang Q, Fu YF (2009) Evaluation of putative reference genes for gene expression normalization in soybean by quantitative real-time RT-PCR. BMC Mol Biol 10:93.
20. Janská A, Hodek J, Svoboda P, Zámečník J, Prášil IT, et al. (2013) The choice of reference gene set for assessing gene expression in barley (*Hordeum vulgare L.*) under low temperature and drought stress. Molecular genetics and genomics 288:639–649.
21. Exposito-Rodriguez M, Borges A, Borges-Perez A, Perez J (2008) Selection of internal control genes for quantitative real-time RT-PCR studies during tomato development process. BMC Plant Biology 8:131.
22. Dubrovina AS, Kiselev KV, Khristenko VS (2013) Expression of calcium-dependent protein kinase (CDPK) genes under abiotic stress conditions in wild-growing grapevine *Vitis amurensis*. J Plant Physiol 170: 1491–1500.
23. Fan C, Ma J, Guo Q, Li X, Wang H, et al. (2013) Selection of reference genes for quantitative real-time PCR in bamboo (*Phyllostachys edulis*). PLoS One 8:e56573.
24. Gantasala NP, Papolu PK, Thakur PK, Kamaraju D, Sreevathsa R, et al. (2013) Selection and validation of reference genes for quantitative gene expression studies by real-time PCR in eggplant (*Solanum melongena L*). BMC Res Notes 6:312.
25. Wang X, Zhang T, Wen Z, Xiao H, Yang Z, et al. (2011) The chromosome number, karyotype and genome size of the desert plant diploid *Reaumuria soongorica* (Pall.) Maxim. Plant Cell Rep 30:955–964.
26. Qian Z-Q, Xu L, Wang Y-L, Yang J, Zhao G-F (2008) Ecological genetics of *Reaumuria soongorica* (Pall.) Maxim. population in the oasis–desert ecotone in Fukang, Xinjiang, and its implications for molecular evolution. Biochemical Systematics and Ecology 36:593–601.
27. Liu Y, Zhang T, Li X, Wang G (2007) Protective mechanism of desiccation tolerance in *Reaumuria soongorica*: leaf abscission and sucrose accumulation in the stem. Sci China C Life Sci 50:15–21.
28. Liu JQ, Qiu MX, Pu JC, Lu ZM (1982) The typical extreme xerophyte-*Reaumuria soongorica* in the desert of China. Act bot sinica 24:485–488.
29. Bai J, Gong CM, Chen K, Kang HM, Wang G (2009). Examination of antioxidative system's responses in the different phases of drought stress and during recovery in desert plant *Reaumuria soongorica* (Pall.) Maxim. Journal of Plant Biology 52:417–425.
30. Liu YB, Wang G, Liu J, Zhao X, Tan HJ, et al. (2007) Anatomical, morphological and metabolic acclimation in the resurrection plant *Reaumuria soongorica* during dehydration and rehydration. Journal of Arid Environments 70:183–194.
31. Liu Y, Zhang T, Li X, Wang G (2007) Protective mechanism of desiccation tolerance in *Reaumuria soongorica*: leaf abscission and sucrose accumulation in the stem. Sci China C Life Sci 50:15–21.
32. Ma J, Chen F, Zhang H, Xia D (2007) Spatial distribution characteristics of stable carbon isotope compositions in desert plant *Reaumuria soongorica* (Pall.) Maxim. Frontiers of Earth Science in China 1: 150–156.
33. Qian Z-Q, Xu L, Wang Y-L, Yang J, Zhao G-F (2008) Ecological genetics of *Reaumuria soongorica* (Pall.) Maxim. population in the oasis–desert ecotone in Fukang, Xinjiang, and its implications for molecular evolution. Biochemical Systematics and Ecology 36:593–601.
34. Shi Y, Yan X, Zhao P, Yin H, Zhao X, et al. (2013) Transcriptomic Analysis of a Tertiary Relict Plant, Extreme Xerophyte *Reaumuria soongorica* to Identify Genes Related to Drought Adaptation. PLoS ONE 8:e63993.
35. Vandesompele J, De Preter K, Pattyn F, Poppe B, Van Roy N, et al. (2002) Accurate normalization of real-time quantitative RT-PCR data by geometric averaging of multiple internal control genes. Genome biology 3:RE-RESEARCH0034.
36. Pfaffl MW, Tichopad A, Prgomet C, Neuvians TP (2004) Determination of stable housekeeping genes, differentially regulated target genes and sample integrity: BestKeeper–Excel-based tool using pair-wise correlations. Biotechnol Lett 26:509–515.
37. Ramakers C, Ruijter JM, Deprez RH, Moorman AF (2003) Assumption-free analysis of quantitative real-time polymerase chain reaction (PCR) data. Neuroscience letters 339:62–66.
38. Ruijter JM, Ramakers C, Hoogaars WM, Karlen Y, Bakker O, et al. (2009) Amplification efficiency: linking baseline and bias in the analysis of quantitative PCR data. Nucleic Acids Res 37:e45.
39. Khodadadi A, Hamidi- Nia M, Ghaforian M, Mohammadi- Asl J (2012) Evaluation of real-time PCR efficiency by the use of two strategies: standard curve and linear regression. Jundishapur Sci Med J 11(1):85–95
40. Chang E, Shi S, Liu J, Cheng T, Xue L, et al. (2012) Selection of reference genes for quantitative gene expression studies in *Platycladus orientalis* (Cupressaceae) Using Real-Time PCR. PLoS ONE 7: e33278.
41. Dhingra A, Portis AR Jr, Daniell H (2004) Enhanced translation of a chloroplast-expressed RbcS gene restores small subunit levels and photosynthesis in nuclear RbcS antisense plants. Proc Natl Acad Sci USA 101:6315–6320
42. Salvucci ME, Portis AR, Ogren WL (1986) Light and CO(2) response of ribulose-1,5-bisphosphate carboxylase/oxygenase activation in *Arabidopsis* leaves. Plant Physiol 80 655–659.
43. Vu JC, Allen LH, Bowes G (1987) Drought stress and elevated CO(2) effects on soybean ribulose bisphosphate carboxylase activity and canopy photosynthetic rates. Plant Physiol 83:573–578.

44. Carmo-Silva AE, Keys AJ, Andralojc PJ, Powers SJ, Arrabaca MC, et al. (2010) rbcL activities, properties, and regulation in three different C4 grasses under drought. J Exp Bot 61:2355–2366.
45. Ginzinger DG (2002) Gene quantification using real-time quantitative PCR: an emerging technology hits the mainstream. Experimental hematology 30:503–512.
46. Bustin SA, Benes V, Nolan T, Pfaffl MW (2005) Quantitative real-time RT-PCR–a perspective. Journal of Molecular Endocrinology 34:597–601.
47. Dang ZH, Zheng LL, Wang J, Gao Z, Wu SB, et al. (2013) Transcriptomic profiling of the salt-stress response in the wild recretohalophyte *Reaumuria trigyna*. BMC Genomics 14:29.
48. Klie M, Debener T (2011) Identification of superior reference genes for data normalisation of expression studies via quantitative PCR in hybrid roses (*Rosa hybrida*). BMC Research Notes 4:518.
49. Rai V, Dey N (2012) Identification of programmed cell death related genes in bamboo. Gene 497:243–248.
50. Rai V, Ghosh JS, Pal A, Dey N (2011) Identification of genes involved in bamboo fiber development. Gene 478:19–27.
51. Xu H, Chen LJ, Qu LJ, Gu HY, Li DZ (2010) Functional conservation of the plant embryonic flower2 gene between bamboo and *Arabidopsis*. Biotechnol Lett 32:1961–1968.
52. Czechowski T, Stitt M, Altmann T, Udvardi M, Scheible W-R (2005) Genome-wide identification and testing of superior reference genes for transcript normalization in *Arabidopsis*. Plant Physiol 139:5–17.
53. Jain M, Nijhawan A, Tyagi AK, Khurana JP (2006) Validation of housekeeping genes as internal control for studying gene expression in rice by quantitative real-time PCR. Biochem Biophys Res Commun 345: 646–651.
54. Jian B, Liu B, Bi Y, Hou W, Wu C, et al. (2008) Validation of internal control for gene expression study in soybean by quantitative real-time PCR. BMC Mol Biol 9:59.

2

Allelic Combinations of Soybean Maturity Loci *E1, E2, E3* and *E4* Result in Diversity of Maturity and Adaptation to Different Latitudes

Bingjun Jiang[1] ©, Haiyang Nan[2,3] ©, Youfei Gao[1] ©, Lili Tang[2], Yanlei Yue[1], Sijia Lu[2,3], Liming Ma[1], Dong Cao[2], Shi Sun[1], Jialin Wang[2], Cunxiang Wu[1], Xiaohui Yuan[2], Wensheng Hou[1], Fanjiang Kong[2]*, Tianfu Han[1]*, Baohui Liu[2]*

1 The National Key Facility for Crop Gene Resources and Genetic Improvement and MOA Key Lab of Soybean Biology (Beijing), Institute of Crop Sciences, The Chinese Academy of Agricultural Sciences, Beijing, China, **2** The Key Laboratory of Soybean Molecular Design Breeding, Northeast Institute of Geography and Agroecology, Chinese Academy of Sciences, Harbin, China, **3** University of Chinese Academy of Sciences, Beijing, China

Abstract

Soybean cultivars are extremely diverse in time to flowering and maturation as a result of various photoperiod sensitivities. The underlying molecular genetic mechanism is not fully clear, however, four maturity loci *E1, E2, E3* and *E4* have been molecularly identified. In this report, cultivars were selected with various photoperiod sensitivities from different ecological zones, which covered almost all maturity groups (MG) from MG 000 to MG VIII and MG X adapted from latitude N 18° to N 53°. They were planted in the field under natural daylength condition (ND) in Beijing, China or in pots under different photoperiod treatments. Maturity-related traits were then investigated. The four *E* maturity loci were genotyped at the molecular level. Our results suggested that these four *E* genes have different impacts on maturity and their allelic variations and combinations determine the diversification of soybean maturity and adaptation to different latitudes. The genetic mechanisms underlying photoperiod sensitivity and adaptation in wild soybean seemed unique from those in cultivated soybean. The allelic combinations and functional molecular markers for the four *E* loci will significantly assist molecular breeding towards high productivity.

Editor: Fan Chen, Institute of Genetics and Developmental Biology, Chinese Academy of Sciences, China

Funding: This work was supported by the National Natural Science Foundation of China (31201232, 31071445, 31171579, 31201222 and 31371643); Open Foundation of Key Laboratory of Soybean Molecular Design Breeding, Chinese Academy of Sciences; "Hundred Talents" Program of Chinese Academy of Sciences; Strategic Action Plan for Science and Technology Innovation of Chinese Academy of Sciences (XDA08030100); Heilongjiang Natural Science Foundation of China (ZD201001, JC201313); China Agriculture Research System (CARS-04) and Chinese Academy of Agricultural Sciences (CAAS) Innovation Project. The funders had no role in study design, data collection and analysis, decision to publish, or preparation of the manuscript.

Competing Interests: The authors have declared that no competing interests exist.

* Email: kongfj@iga.ac.cn (FK); hantianfu@caas.cn (TH); liubh@iga.ac.cn (BL)

© These authors contributed equally to this work.

Introduction

Soybean (*Glycine max* (L.) Merrill) is a short-day crop with high protein and oil contents. Many cultivars were bred with different maturity to adapt to various ecological environments. For the convenience of breeding layout, 13 MGs from MG000 to MGX were classified in North America [1–3]. Chinese soybean researchers divided cultivars into different maturity groups as well [4–5]. Soybean production has increased in America in response to recent increases in global demand and maturity is the key factor determining soybean productivity. Therefore, it is rather important to understand the mechanism of soybean maturity diversity and adaptation.

Flowering and maturity were highly controlled by major genes in soybean. Up to now, nine maturity loci have been identified as *E1–E8* and *J* [1], [6–12]. These loci have different roles under different photoperiods. Wang et al [13] found that long daylength condition (LD) might strengthen while short daylength condition (SD) might weaken these maturity loci. More results and progress of maturity genes was reviewed by Xia et al [14]. Furthermore, four loci were identified at molecular level. *E1* gene was identified as a transcription factor which functions as a flowering repressor with a putative nuclear localization signal and a B3-related domain [15]. *E2* is an orthologue of Arabidopsis flowering gene *GIGANTEA* [16]. *E3* and *E4* are phytochrome genes *GmPhyA3* [17] and *GmPhyA2* [18], respectively. In addition, two homologs of soybean *Flowering Locus T* (*FT*) genes, *GmFT2A* and *GmFT5A* were identified and coordinately regulate flowering [19]. Four identified maturity genes *E1*, *E2*, *E3* and *E4* delay flowering and maturity under LD through down regulating *GmFT2A* and *GmFT5A* [15], [16], [19]. As for other loci, more studies should be done before learning their molecular identities.

Although these four known loci *E1*, *E2*, *E3* and *E4* provide an important key to learn the mechanism of flowering and maturity,

Table 1. Cultivars from North America and their respective maturity group.

Code	PI number	Variety	MG	Code	PI number	Variety	MG
CAAE001	PI548594	Maple Presto	000	**CAAE021**	PI534646	Flyer	IV
CAAE002	PI567787	OAC Vision	000	**CAAE022**	PI598222	TN4-94	IV
CAAE003	PI548648	Canatto	00	**CAAE023**	PI564849	Nathan	V
CAAE005	PI592523	Glacier	00	**CAAE024**	PI572239	Holladay	V
CAAE006	PI629004	MN0201	0	**CAAE025**	PI633609	Lonoke	V
CAAE007	PI596541	Traill	0	**CAAE026**	PI561400	Rhodes	V
CAAE008	PI612764	MN0901	0	**CAAE027**	PI633610	Desha	VI
CAAE009	PI599300	Surge	NA	**CAAE028**	PI592756	Dillon	VI
CAAE010	PI548641	Haroson	I	**CAAE029**	PI617045	NC-Roy	VI
CAAE011	PI614833	NE1900	I	**CAAE030**	PI599333	Musen	VI
CAAE012	PI608438	Titan	I	**CAAE031**	PI531068	Stonewall	VII
CAAE013	PI561858	Holt	II	**CAAE032**	PI595645	Benning	VII
CAAE014	PI567786	OAC Talbot	II	**CAAE033**	PI617041	Santee	VII
CAAE015	PI533655	Burlison	II	**CAAE034**	PI555453	Hagood	VII
CAAE016	PI595926	Athow	III	**CAAE035**	PI603953	Motte	VIII
CAAE017	PI548634	Zane	III	**CAAE036**	PI548970	Foster	VIII
CAAE018	PI593258	Macon	III	**CAAE037**	PI548663	Dowling	VIII
CAAE019	PI578057	Saline	III	**CAAE038**		Jupiter	X
CAAE020	PI614155	NS93-4118	IV				

NA, not available.

Table 2. Cultivars and accessions from China and far-east Russia and their adaption latitudes.

Code	Variety	Latitude	Code	Variety	Latitude
CAAE039	Dengke 2	N 47°–53°	**CAAE064**	Fengchengzaochadou	N 25°–30°
CAAE040	Huajiang 4	N 48°–50°	**CAAE065**	Jin 6606	N 41°–43°
CAAE041	Heihe 27	N 46°–48°	**CAAE066**	Jinzhou 8–14	N 40°–43°
CAAE042	Heihe 3	N 46°–50°	**CAAE071**	Bahong 1	N 39°–42°
CAAE043	Heihe 43	N 46°–48°	**CAAE072**	Mianyanghuangwofeng	N 30°
CAAE044	Suinong 14	N 44°–48°	**CAAE073**	Lüpidou	N 25°–28°
CAAE045	Hefeng 25	N 39°–43°	**CAAE074**	Liuyuezao	N 25°–28°
CAAE047	Jilin 3	N 44°–46°	**CAAE075**	Ruijinxiaohuangdou	N 24°–27°
CAAE049	Jiunong 21	N 42°–46°	**CAAE076**	Edou 2	N 30°–35°
CAAE050	Jilin 30	N 41°–43°	**CAAE077**	Yulindahuangdou	N 23°
CAAE052	Jindou 19	N 35°–40°	**CAAE081**	Qiudou 1	N 25°–30°
CAAE053	Tiefeng 31	N 35°–40°	**CAAE082**	Jiangledaqingdou	N 27°
CAAE054	Jidou 12	N 35°–38°	**CAAE084**	Guixia 1	N 22°–26°
CAAE055	Qihuang 28	N 34°–37°	**CAAE085**	Nandou 12	N 29°–32°
CAAE056	Zhonghuang 13	N 30°–40°	**CAAE086**	Zigongdongdou	N 29°
CAAE057	Xudou 9	N 32°–35°	**CAAE087***	Heiheyesheng	N 50°
CAAE058	Xudou 1	N 32°–35°	**CAAE088***	Bayanyesheng	N 46°
CAAE059	Fengshouhuang	N 35°–38°	**CAAE089***	Baiyangdianyesheng	N 39°
CAAE061	Dandou 2	N 40°–42°	**CAAE090***	Guangxiyesheng	N 23°
CAAE062	Yuejin 4	N 35°–38°	**CAAE091**	Mohe 1	N 53°
CAAE063	Jinda 814	NA	**CAAE092**	Ziweicika 4/75	N 50°

CAAE092 is a cultivar from far east Russia; *indicates wild soybeans: CAAE087, CAAE088, CAAE089 and CAAE090.

we mainly got knowledge based on few cultivars but not on a population level. Population-level knowledge will provide another different view of these four loci's role on maturity and adaptation. Therefore, in this study we selected a set of soybean cultivars which cover 12 maturity groups from MG000 to MGVIII and MGX plus some cultivars with wide range of latitude from N 18° to N 53°. These cultivars were subjected to different photoperiod treatments. Traits of beginning bloom (R1), physiological maturity (R7) and full maturity (R8) were investigated [20]. Maturity loci *E1*, *E2*, *E3* and *E4* were genotyped in the population. Further association analysis was done. The results showed that allelic combinations of these four *E* genes significantly determine the ecological-economical adaption of cultivars although they have different impacts on maturity. In addition, the genetic mechanisms underlying photoperiod sensitivity and adaptation in wild soybean seemed unique from those in cultivated soybean.

Materials and Methods

Soybean cultivars were selected from North America (Table 1), China and Russia (Table 2) [21–23]. Four wild soybean accessions were also included for genotyping only (Table 2), which are CAAE087 (Heiheyesheng) collected in Heihe (N 50°22′, E 127°53′), Heilongjiang, China; CAAE088 (Bayanyesheng) in Bayan (N 46°08′, E 127°39′), Heilongjiang, China; CAAE089 (Baiyangdianyesheng) in Baoding (N 38°51′, E 115°30′), Hebei, China; and CAAE090 (Guangxiyesheng) in Nanning (N 22°48′, E 108°19′), Guangxi, China. They covered 12 maturity groups from MG 000 to MG VIII and MG X and ranged from N 18° to N 53° indicated in Table 1 and Table 2.

Due to seed availability, 59 cultivars were selected for both field and pot experiments, 12 cultivars only in the field experiment, and 4 cultivars only in the pot experiment. For the field experiment, seeds were sowed on May 14th, 2012 in Beijing (N 39°97′, E 116°34′) and maturity-related traits of R1, R7 and R8 were recorded regularly [20]. For the pot experiment, seeds were sowed in 10-liter pots on May 16th 2012 and grown under ND in Beijing. After emergence, the seedlings were thinned until each pot contained five uniform plants. These uniform plants were grown until the unifoliate expanded then treated with different photoperiods (LD, 16 h light/8 h dark; SD, 12 h light/12 h dark; and ND). For SD, the plants were transferred to dark room to shorten the daylength. For LD, incandescent bulbs (50 μmol m^{-2} s^{-2} at the top of plants) with automatic timer controls were used to extend the daylength. Additional details of plant growth and treatments were the same as reported by Wu et al [24]. The days to first flowering of each plant was recorded. Both experiments finished on Oct 15th, 2012. Photoperiod sensitivity (PS) was thus calculated as the following function, where DFF_{LD} is the days to first flowering (R1) from the expansion of the first pair of unifoliates (V1) under LD while DFF_{SD} under SD [25].

$$PS(\%) = (DFF_{LD} - DFF_{SD})/DFF_{LD} \times 100\%$$

Genomic DNA was isolated from soybean unifoliate leaves using TianGen New Plant Genomic DNA Isolation Kit (DP320). Maturity loci *E1*, *E2*, *E3* and *E4* were genotyped using functional allele specific molecular markers [26].

Table 3. The days from VE to R1, R7 and R8 of soybean Cultivars under ND condition at Beijing.

Code	Phenotype			Code	Phenotype		
	VE-R1	VE-R7	VE-R8		VE-R1	VE-R7	VE-R8
CAAE001	20.7±1.7	61.0±3.8	70.9±1.8	**CAAE038**	122.3±1.0	NA	NA
CAAE002	20.0±1.8	63.9±3.4	71.5±1.9	**CAAE039**	20.4±0.8	61.5±4.8	71.7±2.0
CAAE003	20.1±0.4	57.2±1.0	68.0±0.0	**CAAE040**	19.0±0.0	67.0±6.7	76.7±3.4
CAAE005	20.3±1.8	68.4±1.6	73.7±1.0	**CAAE041**	20.1±0.4	69.7±0.5	76.7±1.3
CAAE006	19.9±1.3	69.3±3.0	77.0±2.2	**CAAE042**	20.4±1.5	71.3±1.8	78.3±2.1
CAAE007	20.3±2.0	70.1±2.2	77±2.2	**CAAE043**	19.1±0.5	71.0±2.2	81.3±6.5
CAAE008	22.4±1.6	93.4±10.6	108.5±15	**CAAE044**	21.1±0.7	71.3±2.4	80.7±6.3
CAAE009	21.9±1.2	86.9±2.1	100.7±3.6	**CAAE045**	20.2±0.4	71.3±1.8	78.3±2.1
CAAE010	22.5±1.4	85.3±2.0	90.9±1.6	**CAAE047**	23.1±2.3	83.9±1.3	94.4±4.2
CAAE011	25.8±1.9	81.1±11.1	103.9±5.9	**CAAE049**	23.4±1.2	83.9±9.9	101.4±7.0
CAAE012	21.9±1.4	89.7±1.6	101.3±2.6	**CAAE050**	24.9±1.1	98.6±5.2	109.9±7.3
CAAE013	22.7±1.4	94.2±1.8	109.9±4.7	**CAAE052**	24.5±0.8	99.9±2.8	114.2±2.4
CAAE014	21.6±0.5	92.5±3.0	100.3±3.1	**CAAE053**	23.1±0.3	113.6±3.0	121.7±2.2
CAAE015	25.2±1.9	102.1±7.3	113.9±5.3	**CAAE054**	37.4±0.6	119.8±6.6	127.3±7.0
CAAE016	24.9±1.0	106.3±7.0	118.5±3.5	**CAAE055**	42.7±2.3	127.0±8.0	136.8±5.1
CAAE017	26.9±1.8	117.1±2.3	125.5±1.5	**CAAE056**	36.4±2.0	113.7±6.3	121.1±2.7
CAAE018	24.7±0.6	117.9±1.8	124.9±3.3	**CAAE057**	31.3±1.0	111.4±2.5	119.9±1.4
CAAE019	28.6±1.2	123.4±2.7	131.6±3.3	**CAAE058**	24.2±1.0	95.9±2.8	106.2±5.2
CAAE020	24.2±1.2	116.4±2.8	125.7±1.9	**CAAE059**	25.5±0.5	111.2±3.1	118.7±2.6
CAAE021	26.3±1.3	123.4±3.3	133.4±2.9	**CAAE063**	41.9±2.1	116.0±2.0	122.0±0.0
CAAE022	32.5±4.4	125.1±4.1	134.6±1.4	**CAAE064**	41.3±1.0	101.9±3.7	111.3±4.7
CAAE023	56.1±3.9	131.9±7.8	139.3±2.4	**CAAE065**	27.1±1.2	103.9±9.1	115.7±5.1
CAAE024	50.6±2.6	132.9±6.1	141.0±3.9	**CAAE066**	42.5±1.5	116.0±3.2	126.8±6.3
CAAE025	53.7±1.1	140.6±4.5	142±0.0†	**CAAE071**	49.1±4.0	132.6±6.4	140.0±5.6
CAAE026	54.9±1.9	141.5±4.4	143.2±1.8	**CAAE072**	91.3±2.7	141.0±1.4	NA
CAAE027	61.8±4.6	142.5±4.9	142.0±0.0†	**CAAE073**	54.2±1.1	140.0±4.7	144.0±2.8†
CAAE028	63.3±5.5	141.8±4.3	144.0±1.7†	**CAAE074**	69.0±0.0	125.4±4.6	133.5±4.6
CAAE029	59.0±5.2	142.0±3.1	145.0±0.0†	**CAAE075**	94.0±2.1	NA	NA
CAAE030	75.1±2.0	141.9±3.0	145.0±0.0†	**CAAE076**	65.5±5.5	143.8±1.6	NA
CAAE031	77.7±3.3	142.8±1.1	NA	**CAAE077**	71.8±9.8	NA	NA
CAAE032	70.8±3.2	NA	NA	**CAAE081**	NA	NA	NA
CAAE033	70.4±3.1	145.0±0.0	NA	**CAAE082**	96.8±1.8	NA	NA
CAAE034	77.7±1.6	NA	NA	**CAAE084**	98.3±2.9	NA	NA
CAAE035	75.9±4.4	145.0±0.0	NA	**CAAE085**	95.1±2.6	NA	NA

Table 3. Cont.

Code	Phenotype			Code	Phenotype		
	VE-R1	VE-R7	VE-R8		VE-R1	VE-R7	VE-R8
CAAE036	74.3±4.5	NA	NA	**CAAE086**	NA	NA	NA
CAAE037	69.6±4.6	146.1±0.9	NA				

†Partial plants (<50%) matured. NA, not available.

Ethics Statement

No specific permissions were required for domestic research of the collections of wild soybean accessions in Heihe (N 50°22′, E 127°53′), Bayan (N 46°08′, E 127°39′), Baoding (N 38°51′, E 115°30′) and Nanning (N 22°48′, E 108°19′), China. All the field studies did not involve endangered or protected species.

Results

Soybean cultivars have diverse flowering and maturity dates

Seventy-one cultivars were planted under ND in the field at Beijing. These cultivars showed rich diversity in maturity (Table 3). Two cultivars CAAE081 and CAAE086 failed to flower until the experiment ended. Nine cultivars CAAE032, CAAE034, CAAE036, CAAE038, CAAE075, CAAE077, CAAE082, CAAE084 and CAAE085, flowered but did not reach R7. Six cultivars CAAE031, CAAE033, CAAE035, CAAE037, CAAE072 and CAAE076 did not reach R8 although they flowered and podded. For other cultivars that regularly flowered and matured, the days to R1 from emergence (VE) ranged from 19.0 (CAAE040) to 75.1 (CAAE030), the days to R7 ranged from 57.2 (CAAE003) to 142.5 (CAAE027), and the days to R8 ranged from 68.0 (CAAE003) to 145.0 (CAAE029 and CAAE030). Thus, the range of time to R1, R7 and R8 showed the maturity diversity of these cultivars. The cultivars that reached R1 less than 50 days after emergence could mature before the frost in Beijing, those that reached R1 in greater than 50 days but less than 70 days could mature partially, and the ones that reached R1 after 70 days could hardly mature (Table 3). In Figure 1 and Figure 2, maturity-related traits of R1, R7 and R8 generally increased from early MG to late MG and from high latitude to low latitude.

Soybean cultivars react variously to photoperiod treatments

Sixty-three cultivars were planted in pots. They were treated with different photoperiods after V1. They began flowering in 18.4 to 32.8 days after emergence under SD, in 20.0 to 122.3 days under ND, and in 20.5 to 113.7 under LD, while under LD three cultivars CAAE031 (PI531068, MGVII), CAAE084 (Qiudou 1), and CAAE086 (Zigongdongdou) did not flower (Table 4). The PS was calculated according to the equation [25], which ranged between 10% and 80%. For the three cultivars that did not flower, it could be set at 100%. Thus, CAAE035 (PI603953, MGVIII), CAAE072 (Mianyanghuangwofeng), CAAE085 (Nandou 12), CAAE075 (Ruijinxiaohuangdou) and the former three cultivars are most sensitive to photoperiod (PS>75%). CAAE001 (PI548594, MG000), CAAE002 (PI567787, MG000), CAAE005 (PI592523, MG00), CAAE007 (PI596541, MG0), CAAE042 (Heihe 3) and CAAE091 (Mohe 1) could be classified as photoperiod insensitive because their photoperiod sensitivities are lower than 10%. These data suggested that these soybean cultivars diversify significantly in photoperiod sensitivity. Moreover, as shown in Figure 2, the photoperiod sensitivity of these cultivars generally increased from early MG to late MG and from high to low latitude.

Genotyping soybean cultivars of maturity loci *E1*, *E2*, *E3* and *E4*

Eighty-five cultivars were genotyped at four maturity loci *E1*, *E2*, *E3* and *E4* (Table 4). Thirty-eight cultivars are from North America, which cover 12 maturity groups from MG000 to MGVIII and MGX. Other cultivars are from China except

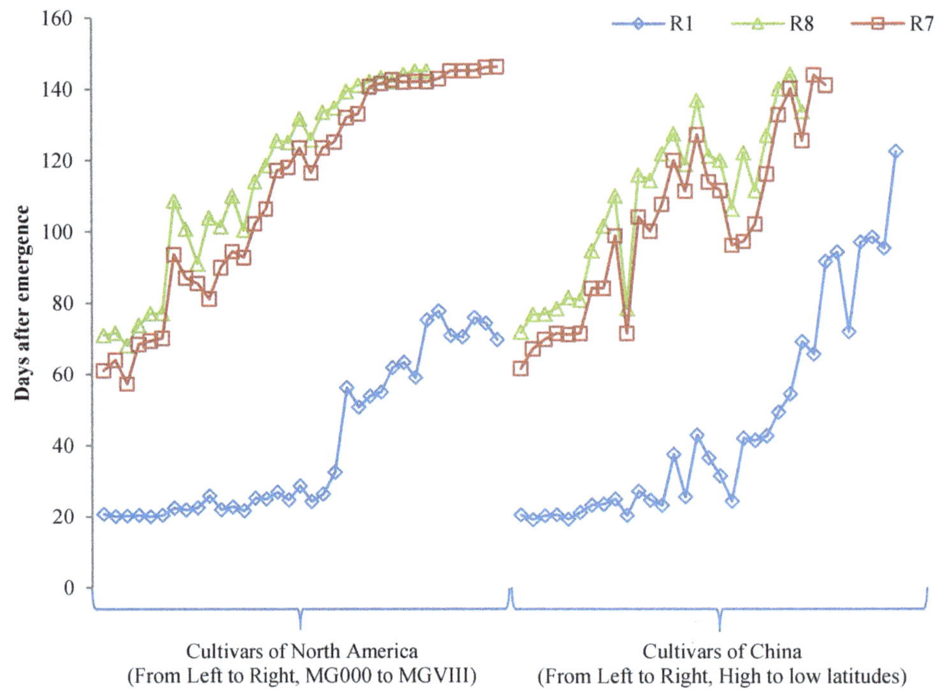

Figure 1. Maturity-related traits of soybean R1, R7 and R8 in the field at Beijing. Left, cultivars of North America, which are approximately sorted by maturity group. Right, cultivars of China, which are sorted roughly by adaption latitude.

CAAE092, which is from Russia (Tables 1 and 2). There are ten genotypes in total in this population (Table 5). The genotypes of *E1/E2/E3/E4* and *E1/e2/E3/E4* are the majority types, which were identified in 28 and 19 cultivars respectively (Table 5). Three genotypes of *e1-as/E2/e3/E4*, *e1-as/e2/e3/e4* and *e1/e2/e3/E4* were identified only in one variety each (Table 5). For the *E1* locus, allele *e1* and *el-as* are always detected in early-maturing cultivars from MG000 to MGIV or from high latitudes adapted cultivars. Moreover, all of the four wild soybeans were *E1/E2/E3/E4* type.

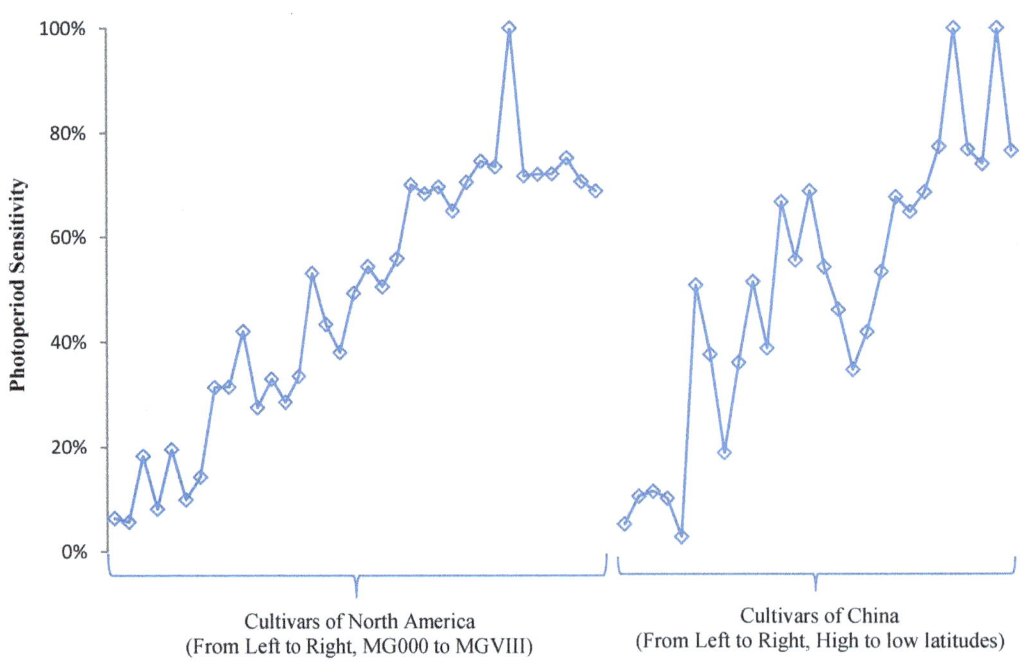

Figure 2. Photoperiod sensitivity of soybean cultivars. Left, cultivars of North America, which are approximately sorted by maturity group. Right: cultivars of China, which are sorted roughly by adaption latitude.

Table 4. The days to R1 from V1 under different photoperiod treatments and the resultant PS at Beijing.

Cultivar	Days from V1 to R1			PS (%)	Cultivar	Days from V1 to R1			PS (%)
	SD	ND	LD			SD	ND	LD	
CAAE001	21.4±1.1	20.4±0.7	22.9±2.4	6.3	**CAAE035**	26.9±1.3	83.2±1.1	108.2±8.9	75.2
CAAE002	22.6±1.1	20.9±1.4	23.9±2.2	5.6	**CAAE036**	26.7±2.4	83.9±1.4	90.6±0.5	70.6
CAAE003	21.4±1.5	24.7±1.7	26.2±3.3	18.2	**CAAE037**	27.6±0.7	73.9±3.1	88.4±0.5	68.8
CAAE005	21.4±2.0	20.0±0.6	23.2±0.9	8.1	**CAAE050**	23.1±0.9	30.2±1.4	37.1±5.0	37.6
CAAE006	21.7±2.1	22.0±2.5	27.0±5.4	19.5	**CAAE053**	23.1±1.7	28.9±2.2	37.8±6.5	38.8
CAAE007	20.6±1.3	20.1±0.6	22.9±1.7	9.8	**CAAE052**	22.9±0.8	31.9±1.9	47.2±3.8	51.5
CAAE008	24.1±3.2	25.5±0.9	28.1±4.1	14.2	**CAAE054**	24.1±1.1	44.1±1.8	54.3±3.0	55.6
CAAE009	22.6±2.7	29.8±1.9	32.9±3.4	31.4	**CAAE057**	27.0±0.8	45.0±2.7	50.2±1.9	46.2
CAAE010	22.8±1.7	28.5±2.1	33.2±1.6	31.4	**CAAE056**	28.0±1.7	49.5±1.4	61.3±3.2	54.3
CAAE011	24.8±1.6	32.5±1.5	42.7±3.4	42.0	**CAAE061**	26.7±3.1	50.3±4.6	54.3±7.1	50.9
CAAE012	25.1±3.0	27.9±2.3	34.6±3.1	27.5	**CAAE076**	27.3±0.8	73.3±1.5	86.9±0.9	68.5
CAAE013	22.2±2.3	28.6±2.1	33.1±2.5	33.0	**CAAE064**	32.8±2.0	45.9±2.4	56.5±2.8	41.9
CAAE014	24.8±1.9	28.7±2.6	34.7±4.7	28.5	**CAAE059**	24.1±1.4	44.8±1.6	77.4±2.0	68.8
CAAE015	26.7±2.1	31.2±2.6	40.1±3.7	33.5	**CAAE084**	30.8±1.9	106.4±0.7	NA	100.0
CAAE016	22.0±1.9	31.2±3.2	46.9±5.2	53.1	**CAAE045**	24.3±2.1	27.6±1.5	30.0±2.6	18.9
CAAE017	26.2±2.1	35.6±6.0	46.3±0.5	43.4	**CAAE041**	21.2±1.7	22.8±0.9	23.6±1.3	10.2
CAAE018	25.1±0.4	31.5±0.8	40.5±1.6	38.0	**CAAE042**	23.0±1.7	24.3±1.4	23.7±1.7	2.8
CAAE019	25.3±2.1	39.4±1.0	49.9±3.1	49.3	**CAAE040**	20.2±1.5	20.4±0.5	22.8±1.6	11.5
CAAE020	21.3±1.6	34.1±6.0	46.7±4.2	54.4	**CAAE065**	26.4±1.4	34.1±2.9	41.3±2.7	36.0
CAAE021	23.7±2.7	36.7±3.1	47.9±3.6	50.5	**CAAE066**	26.5±0.5	48.6±1.3	56.9±1.9	53.4
CAAE022	23.2±2.2	45.2±2.5	52.5±5.0	55.9	**CAAE074**	32.2±1.0	62.9±0.3	91.5±0.9	64.8
CAAE024	25.0±0.8	60.2±1.2	83.3±0.6	70.0	**CAAE073**	26.6±1.2	63.8±5.1	82.0±0.7	67.6
CAAE025	27.9±1.4	67.1±1.4	87.9±1.1	68.3	**CAAE072**	28.2±1.8	92.7±3.3	124.4±0.8	77.3
CAAE026	26.4±1.3	67.2±0.8	86.9±0.8	69.6	**CAAE091**	20.4±2.8	22.9±0.7	21.6±1.2	5.2
CAAE027	29.5±1.0	67.0±0.5	84.0±0.0	64.9	**CAAE085**	29.8±1.7	98.6±1.3	126.8±1.7	76.5
CAAE028	26.3±2.0	68.1±1.3	89.0±2.3	70.5	**CAAE075**	28.9±1.4	98.3±1.4	124.5±1.2	76.8
CAAE029	24.1±0.8	82.1±3.3	94.7±0.5	74.5	**CAAE058**	23.9±1.4	30.9±1.9	36.6±7.0	34.7
CAAE030	25.0±0.8	83.8±1.4	94.1±0.6	73.4	**CAAE077**	29.6±2.8	96.4±0.5	113.7±1.0	74.0
CAAE031	24.7±0.8	83.6±0.5	NA	100.0	**CAAE062**	27.6±0.9	55.3±2.3	83.0±1.4	66.7
CAAE032	25.1±0.7	72.6±1.3	88.6±1.1	71.7	**CAAE092**	18.4±2.0	20.9±1.6	20.5±1.8	10.6
CAAE033	24.9±0.4	73.9±2.2	88.9±1.9	72.0	**CAAE086**	32.8±0.8	122.3±0.8	NA	100.0
CAAE034	24.9±0.6	84.6±1.1	89.0±1.7	72.0					

NA, not available.

Table 5. Genotype of soybean cultivars at four maturity loci *E1, E2, E3* and *E4.*

Genotype				Num	Variety
E1	**E2**	**E3**	**E4**		
E1	*E2*	*E3*	*E4*	28	CAAE023 (MGV), CAAE024 (MGV), CAAE025 (MGV), CAAE026 (MGV), CAAE027 (MGVI), CAAE028 (MGVI), CAAE029 (MGVI), CAAE030 (MGVI), CAAE031 (MGVII), CAAE032 (MGVII), CAAE033 (MGVII), CAAE034 (MGVII), CAAE035 (MG VIII), CAAE036 (MG VIII), CAAE037 (MG VIII), CAAE038 (MGX), CAAE071, CAAE072, CAAE075, CAAE081, CAAE082, CAAE084, CAAE085, CAAE086, CAAE087, CAAE088, CAAE089, CAAE090
E1	*e2*	*E3*	*E4*	19	CAAE047, CAAE049, CAAE050, CAAE052, CAAE054, CAAE055, CAAE056, CAAE057, CAAE058, CAAE061, CAAE062, CAAE063, CAAE064, CAAE065, CAAE066, CAAE073, CAAE074, CAAE076, CAAE077
E1	*e2*	*e3*	*E4*	3	CAAE044, CAAE045, CAAE059
e1-as	*E2*	*E3*	*E4*	8	CAAE016 (MGIII), CAAE017 (MGIII), CAAE018 (MGIII), CAAE019 (MGIII), CAAE020 (MGIV), CAAE021 (MGIV), CAAE022 (MGIV), CAAE053
e1-as	*E2*	*e3*	*E4*	1	CAAE011 (MGI)
e1-as	*e2*	*E3*	*E4*	8	CAAE003 (MG00), CAAE008 (MG0), CAAE009, CAAE010 (MGI), CAAE012 (MGI), CAAE013 (MGII), CAAE014 (MGII), CAAE015 (MGII)
e1-as	*e2*	*e3*	*E4*	7	CAAE005 (MG00), CAAE006 (MG0), CAAE007 (MG0), CAAE040, CAAE041, CAAE042, CAAE043
e1-as	*e2*	*e3*	*e4*	1	CAAE039
e1	*e2*	*e3*	*E4*	1	CAAE092
e1	*e2*	*e3*	*e4*	3	CAAE001 (MG000), CAAE002 (MG000), CAAE091

Maturity loci *E1, E2, E3* and *E4* have different impacts on maturity and photoperiod response

In general, recessive alleles *e1*, *el-as*, *e2*, *e3* and *e4* promoted flowering and maturity but with different impacts (Figure 3). The allele *e4* was detected in only four cultivars (CAAE039 with the genotype *e1-as/e2/e3/e4*, and CAAE001, CAAE002 and CAAE091 with the genotype of *e1/e2/e3/e4*) in the population. The four cultivars with recessive *e4* alleles were adapted to high latitude and showed photoperiod insensitivity suggesting the importance of the *e4* allele for high latitude adaptation. The cultivars with the allele *e1*, *e1-as* or *e2* exhibited a narrower range of the days from VE to R1 (VE-R1) than that of the days from VE to R7 (VE-R7) and that of the days from R1 to R7 (R1–R7) (Figure 3). In contrast, the cultivars with the allele *e3* or *e4* showed a consistently narrow range although some outliers existed (Figure 3). Moreover, these four recessive alleles promoted flowering under different photoperiod conditions (Figure 4), and the cultivars with more recessive alleles of *e1*, *e1-as*, *e2*, *e3* and *e4* had a lower PS during photoperiod treatments (Figure 5).

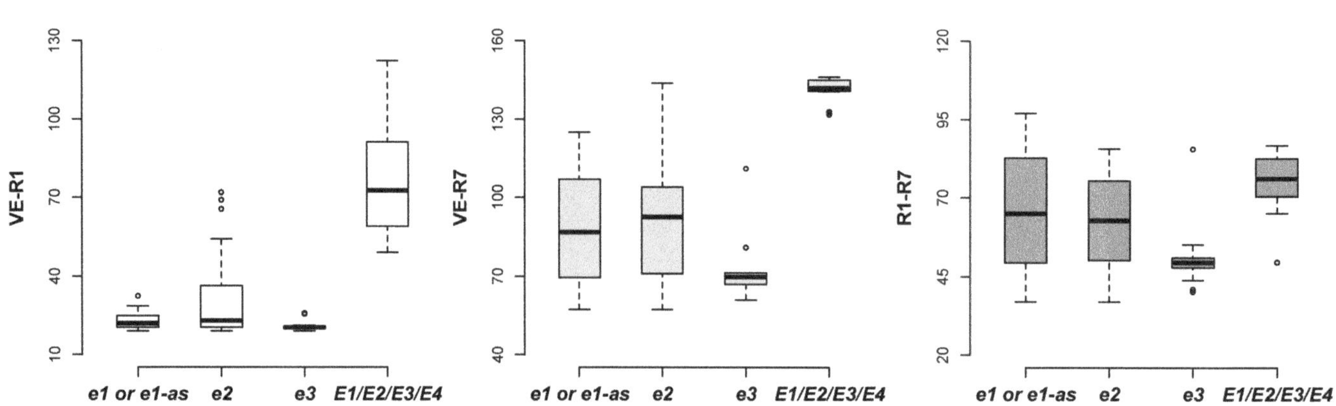

Figure 3. Quartile box plots showing days between the stages of VE, R1 and R7. Circles show outliers.

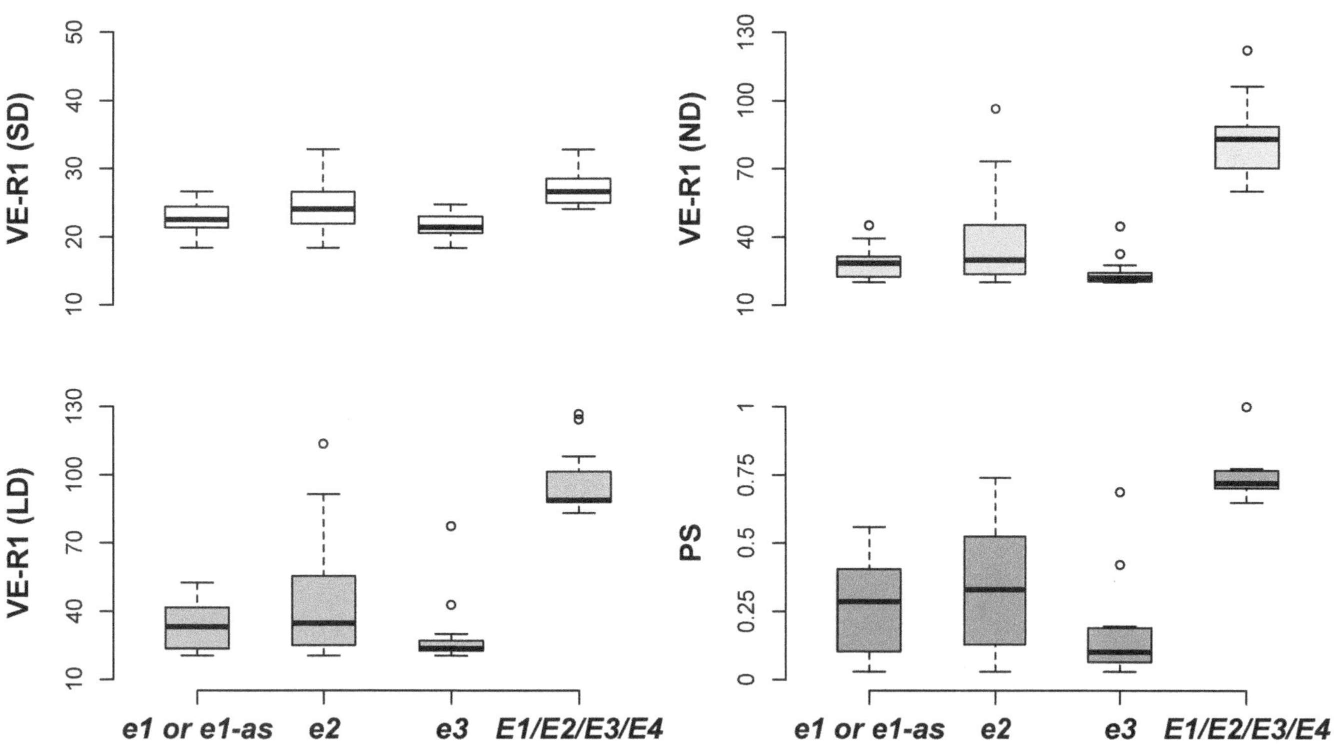

Figure 4. Quartile box plots showing days between the stages of VE and R1 under different photoperiod conditions and PS. Circles show outliers.

Discussion

Soybean cultivars from different maturity groups show diversity in flowering, maturity and photoperiod sensitivity

The tested soybean cultivars were selected from North America, China and Russia. Some of them covered from MG000 to MGVIII and MGX, almost all of the total 13 MG [2] and the others were collected from N 18° to N 53° to cover the wide range of latitude, which represents the main soybean producing area in China. In addition, some wild soybean accessions were also included. Thus, the population of lines used here should exhibit the diversity of maturity not only in phenotype but also in genotype. The field experiment under ND provided strong evidence. Some cultivars failed to flower (R1), some could not reach pod yellowing (R7) and some could not reach full maturity (R8) (Table 3). Even for those cultivars that flowered and matured, the days to first flowering, physiological maturity and full maturity varied significantly (Table 3). What's more, in the experiment of photoperiod treatments, the population also showed diversity of photoperiod sensitivity as expected. The days to first flowering varied between different photoperiod treatments and between different cultivars. The photoperiod sensitivity also varied from 10% to 80% (Table 4). Six cultivars were classified as photoperiod insensitive: CAAE001 (PI548594, MG000, *e1/e2/e3/e4*), CAAE002 (PI567787, MG000, *e1/e2/e3/e4*), CAAE005 (PI592523, MG00, *e1-as/e2/e3/E4*), CAAE007 (PI596541, MG0, *e1-as/e2/e3/E4*), CAAE042 (Heihe 3, *e1-as/e2/e3/E4*) and CAAE091 (Mohe 1, *e1/e2/e3/e4*). These results were consistent with previous report that soybean photoperiod insensitivity was at least conditioned by three genetic mechanisms according to allelic combinations of *E1*, *E2* and *E4*: *e3/e4*; *e1/e3 or e1/e4* and *e1-as/e3/E4*. In the genetic mechanism of *e1-as/e3/E4*, novel unidentified gene/genes participated in photoperiod insensitivity [26], [27].

E genes have different impacts on flowering and maturation

Soybean, as a short-day crop, has many cultivars with diversified maturity structure. *E1*, *E2* and *E3* are involved with different impacts. In the field experiment under ND, the recessive allele *e1*, *el-as* and *e2* significantly narrowed the variation of VE-R1 more than that of VE-R7 and R1–R7 (Figure 3), suggesting that *E1* and *E2* genes have significant impact on pre-flowering development other than post-flowering responses. However, the loci *E3* and *E4* might function not only in pre-flowering development but also in post-flowering development, indicated by the narrow variation in VE-R1, VE-R7 and R1–R7 of cultivars with *e3* and e4 alleles (Figure 3). This result is consistent with Xu et al [26]. that *E3* and *E4* respond not only to pre-flowering but also to post-flowering by increasing pod filling duration, number of nods and pod numbers by up-regulating the expression of growth habit gene *Dt1*. This result implies the significance of *E3* and *E4* loci for molecular genetic breeding to increase soybean productivity. Moreover, the outliers suggested that the disfunction of *E3* and *E4* might be interrupted by other genes. Similarly, the loci *E1* and *E3* are related with photoperiod sensitivity (Figure 4).

For the genotype of *E1/E2/E3/E4*, the photoperiod sensitivity was mostly above 70%. Compared with genotypes *E1/E2/E3/E4*, *e1-as/E2/E3/E4*, *e1-as/e2/E3/E4* and *e1-as/e2/e3/E4* (Figure 5), the photoperiod sensitivity decreased with the numbers of recessive

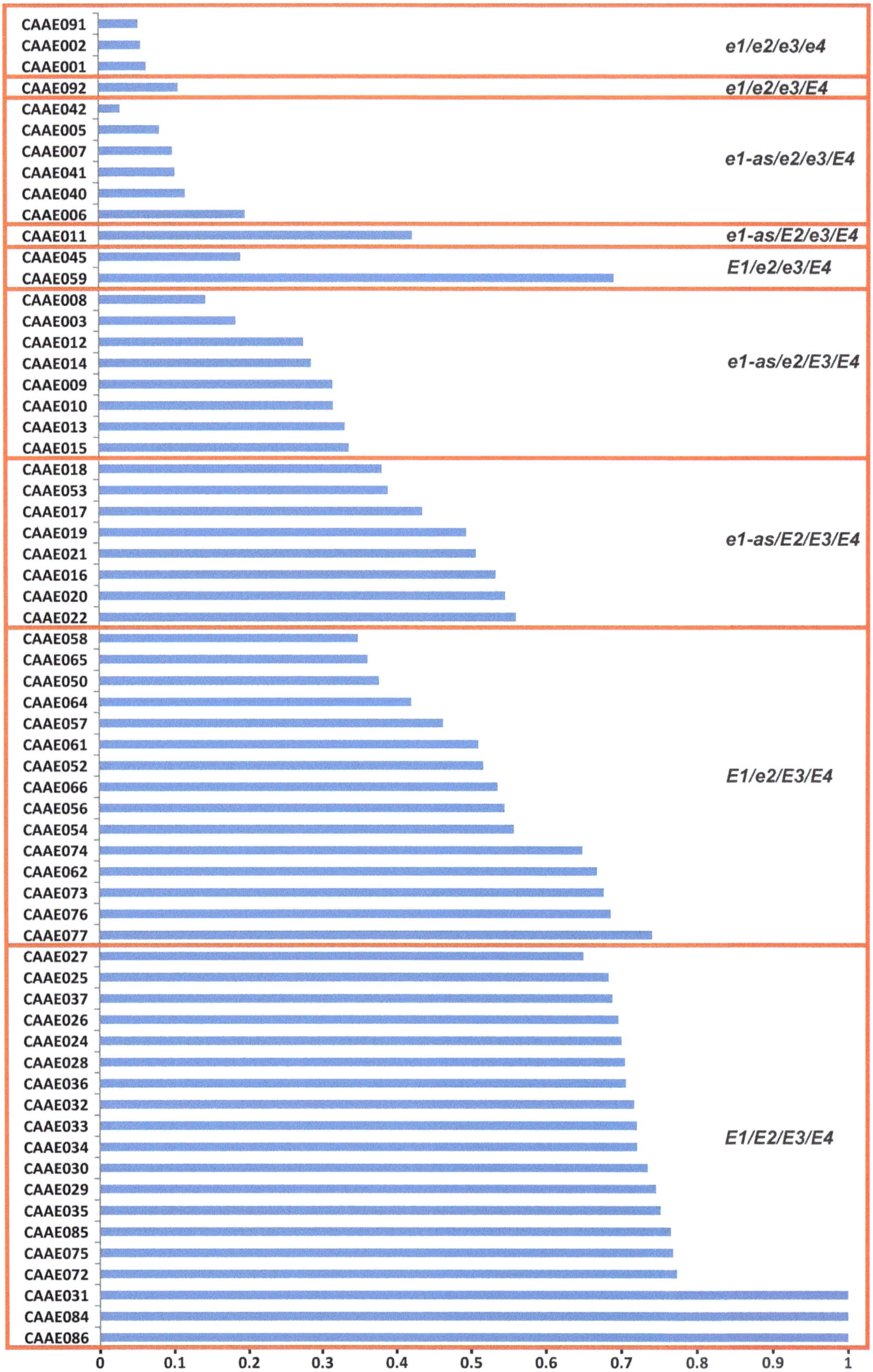

Figure 5. Photoperiod sensitivity grouped by *E* genotypes.

alleles. While comparing *E1/e2/E3/E4* and *e1-as/E2/E3/E4* with *E1E2E3E4*, it was suggested that *E1* plays a more important role than *E2* because *e1* decreased the photoperiod sensitivity more significantly and narrowed its range. These results further proved that soybean photoperiod insensitivity was involved by four maturity loci *E1, E3, E4* and *E7* while *E2* locus was not involved [28], [29]. In these tested cultivars of China, *E1/e2/E3/E4* is much more abundant. It ranged from N 18° -N 42° while *E1/E2/E3/E4* was distributed south of N 39° except for the four wild soybeans (CAAE087, Heiheyesheng; CAAE088, Bayanyesheng; CAAE089, Baiyangdianyesheng; and CAAE090, Guangxiyesheng). For *e1* or *el-as* alleles, the associated cultivars where mostly located above N 45° except for CAAE053 (Tiefeng 31, N 35°–40°). Thus, to improve the ecological adaptability of cultivars, the *E1* gene must function less because of its most strong impact on delaying maturity while other *E* genes might become important in adaptation. Unlike cultivated soybean, two wild soybean accessions, CAAE087 (Heiheyesheng) and CAAE088 (Bayanyesheng) adapted north of N 46° in China where cultivars generally had less photoperiod sensitivity where genotyped as *E1/E2/E3/E4*. It is greatly important for wild soybean to adapt and survive during season alteration. The genetic mechanisms underlying photoperiod insensitivity and adaptation in wild soybean were therefore unique from those in cultivated soybean.

Allelic combinations of *E* genes determine maturity groups

In the population analyzed here, MGV to MGX have the same genotype at *E1/E2/E3/E4*. It suggests that in these maturity groups, other unknown maturity genes should be involved in the determination of Mature Group. MGIII to MGIV are mainly genotyped as *e1-as/E2/E3/E4*, and MGII is *e1-as/e2/E3/E4*. For MGII to MGX, each group has one genotype. MGI has the genotypes of *e1-as/e2/E3/E4* and *e1-as/E2/e3/E4*; MG0 and MG00 both have the genotypes of *e1-as/e2/E3/E4* and *e1-as/e2/e3/E4*; and MG000 has *e1/e2/e3/e4*. From MG000 to MGI, each maturity group has two genotypes, which means that photoperiod is the first key factor in these region to determine maturity group. However, the more recessive alleles at *E* genes, the earlier cultivars mature. Due to the limited number of cultivars used in each MG, the maturity genotypes for each group may be underestimated and additional genotypes for each MG may be identified with a larger sample of cultivars. Although it is not possible to enumerate all genotypes of a given maturity group, this limited sample of cultivars showed that allelic combinations of *E* genes determine maturity groups in general.

Conclusions

The *E* genes (*E1, E2, E3* and *E4*) have different roles in maturity and photoperiod sensitivity and their allelic combinations determine maturity group and adaptation to different latitude.

Acknowledgments

We thank Randall L. Nelson (Soybean/Maize Germplasm, Pathology and Genetics Research Unit, USDA and Department of Crop Sciences, University of Illinois, Urbana, Illinois, USA), Lijuan Qiu and Zhangxiong Liu (Institute of Crop Sciences, Chinese Agricultural Academy of Sciences, China) for providing some soybean germplasm. We also appreciate Dr. Elroy R. Cober (Eastern Cereal and Oilseed Research Centre, Agriculture and Agri-Food Canada, Canada) and Dr. Jun Abe (Research Faculty of Agriculture, Hokkaido University) for proofreading of the manuscript.

Author Contributions

Conceived and designed the experiments: BJ FK TH BL. Performed the experiments: BJ HN YG LT YY SL LM JW SS DC. Analyzed the data: BJ HN YG CW XY WH. Contributed reagents/materials/analysis tools: FK TH BL. Contributed to the writing of the manuscript: BJ FK.

References

1. Cober ER, Voldeng HD (2001) A new soybean maturity and photoperiod-sensitivity locus linked to *E1* and *T*. Crop Sci. 41: 698–701.
2. Hartwig E (1970) Growth and reproductive characteristics of soybeans [*Glycine max* (L) Merr.] grown under short-day conditions. Trop. Sci. 12: 47–53.
3. Zhang L, Kyei-Boahen S, Zhang J, Zhang M, Freeland T, et al. (2007) Modifications of optimum adaptation zones for soybean maturity groups in the USA. Crop Manage. http://dx. doi. org/10.1094/ CM-2007-0927-01-RS.
4. Hao G, Chen X, Pu M (1992) Maturity groups of soybean cultivars in China. Acta Agr. Sin. 18: 275–281.
5. Wang G (1981) The research about the ecological division of soybean cultivars in China. Sci. Agr. Sin.: 39–46.
6. Bernard R (1971) Two major genes for time of flowering and maturity in soybeans. Crop sci. 11: 242–244.
7. Bonato ER, Vello NA (1999) *E6*, a dominant gene conditioning early flowering and maturity in soybeans. Genet. Mol. Biol. 22: 229–232.
8. Buzzell R (1971) Inheritance of a soybean flowering response to fluorescent-daylength conditions. Can. J. Genet. Cytol. 13: 703–707.
9. Buzzell R, Voldeng HD (1980) Inheritance of insensitivity to long daylength. Soyb. Genet. Newsl. 7: 26–29.
10. Cober ER, Molnar SJ, Charette M, Voldeng HD (2010) A new locus for early maturity in soybean. Crop sci. 50: 524–527.
11. McBlain B, Bernard R (1987) A new gene affecting the time of flowering and maturity in soybeans. J. Hered. 78: 160–162.
12. Ray JD, Hinson K, Mankono J, Malo MF (1995) Genetic control of a long-juvenile trait in soybean. Crop sci. 35: 1001–1006.
13. Wang Y, Wu C, Zhang X, Wang Y, Han TF (2008) Effects of soybean major maturity genes under different photoperiods. Acta Agr. Sin. 34: 1160–1168.
14. Xia Z, Zhai H, Liu B, Kong F, Yuan X, et al. (2012) Molecular identification of genes controlling flowering time, maturity, and photoperiod response in soybean. Plant Syst. Evol.: 1217–1227.
15. Xia Z, Watanabe S, Yamada T, Tsubokura Y, Nakashima H, et al. (2012) Positional cloning and characterization reveal the molecular basis for soybean maturity locus *E1* that regulates photoperiodic flowering. Proc. Natl. Acad. Sci. 109: E2155–E2164.
16. Watanabe S, Xia Z, Hideshima R, Tsubokura Y, Sato S, et al. (2011) A map-based cloning strategy employing a residual heterozygous line reveals that the *GIGANTEA* gene is involved in soybean maturity and flowering. Genetics 188: 395–407.
17. Watanabe S, Hideshima R, Xia Z, Tsubokura Y, Sato S, et al. (2009) Map-based cloning of the gene associated with the soybean maturity locus *E3*. Genetics 182: 1251–1262.
18. Liu B, Kanazawa A, Matsumura H, Takahashi R, Harada K, et al. (2008) Genetic redundancy in soybean photoresponses associated with duplication of the *phytochrome A* gene. Genetics 180: 995–1007.
19. Kong F, Liu B, Xia Z, Sato S, Kim BM, et al. (2010) Two coordinately regulated homologs of *FLOWERING LOCUS T* are involved in the control of photoperiodic flowering in soybean. Plant physiol. 154: 1220–1231.
20. Fehr W, Caviness C, Burmood D, Pennington J (1971) Stage of development descriptions for soybeans, *Glycine max* (L.) Merrill. Crop sci. 11: 929–931.
21. Jilin Academy of Agricultural Sciences (1985) Soybean cultivars in China. 1 ed. Agriculture Press, Beijing.
22. Soybean Institute, Jilin Academy of Agricultural Sciences (1993) Soybean cultivars in China (1978–1992). 1 ed. Agr. Press, Beijing.
23. Li WD, Zhang MC (2006) Huang-Huai-Hai summer-sowed soybeans and their parameters. 1 ed. China Agr. Sci. Technol. Press, Beijing.
24. Wu C, Ma Q, Yam KM, Cheung MY, Xu Y, et al. (2006) In situ expression of the *GmNMH7* gene is photoperiod-dependent in a unique soybean (*Glycine max* [L.] Merr.) flowering reversion system. Planta 223: 725–735.
25. Fei Z, Wu C, Sun H, Hou W, Zhang B, et al. (2009) Identification of photothermal responses in soybean by integrating photoperiod treatments with planting-date experiments. Acta Agr. Sin.: 1525–1531.
26. Xu M, Xu Z, Liu B, Kong F, Tsubokura Y, et al. (2013) Genetic variation in four maturity genes affects photoperiod insensitivity and *PHYA*-regulated post-flowering responses of soybean. BMC Plant Biol. 13: 91.
27. Tsubokura Y, Matsumura H, Xu M, Liu B, Nakashima H, et al. (2013) Genetic variation in soybean at the maturity locus *E4* is involved in adaptation to long days at high latitudes. Agronomy 3: 117–134.
28. Cober ER, Tanner J, Voldeng HD (1996) Genetic control of photoperiod response in early-maturing, near-isogenic soybean lines. Crop sci. 36: 601–605.
29. Cober ER, Tanner J, Voldeng HD (1996) Soybean photoperiod-sensitivity loci respond differentially to light quality. Crop sci. 36: 606–610.

A Soybean C2H2-Type Zinc Finger Gene *GmZF1* Enhanced Cold Tolerance in Transgenic *Arabidopsis*

Guo-Hong Yu[1¶], Lin-Lin Jiang[1¶], Xue-Feng Ma[1,2], Zhao-Shi Xu[3], Meng-Meng Liu[1], Shu-Guang Shan[1], Xian-Guo Cheng[1*]

1 Key Lab. of Plant Nutrition and Fertilizer, Ministry of Agriculture, Institute of Agricultural Resources and Regional Planning, Chinese Academy of Agricultural Sciences, Beijing, China, **2** Institute of Agro-Products Processing Science and Technology, Chinese Academy of Agricultural Sciences, Beijing, China, **3** Institute of Crop Science, Chinese Academy of Agricultural Sciences (CAAS)/National Key Facility for Crop Gene Resources and Genetic Improvement, Key Laboratory of Biology and Genetic Improvement of Triticeae Crops, Ministry of Agriculture, Beijing, China

Abstract

Zinc finger proteins were involved in response to different environmental stresses in plant species. A typical Cys2/His2-type (C2H2-type) zinc finger gene *GmZF1* from soybean was isolated and was composed of 172 amino acids containing two conserved C2H2-type zinc finger domains. Phylogenetic analysis showed that GmZF1 was clustered on the same branch with six C2H2-type ZFPs from dicotyledonous plants excepting for GsZFP1, and distinguished those from monocotyledon species. The GmZF1 protein was localized at the nucleus, and has specific binding activity with EP1S core sequence, and nucleotide mutation in the core sequence of *EPSPS* promoter changed the binding ability between GmZF1 protein and core DNA element, implying that two amino acid residues, G and C boxed in core sequence TGACAGTGTCA possibly play positive regulation role in recognizing DNA-binding sites in GmZF1 proteins. High accumulation of *GmZF1* mRNA induced by exogenous ABA suggested that *GmZF1* was involved in an ABA-dependent signal transduction pathway. Over-expression of *GmZF1* significantly improved the contents of proline and soluble sugar and decreased the MDA contents in the transgenic lines exposed to cold stress, indicating that transgenic *Arabidopsis* carrying *GmZF1* gene have adaptive mechanisms to cold stress. Over-expression of *GmZF1* also increased the expression of cold-regulated *cor6.6* gene by probably recognizing protein-DNA binding sites, suggesting that *GmZF1* from soybean could enhance the tolerance of *Arabidopsis* to cold stress by regulating expression of cold-regulation gene in the transgenic *Arabidopsis*.

Editor: Fan Chen, Institute of Genetics and Developmental Biology, Chinese Academy of Sciences, China

Funding: This work was supported by the National Key Project for Cultivation of New Varieties of Genetically Modified Organisms (2013ZX08002-005). The funders had no role in study design, data collection and analysis, decision to publish, or preparation of the manuscript.

Competing Interests: The authors have declared that no competing interests exist.

* Email: chengxianguo@caas.cn

¶ These authors contributed equally to this work.

Introduction

Plants are usually exposed to various environmental stress factors affecting plant growth and crop productivity, such as drought, high salt and low temperature. As an adaptive response, the plants have developed a resistance mechanism to abiotic stresses to achieve an optimal adaptation [1]. It was reported that a number of transcription factor gene from different plants was induced or repressed during the responses and acclimations [2]. For example, some typical transcription factors, AP2/ERF, bZIP, NAC, MYB, MYC, WRKY and zinc finger proteins have been well characterized, and confirmed to be involved in stress response via transcriptional regulation modulation [3,4,5], and these transcription factors usually have specific structure and/or conserved domains functioning in the adaptation to abiotic stresses [6,7].

The first zinc finger transcription factor IIIA (TFIIIA) was recognized as a repeated zinc-binding motif in *Xenopus*, and is a classical Cys2/His2-type (C2H2-type) zinc finger gene [6]. The C2H2-type zinc finger has been well characterized in the eukaryotic transcription factors having specific $CX_{2-4}CX_3FX_5LX_2HX_{3-5}H$ structure (X: any amino acid; number: amino acid amounts), and this conserved motif contains two pairs of specific Cys and His residues which form a tetrahedral structure with a zinc ion [7]. In plants, the C2H2-type zinc finger has a highly conserved QALGGH motif in a putative DNA-contacting surface, which is unique in zinc-finger proteins from plants [8,9]. *In vitro* analysis revealed that the conserved QALGGH motif in plants plays a critical role in DNA binding activity, and each amino acid residue in the motif seems to be essential for maintaining effective binding between DNA and C2H2-type zinc finger proteins [10,11], and these specific domains conferred that zinc-finger proteins play important regulatory role by recognizing the target sequences and regulating expression of target gene in a plant-specific manner [1,8,12].

As a first C2H2-type zinc finger protein in plants, ZPT2-1 was found in petunia (*Petunia hybrid*), and identified to interact with the specific DNA sequences in the promoter region of 5-enolpyruvylshikimate-3-phosphate synthase (EPSPS) [1,13]. Thereafter, several TFIIIA-type zinc finger proteins have been consecutively reported [1,8], and these C2H2-type zinc finger genes paly important regulation roles in responses to various abiotic stresses [8,9,14]. In petunia, the TFIIIA-type zinc finger genes, *ZPT2-2* and *ZPT2-3*, were regulated by cold and/or drought, and over-expression of *ZPT2-3* gene in the transgenic petunia increased the tolerance of plants to drought stress [15,16]. *STZ* is one of the *ZPT2*-related genes in *Arabidopsis*, and was isolated by complementation of the salt-sensitive phenotype with a yeast calcineurin mutant [17], and the expression of *STZ* gene has been identified to be responsive to drought, salt, cold and abscisic acid (ABA), and that constitutive expression of *STZ* resulted in a suppression in growth, and accompanied an enhancement of plants adaptation to drought and osmotic stresses [18,19,20].

To date, few zinc finger proteins from cultivar soybean (*Glycine max*) have been isolated and characterized. *SCOF-1,* a typical C2H2-type zinc finger gene, was isolated from soybean and was confirmed to have a positive role in regulating the expression of cold-regulated gene and enhancing cold tolerance in transgenic plants [21]. *GsZFP1*, a zinc finger transcription factor lacking typical QALGGH motif was isolated from wild soybean (*Glycine soja* L. G07256), and transformed into *Arabidopsis*, and confirmed to play a crucial role in withstanding cold and drought stresses [22], indicating that the presence of QALGGH motifs in the C2H2 zinc finger *GsZFP1* gene from wild soybean seems to be not a crucial element in functioning during adaptation to abitotic stress. In our study, a novel C2H2-type zinc finger gene, *GmZF1* was isolated from cultivar soybean, and identified to have two typical conserved QALGGH motifs, and is significantly different from *GsZFP1* gene in wild soybean [22]. To our knowledge, the understanding on the function of zinc finger C2H2 from cultivar soybean in responses to cold stress is still limited. Therefore, based on the structure analysis and *in vitro* identification of *GmZF1* gene, we transformed the *GmZF1* gene into *Arabidopsis* plants and identified three homozygous lines which were subsequently used to observe the function effects of *GmZF1* gene in responses to cold stress and exogenous ABA. Our data showed that *GmZF1* from cultivar soybean was induced by ABA, and over-expression of *GmZF1* gene significantly enhanced the tolerance of the transgenic *Arabidopsis* to low temperature stress.

Results

Characteristics of the *GmZF1* gene

The *GmZF1* cDNA is composed of 765-bp nucleotides, and eocodes a predicted protein of 172 amino acids with a calculated molecular mass of 19.2 kDa, and flanked by a 97 bp fragment at 5′ end and a 149 bp fragment at 3′ end at two untranslating regions, respectively (accession number DQ055134). Electrophoresis analysis showed that PCR product from the genomic DNA of soybean seedlings has the same size as that from the cDNA length generated by RT-PCR using total RNA as remplate, and sequencing also showed that the generated *GmZF1* clones from gDNA and cDNA have identical sequence, indicating that *GmZF1* gene has no intron in the genomic DNA. A homology search against the GenBank database showed that GmZF1 is a homolog of C2H2-type zinc finger proteins (ZFPs), and has two typical C2/H2 type zinc finger domains ($CX_2CX_3FX_3QALGGHX_3H$) (Fig. 1a). GmZF1 has 57%, 48%, 47% and 46% homology with the C2H2-type ZFPs from *Medicago truncatula*, Petunia, *Arabidopsis* and rice, respectively, and no highly homology was observed between these genes from different plants, suggesting that *GmZF1* belongs to a novel subfamily member of C2H2 zinc finger. Like C2H2 zinc finger genes from four kinds of plants mentioned above, the GmZF1 protein also has plant-specific QALGGH motifs as well as a conserved motif L-box with more Leu residues, and contains a short hydrophobic region with a highly conserved DLN box near the C-terminus of GmZF1 (Fig. 1a), which may function as a transcription repression domain [16,18]. It is noteworthy that GmZF1 have two conserved domains with QALGGH motifs, and is significantly different from *GsZFP1* isolated from wild soybean, the later only has one C2H2 domain lacking QALGGH motif, and the number of amino acid residues between two His (H) residues is composed of four amino acids, which is different from GmZF1 having three amino acids between two His amino acids (as indicated by # and * respectively in Fig. 1a), and no DLN box was found at the C-terminus of GsZFP1 [22].

For profiling the differences in genetic characteristics between C2H2 family members from different plants, a systematic phylogenetic analysis was carried out. The results showed that GmZF1 was clustered on the same branch with six C2H2-type ZFPs from dicotyledonous plants, and distinguished them from monocotyledon species (Fig. 1b), indicating that dicotyledonous plants and monocotyledon plants have significant difference in genetic characteristics of C2H2-type genes, and probably play different regulation roles in plant responses to abiotic stresses. What's worthy reminding is that *GmZF1* from cultivar soybean and *GsZFP1* from wild soybean respectively belong to two different subfamily members in C2H2 zinc finger super family.

Analysis for SDS-PAGE and DNA-binding activity of GmZF1

Analysis of SDS-PAGE indicated that the GmZF1::GST fusion protein with 45.2 kD molecular weight was successfully expressed in *E. coli* strain (Fig. 2a), and the size of expressed protein is identical with the predicted molecular weight of target protein in the fusion vector vector pGEX-4 T-1. As a DNA-binding motif, a C2H2-type zinc-finger domain in the GmZF1 has been identified in many transcription factors. In our study, we observed that the transcription factors containing the zinc finger motif could specifically bind to EP1S core sequence (TGACAGTGTCA), which was originally identified as a *cis*-element within the *EPSPS* gene promoter in petunia [13,23]. For identifying DNA-binding ability of GmZF1, a gel-shift assay was performed *in vitro* using the procedure as described in the method. Data showed that GmZF1 protein was expressed as fusion proteins with GST in *E. coli.* To better understand the molecular mechanism of specific-binding between the GmZF1 proteins and EP1S core sequences, a series of synthesis probes including wild type and mutants was prepared as the following descriptions. Briefly, the EP1S core sequence (wild-type, E1) and two probes (E2: bases substitution, E3: with four bases substitution) in the repeated sequences were prepared (Fig. 2b), and fusion protein was purified for gel shift assay. Assay showed that the GmZF1::GST fusion protein strongly bound to E1 and E2 probes, but weakly binding to E3 probe (Fig. 2c), no complex was found in reaction solution containing GST protein and probes, suggesting that GmZF1 protein could bind specifically to the EP1S core sequence *in vitro*, and two amino acid residues, G and C boxed in core sequence of TGACAGTGTCA probably play key role in recognizing DNA-binding sites in GmZF1 proteins.

Subcellular localization of GmZF1

To identify the subcellular localization of GmZF1, we fused the full length of GmZF1 to GFP vector under control of the constitutive 35S promoter. Both the recombinant DNA constructs encoding a GmZF1::GFP fusion protein and a GFP protein were respectively introduced into *Arabidopsis* protoplast cells.

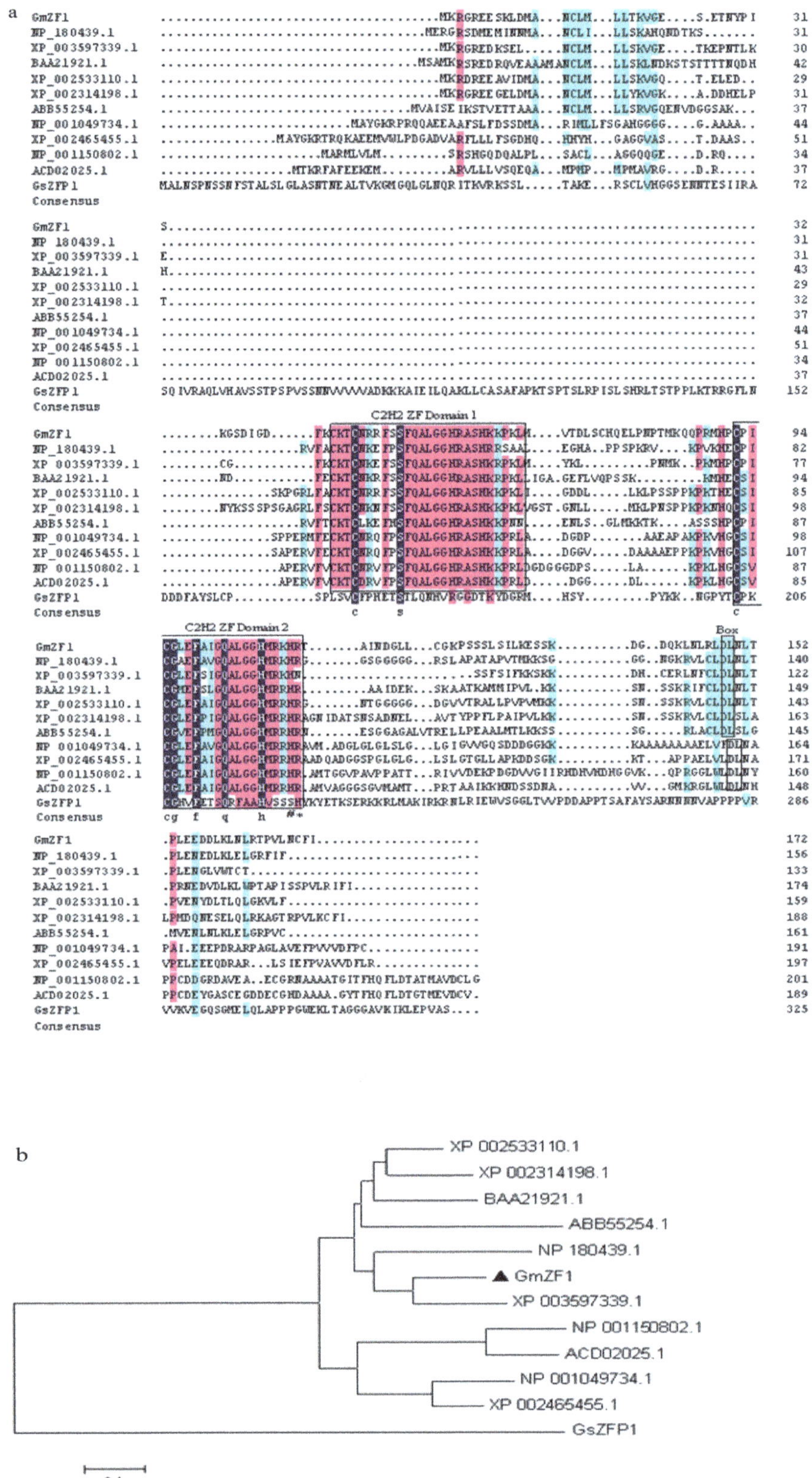

Figure 1. Sequence characteristic of GmZF1 deduced amino acids. a: Multiple alignment of GmZF1 amino acid sequences with XP_003597339.1 (*Medicago truncatula*), BAA21921.1 (*Petunia hybrid*)), NP_180439.1 (*Arabidopsis*) and NP_001049734.1 (*Oryza sativa*). The two zinc finger domains and the conserved DLN amino residues are boxed. b: Phylogenetic analysis of GmZF1 and related proteins. A mid-point rooted neighbor-joining phylogeny was constructed by 12 amino acid sequences from diverse organisms. Excepting for GmZF1, the other C2H2 type zinc finger proteins were demonstrated by XP_002533110.1 (*Ricinus communis*), XP_003597339.1 (*M. truncatula*), BAA21921.1 (*Petunia*), XP_002314198.1 (*Populus*), NP_180439.1 (*Arabidopsis*), ABB55254.1 (*Brassica*), NP_001049734.1 (*Oryza sativa*), XP_002465455.*1* (*Broomcorn*), NP_001150802.1 (*Zea mays*), *ACD02025*.1 (*Triticum aestivum*) and GsZFP1 (*Glycine sojia*. L).

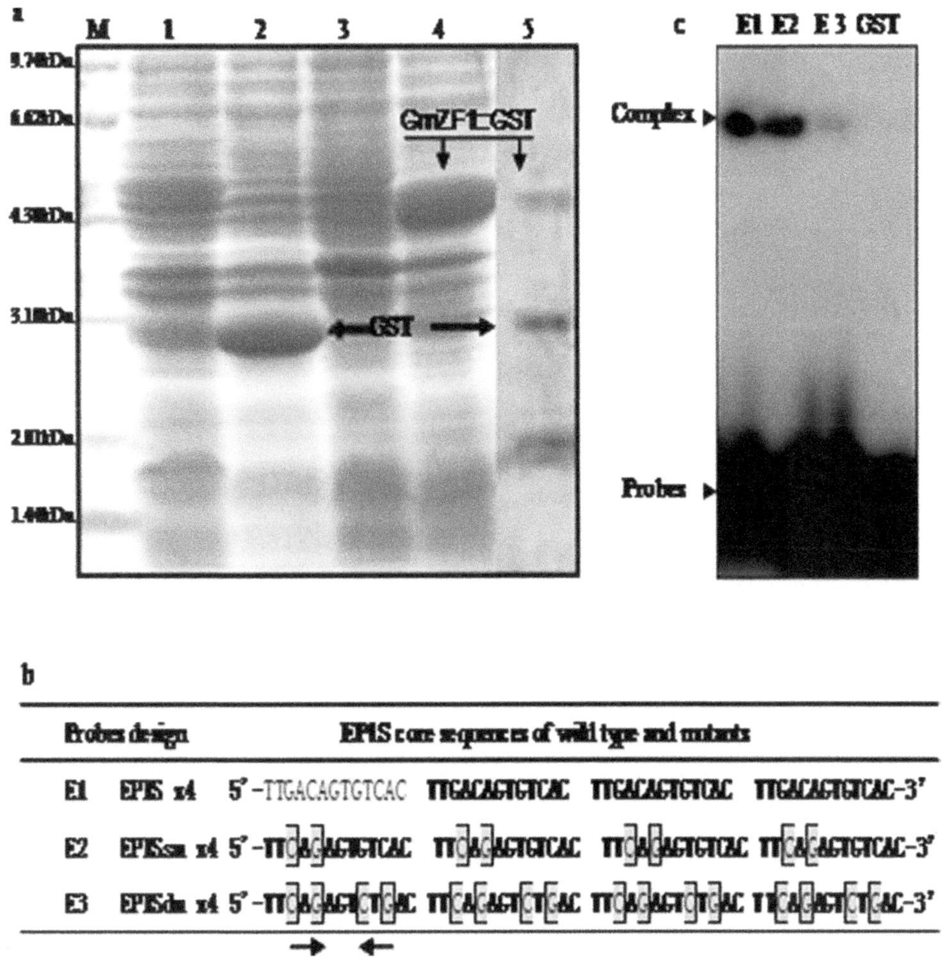

Figure 2. SDS PAGE induction expression and gel shift assay for DNA-binding activities *in vitro*. a: Induction expression and purification of target proteins, GmZF1::GST fusion protein and GST protein. Lane 1-uninduced strains containing pGEX4T-1 vector; Lane 2-induced strains containing pGEX4T-1; Lane-3-uninduced strains containing GmZF1::GST fusion; Lane 4-induced strains containing GmZF1::GST fusion; Lane 5-purified target protein as indicated by arrows. b: Probe design and composition. Nucleotide sequences of EP1S and mutated EP1S (E1, E2 and E3) probes. The nucleotide mutations in the EP1S core motif for each probe are boxed. c: The gel shift assay was performed in a solution containing 0.2 µg GmZF1::GST proteins or GST proteins and the probes ^{32}P-labeled EP1S (E1) or mutant EP1S (E2 and E3), respectively. The GmZF1::EP1S complex and free probes are indicated by arrows.

Localization of the GmZF1::GFP fusion protein was visualized exclusively in the nucleus (Fig. 3b), whereas the control GFP (35S::GFP) was distributed throughout the protoplast cells (Fig. 3a), demonstrating that GmZF1 is a nuclear protein possibly functioning as a transcriptional activator.

Expression of *GmZF1* in response to stress

To examine the transcription profile of *GmZF1* in soybean under cold stress, the expression of *GmZF1* in soybean was detected using semi-quantitative RT-PCR. The result showed that *GmZF1* was weakly expressed in young seedlings, and induced by low temperature and reached a maximum at 5 h after 4°C stress (Fig. 3c). Data also showed that *GmZF1* gene could be induced by exogenous ABA, and seems to have similar expression pattern under cold stress, and maintained at a high expression level when the seedlings were kept at ABA treatment (Fig. 3d), suggesting that *GmZF1* might be involved in plant responses to cold stress via an ABA-dependent pathway.

Phenotype changes of plants exposed to cold stress

To observe the difference in germination and survival rates between the wild type and the transgenic *Arabidopsis*, two independent culture experiments was performed at 22°C and at 4°C, respectively (Fig. 4a, b) Fig. 4a showed that both wild type and transgenic lines almost exhibited identical germination rates under favorable condition, but the survival rate of transgenic line OE1 was significantly higher than that of wild types under cold stress when wild type and transgenic lines of one-week-old seedlings were exposed to 4°C for one week (Fig. 4e). For observing the phenotypic changes of wild type and transgenic *Arabidopsis* under cold stress, the seedlings of two-week-old were transferred onto the MS medium, and cultured at 22°C (Fig. 4c) and 4°C (Fig. 4d) for one week, respectively (note: white bacterial plaques appeared in the mediums because of possible contaminant from culture processes). Data showed that transgenic lines showed a better phenotypic characteristic in withstanding cold stress compared to wild types (Fig. 4d), because both fresh weight and root lengths from the transgenic lines were significantly increased comparing to the wild types (Fig. 4f, g). These data suggested that *GmZF1* in the transgenic *Arabidopsis* could enhance the tolerance

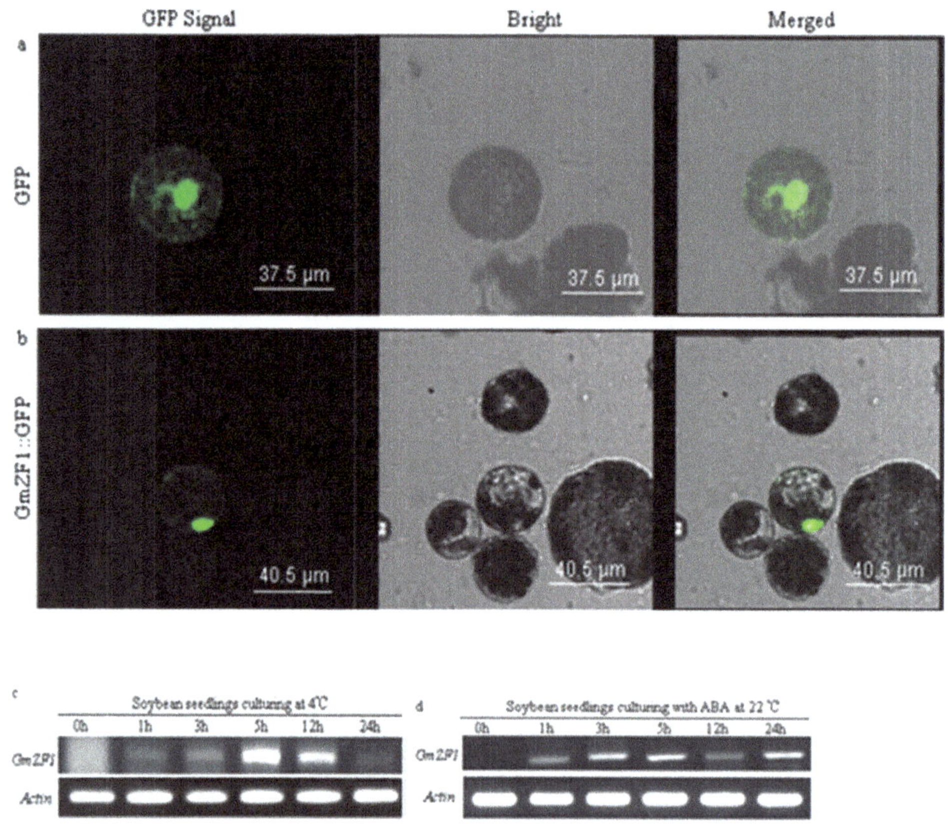

Figure 3. Subcellular localization of the GmZF1 proteins and mRNA accumulation of *GmZF1* gene in soybean seedlings respectively exposed to cold and ABA. a: Images expressing h16318-GFP in protoplast cells; b: images expressing the GmZF1::GFP fusion protein, fluorescent-field illumination was used to examine GFP signal (left); followed by bright-field illumination (middle) and confocal microscopy (right) for an merged image of bright and fluorescent illumination. c: the relative accumulation of *GmZF1* mRNA in soybean seedlings of 10-day-old exposed to 4°C at different time points; d: the accumulation of GmZF1 mRNA in soybean seedlings of 10-day-old exposed to ABA of 200 μM at different time points.

plants to the cold stress. *GmZF1* and *GsZFP1* [22] seems to play similar roles in acclimating the tolerance of plants to cold stress, although these two genes have differences in the structure, especially in the conserved domains of C2H2 zinc finger gene, suggesting that different zinc finger genes in the same super family could demonstrate an uniformity in gene functions even though they have the difference in the genetic characteristics, and this seems to imply that the diversity genetic of genes is an essential factor in regulating the expression of target genes related to abiotic stresses. Therefore, to better elucidate the molecular regulation mechanism of *GmZF1* from cultivar and *GsZFP1* from wild soybean in responding to cold stress, further investigations is necessary.

Over-expression of *GmZF1* led to physiological changes in the transgenic lines under cold stress

To further characterize the function of *GmZF1*, we generated *Arabidopsis* transgenic plants carrying *35S::GmZF1*, in which *GmZF1* was driven by the CaMV35S promoter. Total 22 transformed *Arabidopsis* plants with kanamycin-resistant were obtained, and three independent homozygous transgenic lines (OE1, OE4 and OE7) carrying *GmZF1* were continuously selected by the PCR analysis until T3 transgenic *Arabidopsis* lines. For cold stress, 5-week-old wild-type and transgenic lines were transferred to the incubator at 4°C for two days, and morphological observation showed that the transgenic lines were not significantly inhibited in growth comparing to wild type in spite of being a short term cold stress, and the growth of the transgenic lines carrying *GmZF1* gene were slightly damaged comparing to wild type (Fig. 5a, b). However, three transgenic lines have remarkable physiological changes in accumulation of the proline, soluble sugar and MDA in the leaves comparing to the wild type after cold stress. Cold stress enhanced accumulation of proline and soluble sugar in the wild type and transgenic lines, but the increase times in the contents of proline in the transgenic lines (OE1, 1.3-fold; OE4, 1.3-fold and OE7, 1.6-fold) were significantly higher than that in the wild type (1.1-fold) (Fig. 5c). Similarly, the increase times in the content of soluble sugar in the leaves from OE1, OE4 and OE7 (1.7-fold, 1.9-fold and 2.2-fold) were also higher than that in the wild type (1.4-fold) after cold stress for two days (Fig. 5d). In contrast, the contents of MDA in the transgenic lines (OE1, 0.012 $\mu mol \cdot g^{-1}$; OE4, 0.006 $\mu mol \cdot g^{-1}$ and OE7, 0.015 $\mu mol \cdot g^{-1}$) were decreased comparing to the wild type (0.019 $\mu mol \cdot g^{-1}$) after cold stress (Fig. 5e), demonstrating that over-expression of *GmZF1* reduced oxidative damage of membrane lipid peroxidation, and further maintained osmotic balance in the plant cells.

To observe the physiological effects of exogenous ABA supply on the transgenic *Arabidopsis* lines, the leaves from the wild type and the transgenic *Arabidopsis* lines exposed to ABA solution was sampled, and used for the determination in the content changes of praline, soluble sugar and MDA. Data showed that the contents of proline in the wild type were almost identical with that in the transgenic lines before ABA treatment. However, the content of

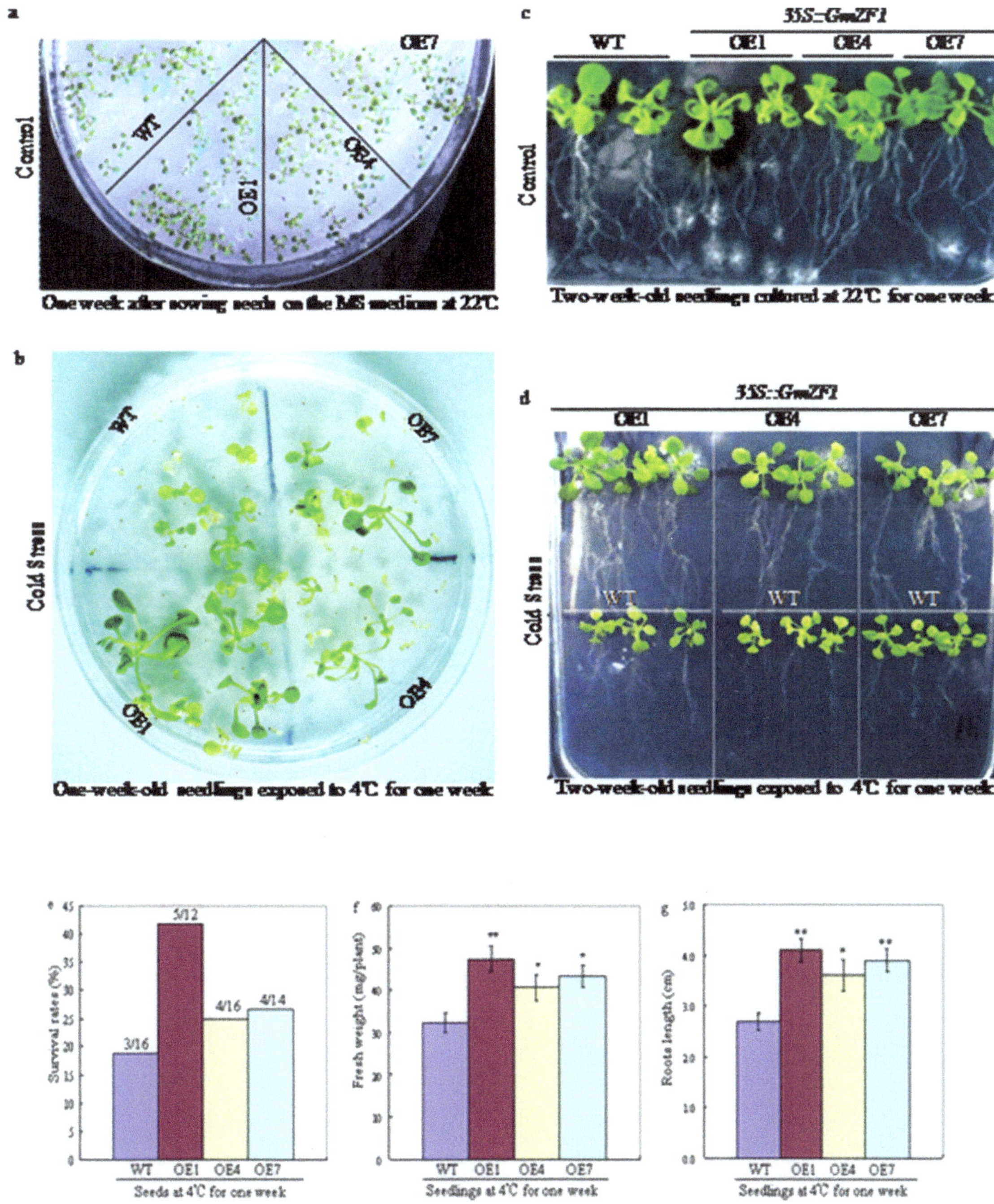

Figure 4. Morphological identification on wild type and transgenic *Arabidopsis* exposed to cold stress at 4°C on the MS medium. a: *Arabidopsis* seedlings of one-week-old after culturing at 22°C; c: *Arabidopsis* seedlings of two-week-old cultured at 22°C for one week; b and d: *Arabidopsis* seedlings of one-week-old cultured at 4°C for one week or two weeks, respectively; e: the survival rates of *Arabidopsis* seedlings exposed to 4°C for one week; f: the fresh weight per plant of two-week-old *Arabidopsis* after culturing at 4°C for one week; g: the roots length of two-week-old *Arabidopsis* seedlings after culturing at 4°C. WT-wild type; OE1, OE4 and OE7- transgenic lines.

proline in the transgenic lines (OE1, 142.4 ng·mg^{-1}; OE4, 427.2 ng·mg^{-1} and OE7, 412.9 ng·mg^{-1}) were significantly higher than that in the wild type (125.7 ng·mg^{-1}) after application of ABA (Fig. 6c). In addition, ABA treatment also enhanced the levels of soluble sugar in the wild type and the transgenic lines, but the increase times in the content of soluble sugar in the transgenic lines (OE1, 1.5-fold; OE4, 2.1-fold and OE7, 2.0-fold) were higher than that in the wild type (1.4-fold) after two days by ABA treatment for (Fig. 6d). The contents of MDA in the transgenic lines (OE1, 0.012 μmol·g^{-1}; OE4, 0.009 μmol·g^{-1} and OE7, 0.011 μmol·g^{-1}) were reduced comparing to the wild type (0.013 μmol·g^{-1}) after ABA treatment, whereas the contents of MDA in the transgenic lines (OE1, 0.009 μmol·g^{-1}; OE4, 0.009 μmol·g^{-1} and OE7, 0.008 μmol·g^{-1}) were slightly higher than that in the wild type (0.007 μmol·g^{-1}) before ABA treatment (Fig. 6e).

The induction of numerous stress-responsive genes is a hallmark of stress acclimation in plants. To elucidate the molecular

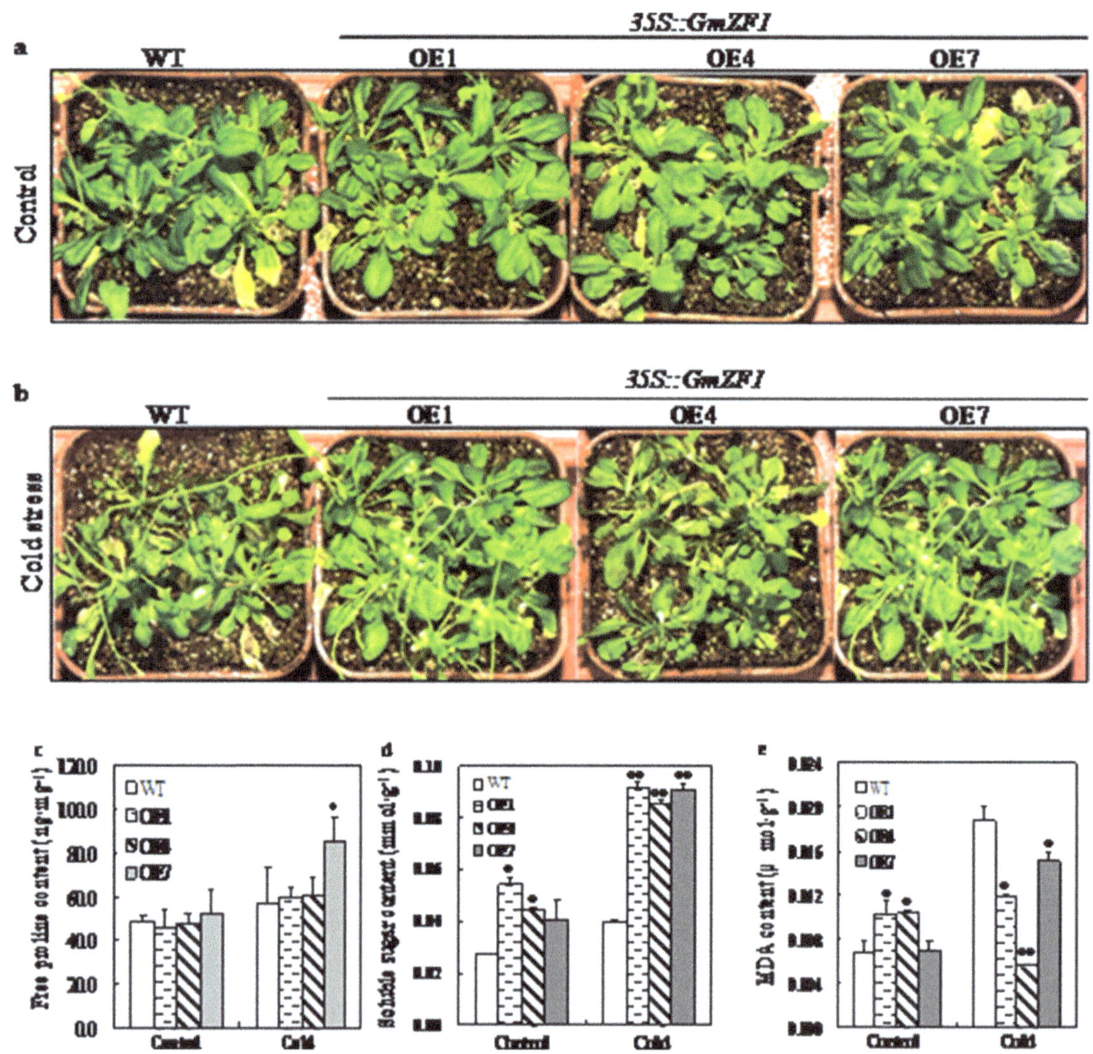

Figure 5. Phenotypic and physiological changes in wild type and transgenic *Arabidopsis* lines under cold stress. a: Phenotypes of wild type and transgenic lines of 5-week-old seedlings before cold stress (Control). b: Phenotypes of wild type and transgenic lines of 5-week-old seedlings exposed to 4°C for one week (Cold stress). c, d and e respectively represent the content changes of free proline, soluble sugar and MDA in the leaves of wild type and in transgenic lines before (Control) and after cold stress (Cold). WT-wild type; OE1, OE4 and OE7-transgenic lines; single * and duble ** respectively means significant difference at 0.05 and 0.01 levels.

mechanism of *GmZF1* in responding to cold, both wild type and transgenic seedlings of *Arabidopsis* were incubated at 4°C or by ABA for 48 hours to examine the expression of cold-responsive genes *cor6.6* at different time points by real-time PCR analysis. As shown in Fig. 7, both wild type and transgenic lines (OE1, OE4 and OE7) exhibited a similar levels in the expression of *cor6.6* before cold treatment, and the OE1 line began to appear significant up-regulations in *cor6.6* expressions in the roots after 12 hours cold treatment, while the OE7 line significantly increased the expression of *cor6.6* gene in the leaves after 12 h cold stress (Fig. 7a, b) When the seedlings of *Arabidopsis* were exposed to exogenous ABA, the expressions of *cor6.6* gene in the roots of the OE1 and OE7 lines were significantly induced after 24 hours (Fig. 7c), and the all transgenic lines (OE1, OE4 and OE7) exhibited remarkable increase in the expression of *cor6.6* gene in the leaves after 12 hours (Fig. 7d). Although both the wild type and transgenic lines at some time points did not exhibit significant differences in the expression of *cor6.6*, but the transgenic lines, OE1, OE4 and OE7 still showed higher transcript levels of *cor6.6* gene than the wild type under cold stress or exogenous ABA supply (Fig. 7b, c and d), suggesting that *GmZF1* may play an important role in activating cold-resistance genes in the transgenic *Arabidopsis* responding to cold stress through an signal pathway depending ABA.

Fig. 7e, f showed that the expressions of *GmZF1* gene in the roots and in the leaves of transgenic lines significantly increased after exposing the plants to cold stress for 6 hours, and this result seems to be identical with that in soybean seedlings exposed to cold stress or ABA (Fig. 3c, d). Fig. 7e showed that the expression of *GmZF1* gene in the roots of OE4 and OE7 was almost identical, but the OE7 line exhibited obvious increase in the expression of *GmZF1* gene in the leaves after 6 hours cold stress relative to other transgenic lines, and then maintained at relative stable levels at other time points (Fig. 7f). In addition, accumulation pattern of *GmZF1* gene in the transgenic *Arabidopsis* lines treated by exogenous ABA also showed that *GmZF1* gene in the leaves of the transgenic *Arabidopsis* lines was strongly induced by ABA, and the accumulation levels *GmZF1* gene in the roots of OE1 line were

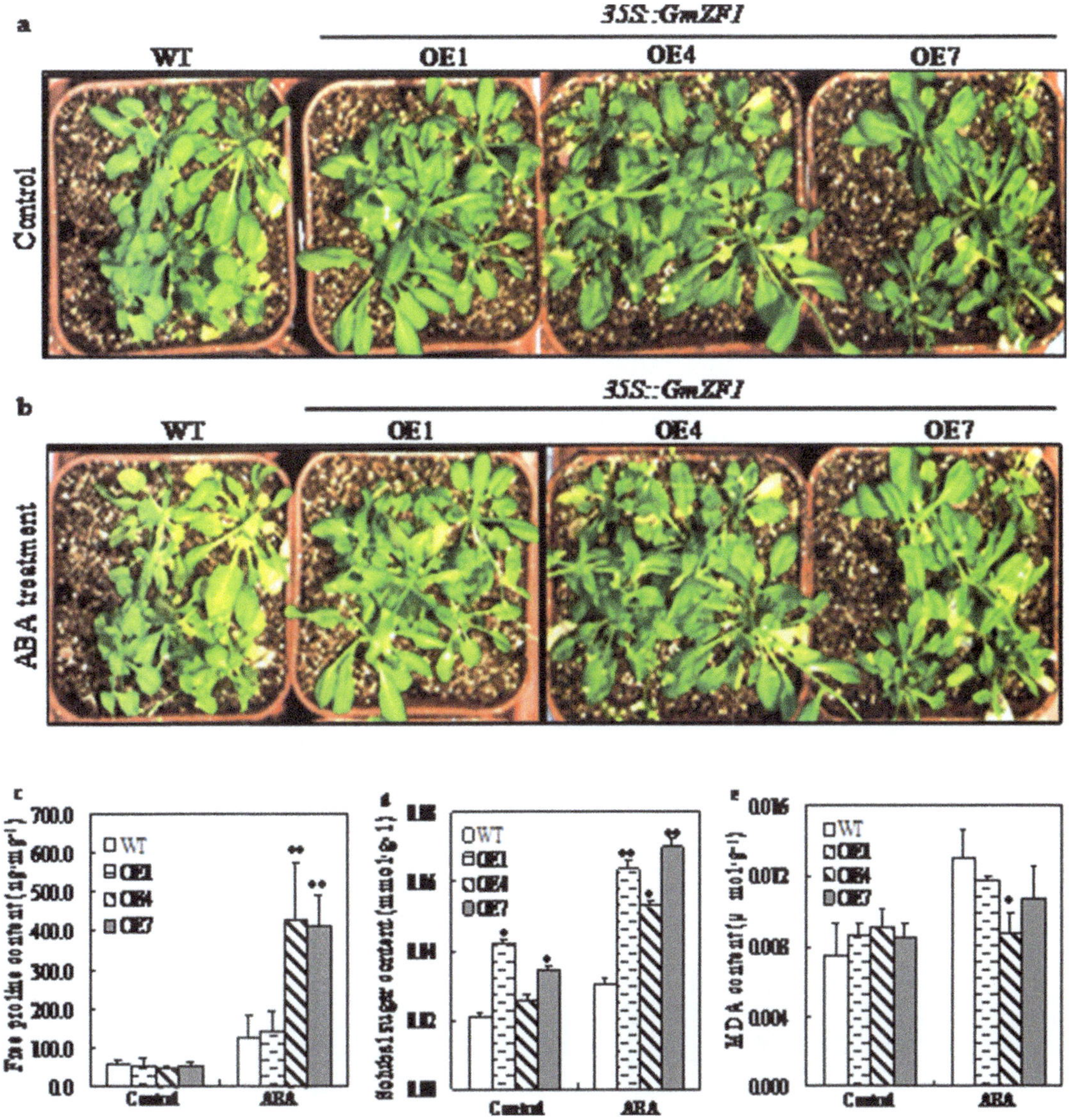

Figure 6. Phenotypic and physiological changes in wild type and transgenic *Arabidopsis* lines exposed to ABA. a: Phenotypes of wild type and transgenic lines of 5-week-old seedlings before ABA treatment. b: Phenotypes of wild type and transgenic lines of 5-week-old seedlings exposed to 200 μM ABA for one week. c, d and e respectively represent the content changes of free proline, soluble sugar and MDA in the leaves of wild type and transgenic lines before (Control) and after ABA treatment (ABA). WT-wild type; OE1, OE4 and OE7-transgenic lines; single * and duble ** respectively means significant difference at 0.05 and at 0.01 levels.

increased significantlyafter exposing to ABA for 24 hours (Fig. 7g) hours, and the accumulation of *GmZF1* mRNA in the leaves of OE the transgenic lines exhibited strong induction by ABA after 6 hours, and only OE7 line has almost identical induction accumulation patterns of *GmZF1* mRNA by ABA, and the induction accumulation seems to be inhibited after 48 hours (Fig. 7h). However, the accumulation of *GmZF1* mRNA in the OE7 transgenic line was little after exposing to ABA supply compared to the other transgenic lines, indicating that different transgenic lines exhibited different accumulation patterns in expressions of *GmZF1* gene in the plant organs, and seems to imply that accumulation patterns of *GmZF1* mRNA in the transgenic plants probably are related to the positions and copy numbers of *GmZF1* gene inserting in the genome of *Arabidopsis*.

Discussion

The C2H2 zinc-finger transcription factors were usually thought to be involved in plant development and have various adaptive responses to the environment stress [10,24]. Although the roles of some C2H2 zinc-finger transcription factors have been identified to be related to stress and developmental processes, the functions of C2H2 ZFPs from soybean involved in stress response are largely unknown [22].

In this study, as a novel C2H2 zinc finger protein gene, *GmZF1* from soybean was characterized. Sequence analysis revealed that the GmZF1 had high identity with other C2H2-type ZFPs, and shared two zinc finger motifs containing a conserved plant-specific QALGGH amino acid sequence which was proved to be critical for DNA-binding activity [11]. Based on the present data, we predicted that the binding activity of GmZF1 to EP1S core elements was probably mediated by the QALGGH sequence.

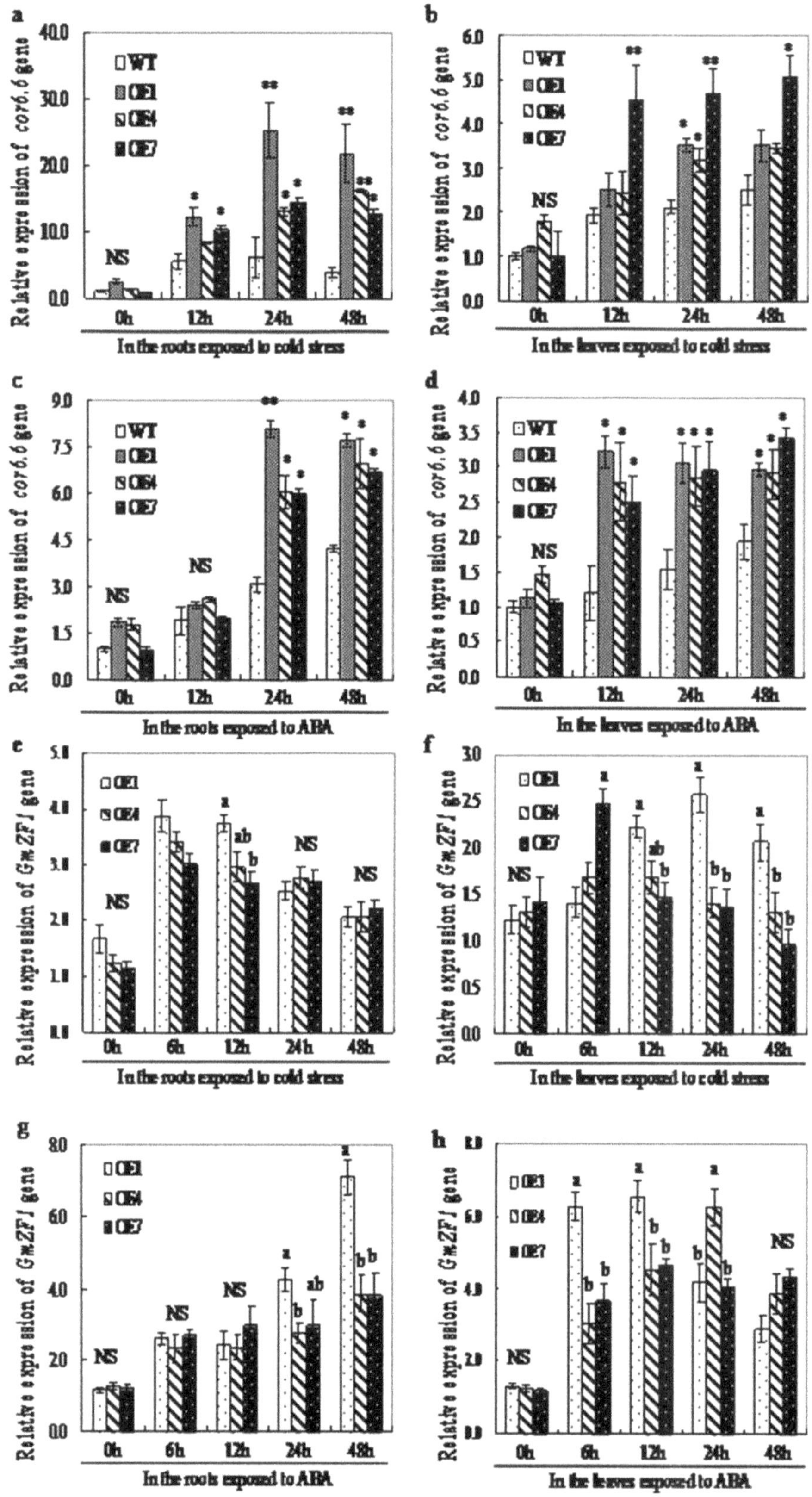

Figure 7. Responsive expression of *cor6.6* gene and relative accumulation of *GmZF1* mRNA in wild type and transgenic lines both under cold stress and ABA supply. a and b respectively represent the relative expressions of *cor6.6* gene in the roots and leaves of plants exposed to cold stress; c and d respectively represent the relative expressions of *cor6.6* gene in the roots and leaves of plants exposed to ABA, and the expression level of *cor6.6* gene in unstressed wild type was set at 1.0 in the transgenic *Arabidopsis* lines. e: the relative expressions of *GmZF1* gene in the roots; f: the relative expressions of *GmZF1* gene in the leaves of transgenic plants exposed to cold stress; g and h respectively represent the relative expressions of *GmZF1* gene in the roots and leaves of transgenic plants exposed to ABA. Error bars indicate the standard deviation; NS means

no significant differences; lowercase letters labeled on the columns mean statistical difference at 0.05 level ce at 0.05 level; single * and duble ** respectively means significant difference at 0.05 and at 0.01 levels; WT-wild type; OE1, OE4 and OE7-transgenic lines.

Additionally, the C-terminus of *GmZF1* gene contains typical Leu-rich L-box and DLN-box which play roles in protein interactions or in maintaining the folded structure [19]. The DLN-box was thought to function in transcriptional repression. As reported previously, the DLN-boxes in the zinc finger proteins ZPT2–3 from petunia [16] and STZ/ZAT10 from *Arabidopsis* exhibited repression roles in transcription activities [18]. However, some zinc finger proteins containing the DLN-box were involved in transcriptional activation, such as ThZF1 from *Thellungiella halophila* [25] and CaZF from chickpea [26]. Subcellular localization analysis revealed that GmZF1 localized at nuclei (Fig. 3), implying that GmZF1, like other ZFPs from plant TFIIIA-type, functions as a transcription factor in plant cells and may play an important role in signaling pathway in soybean under abiotic stress.

As a key regulator, ABA plays an important role in signaling pathways under stresses, such as drought, low temperature and osmotic stress, and induces the expressions of a number of genes that respond to abiotic stress [30,31]. For example, the expression of *AZF2* gene from *Arabidopsis* was strongly induced following ABA treatment under drought and salt stress [1,12,18]. However, it was reported that some C2H2-type zinc finger genes are induced by dehydration and cold stress, but do not respond to exogenous ABA [2,32,33], suggesting that two kinds of signaling transduction pathways, ABA-independent pathway and ABA-dependent pathway play different regulation roles in responding to abiotic stresses, repectively, and that the initial stress signal was converted into cellular responses [34]. In this study, our data proved that *GmZF1* was involved in plant responses to cold stress through an ABA-dependent signal transduction pathway since the expression of *GmZF1* in soybean seedlings was clearly induced by ABA (Fig. 4). In addition, accumulation pattern of *GmZF1* gene in the transgenic *Arabidopsis* lines treated by exogenous ABA also showed that over-expression of *GmZF1* gene in *Arabidopsis* was strongly induced by ABA.

Some of the C2H2-type ZFPs from different plant species are confirmed to play regulatory roles in stress responses, such as AZF2 and STZ in Arabidopsis [12,17,27], SCOF-1 in soybean [21], GsZFP1 in wild soybean [22], StZFP1 in potato [9], DST in rice [28] and TaCHP in wheat [29]. Expression analysis revealed that GmZF1 was clearly induced by cold stress (Fig. 4), suggesting that *GmZF1* might be involved in plant responses to cold stress in soybean. Transgenic *Arabidopsis* plants over-expressing *GmZF1* were evaluated for the involvement of *GmZF1* gene in cold tolerance of plants. Although no morphological differences between the transgenic lines and the wild type was observed, the remarkable differences in accumulation of free proline, soluble sugar and MDA were confirmed after cold and ABA treatment (Fig. 5c–e and Fig. 6c–e). Accumulation of proline by stress-induced has been observed in many plant species, and functions as an osmo-protectant coping with stress [32,35,36]. For example, the transgenic tobacco plants over-expressing *GmDREB3* could accumulate much free proline than the wild-type plants after drought stress treatment for 16 d [37]. In plants, soluble sugar has been shown to fulfill a dual role as both metabolites and as signaling molecules [38,39] that may play important roles in the mechanisms of plant responding to the stress [40,41]. In our study, the accumulations of proline and soluble sugar in the wild type and transgenic lines were improved, and increase times in the contents of proline and soluble sugar in the transgenic lines were higher than that in wild type during cold and ABA treatment (Fig. 5 and Fig. 6), proving that the accumulation of soluble sugars has been occurred in many plant species during cold acclimation [42]. Based on our investigations, both wild type and transgenic *Arabidopsis* lines basically have no significant differences in germination rates. However, the survival rates from these two phenotypes was remarkable when these two seedlings were exposed to cold stress, the survival rates of transgenic *Arabidopsis* lines cultured at 4°C significantly increased after one week comparing to the wild types (Fig. 4 a,b), indicating that *GmZF1* gene could play regulation role during the plant growth and development. Our study showed that over-expression of *GmZF1* gene significantly enhanced the transcription levels of *cor6.6* gene responding to cold stress and ABA supply, and increased the accumulation of proline, soluble sugar in the transgenic *Arabidopsis*, and reduced the content of MDA in the transgenic lines comparing to the wild type, and significantly improved the tolerance of the transgenic lines exposed to cold stress and exogenous ABA (Fig. 5e; Fig. 6e). In conclusion, over-expression of *GmZF1* resulted in an enhancement in cold tolerance of the transgenic *Arabidopsis* by activating transcription of *cor6.6* and/or cold resistance-genes (Fig. 7) as well as accumulation changes of proline, soluble sugar and MDA [43–45]. Bioinformatics analysis fund that the promoter with 1955 bp of *cor6.6* gene contains 7×G<u>A</u>C<u>A</u> and 6×G<u>T</u>C<u>A</u> repeat in core element TTGA-CAGTGTCAC, respectively (NCBI Reference Sequence in *Arbidopsis thaliana* chromosome 5: NC_003076.8-AT5G15950/NM_121600.3/NP_197099.1), and these two core elements possibly play key role in recognizing DNA-binding sites in target proteins, because the mutation of G and C in the two boxed core elements could change the ability of GmZf1 protein binding to DNA, especially the mutation of G and C in the GCTA box (Fig. 3c). Therefore, *GmZF1* could play regulation role by activating the transcription of *cor6.6* cold-regulated gene in *Arabidopsis*, and led to an enhancement in the tolerance of transgenic plants to cold stress.

Materials and Methods

Plant materials and growth conditions

Soybean seedlings (*Glycine max*) were grown in a growth chamber at 24°C with 60% relative humidity under 16 h light and 8 h darkness. *Arabidopsis* plants (genotype Colombia) were grown in a controlled environmental chamber at 22°C and 70% humidity with a 14 h light/10 h darkness cycle under normal light intensity (150 $Em^{-2} \cdot s^{-1}$). T_1 seeds were sterilized and planted on the MS medium containing kanamycin of 50 $\mu g \cdot mL^{-1}$ for the selection of transgenic plants. After continuously screening on the MS medium containing kanamycin of 50 $\mu g \cdot mL^{-1}$ for two times and the generated T3 seedlings of transgenic *Arabidopsis* lines were transplanted to pots or MS medium for further investigations.

Isolation and sequence analysis of *GmZF1*

The zinc finger gene *GmZF1* was cloned from soybean based on the expressed sequence tag (EST) database of soybean. Total RNA was extracted from soybean seedlings using RNA Prep. Pure plant kit (Tiangen Biological Company, Beijing), and cDNA synthesis was performed by a reverse transcription kit (TaKaRa Dalian BioCompany). The entire *GmZF1* cDNA of was obtained by PCR using the specific primers (F: 5′-AGAGGAAACTAGCTAGGG-CACTTC-3′ and R: 5′-CCCGAGAACTAAGAAGTTTCG-TATT-3′). The deduced amino acid sequences of soybean GmZF1 zinc finger protein were matched by a protein blast procedure (http://www.ncbi.nlm.nih.gov/BLAST/). BioEdit version 7.0 software was used to multiple sequence alignments. Phylogenetic analysis was carried out by MEGA version 4.0 with adopting position correction distance using a bootstrap replicate number of 1000.

Induction expression of *GmZF1* fusion protein and electrophoretic mobility shift assay

An entire 516-bp *GmZF1* fragment containing the DNA binding domain was amplified by PCR using a pair of primers F and R (F: 5′-ACAACTCGAGATGAAGAGAGGCAGAGAA-3′ and R: 5′-AGACGAATTCAATGAAACAATTGAGCAC-3′), and was subcloned onto the pGEX4 T-1 vector by inserting at two specific sites, *EcoR*I and *Xho*I (Amersham Biosciences), and the recombinant pGEX-4 T-1 plasmids were identified by sequencing, and transferred into *Escherichia coli* BL21 cells (Amersham Biosciences). The induction of target fusion proteins was performed by adding 0.8 mM IPTG (Isopropyl-β-d-thiogalactopyranoside) into the culture of *E. coli* strains, and the cultures carrying GmZF1::GST fusion proteins was incubated for 5 hours at 37°C with 250 rpm The bacterial cells were shattered by ultrasonication in phosphate-buffered saline (PBS) and centrifuged at 11000 *g* for 10 min to remove insoluble cell debris, and then the supernates were collected, and applied on the gel containing 12% ployacrylamide for electrophoresis detection of fusion protein by the method of SDS-PAGE described in the book [46]. The GST::GmZF1 fusion protein was purified using a glutathione-Sepharose 4B column (Amersham Biosciences) according to the manufacturer's instructions. The 52 bp DNA fragment containing four copies of wild-type or mutant EP1S core sequences were labeled by 25 $\mu Ci \cdot \mu L^{-1}$ of γ-^{32}P-dATP (Amersham Biosciences). The DNA-binding reaction was performed in a 20 μL binding buffer containing ^{32}P-labelled probe, glycerine and purified fusion protein [24], and were subjected to electrophoresis with the gel containing 0.53 Tris-borate-EDTA, 6% polyacrylamide, and the gels were dried and visualized by autoradiography.

Subcellular localization of *GmZF1* gene

In brief, an entire *GmZF1* cDNA fragment was amplified by PCR procedure with a pair of specific primers (F: 5′-AACACTG-CAGATGAAGAGAGGCAGAGA-3′), and R: 5′-CGGGATC-CAATGAAACAATTGAGCAC-3′.). Based on the specific sites in the MCS of GFP vector, two specific sites, a *Pst*I and *BamH*I were respectively inserted into the 5′ ends of the primers for subcloning requirement. The digested PCR fragment was inserted into the multiple cloning sites (MCS) in the GFP vector carrying a GFP protein driven by a 35 S promoter, and this construct carrying 35 S::GFP or 35 S::GmZF1::GFP were introduced into *Arabidopsis* protoplast cells, which was prepared by young seedlings of *Arabidopsis,* and the transformation procedures were performed by the method as described previously [47,48]. The images were captured using a Nikon Eclipse TE2000-U microscope (Nikon).

Transformation of *Arabidopsis*

A pair of specific primers, the *GmZF1* cDNA fragment was generated by PCR procedure using the specific primers (F: 5′-AACACCATGGCCATGAAGAGAGGCAGAGA-3′, and R: 5′-AGACACTAGTAATGAAACAATTGAGCAC-3′. The generated PCR product directly was inserted onto the pCAMBIA1304 vector using a pair of specific restriction sites (*Nco*I and *Spe*I), and an over-expression vector p35 S::GmZF1 was constructed, and introduced into *Agrobacterium tumefaciens* strain C58C1. The flowering *Arabidopsis* seedlings were used for genetic transformation at 25°C by the method of floral-dip. T3 transgenic *Arabidopsis* plants were continuously cultured in the medium containing kanamycin of 45 $mg \cdot L^{-1}$ and three positive transgenic *Arabidopsis* were identified and the T3 seeds of transgenic *Arabidopsis* lines and wild type *Arabidopsis* were sowed on MS medium based on the requirements of experiment, and the resulting seedlings were accordingly transferred into the pots and MS medium for cold stress and ABA treatment.

Stress treatments

Cold stress was performed by incubating soybean seedlings of 10-day-old at 4°C for 24 hours and *Arabidopsis* seedlings of two-week-old seedlings were incubated at 4°C for 48 hours. ABA treatment was respectively carried out by completely spraying 200 $\mu mol \cdot L^{-1}$ ABA onto the 10-day-old soybean seedlings and two-week-old *Arabidopsis* seedlings Soybean leaves from the same position were respectively sampled at 0, 1, 3, 5, 12 and 24 hours, and *Arabidopsis* leaves were collected respectively at 0, 6, 12, 24 and 48 hours. Equal amounts of seeds from wild type and transgenic lines were sowed on MS medium at 22°C (Control), and the generated seedlings of one-week-old *Arabidopsis* were cultured at 22°C and at 4°C (Cold stress), respectively, for one week in two independent growth chambers, and used for investigation in survival rates. For evaluating the changes of Arabidopsis biomass and root development under cold stress, two-week-old seedlings of *Arabidopsis* from control culture were transplanted on the MS medium, and cultured respectively at 22°C (Control) and at 4°C (Cold stress) for one week in two independent growth chambers, and biological fresh weights per plant and root lengths from two treatments were statistically investigated based on the three replicates.

Gene expression analysis

Based on the equal amounts of leaves RNA from each time point, the expression pattern of *GmZF1* gene in soybean was characterized using semi-quantitative RT-PCR by GmF and GmR primers, and accumulation analysis of *GmZF1* mRNA in the transgenic *Arabidopsis* lines was performed using the specific primers (GmF: 5′-ATGAAGAGAGGCAGAGAA-3′ and GmR: 5′-AATGAAACAATTGAGCAC-3′) by quantitative real-time PCR (qPCR) For understanding the expression profile of GmZF1 gene and cold-responsive marker gene *cor6.6* in the transgenic *Arabidopsis* under low temperature stress and exogenous ABA supply, the wild type Arabidopsis and transgenic *Arabidopsis* lines carrying GmZF1 gene were incubated at 4°C for 48 hours in darkness. All samples including the roots and leaves were collected with three biological replicates at the designated time intervals after cold and ABA treatment, and were quickly frozen in liquid nitrogen and stored at 80°C for total RNA isolation. As a reference gene in *Arabidopsis*, actin was used as an internal maker, and amplified by the specific primers [(F 5′-AAGTATCCTATT-GAGCATGGTGTTG-3′; R 5′- CTGGCGTACAAGGAGAG-A-3′), (accession number: AEE76148)], and the *cor6.6* gene was amplified by the primers (F 5′- ATGTCAGAGACCAACAA-GAATG-3′; R5′- CTTGTTCAGGCCGGTCTTG-3′) (accession number: CAA38894) with qPCR, which was performed using a SYBR premix Ex Taq kit (TaKaRa) according to the manufacturer's instructions on a 7900HT Real-time PCR system.

Accumulation analysis of free proline, soluble sugar and malondialdehyde (MDA)

0.5 g fresh leaves from each treatment of the wild-type and the transgenic *Arabidopsis* lines were respectively collected at different time point, and respectively were treated according to requirements of experiment and used for measurements of free praline, soluble sugar and MDA. The contents of free proline were measured by the previous procedure [40,49,50]. Measurement of MDA content in *Arabidopsis* leaves was determined as described previously [51,52], and the content of soluble sugar was measured with the method as described by [53].

Author Contributions

Conceived and designed the experiments: XGC ZSX. Performed the experiments: GHY LLJ XFM MML SGS. Analyzed the data: ZSX. Contributed reagents/materials/analysis tools: GHY. Wrote the paper: XGC GHY ZSX.

References

1. Kodaira KS, Qin F, Tran LSP, Maruyama K, Kidokoro S, et al. (2011) *Arabidopsis* Cys2/His2 zinc-finger proteins AZF1 and AZF2 negatively regulate abscisic acid-repressive and auxin-inducible genes under abiotic stress conditions. Plant Physiol 157: 742–756.
2. Yamaguchi-Shinozaki K, Shinozaki K (2006) Transcriptional regulatory networks in cellular responses and tolerance to dehydration and cold stresses. Annu Rev Plant Biol 57: 781–803.
3. Umezawa T, Fujita M, Fujita Y, Yamaguchi-Shinozaki K, Shinozaki K (2006) Engineering drought tolerance in plants: discovering and tailoring genes to unlock the future. Curr Opin Biotechnol 17: 113–122.
4. Zhang G, Chen M, Chen X, Xu Z, Guan S, et al. (2008) Phylogeny, gene structures, and expression patterns of the ERF gene family in soybean (*Glycine max*). J Exp Bot 59: 4095–4107.
5. Xu ZS, Chen M, Li LC, Ma YZ (2011) Functions and application of the AP2/ERF transcription factor family in crop improvement. J Integr Plant Biol 53: 570–585.
6. Laity JH, Lee BM and Wright PE (2001) Zinc finger proteins: new insights into structural and functional diversity. Curr Opin Struct Biol 11: 39–46.
7. CO Pabo, Peisach E, Grant RA (2001) Design and selection of novel Cys2His2 zinc finger proteins. Annu Rev Biochem 70: 313–340.
8. Takatsuji H (1999) Zinc-finger proteins: the classical zinc finger emerges in contemporary plant science. Plant Mol Biol 39: 1073–1078.
9. Tian ZD, Zhang Y, Liu J, Xie CH (2010) Novel potato C2H2-type zinc finger protein gene, *StZFP1*, which responds to biotic and abiotic stress, plays a role in salt tolerance. Plant Biol 12: 689–697.
10. Ciftci-Yilmaz S, Mittler R (2008) The zinc finger network of plants. Cell Mol Life Sci 65: 1150–1160.
11. Kubo K, Sakamoto A, Kobayashi A, Rybka Z, Kanno Y, et al. (1998) Cys2/His2 zinc-finger protein family of petunia: evolution and general mechanism of target-sequence recognition. Nucleic Acids Res 26: 608–615.
12. Chen Y, Sun A, Wang M, Zhu Z, Ouwerkerk PB (2013) Functions of the CCCH type zinc finger protein OsGZF1 in regulation of the seed storage protein GluB-1 from rice. Plant Mol Biol DOI 10.1007/s11103-013-0158-5.
13. Takatsuji H, Mori M, Benfey PN, Ren L, Chua NH (1992) Characterization of a zinc finger DNA-binding protein expressed specifically in Petunia petals and seedlings. EMBO J 11: 241–249.
14. Liu P, Xu ZS, Pan-Pan L, Hu D, Chen M, et al. (2013) A wheat plasma membrane-localized *PI4K* gene possessing threonine autophophorylation activity confers tolerance to drought and salt in *Arabidopsis*. J Exp Bot 64: 2915–2927.
15. Van Der Krol AR, Van Poecke RMP, Vorst OFJ, Voogd C, Van Leeuwen W, et al. (1999) Developmental and wound-, cold-, desiccation-, ultraviolet-b-stress-induced modulations in the expression of the Petunia Zinc Finger Transcription Factor Gene *ZPT2–2*. Plant Physiol 121: 1153–1162.
16. Sugano S, Kaminaka H, Rybka Z, Catala R, Salinas J, et al. (2003) Stress responsive zinc finger gene *ZPT2–3* plays a role in drought tolerance in petunia. Plant J 36: 830–841.
17. Lippuner V, Cyert MS, Gasser CS (1996) Two classes of plant cDNA clones differentially complement yeast calcineurin mutants and increase salt tolerance of wild-type yeast. J Biol Chem 271: 12859–12866.
18. Sakamoto H, Maruyama K, Sakuma Y, Meshi T, Iwabuchi M, et al. (2004) *Arabidopsis* Cys2/His2-type zinc-finger proteins function as transcription repressors under drought, cold, and high-salinity stress conditions. Plant Physiol 136: 2734–2746.
19. Sakamoto H, Araki T, Meshi T, Iwabuchi M (2000) Expression of a subset of the Arabidopsis Cys (2)/His (2)-type zinc-finger protein gene family under water stress. Gene 248: 23–32.
20. Mittler R, Kim YS, Song L, Coutu J, Coutu A, et al. (2006) Gain-and loss-of-function mutations in Zat10 enhance the tolerance of plants to abiotic stress. FEBS Lett 580: 6537–6542.
21. Kim JC, Lee SH, Cheong YH, Yoo CM, Lee SI, et al. (2001) A novel cold-inducible zinc finger protein from soybean, SCOF-1, enhances cold tolerance in transgenic plants. Plant J 25: 247–259.
22. Luo X, Bai X, Zhu D, Li Y, Ji W, et al. (2012) GsZFP1, a new Cys2/His2-type zinc-finger protein, is a positive regulator of plant tolerance to cold and drought stress. Planta 235: 1141–1155.
23. Takatsuji H, Nakamura N, Katsumoto Y (1994) A new family of zinc finger proteins in petunia: structure, DNA sequence recognition, and floral organ-specific expression. Plant Cell 6: 947–958.
24. Lu X, Li Y, Su Y, Liang Q, Meng H, et al. (2012) An *Arabidopsis* gene encoding a C2H2-domain protein with alternatively spliced transcripts is essential for endosperm development. J Exp Bot 63: 5935–5944.
25. Xu SM, Wang XC, Chen J (2007) Zinc finger protein 1 (ThZF1) from salt cress (*Thellungiella halophila*) is a Cys-2/His-2-type transcription factor involved in drought and salt stress. Plant Cell Rep 26: 497–506.
26. Jain D, Roy N, Chattopadhyay D (2009) *CaZF*, a plant transcription factor functions through and parallel to HOG and calcineurin pathways in *Saccharomyces cerevisiae* to provide osmotolerance. PLoS ONE 4: e5154.
27. Nakai Y, Nakahira Y, Sumida H, Takebayashi K, Nagasawa Y, et al. (2013) Vascular plant one-zinc-finger protein 1/2 transcription factors regulate abiotic and biotic stress responses in Arabidopsis. Plant J 73: 761–775.
28. Huang XY, Chao DY, Gao JP, Zhu MZ, Shi M, et al. (2009) A previously unknown zinc finger protein, DST, regulates drought and salt tolerance in rice via stomatal aperture control. Genes & Development 23: 1805–1817.
29. Li C, Lv J, Zhao X, Ai X, Zhu X, et al. (2010) *TaCHP*: a wheat zinc finger protein gene down-regulated by abscisic acid and salinity stress plays a positive role in stress tolerance. Plant Physiol 154: 211–221.
30. Xiong L, Schumaker KS, Zhu JK (2002) Cell signaling during cold, drought, and salt stress. Plant Cell 14: 165–183.
31. Finkelstein RR, Gampala SSL, Rock CD (2002) Abscisic acid signaling in seeds and seedlings. Plant Cell 14: 15–45.
32. Davletova S, Schlauch K, Coutu J, Mittler R (2005) The zinc-finger protein Zat12 plays a central role in reactive oxygen and abiotic stress signaling in *Arabidopsis*. Plant Physiol 139: 847–856.
33. Zhu JK (2002) Salt and drought stress signal transduction in plants. Annu Rev Plant Biol 53: 247–273.
34. Shinozaki K, Yamaguchi-Shinozaki K, Seki M (2003) Regulatory network of gene expression in the drought and cold stress responses. Curr Opin Plant Biol 6: 410–417.
35. Igarashi Y, Yoshiba Y, Sanada Y, Yamaguchi-Shinozaki K, Wada K, et al. (1997) Characterization of the gene for 1-pyrroline-5-carboxylate synthetase and correlation between the expression of the gene and salt tolerance in Oryza sativa L. Plant Mol Biol 33: 857–865.
36. Dobrá J, Vanková R, Havlová M, Burman AJ, Libus J, et al. (2011) Tobacco leaves and roots differ in the expression of proline metabolism-related genes in the course of drought stress and subsequent recovery. J Plant Physiol 168: 1588–1597.
37. Chen M, Xu Z, Xia L, Li L, Cheng X, et al. (2009) Cold-induced modulation and functional analyses of the DRE-binding transcription factor gene, *GmDREB3*, in soybean (*Glycine max*). J Exp Bot 60: 121–135.
38. Bolouri Moghaddam MR, Le Roy K, Xiang L, Rolland F, Van den Ende W (2010) Sugar signaling and antioxidant network connections in plant cells. FEBS J 277: 2022–2037.
39. Smeekens S, Ma J, Hanson J, Rolland F (2010) Sugar signals and molecular networks controlling plant growth. Curr Opin Plant Biol 13: 273–278.
40. Sperdouli I, Moustakas M (2012) Interaction of proline, sugars, and anthocyanins during photosynthetic acclimation of *Arabidopsis thaliana* to drought stress. J Plant Physiol 169: 577–585.
41. Ramel F, Sulmon C, Gouesbet G, Couéel I (2009) Natural variation reveals relationships between pre-stress carbohydrate nutritional status and subsequent responses to xenobiotic and oxidative stress in *Arabidopsis thaliana*. Anna Bot 104: 1323–1337.
42. Guy CL (1990) Cold Acclimation and freezing stress tolerance: Role of protein metabolism. Annu Rev Plant Biol 41: 187–223.
43. Giannakoula A, Moustakas M, Mylona P, Papadakis I, Yupsanis T (2007) Aluminium tolerance in maize is correlated with increased levels of mineral nutrients, carbohydrates and proline and decreased levels of lipid peroxidation and Al accumulation. J Plant Physiol 165: 385–396.
44. Zhan X, Wang B, Li H, Liu R, Kalia RK, et al. (2012) *Arabidopsis* proline-rich protein important for development and abiotic stress tolerance is involved in microRNA biogenesis. Proc Natl Acad Sci USA 109: 18198–203.
45. Xu ZS, Xia LQ, Chen M, Cheng XG, Zhang RY, et al. (2007) Isolation and molecular characterization of the *Triticum aestivum* L. Ethylene-responsive

factor 1 (*TaERF1*) that increases multiple stress tolerance. Plant Mol Biol 65: 719–732.
46. Davis LG, Kuehl WM, Battey JF (1994) Section, 11-3: Acidic guanidine isothiocyanate/phenol/chlororm extraction for isolation of RNA; Section, 17-1: Electrophoresis of protein on sodium dodecy sulfate polyacrylamide gels. In Basic methods in Molecular Biology, 2nd ed.1994, 335–338, 661–668, Appleton & Lange, USA.
47. Han MJ, Jung KH, Yi G, Lee DY, An G (2006) Rice Immature Pollen 1 (RIP1) is a regulator of late pollen development. Plant Cell Physiol. 47: 1457–1472.
48. Liu XM, Nguyen XC, Kim KE, Han HJ, Yoo J, et al. (2013) Phosphorylation of the zinc finger transcriptional regulator *ZAT6* by *MPK6* regulates *Arabidopsis* seed germination under salt and osmotic stress. Biochem Biophys Res Commun 430: 1054–1059.
49. Giannakoula A, Moustakas M, Syros T, Yupsanis T (2012) Aluminium stress induces upregulation of an efficient antioxidant system in the Al-tolerant maize line but not in the Al-sensitive line. Environ Exp Bot 67: 487–494.
50. Zhan X, Wang B, Li H, Liu R, Kalia RK, et al. (2012) Arabidopsis proline-rich protein important for development and abiotic stress tolerance is involved in microRNA biogenesis. Proc Natl Acad Sci USA 109: 18198–203.
51. Yue Y, Zhang M, Zhang J, Tian XL, Duan LS, et al. (2012) Overexpression of the *AtLOS5* gene increased abscisic acid level and drought tolerance in transgenic cotton. J Exp Bot 63: 3741–3748.
52. Ai L, Li ZH, Xie ZX, Tian XL, Eneji AE, et al. (2008) Coronatine alleviates polyethylene Glycol-induced water stress in Two Rice (*Oryza sativa L.*) cultivars. J Agronomy and Crop Sci 194: 360–368.
53. Taji T, Ohsumi C, Iuchi S, Seki M, Kasuga M, et al. (2002) Important roles of drought-and cold-inducible genes for galactinol synthase in stress tolerance in *Arabidopsis thaliana*. Plant J 29: 417–426.

4

Comparison of Soil Quality Index Using Three Methods

Atanu Mukherjee*, Rattan Lal

Carbon Management and Sequestration Center, School of Environment and Natural Resources, The Ohio State University, Columbus, Ohio, United States of America

Abstract

Assessment of management-induced changes in soil quality is important to sustaining high crop yield. A large diversity of cultivated soils necessitate identification development of an appropriate soil quality index (SQI) based on relative soil properties and crop yield. Whereas numerous attempts have been made to estimate SQI for major soils across the World, there is no standard method established and thus, a strong need exists for developing a user-friendly and credible SQI through comparison of various available methods. Therefore, the objective of this article is to compare three widely used methods to estimate SQI using the data collected from 72 soil samples from three on-farm study sites in Ohio. Additionally, challenge lies in establishing a correlation between crop yield versus SQI calculated either depth wise or in combination of soil layers as standard methodology is not yet available and was not given much attention to date. Predominant soils of the study included one organic (Mc), and two mineral (CrB, Ko) soils. Three methods used to estimate SQI were: (i) simple additive SQI (SQI-1), (ii) weighted additive SQI (SQI-2), and (iii) statistically modeled SQI (SQI-3) based on principal component analysis (PCA). The SQI varied between treatments and soil types and ranged between 0–0.9 (1 being the maximum SQI). In general, SQIs did not significantly differ at depths under any method suggesting that soil quality did not significantly differ for different depths at the studied sites. Additionally, data indicate that SQI-3 was most strongly correlated with crop yield, the correlation coefficient ranged between 0.74–0.78. All three SQIs were significantly correlated ($r = 0.92$–0.97) to each other and with crop yield ($r = 0.65$–0.79). Separate analyses by crop variety revealed that correlation was low indicating that some key aspects of soil quality related to crop response are important requirements for estimating SQI.

Editor: Upendra M. Sainju, Agricultural Research Service, United States of America

Funding: This research was part of a collaborative project supported by the USDA-NIFA, award number: 2011-68002-30190, "Cropping Systems Coordinated Agricultural Project (CAP): Climate Change, Mitigation, and Adaptation in Corn-based Cropping Systems" (sustainablecorn.org). The funders had no role in study design, data collection and analysis, decision to publish, or preparation of the manuscript.

Competing Interests: The authors have declared that no competing interests exist.

* Email: mukherjee.70@osu.edu

Introduction

A wide range of agricultural soils represents diversely managed arable lands while the main goal to improve soil quality, crop yield, and reduce the ecological foot print. Soil quality is defined as the soil's capacity to function within natural or managed ecosystem boundaries and to sustain plant productivity while reducing soil degradation [1–4]. As soil quality is a complex functional concept and cannot be measured directly in the field or laboratory [5] but can only be inferred from soil characteristics [6], a range of soil parameters or indicators has been identified to estimate soil quality. However, soil quality is often related to the management goal and practices as well to soil characteristics. Thus, a mathematical or statistical framework was put forward in early 1990s to estimate soil quality index (SQI) [1,3,4]. The SQI was assessed so that the management goals are not only focused on productivity per se, which may result in soil degradation [7], but also on environmental issues. Thus, an appropriate SQI may have three component goals: environmental quality, agronomic sustainability, and socio-economic viability [8].

Estimation of SQI is a complex process and difficult task [9], especially when linked with several functional goals. Yet a considerable progress has been made towards estimating SQI across a number of soil types and management practices [8–14]. Most studies indexed soil quality employing only one method, with a few exceptions [8,15,16]. As computation of SQI is difficult, a strong need exists for developing a user-friendly and credible SQI through comparison of various available methods. Thus, the objective of this study was to compare SQIs computed by three methods which are conceptually different from each other. The study is based on the hypothesis that SQIs computed from three methods have similar relationship with crop yields. Data are scarce on validation of SQI against crop yield as most studies focused on the environmental aspects of the soil as end point variable and that SQI of various soil layers has not been computed [13,16–18]. Nevertheless, there is a challenge in validating SQI against crop yield as SQI computed only from surface soils (0–10 and/or 0–20 cm) may not evaluate realistic relationship between soil quality and crop yield because root system can extend to deeper layers [19]. Thus, the other objective of this study was to compute SQI from multiple depths of the soil to examine its relationship with crop yield.

Materials and Methods

Soil sampling and analyses

Soil samples from the field (on-farm) were collected from Logan county, Ohio that included an organic (Martisco Variant silt loam: Mc, organic parent material, >26% organic matter, 40°25′12.4″N, 83°40′55.9″W), and a mineral soil (Crosby silt loam: CrB: *Fine, mixed, active, mesic Aeric Epiaqualfs*, sand: 23%, silt: 35%, clay: 42%, 40°24′52.4″N, 83°39′21.9″W), and from Franklin county, Ohio that included a Kokomo silty clay loam (Ko: *Fine, mixed, superactive, mesic Typic Argiaquolls*, sand: 9%, silt: 62%, clay: 29%, 40°00′41.2″N, 83°12′23.1″W) in May, 2013. These locations were farmer-owned field sites and before collecting the soils from specific sites permissions of the farmers were obtained from the land-owners or farmers. Initial permission was granted through The Ohio State University's extension managers and thereafter communication was established directly by the primary author. Additionally, no endangered or protected species were involved in the current study sites, and thus no such permission was required from any other regulatory agencies. The properties of the soils and other management details were presented elsewhere [20,21] and briefly presented in the discussion section. Soil samples were obtained from 0–10, 10–20, 20–40, and 40–60 cm depths under no-tillage (NT) and conventional tillage (CT) practices for Mc and CrB soil series and under NT for one year cover crop (CC) and no-cover crop (NCC) for Ko series. The organic soil (Mc) has been under practice of CT corn (*Zea mays*) for 10 years which was compared with the adjacent grassland soil (termed as NT). The mineral soil (CrB) was under NT and recent introduction (one year) of CT corn practices. Both soils received fertilizer (N:P:K as 5.5:26:30) with the rates of 336 (30, 142 and 164 kg ha^{-1} of N, P, and K, respectively) and 280 (25, 118 and 137 kg ha^{-1} of N, P, and K, respectively) kg ha^{-1} for corn, for Mc and CrB soils, respectively. The other site (Ko) was under NT corn and soybean (*Glycin max L.*) annual rotation. A mixture of pea (*Pisum sativum L.*) and turnip (*Brassica rapa L.*) cover crops with the seeding rate of 140 kg ha^{-1} was seeded prior to growing soybean in 2013. Roundup-ready soybean was seeded on 20 April, 2013 and harvested on 30 September, 2013. A row spacing of 38 cm was used during planting for corn and soybean. Soil core (inside diameter: 5.3 cm, height: 5.9 cm) samples were collected manually using soil sampler before planting (7 May, 2013) from four depths (0–10, 10–20, 20–40, and 40–60) for determination of bulk density (BD) by dividing dry mass of soil in the core by core volume [22] and hydrologic properties. Undisturbed soil cores were used to determine the water retention at field capacity (0.033 MPa), while sieved samples (<2-mm size) were used to determine the permanent wilting point (1.5 MPa). The potential available water capacity (AWC) of the soil was calculated as the difference in volumetric water content at 0.033 and 1.5 MPa moisture potentials. The water content of a soil layer was calculated by multiplying thickness of the layer with the volumetric water capacity. A minimum of three field penetration resistance (PR) measurements for soil's mechanical strength were made for each depth using a CP40II cone penetrometer [23]. Bulk samples were obtained to measure the aggregate size distributions and fraction of water stable aggregates (WSA) using the wet sieving method [24]. Five sieves of diameter sizes 4.75, 2, 1, 0.5, and 0.25 mm openings were placed into a Yoder apparatus. Air-dried soils (5–8 mm, 51 g) were slowly wetted by capillarity action by adjusting the water level in the container so that the base of the top sieve just touched the water. The sieve combination was oscillated mechanically in the water at 60 oscillations per minute for 30 min. Aggregates retained in the sieves were transferred to glass beakers and the weight of each of five fractions was measured after drying at 60°C overnight. The data were used to compute WSA, mean weight diameter (MWD), and geometric mean diameter (GMD) [25]. Soil water retention at matrix potentials of 0.033 and 1.5 MPa were measured using a pressure plate apparatus [26]. The pH and electrical conductivity (EC) of soil were determined in 1:2 soil:water ratio slurry using a Thermo-scientific Orion Star Series pH/Conductivity Meter. Concentrations of soil organic C and total N were determined using an Elemental analyzer (Vario Max, Elementar Americas, Inc., Germany) by the dry combustion method (900°C) after grinding subsamples to 0.25 mm. Total C and N stocks were calculated by multiplying the respective elemental concentrations by BD and the thickness of the soil layer [27].

Soil quality index (SQI) calculations

All the SQI methods involved a set of 72 soil samples and a number of soil quality indicators as parameters. The 13 parameters used for developing SQIs were pH, EC, BD, WSA, GMD, MWD, PR, SOC concentration, N concentration, C-Stock, N-Stock, AWC, and soil water content. As some of these parameters were synthesized and redundant, only nine parameters were chosen omitting MWD, C and N concentrations, and water content from the SQI-1 and SQI-2 calculations to avoid redundancy. However, SQI-3 was not a primarily additive approach and redundancy of the parameters was eliminated through the processes of elimination as prescribed before. Thus, all 13 parameters were initially included in SQI-3 model and only four non-redundant soil parameters with maximum variations in the dataset were finally retained in the model as described in the following section. The average and standard deviation values of these parameters grouped by soil types are presented in Table 1. Under the proposed framework an ideal soil would have SQI value of 1 for the highest quality soil and 0 for the severely degraded soil [1–4].

Simple additive SQI (SQI-1)

Simple additive SQI was estimated following the method outlined by Amacher et al. [28]. In this method, soil parameters were given threshold values based primarily on the literature review and expert opinion of the authors. The threshold levels, interpretations, and associated unitless soil index score values are listed in Table 2. The individual index values were then summed up to obtain a total SQI:

$$\sum SQI = \sum \textit{Individual soil parameter index values} \quad (1)$$

The scaled SQI (SQI-1) of individual soil, was computed by Eq. 2:

$$SQI-1 = (\sum SQI - SQI_{Min})/(SQI_{Max} - SQI_{Min}), \quad (2)$$

whereas, SQI_{Min} = Minimum value of SQI, and SQI_{Max} = Maximum value of SQI from the total dataset.

Weighted additive SQI (SQI-2)

In this approach, each soil parameter was first assigned unitless score ranging from o to 1 by employing linear scoring functions [8]. Non-linear scoring functions were avoided because of their lower capacity of predicting the end point variable or crop yield [8]. Soil parameters were divided into groups based on three mathematical algorithm functions: (a) 'more is better' (e.g., WSA, GMD, C-Stock, N-Stock, and AWC) (b) 'less is better' (e.g., BD, PR), and (c) 'optimum' (e.g., pH and EC). 'Optimum' properties are those which have positive influence up to a certain level beyond which the influence could be considered detrimental [13]. For 'more is better' parameters, each observation was divided by the highest observed value of the entire dataset so that the highest observed value would have a score of 1; for 'less is better'

Table 1. Descriptive statistics of all the soil indicators collected from four depths under three soil types used to estimate SQI.

	0–10 cm		10–20 cm		20–40 cm		40–60 cm	
Ko								
	Mean	SD	Mean	SD	Mean	SD	Mean	SD
pH	6.0	0.5	6.2	0.6	6.4	0.6	6.4	0.6
EC (μs cm^{-1})	190.6	151.1	165.6	84.2	151.7	71.8	175.9	92.2
BD (Mg m^{-3})	1.4	0.2	1.5	0.1	1.6	0.1	1.6	0.1
WSA %	73.1	14.8	79.5	12.0	73.3	22.5	84.7	5.6
GMD (mm)	1.4	0.4	1.6	0.4	1.4	0.3	1.5	0.3
MWD (mm)	2.6	1.7	3.1	1.8	2.3	1.7	3.4	1.1
PR (Mpa)	2.2	0.3	2.4	0.2	2.3	0.2	2.2	0.4
N (%)	0.2	0.0	0.2	0.0	0.2	0.0	0.2	0.0
SOC (%)	2.4	0.2	2.1	0.2	2.0	0.2	2.0	0.3
N-Stock (Mg/ha)	3.5	0.5	3.0	0.3	6.4	0.6	6.4	1.1
C-Stock (Mg/ha)	33.7	3.7	30.6	2.8	62.5	6.8	63.7	7.2
AWC (%)	38.5	4.1	36.7	2.1	35.8	6.4	39.9	5.8
Water content (cm)	3.8	0.4	3.7	0.2	7.2	1.3	7.9	1.0
CrB								
	0–10 cm		10–20 cm		20–40 cm		40–60 cm	
	Mean	SD	Mean	SD	Mean	SD	Mean	SD
pH	7.4	0.2	7.4	0.2	7.4	0.2	7.4	0.1
EC (μs cm^{-1})	250.3	44.6	217.8	26.8	212.3	31.8	223.5	45.0
BD (Mg m^{-3})	1.5	0.2	1.5	0.1	1.5	0.3	1.5	0.3
WSA %	82.4	4.4	76.4	9.7	76.5	9.5	76.5	17.7
GMD (mm)	1.2	0.1	1.1	0.2	1.1	0.2	1.2	0.3
MWD (mm)	2.0	0.5	1.7	0.8	1.5	0.7	2.0	1.1
PR (Mpa)	2.4	0.2	2.5	0.5	2.6	0.2	2.9	0.7
N (%)	0.2	0.0	0.2	0.0	0.2	0.0	0.2	0.0
SOC (%)	2.4	0.3	2.1	0.4	1.8	0.4	1.8	0.5
N-Stock (Mg/ha)	3.2	0.2	3.1	0.3	5.2	1.6	5.1	2.2
C-Stock (Mg/ha)	34.3	3.2	32.2	4.1	53.8	16.8	54.6	24.4
AWC (%)	17.8	9.0	10.1	6.8	10.7	8.2	9.0	3.2
Water content (cm)	1.8	0.9	1.0	0.7	2.1	1.6	1.8	0.6
Mc								

Table 1. Cont.

Ko								
	0–10 cm		10–20 cm		20–40 cm		40–60 cm	
	Mean	SD	Mean	SD	Mean	SD	Mean	SD
	0–10 cm		10–20 cm		20–40 cm		40–60 cm	
	Mean	SD	Mean	SD	Mean	SD	Mean	SD
pH	7.4	0.1	7.4	0.1	7.4	0.1	7.5	0.1
EC (μs cm^{-1})	616.3	40.1	570.0	92.8	633.2	126.4	599.2	129.2
BD (Mg m^{-3})	0.6	0.0	0.7	0.0	0.6	0.1	0.6	0.1
WSA %	91.2	0.5	91.0	2.5	89.7	2.5	82.5	8.7
GMD (mm)	2.1	0.0	2.2	0.0	2.2	0.0	2.2	0.1
MWD (mm)	5.5	0.1	5.5	0.2	5.7	0.4	5.0	0.5
PR (Mpa)	1.3	0.1	1.5	0.3	1.6	0.3	1.9	0.1
N (%)	1.0	0.1	0.9	0.1	0.9	0.1	1.0	0.1
SOC (%)	15.1	1.1	15.0	0.8	14.9	0.9	15.9	0.6
N-Stock (Mg/ha)	6.1	0.8	6.3	0.7	11.2	1.7	12.6	1.9
C-Stock (Mg/ha)	95.7	11.7	102.8	8.8	186.6	36.2	196.6	28.0
AWC (%)	248.3	151.3	84.8	42.4	48.2	13.8	36.1	22.1
Water content (cm)	24.8	15.1	8.5	4.2	9.6	2.8	7.2	4.4

Abbreviations: SD: standard deviation, EC: electrical conductivity, BD: bulk density, WSA: water stable aggregates, GMD: geometrical mean diameter, MWD: mean weight diameter, PR: penetration resistance, N: nitrogen, SOC: soil organic carbon, AWC: available water capacity.

Table 2. Soil indicators, threshold values, interpretations and scores.

Indicators	Range	Interpretation	Score	Reference
pH	5.5–7.2	Slightly acidic to neutral: Optimum for plant growth	2	[28]
	>7.2<8.0	Slightly to moderately alkaline: Preferred by some plants, possible P and some metal deficiencies	1	
EC (us/cm)	<200	Low salt level	0	[42]
	200–500	Optimum salt level for plants	1	
	>500	High salt level, adverse effect likely	0	
BD (Mg/m^3)	<1.0	High organic soil, supports plant roots	2	[28,45–47]
	1.0–1.5	Adverse effects unlikely	1	
	>1.5	Adverse effects likely	0	
WSA (%)	<50	Infiltration and soil erosion problems likely	0	[30,42]
	50–70	Moderate constraints	1	
	70–90	Good soil	2	
	>90	Excellent soil	3	
GMD (mm)	<1.0	Infiltration and soil erosion problems likely	0	[30,42]
	1–2	Moderate limitations	1	
	>2	No limitation	2	
MWD (mm)	<1.0	Infiltration and soil erosion problems likely	0	[30,42]
	1–2	Moderate limitations	1	
	2–5	Slight limitations	2	
	>5.0	No limitation	3	
PR (Mpa)	1–2	Adverse effect on plant root unlikely	2	[42,48]
	2–3	Moderate adverse effect on plant root	1	
	>3.0	Severe adverse effect on plant roots	0	
N (%)	0.2–0.3	Moderate limitation	1	[41,42]
	>0.3	Slight to no limitation	2	
SOC (%)	2–3	Moderate limitation	1	[41,42]
	>3.0	Slight to no limitation	2	
N-Stock (Mg/ha)	<5.0	N deficient	1	Authors' opinion
	5–10	Moderate to optimum N level	2	
	>10.0	N-rich soil	3	
C-Stock (Mg/ha)	<50.0	C deficient	1	Authors' opinion
	50–100	Moderate to optimum C level	2	
	>100	C-rich soil	3	
AWC (%)	<20	Water-stress to plants	0	[31,42]
	20–50	Moderate water availability	1	
	>50	Good water capacity for plants	2	
Water content(cm)	<5.0	Water-stress to plants	0	Authors' opinion
	5–10	Moderate water availability	1	
	>10	Good water capacity for plants	2	

Abbreviations are same as Table 1.

parameters, the lowest observed value in the entire dataset was divided by each observation so that the lowest observed value received a score of 1; and 'optimum' parameters were scored up to a threshold value as 'more is better', and thereafter above the threshold values were scored as 'less is better' [8,29]. For example, pHs up to 5.5–7.2 were scored as 'more is better', and pHs >7.2 were scored as 'less is better'.

After normalizing soil parameters, the scores were integrated into a single index value for each soil using a weighted additive approach initially suggested by Karlen and Stott [4], but modified later by Fernandes et al. [13]. The following weighted additive function was used for development of SQI-2 (Eq. 3):

$$SQI-2 = [(\text{Weight1}) * \text{RDC}] + [(\text{Weight2}) * \text{WSC}] + [(\text{Weight3}) * \text{NSC}], \quad (3)$$

where, RDC (root development capacity) is the rating for the soil's ability to allow plant root development, WSC (water storage capacity) is the rating for the soil's ability to store water, NSC

Table 3. Model of SQI-2.

Soil function	Weight	Soil Indicators	Sub-weight	Scaled score	B×C	$\sum$ B×C	D×A	%	SQI
	A		B	C		D			
RDC	0.4	BD	0.35	0.41	0.14	0.46	0.18	52.94	0.34
		PR	0.35	0.52	0.18				
		WSA	0.20	0.54	0.11				
		GMD	0.10	0.34	0.03				
WSC	0.2	AWC	1.00	0.06	0.06	0.06	0.01	2.94	
NSC	0.4	pH	0.30	0.79	0.24	0.38	0.15	44.12	
		EC	0.30	0.25	0.08				
		C-stock	0.20	0.11	0.02				
		N-stock	0.20	0.18	0.04				

Abbreviations: RDC: root development capacity, WSC: water storage capacity, NSC: nutrient storage capacity; all other abbreviations are same as Table 1.

(nutrient supply capacity) is the rating for the soil's ability to supply nutrients. Weight 1, 2 and 3 are the respective numerical weights for each soil function (RDC, WSC, and NSC). The numerical weights were assigned to each soil function according to their importance in fulfilling the management goal(s) of maintaining soil quality. While the summation of all these numerical weights (Weight 1, 2 and 3) are supposed to be 1 and distributed evenly as 0.33, 0.33, and 0.34 [4], however, a modified approach was taken for the current study following Fernandes et al. [13], where lower weightage was given to the functional attribute which had lower number of representative indicators in the model. Similar to Fernandes et al. [13] weight values were arbitrarily chosen for RDC, WSC, and NSC as 0.4, 0.2, and 0.4 in the current study. The WSC received lower weight value than others because of low number of representative indicators (only AWC) [13] among all included parameters.

The soil parameters or indicators selected for (i) RDC were BD, PR, WSA, and GMD, (ii) WSC was AWC and (iii) NSC were pH, EC, C-Stock, and N-Stock. Within this network, the sub-weight values were given to each indicator based on their importance under the particular soil functional property, field versus laboratory measurements and scope of redundancy. The sub-weight values of different soil indicators or parameters were summed up to 1 under each soil functional property [13]. The rank of the subweight values were as follows: field measurement indicators (BD, PR) > laboratory measurement indicators (WSA, AWC, pH, EC) >> synthesized parameters (GMD, C-Stock, and N-Stock). Field indicators were given the maximum weightage as these were the most representative of the soil's natural conditions and synthesized parameters received the lowest weightage to avoid data redundancy in the model. Thereafter, scaled SQI (SQI-2) of individual soil was computed by Eq. 2. An example of this model of development of SQI is shown in Table 3.

Statistically modeled SQI (SQI-3)

A statistics-based model was used to estimate SQI using principal component analysis (PCA) [8,12,14,18,30,31]. The PCA-model is used to create a minimum data set (MDS) to reduce the indicator load in the model and avoid data redundancy [8]. The main difference between the first two versus the PCA method is that the first two rely mainly on subjective expert opinion and literature review, while the PCA method is more objective of using a number of statistical tools (multiple correlation, factor and cluster analyses) which could avoid any biasness and data redundancy by choosing an MDS using mathematical formulae [8,32]. A relative advantage and disadvantage of each SQI method is explained in the later section. The preliminary function of PCA is to reduce the dimensionality of the entire data set consisting of a large number of interrelated variables, while retaining as much as possible of the variations present in the data set. This is achieved by transformation to a new set of variables, the principal components (PCs), which are uncorrelated, and ordered so that the first few retain most of the variation present in all of the original variables [33,34]. In other words, the PCA method was chosen as a data reduction tool to select the most appropriate indicator(s) to represent and estimate SQI [18].

All the original observations (untransformed data) of each soil were included in the PCA model using SPSS, version 21.0 [35]. The PCs with high eigenvalues represented the maximum variation in the dataset [8,12,14,18,30,31]. While most studies have assumed to examine PCs with eigenvalue >1.0 following Kaiser [36], the present study had the third and fourth components with eigenvalues of 0.98 and 0.89 and variances >5% (Table 4, Fig. 1). These were examined with first two PCs (eigenvalues 8.23, and 1.67) as

Table 4. Results of principal component analyses (PCA).

	PC-1	PC-2	PC-3	PC-4
Eigenvalues	8.23	1.67	0.98	0.89
% Variance	63.34	12.87	7.56	6.82
Cumulative variance	63.3	76.2	83.8	90.6
Eigen vectors or factor loading				
pH	0.423	0.351	−0.378	**0.711**
EC	**0.925**	0.113	−0.153	0.147
BD	**−0.881**	−0.034	0.212	0.002
WSA	0.590	−0.020	**0.669**	0.343
GMD	**0.927**	0.024	0.276	−0.024
MWD	**0.904**	0.001	0.342	0.046
PR	−0.770	0.268	−0.131	0.092
SOC	**0.975**	0.073	−0.109	−0.025
N	**0.972**	0.050	−0.103	−0.053
C-Stock	0.831	0.412	−0.107	−0.268
N-Stock	0.722	0.440	−0.067	−0.385
AWC	0.555	**−0.782**	−0.189	0.057
Water content	0.624	−0.693	−0.199	−0.066

Abbreviations are same as Table 1; PC: principal component; bold values under each component are highly weighted and underlined bold values are selected in minimum data set.

prescribed for the cases of fewer than three components with eigenvalue >1.0 [8,37]. Under a given PC, each variable had corresponding eigenvector weight value or factor loading (Table 2). Only the 'highly weighted' variables were retained to include in the MDS (Table 2). The 'highly weighted' variables were defined as the highest weighted variable under a certain PC and absolute factor loading value within 10% of the highest values under the same PC [37]. Thus, the bold-face values (Table 4: EC, BD, GMD, MWD, SOC, and N for PC-1, AWC for PC-2, WSA for PC-3, and pH for PC-4) were considered highly weighted eigenvectors and therefore were initially selected in the MDS. However, when more than one variable was retained under a particular PC, multivariate correlation matrix (Table 5) was used to determine the correlation coefficients between the parameters [8,12]. If the parameters were significantly correlated ($r>0.60$, $p<0.05$), then the one with the highest loading factor was retained in the MDS and all others were eliminated from the MDS to avoid redundancy. Following this procedure, except SOC, all other eigenvectors from PC-1 were eliminated from MDS due to high and significant correlation between each other (Table 5). The non-correlated parameters under a particular PC were considered important and retained in the MDS [10,12]. All the bold-faced and underlined soil parameters in Table 4 were selected in the final MDS.

After selection of parameters for the MDS, all selected observations were transformed using linear scoring functions (less is better, more is better and optimum) as described in SQI-2 method. Once the selected observations were transformed in numerical scores (ranged 0–1), a weighted additive approach was used to integrate them into indices for each soil [8,12]. Each PC explained a certain amount of variation in the dataset (Table 4), which was divided by the maximum total variation of the all PCs selected for the MDS to get a certain weightage value under a particular PC [8]. For example, the % variance (63.3) was divided by total cumulative variance (90.6) to obtain the weight value of 0.7 for PC-1 (Table 4). Thereafter, the weighted additive SQI was computed using Eq. 4:

$$SQI-3\,(PCA)=\sum Weight * Individual\,oil\,parameter\,score \quad (4)$$

Validation of SQI

The SQIs estimated from three different methods were validated against soybean (*Glycin max*) and corn (*Zea mays*) yield data of that particular year of collection of soil and crops by computing correlation coefficients [8]. The SQI values were also compared through pearson correlation to understand the effectiveness of each other.

Statistical analyses

All values are presented as means ± standard deviations of three field or laboratory measurements. Significant differences between treatments were analyzed using Tukey's test in PROC GLM in SAS version 9.2 [38]. Treatment differences were deemed significant at $p<0.05$. The PCA was performed in SPSS version 21 [35]. Descriptive statistics and linear regressions were computed in Microsoft Excel [39] and all the figures were obtained using Sigmaplot Version 12.0 [40].

Results and Discussions

Most data of three farm sites have been presented and the detailed discussions of soil type, crop and management effects on soil properties are discussed elsewhere [20,21]. The mean and

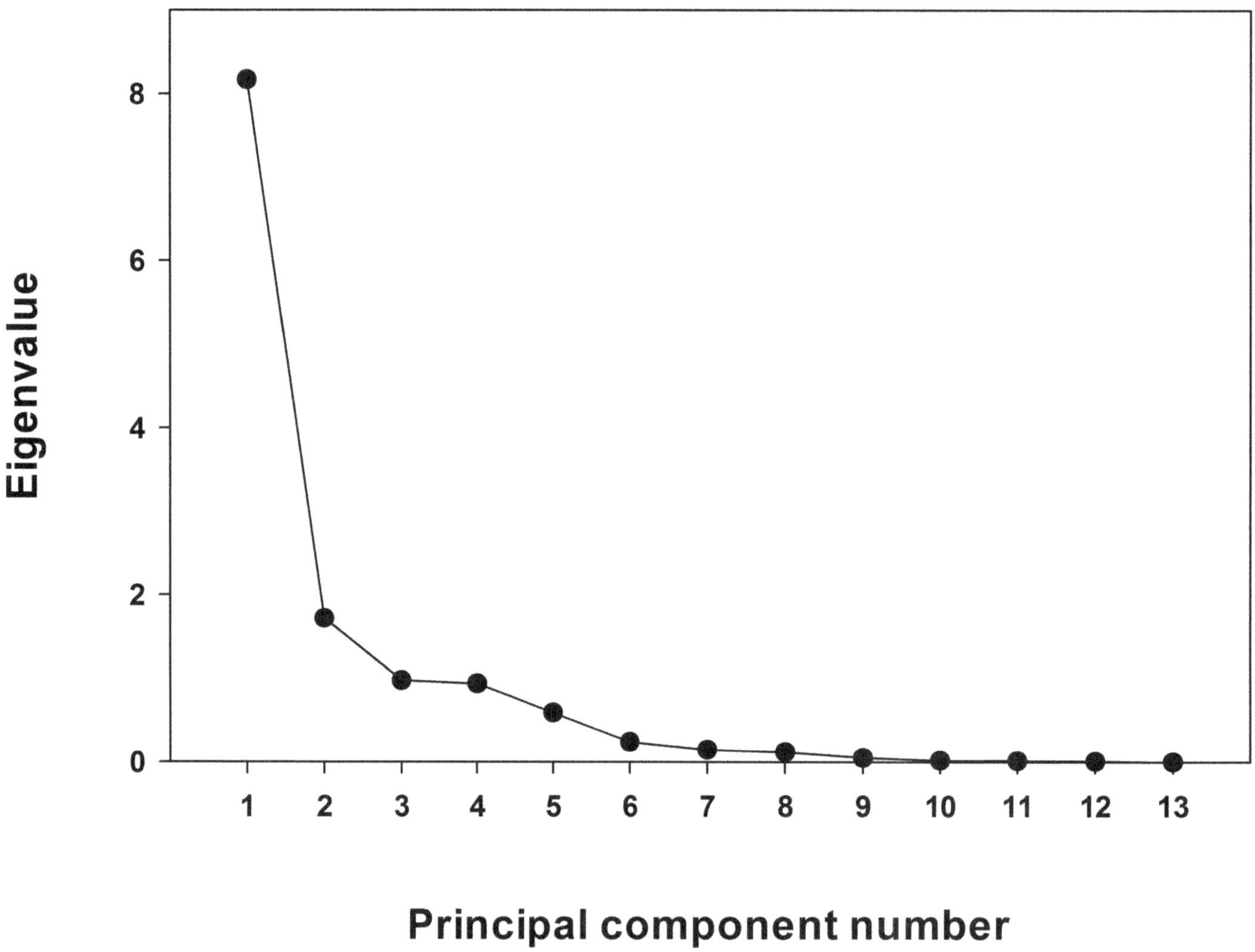

Figure 1. Scree plot of principal component analyses.

standard deviation values of the soil parameters by depth under three soil types are listed in Table 1. Most properties of mineral soils (Ko and CrB) were significantly different compared to those of the organic soil (Mc) and except for AWC and water content two mineral soils' characteristics were similar across soil profile up to 60 cm depth (Table 1). The water characteristics of Ko soil were significantly higher than those of CrB. On the other hand, organic soil (Mc) had much improved (lower BD and PR, and higher WSA, SOC, N concentrations and stock, AWC and water content) soil properties compared to those of the mineral counterparts in all four different depths (Ko and CrB). Impacts of all these parameters on overall soil quality in four soil depths calculated by three indexing methods are discussed in the following sections.

Under the framework of SQI-1, nine soil quality indicators were integrated numerically after scoring them primarily using the information from literature review. However, scoring data on some physicochemical indicators (e.g., C and N stocks) is scarce in the literature and thus, authors' opinion was used in those cases. Further, different studies used different scoring on the same indicator based on soil type, and management goals. For an example, while Amacher and Perry [28] had eight different range classes and scores for pH, Feiza et al. [41] had only four classes and scores available for pH. Thus, scoring on indicators in the current study (Table 2) had importance on experts' opinion although the knowledge was primarily based on available literature and authors' experience [28,42]. The high variability in the observations (Table 1) was reflected in the SQI-1 values. There was no statistical difference ($p>0.05$) across treatments (NCC versus CC, and CT versus NT) and depth under specific soil type (Table 6). Overall performance of indexing methods was determined from averaging SQI values obtained from calculations for each depth and presented in Fig. 2. The, SQI-1 was significantly higher in the following order: Mc>Ko>CrB (Fig. 2), indicating that soil quality was influenced more by soil type than management and sampling depth under the SQI-1 approach.

Overall percentage of various soil functional influences in SQI-2 is presented in Fig. 3 and depthwise percentage of influence in SQI-2 is presented in Fig. S1. Generally, mineral soils (Ko and CrB) did not have any depthwise significant changes for WSC but 0–10 cm layer of organic (Mc) soil had significantly higher influence of WSC in SQI-2 compared to lower depths. On the other hand, while both Ko and Mc had significantly higher influence of RDC in SQI-2 in the upper layers than the lower depths, the opposite trend was found for the NSC (Fig. S1). In the

Table 5. Pearson correlation coefficients (r) for all soil indicators.

Correlation (r) matrix	pH	EC	BD	WSA	GMD	MWD	PR	N	SOC	N-Stock	C-Stock	AWC
EC	0.547											
BD	−0.394	−0.88										
WSA	0.248	0.483	−0.338									
GMD	0.279	0.8	−0.762	0.673								
MWD	0.286	0.782	−0.725	0.734	0.975							
PR	−0.129	−0.655	0.684	−0.456	−0.71	−0.687						
N	0.405	0.922	−0.908	0.47	0.877	0.835	−0.737					
SOC	0.438	0.93	−0.914	0.473	0.881	0.842	−0.731	0.995				
N-Stock	0.271	0.632	−0.551	0.32	0.642	0.595	−0.396	0.709	0.707			
C-Stock	0.382	0.764	−0.702	0.367	0.742	0.695	−0.489	0.822	0.833	0.965		
AWC	0.091	0.447	−0.449	0.275	0.439	0.441	−0.525	0.504	0.489	0.106	0.181	
Water content (cm)	0.092	0.487	−0.485	0.282	0.501	0.496	−0.557	0.558	0.546	0.28	0.333	0.967
Significance level												
EC	<0.001											
BD	<0.001	<0.001										
WSA	0.018	<0.001	0.002									
GMD	0.009	<0.001	<0.001	<0.001								
MWD	0.007	<0.001	<0.001	<0.001	<0.001							
PR	0.141	<0.001	<0.001	<0.001	<0.001	<0.001						
N	<0.001	<0.001	<0.001	<0.001	<0.001	<0.001	<0.001					
SOC	<0.001	<0.001	<0.001	<0.001	<0.001	<0.001	<0.001	<0.001				
N-Stock	0.011	<0.001	<0.001	0.004	<0.001	<0.001	<0.001	<0.001	<0.001			
C-Stock	<0.001	<0.001	<0.001	0.001	<0.001	<0.001	<0.001	<0.001	<0.001	<0.001		
AWC	0.224	<0.001	<0.001	0.011	<0.001	<0.001	<0.001	<0.001	<0.001	0.195	0.07	
Water content (cm)	0.22	<0.001	<0.001	0.009	<0.001	<0.001	<0.001	<0.001	<0.001	0.01	0.003	<0.001

Abbreviations are same as Table 1.

Table 6. Treatment effects on soil quality index (SQI) under three soil types (Ko: Kokomo, CrB: Crosby silty loam, Mc: Muck) and four soil depths.

Treatments									
Ko		NCC	CC	NCC	CC	NCC	CC	NCC	CC
		0–10 cm		10–20 cm		20–40 cm		40–60 cm	
	SQI-1	0.36a	0.36a	0.40a	0.40a	0.45a	0.48a	0.48a	0.50a
	SQI-2	0.40a	0.37a	0.39a	0.36a	0.40a	0.36a	0.43a	0.38a
	SQI-3	0.21a	0.23a	0.20a	0.22a	0.19a	0.20a	0.21a	0.21a
CrB		CT	NT	CT	NT	CT	NT	CT	NT
		0–10 cm		10–20 cm		20–40 cm		40–60 cm	
	SQI-1	0.36a	0.38a	0.36a	0.17a	0.45a	0.29a	0.38a	0.26a
	SQI-2	0.42a	0.40b	0.40a	0.37a	0.41a	0.40a	0.41a	0.40a
	SQI-3	0.26a	0.24a	0.24a	0.22a	0.22a	0.22a	0.22a	0.21a
Mc		CT	NT	CT	NT	CT	NT	CT	NT
		0–10 cm		10–20 cm		20–40 cm		40–60 cm	
	SQI-1	0.88a	0.88a	0.90a	0.90a	0.90a	0.90a	0.90a	0.81a
	SQI-2	0.70a	0.72a	0.65a	0.61a	0.65a	0.66a	0.67a	0.65a
	SQI-3	0.85a	0.85a	0.82a	0.78a	0.78a	0.77a	0.82a	0.81a

Abbreviations: NCC: no cover crop, CC: cover crop, CT: conventional tillage, NT: no tillage.
Different letters under specific row and depth indicate significant differences at $p<0.05$ level, underline value is significant at $p<0.1$ level.

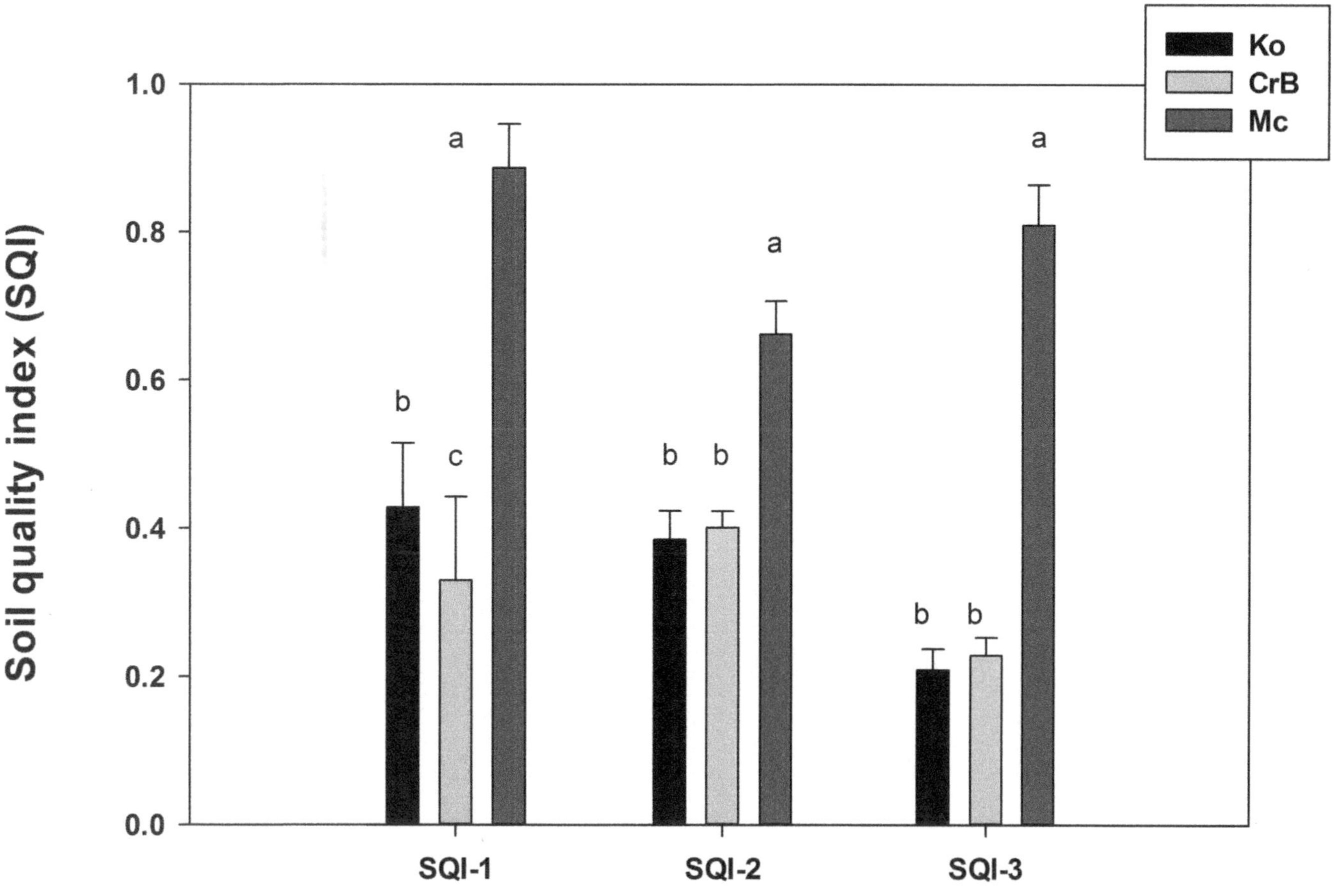

Figure 2. Overall soil quality index (SQI) values under different soil types. Different letters indicate significant differences at $p<0.05$ level for particular indexing method.

general composition of SQI-2, the main contributors were RDC, which ranged from 39–57% and NSC ranged from 35–59%. However, the WSC was relatively small and ranged merely from 0.1–12% (Fig. 3). The relative influence of root development functions in SQI-2 has been significantly higher in Ko than CrB, and Mc by 8 and 9%, respectively and the same for nutrient storage functions in SQI-2 has been significantly higher in CrB than Ko and Mc by 13 and 8%, respectively (Fig. 3). On the other hand, Mc soil had significantly higher water storage functional contribution to SQI-2 by 53 and 500%, respectively, compared to that of mineral (Ko and CrB) soils (Figs. 3). Thus, while the mineral soil quality was more sensitive to the functional attributes dedicated to root development and nutrient storage, muck soil quality was more influenced by soil attributes related to water storage under the scheme of SQI-2. Further, there was no significant effect on management practices on %RDC and %NSC to SQI-2 in different soil/management/crop combinations (Fig. 4). Note that CC/NCC management was used on field cultivated to soybean but CT/NT was used to soil under corn. However, contribution of WSC to SQI-2 for the case of CrB soil under NT management was significantly higher (178%) than that of CT practice (Fig. 4).

Correlation coefficients of SQIs to each other and depthwise SQIs and dry grain and straw are given in Table 7. Crop yield is obviously related to soil quality but soil quality in different layers may not be same and can influence crop yield accordingly. Data on SQI variation for different soil layers are scarce and thus no comparison to the data presented in the current study (Table 6) was possible. However, while numerical values of SQI were almost unchangeable for different depths at Mc site, SQI generally variable with depth in Ko and CrB sites (Table 6). One goal of this study was to validate SQI against crop yield as only a few studies have done so with some exceptions [8,14]. Challenge lies in establishing a correlation between crop yield versus SQI calculated either depthwise or in combination of soil layers as standard methodology is not yet available and was not given much attention to date. For example, one can have only one crop yield value for multiple soil layers and thus to date crop yield has been correlated only against SQI calculated from surface layer. Low (ranged from 0.11–0.32) and high (ranged from 0.83–0.96) correlation coefficients of SQI and crop yield across a number of soil/crop combinations were reported [8,31]. The reasons attributed to low correlation between yield and SQI were: (i) the assumptions that the indicators considered good for a soil quality may not always lead to desired outcomes and flexibility in scoring functions needed [8], (ii) indicators which may not directly relate to crop performance were given more weightage by the method employed, or (iii) some key soil indicators were not included in the MDS due to the design of the study [9,43]. Nevertheless, in order to evaluate the index performance in the current study, SQIs were correlated against end point variable (i.e., yield) at variable depths (Table 7). In the present study, depth of soil sampling had

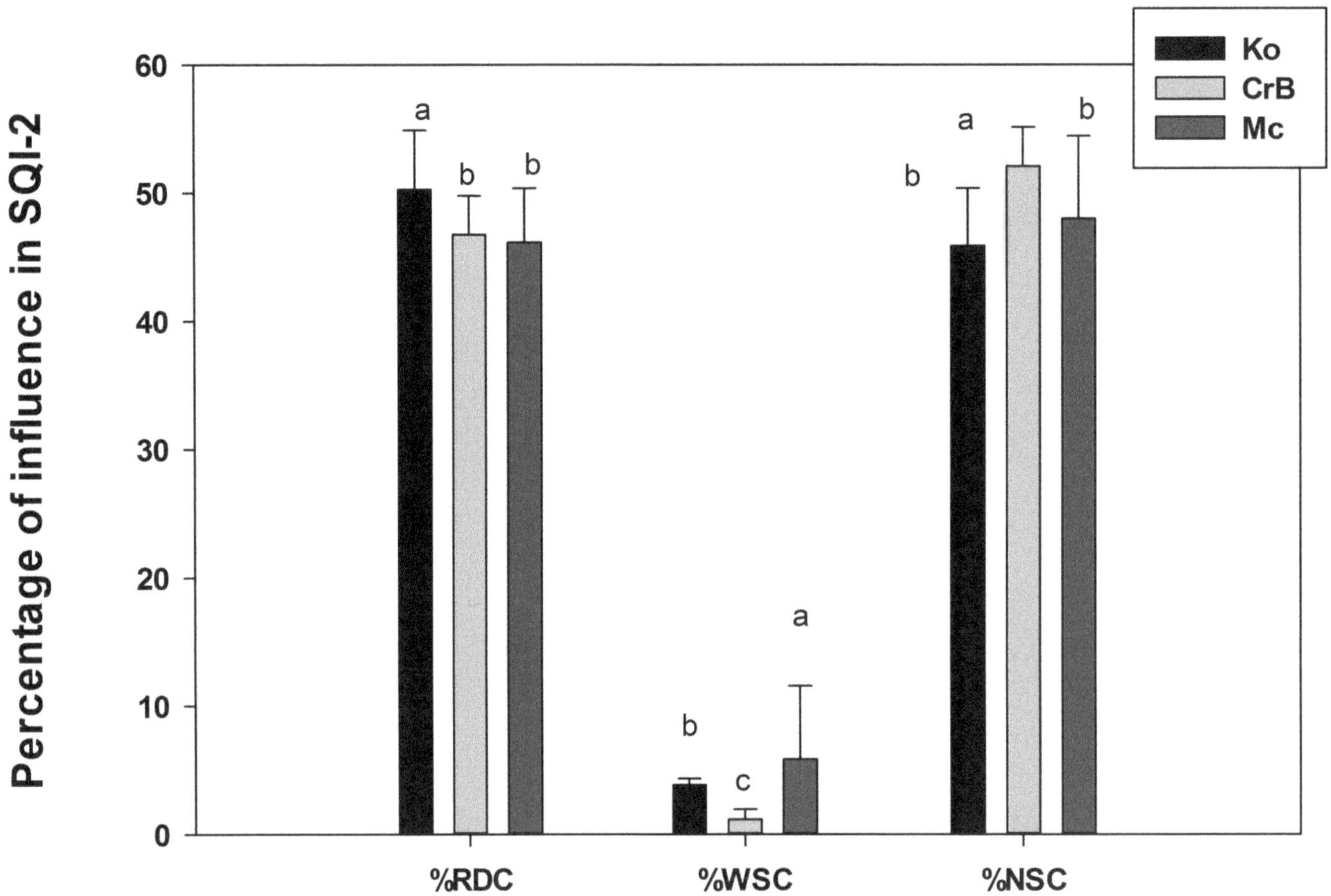

Figure 3. Percentage contribution of each soil function in SQI-2 under different soils. Different letters indicate significant differences at p<0.05 level for particular soil function. Abbreviations: RDC: root development capacity, WSC: water storage capacity, NSC: nutrient storage capacity.

no significant effect in SQI values (Table 6) and thus the individual SQI values were integrated by combining 0–20 cm and 0–60 cm profiles by averaging 0–10, and 10–20 cm for obtaining values for 0–20 cm depth and 0–10, 10–20, 20–40 and 40–60 cm for obtaining values for 0–60 cm depths and correlated the combined values with crop yield (Table 7). In the present study, grain and biomass yields were all significantly ($p<0.05$, r = 0.65–0.79) correlated under any SQI methods employed indicating that either of these methods was successful in predicting crop yield (Table 7). However, correlation coefficients in Table 7 are derived from the complete datasets including both corn and soybean yield but the correlation was considerably low (r<0.20 for soybean grain and r<0.60 for corn grain; data not shown), when dataset was further divided into corn and soybean. As discussed above, low correlation of SQI and crop yield was observed before and one of the listed reasons was some key soil indicators were not included in the dataset as per the experimental design and objective [8]. Crop yield is invariably related to soil fertility status and thus SQI may not always lead to high correlation with yield if some of the related characterizations are not observed as identified before [9,43] which is supported by the data of the current study (Table 7). Thus, the low correlation coefficient values may be because soil fertility (ion exchange capacities, macro and micro nutrients) and microbial (microbial biomass, microbial C, and N, and soil respiration) aspects were not monitored in the current studied sites, especially in the case of soybean grain yield. Nevertheless, relatively higher correlation coefficient was observed in the 0–10 cm soil layer than the deeper soil profile (Table 7) probably due to higher plant root density in the top layer of the soil [44].

Additionally, SQI computed from different methods were also highly correlated with each other (Table 7) suggesting that (i) a relatively easy and user-friendly SQI (SQI-1) can be computed to evaluate and compare soil quality, which is similarly useful to other approaches, (ii) giving certain appropriate weightage on scores (SQI-2) could be similarly useful in predicting particular soil quality and (ii) a minimum group of carefully selected soil indicators (SQI-3) may be used to evaluate soil quality across various soil types and management practices. As all three SQI methods were significantly correlated to each other, it is difficult to conclude which one is the best approach, however, it highly depends on factors such as design of the study, choice of soil parameters included in the model to compute SQI and end point variable (environmental aspects versus crop yield). The advantage of using SQI-1 is that the soil quality could be assessed after measuring any number (low to high) of soil parameters and this procedure is relatively easier compared to other methods as the scoring requires literature review and expert opinions only. The disadvantage of SQI-1 is that it is subjective and relies mainly on researcher's point of view. On the other hand, advantage of SQI-2 is that it includes weightage based on the design of the study, system or the dataset to offset the subjectivity of the approach

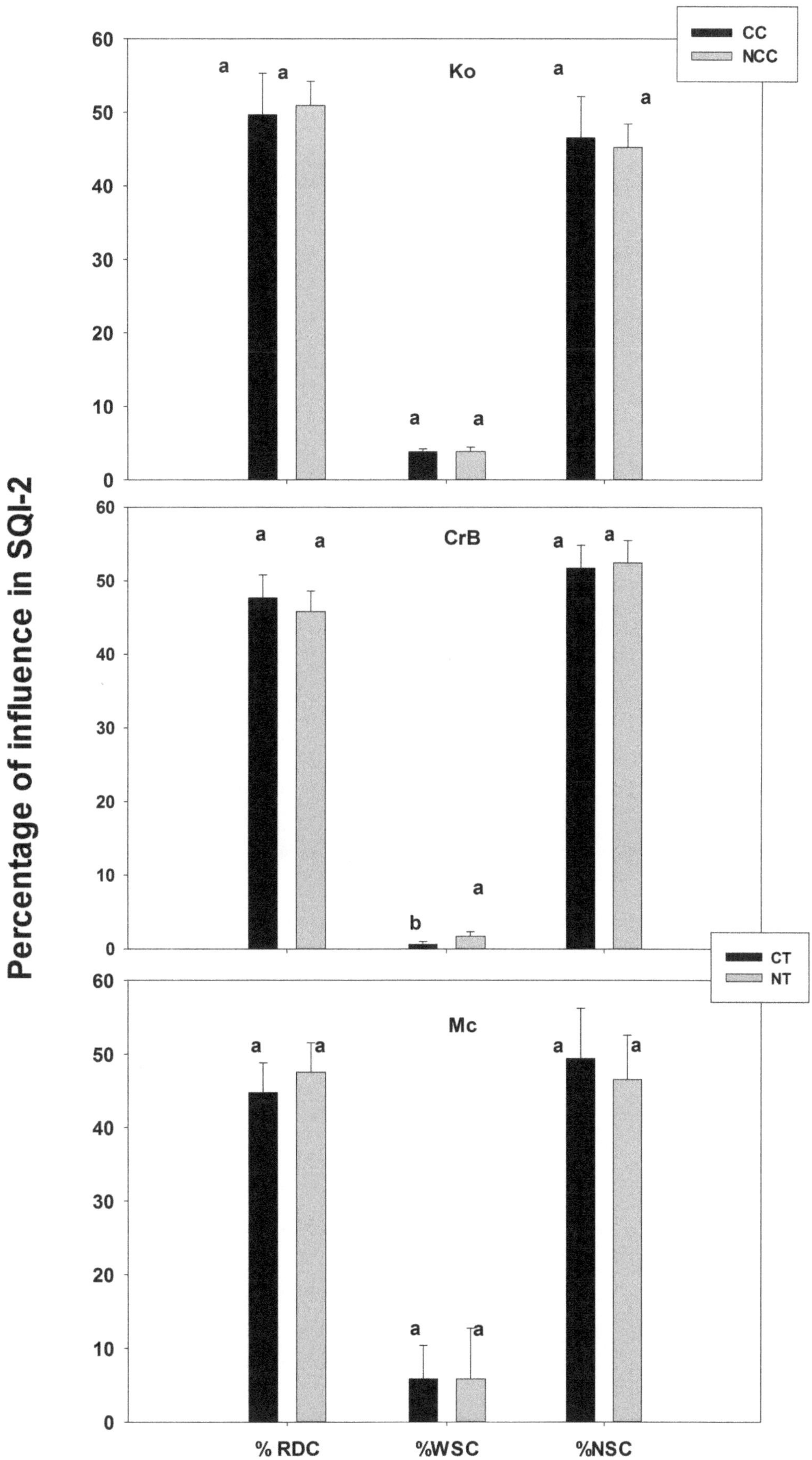
CC
NCC
Ko
CrB
CT
NT
Mc
Percentage of influence in SQI-2
% RDC
%WSC
%NSC
0
10
20
30
40
50
60
a
b

Figure 4. Percentage contribution of each soil function in SQI-2 under various management practices in different soil. Different letters indicate significant differences at p<0.05 level for particular soil function. Abbreviations: RDC: root development capacity, WSC: water storage capacity, NSC: nutrient storage capacity; NCC: no cover crop, CC: cover crop, CT: conventional tillage, NT: no tillage.

present in SQI-2. However, the disadvantage of SQI-2 is that it requires multiple numbers of soil parameters under different soil functional systems which may be expensive and time consuming in practical case. The SQI-3 is advantageous in the aspect of its ability to predict soil quality based on a reduced dataset with low number of soil parameters. Additionally, it is mostly objective approach as the statistical procedure would select a low number of soil parameters needed to calculate SQI based on the variances present in the whole dataset. So, in a long-term aspect within a particular soil/crop system, SQI-3 can be used effectively once it evaluated the most influential soil parameters required to assess soil quality of a particular soil/crop/management scenario. However, current data suggest that relative higher SQI values were obtained for Ko and Mc by SQI-1 than SQI-2 or SQI-3 and for CrB by SQI-2 than SQI-1 or SQI-3 (Fig. 2). Although SQI-1 and SQI-2 had higher values than SQI-3 in different scenarios, however, based on the current experimental design, and selected soil parameters SQI-3 successfully predicted crop yield relatively higher [Table 7 and also when grouped by crops (data not shown)] than SQI-1 and 2. Depth had little influence in any SQIs in the present study and additional chemical and biological parameters are also needed to strengthen the validity of SQI. Thus, SQI-3 appears to be the best method among the three under long-term scenario, especially due to its objective approach, relative higher correlations with crop yield and lower number of indicator selection which is more cost and time-effective over time than other approaches.

Conclusions

The data presented support the following conclusions: The SQI calculated by three different methods indicated that studied muck soil has significantly higher soil quality than that of mineral soils under on-farm conditions. The SQI was affected more by management and soil type than by depths in the studied on-farm sites. The SQI computed using three established methods were all significantly correlated to each other indicating that relatively easy and user-friendly SQI (SQI-1) can be similarly useful to evaluate soil quality, appropriate weightage on scores can predict soil's quality (SQI-2) with high performance but requires a number of soil parameters under different soil functional components and carefully chosen MDS with small numbers of soil variables (SQI-3) may adequately evaluate its quality. All three SQIs were highly and significantly correlated with crop yield. In general, however, analyses by crop type (corn and soybean) revealed that correlation was low ($r<0.20$ or 0.60 for soybean or corn grain yield, respectively; data not shown) suggesting choice of crop-specific and key soil parameters used in computing SQI. Under the current experimental on-farm conditions SQI-1 (for Ko and Mc) and SQI-2 (for CrB) values were higher than SQI-3, however, SQI-3 can be regarded as the best and easiest model given its relatively higher success to predict crop yield and objectivity approach with lower number of indicator selection ability which should be regarded as a relatively less expensive procedure over time compared to SQI-1 and 2. In addition, in order to effectively predict particular crop yield one must include soil fertility and microbial parameters in the model of SQI.

Table 7. Pearson correlation coefficients (r) of soil quality index (SQI) versus crop yield and correlations between different SQI values which were averaged up to a certain depth; all numbers are significant at $p<0.05$ level.

SQI-1	**Dry Grain**	**Dry Straw**
0–60 cm	0.65	0.67
0–20 cm	0.71	0.71
0–10 cm	0.73	0.71
SQI-2	**Dry Grain**	**Dry Straw**
0–60 cm	0.75	0.73
0–20 cm	0.75	0.72
0–10 cm	0.79	0.75
SQI-3	**Dry Grain**	**Dry Straw**
0–60 cm	0.76	0.74
0–20 cm	0.76	0.74
0–10 cm	0.78	0.76
	SQI-1	**SQI-2**
SQI-2	0.92	
SQI-3	0.93	0.97

Supporting Information

Figure S1 Percentage contribution of each soil function in SQI-2 under different soils in four soil layers; different letters indicate significant differences at p<0.05 level for particular soil function. Abbreviations: RDC: root development capacity, WSC: water storage capacity, NSC: nutrient storage capacity.

Acknowledgments

Authors would like to thank Dr. Shiguo Jiang for his help with PCA and Mr. Basant Rimal for his assistance in the laboratory. Authors would also like to thank three anonymous reviewers and an academic editor for their constructive suggestions to improve the manuscript.

Author Contributions

Conceived and designed the experiments: AM RL. Performed the experiments: AM. Analyzed the data: AM. Contributed reagents/materials/analysis tools: AM. Contributed to the writing of the manuscript: AM RL.

References

1. Doran J, Coleman D, Bezdicek D, Stewart B (1994) A framework for evaluating physical and chemical indicators of soil quality.
2. Doran JW, Parkin TB (1994) Defining and assessing soil quality. SSSA special publication 35: 3–3.
3. Karlen DL, Mausbach MJ, Doran JW, Cline RG, Harris RF, et al. (1997) Soil Quality: A Concept, Definition, and Framework for Evaluation (A Guest Editorial). Soil Sci Soc Am J 61: 4–10.
4. Karlen DL, Stott DE (1994) A framework for evaluating physical and chemical indicators of soil quality. In: Doran JW, Coleman DC, Bezdicek DF, Stewart BA, editors. Defining soil quality for a sustainable environment. Madison, WI: Soil Science Society of America. pp. 53–72.
5. Stocking MA (2003) Tropical Soils and Food Security: The Next 50 Years. Science 302: 1356–1359.
6. Diack M, Stott D (2001) Development of a soil quality index for the Chalmers Silty Clay Loam from the Midwest USA. Purdue University: USDA-ARS National Soil Erosion Research Laboratory. pp. 550–555.
7. Larson WE, Pierce FJ (1991) Conservation and enhancement of soil quality. Evaluation of Sustainable Land Management in the Developing World. International Board for Soil Research and Management, Bangkok, Thailand.
8. Andrews S, Karlen D, Mitchell J (2002) A comparison of soil quality indexing methods for vegetable production systems in Northern California. Agriculture, Ecosystems & Environment 90: 25–45.
9. Bhardwaj AK, Jasrotia P, Hamilton SK, Robertson GP (2011) Ecological management of intensively cropped agro-ecosystems improves soil quality with sustained productivity. Agriculture Ecosystems & Environment 140: 419–429.
10. Andrews SS, Carroll CR (2001) Designing a soil quality assessment tool for sustainable agroecosystem management. Ecological Applications 11: 1573–1585.
11. Andrews SS, Karlen DL, Cambardella CA (2004) The soil management assessment framework: A quantitative soil quality evaluation method. Soil Science Society of America Journal 68: 1945–1962.
12. Andrews SS, Mitchell JP, Mancinelli R, Karlen DL, Hartz TK, et al. (2002) On-Farm Assessment of Soil Quality in California's Central Valley. Agron J 94: 12–23.
13. Fernandes JC, Gamero CA, Rodrigues JGL, Mirás-Avalos JM (2011) Determination of the quality index of a Paleudult under sunflower culture and different management systems. Soil and Tillage Research 112: 167–174.
14. Mandal UK, Ramachandran K, Sharma K, Satyam B, Venkanna K, et al. (2011) Assessing Soil Quality in a Semiarid Tropical Watershed Using a Geographic Information System. Soil Science Society of America Journal 75: 1144–1160.
15. Zobeck TM, Halvorson AD, Wienhold B, Acosta-Martinez V, Karlen DL (2008) Comparison of two soil quality indexes to evaluate cropping systems in northern Colorado. Journal of Soil and Water Conservation 63: 329–338.
16. Zornoza R, Mataix-Solera J, Guerrero C, Arcenegui V, Mataix-Beneyto J, et al. (2008) Validating the effectiveness and sensitivity of two soil quality indices based on natural forest soils under Mediterranean conditions. Soil Biology and Biochemistry 40: 2079–2087.
17. Jokela W, Posner J, Hedtcke J, Balser T, Read H (2011) Midwest Cropping System Effects on Soil Properties and on a Soil Quality Index. Agronomy Journal 103: 1552–1562.
18. Navas M, Benito M, Rodriguez I, Masaguer A (2011) Effect of five forage legume covers on soil quality at the Eastern plains of Venezuela. Applied Soil Ecology 49: 242–249.
19. Brady NC, Weil RR (1984) The nature and properties of soils New York: Macmillan.
20. Mukherjee A, Lal R (2014) Long and short-term tillage effects on quality of organic and mineral soils under on-farm conditions in Ohio, Submitted. Soil and Tillage Research.
21. Mukherjee A, Lal R (2014) Short-term effects of cover cropping on quality of a Typic Argiaquolls in Central Ohio, Submitted.
22. Grossman RB, Reinsch TG (2002) Bulk density and linear extensibility. In: Dane JH, Topp GC, editors. Methods of soil analysis, Part 4: Physical methods. Madison, WI: SSSA. pp. 201–254.
23. Herrick JE, Jones TL (2002) A dynamic cone penetrometer for measuring soil penetration resistance. Soil Science Society of America Journal 66: 1320–1324.
24. Yoder RE (1936) A direct method of aggregate analysis of soils and a study of the physical nature of erosion losses. J Am Soc Agron 28: 337–351.
25. Youker RE, McGuinness JL (1957) A short method of obtaining mean weight-diameter values of aggregate analyses of soils. Soil Science 83: 291–294.
26. Dane JH, Hopmans JH (2002) Water retention and storage. In: Dane JH, Topp GC, editors. Methods of soil analysis Part 4, SSSA Book Ser 5. Madison, WI.: SSSA. pp. 671–717.
27. Chhabra A, Palria S, Dadhwal VK (2003) Soil organic carbon pool in Indian forests. Forest Ecology and Management 173: 187–199.
28. Amacher MC, Perry CH (2007) Soil vital signs: A new Soil Quality Index (SQI) for assessing forest soil health.
29. Liebig MA, Varvel G, Doran J (2001) A simple performance-based index for assessing multiple agroecosystem functions. Agronomy Journal 93: 313–318.
30. Li Q, Xu MX, Liu GB, Zhao YG, Tuo DF (2013) Cumulative effects of a 17-year chemical fertilization on the soil quality of cropping system in the Loess Hilly Region, China. Journal of Plant Nutrition and Soil Science 176: 249–259.
31. Masto RE, Chhonkar PK, Singh D, Patra AK (2008) Alternative soil quality indices for evaluating the effect of intensive cropping, fertilisation and manuring for 31 years in the semi-arid soils of India. Environmental monitoring and assessment 136: 419–435.
32. Doran JW, Parkin TB (1996) Quantitative indicators of soil quality: a minimum data set. SSSA special publication 49: 25–38.
33. Dunteman GH (1989) Principal components analysis: Sage.
34. Jolliffe I (2005) Principal component analysis: Wiley Online Library.
35. SPSS (2014) IBM SPSS, Version 21.0, Chicago, USA.
36. Kaiser HF (1960) The application of electronic computers to factor analysis. Educational and Psychological Measurement 20: 141–151.
37. Wander MM, Bollero GA (1999) Soil quality assessment of tillage impacts in Illinois. Soil Science Society of America Journal 63: 961–971.
38. SAS (2012) SAS Institute Inc.; SAS version 9.2.
39. Microsoft (2007) Microsoft Excel. Redmond, Washington.
40. Sigmaplot (2012) Scientific Software Solutions Internationals; Sigmaplot version 12.
41. Feiza V, Feiziene D, Kadziene G, Lazauskas S, Deveikyte I, et al. (2011) Soil state in the 11th year of three tillage systems application on a cambisol. Journal of Food, Agriculture & Environment 9: 1088–1095.
42. Lal R (1994) Methods and guidelines for assessing sustainable use of soil and water resources in the tropics; Washington D.C.: USDA/SMSS Technical Monograph 21.
43. Zornoza R, Mataix-Solera J, Guerrero C, Arcenegui V, García-Orenes F, et al. (2007) Evaluation of soil quality using multiple lineal regression based on physical, chemical and biochemical properties. Science of The Total Environment 378: 233–237.
44. Maurya PR, Lal R (1980) Effects of No-Tillage and Ploughing on Roots of Maize and Leguminous Crops. Experimental Agriculture 16: 185–193.
45. Brzezinska M, Sokolowska Z, Alekseeva T, Alekseev A, Hajnos M, et al. (2011) Some characteristics of organic soils irigated with municipal wastewater. Land Degradation & Development 22: 586–595.
46. Sokoowska Z (2011) Effect of phosphates on dissolved organic matter release from peat-muck soils.
47. Walczak R, Rovdan E, Witkowska-Walczak B (2002) Water retention characteristics of peat and sand mixtures. International Agrophysics 16: 161–166.
48. Carter MR (2006) Quality critical limits and standardization. In: Lal R, editor. Encyclopedia of Soil Science. New York: Marcel Dekker. pp. 1412–1415.

5

Development and Evaluation of SoySNP50K, a High-Density Genotyping Array for Soybean

Qijian Song[1], David L. Hyten[1¤], Gaofeng Jia[1], Charles V. Quigley[1], Edward W. Fickus[1], Randall L. Nelson[2], Perry B. Cregan[1]*

1 Soybean Genomics and Improvement Laboratory, Agricultural Research Service, United States Department of Agriculture, Beltsville, Maryland, United States of America, **2** Pathology and Genetics Research Unit and Department of Crop Sciences, Soybean/Maize Germplasm, Agricultural Research Service, United States Department of Agriculture, University of Illinois, Urbana, Illinois, United States of America

Abstract

The objective of this research was to identify single nucleotide polymorphisms (SNPs) and to develop an Illumina Infinium BeadChip that contained over 50,000 SNPs from soybean (*Glycine max* L. Merr.). A total of 498,921,777 reads 35–45 bp in length were obtained from DNA sequence analysis of reduced representation libraries from several soybean accessions which included six cultivated and two wild soybean (*G. soja* Sieb. et Zucc.) genotypes. These reads were mapped to the soybean whole genome sequence and 209,903 SNPs were identified. After applying several filters, a total of 146,161 of the 209,903 SNPs were determined to be ideal candidates for Illumina Infinium II BeadChip design. To equalize the distance between selected SNPs, increase assay success rate, and minimize the number of SNPs with low minor allele frequency, an iteration algorithm based on a selection index was developed and used to select 60,800 SNPs for Infinium BeadChip design. Of the 60,800 SNPs, 50,701 were targeted to euchromatic regions and 10,000 to heterochromatic regions of the 20 soybean chromosomes. In addition, 99 SNPs were targeted to unanchored sequence scaffolds. Of the 60,800 SNPs, a total of 52,041 passed Illumina's manufacturing phase to produce the SoySNP50K iSelect BeadChip. Validation of the SoySNP50K chip with 96 landrace genotypes, 96 elite cultivars and 96 wild soybean accessions showed that 47,337 SNPs were polymorphic and generated successful SNP allele calls. In addition, 40,841 of the 47,337 SNPs (86%) had minor allele frequencies ≥10% among the landraces, elite cultivars and the wild soybean accessions. A total of 620 and 42 candidate regions which may be associated with domestication and recent selection were identified, respectively. The SoySNP50K iSelect SNP beadchip will be a powerful tool for characterizing soybean genetic diversity and linkage disequilibrium, and for constructing high resolution linkage maps to improve the soybean whole genome sequence assembly.

Editor: Tianzhen Zhang, Nanjing Agricultural University, China

Funding: This study was partially funded by the United Soybean Board (http://www.unitedsoybean.org/) as part of Project #8265 entitled: Whole Genome Analysis of the USDA Soybean Germplasm Collection and Applications for New Gene Discovery (50,000 SNPs). The United Soybean Board has read the manuscript and has indicated their approval of the manuscript. The funders had no role in study design, data collection and analysis, decision to publish, or preparation of the manuscript.

Competing Interests: The authors have declared that no competing interests exist.

* E-mail: Perry.Cregan@ars.usda.gov

¤ Current address: Pioneer Hi-Bred International Inc, Johnston, Iowa, United States of America

Introduction

Even with the extensive linkage disequilibrium (LD) in the soybean [*Glycine max* (L). Merr.] genome, a large number of molecular markers are required to cover the genome for the purpose of genome-wide genetic analysis. Single nucleotide polymorphisms (SNPs) are the most abundant form of DNA polymorphism in eukaryotic genomes [1], and are suitable for the development of high-throughput, easy-to-automate genotyping methods because most SNPs have only two alleles, thereby simplifying genotyping approaches and analysis [2–6]. Previous procedures of SNP discovery in plants involved primer selection, PCR amplification and sequencing of amplicons from a small set of diverse genotypes [7–9]. This approach is time-consuming and expensive. With the advent of second generation sequencing technology, large amounts of DNA sequence data can be generated in a short period of time. Coupled with the availability of the soybean whole genome sequence [10], large numbers of sequence variants can be efficiently identified via the alignment of the short sequence reads from diverse genotypes to the whole genome sequence [11–13]. High throughput SNP discovery is a necessity in light of the availability of high throughput SNP assay systems such as the Infinium platform from Illumina that allows the assay of large numbers of SNPs per DNA sample in parallel on a single silicon slide (http://www.illumina.com/). Thus, the objective of this research was to efficiently identify large numbers of SNPs in soybean using advanced sequencing technologies and SNP identification pipelines; to develop a custom high resolution Illumina Infinium BeadChip containing SNPs with high minor allele frequency, even distribution across the euchromatic and heterochromatic regions of the soybean genome; to evaluate the performance of the chip in a set of diverse landrace and elite soybean genotypes as well as in a diverse set of wild soybean [*G. soja* (Sieb. et Zucc.)] genotypes and to identify candidate regions that may be associated with domestication and recent selection imposed by plant breeding.

Table 1. Illumina GAII DNA sequence analysis of eight genotypes and mixed DNA.

Genotypes	Number of lanes	Number of reads	Total number of bases
PI 468916 and PI 479752	23	120,581,402	4,258,255,310
Essex	18	166,829,261	6,473,293,963
Evans	4	11,638,328	360,108,084
Archer	3	7,759,202	277,664,148
Minsoy	3	26,101,751	881,054,603
Noir 1	3	39,234,960	1,404,545,136
Peking	2	35,012,819	1,237,429,040
Mixed DNA	15	91,764,054	3,421,615,732
Total	71	498,921,777	18,313,966,016

Results

SNP Discovery and Randomness of SNP Distribution in the Soybean Genome

The total number of short reads and total bases obtained from the eight genotypes were 498,921,777 and 18,313 Mb, respectively and the total bases obtained from the eight genotypes ranged from 277 Mb (0.3×) to 6,473 Mb (6.8×) (Table 1). Following alignment of the short reads to the Williams 82 (Glyma1.01) genome sequence, 209,903 SNPs were identified by MAQ and/or CASAVA. The average distance between adjacent SNPs was 4.5 kb. Only 18, 193 and 1575 of the 209,863 gaps between adjacent SNPs had sequence length >200 kb, 100 kb and 50 kb, respectively and the largest gap was 328 kb. Thus, in general, the identified SNPs were fairly well distributed throughout the soybean genome.

SNP Filtering

Of the 209,903 SNPs, a total of 28,891 were A vs. T or G vs. C SNPs and thus, the remaining 188,272 SNPs were a potential source for Infinium II assay (duel color channel, single bead). After eliminating SNPs with N's or ambiguity codes in the 60 nt of flanking sequence and SNPs residing within 25 nt of another SNP, a total of 150,171 of the 188,272 SNPs for Infinium II assay remained. After elimination of the SNPs with flanking sequences that were not unique in the genome, a total of 146,161 SNPs, including 785 SNPs from unanchored scaffolds, was retained and these SNPs were ideal candidates for Infinium II assay design.

Initial SNP Verification and Calculation of Priority Scores

Of the 146,161 SNPs, a total of 23,721, 50,229, 58,281 and 13,930 were assigned to priority groups of A, B, C, and D, respectively. Initial validation of 767 SNPs via Sanger sequencing gave validation rates of 100%, 83%, 77% and 67% for the SNPs with the A, B, C, and D scores, respectively (Table 2). These rates were used as priority scores (R_i) for the corresponding priority groups for calculation of the SNP selection index.

Selection of a Set of 60,800 SNPs

Because Illumina only allowed 60,800 SNPs in the bead pool at the time the chip was developed, a total of 60,701 SNPs were selected from the 20 chromosomes (Gm01–Gm20) and 99 SNPs were selected from 71 unanchored sequence scaffolds (Table S1). Of the 60,701 SNPs, a total of 50,701were selected from the euchromatic regions and 10,000 SNPs were selected from heterochromatic regions with approximately a 5:1 ratio. Thus, the average density of SNPs was approximately 110.5 SNPs per Mb and 20.4 SNPs per Mb in euchromatic and heterochromatic regions, respectively (Table 3). The number of SNPs chosen from each chromosome was the product of the chromosome sequence length by the average SNP density in the euchromatic and heterochromatic regions, respectively. The SNPs along each chromosome were then selected with the iteration algorithm. Genetic mapping of the 99 SNPs from the 71 unanchored sequence scaffolds is expected to anchor 6.5 Mb of the 23 Mb of sequence that was not anchored in the Glyma1.01 build.

Index Score and Evenness of Spacing between the 60,800 Selected SNPs

After SNP selection, the SNP selection index score of the 60,800 selected SNPs was increased versus the source set of 146,161 SNPs. The average SNP selection index score of the selected SNPs was 0.69 vs. 0.60 for the complete set of 146,161 SNPs. Of the 146,161 SNPs, 73,084 were located in euchromatic DNA and 72,272 were located in heterochromatic regions. A total of 50,701 SNPs were selected from the euchromatic regions and 10,000 were selected from the heterochromatic regions (Table 3). In the pre-selected set, the average distance between SNPs was 6.3 kb and 6.8 kb with a standard deviation of 10.7 kb and 14.9 kb in the euchromatic and heterochromatic regions, respectively. Of the 73,084 SNPs identified in the euchromatic regions, 21,721 (30%) were at distances of ≤2 kb between adjacent SNPs. In the heterochromatic regions, a total of 58,766 SNPs were at distances of ≤10 kb, which was 81% of the 72,272 SNPs in the heterochromatic regions. Of these, 62,272 were eliminated via the application of the iteration algorithm. After elimination of SNPs via the application of the iteration algorithm, the result was an average distance between adjacent SNPs in the selected set of 60,800 SNPs of 9.1 kb and 49.1 kb with a standard deviation of 9.1 kb and 33.3 kb in the euchromatic and heterochromatic regions, respectively. Thus, the uniformity of the distance between selected SNPs was enhanced in the euchromatic regions and clustered SNPs were greatly reduced in the heterochromatic regions as a result of SNP selection via the application of the iteration algorithm (Fig. 1).

Table 2. Validation rate of SNPs based on the Sanger sequence analysis of a random set of 767 SNP loci.

Priority class	Total number of loci	Number of loci with null or multiple amplicons	Number of loci with good sequence	Number of SNPs validated	Number of SNPs not validated	Validation rate (%)
A	89	17	72	72	0	100
B	233	43	190	158	32	83
C	390	73	317	244	73	77
D	55	9	46	31	15	67

Distribution of SNPs in Genic Regions

A total of 16,381 (27%) of the 60,800 SNPs resided in genes and, of these, 6,820 SNPs were located in coding DNA sequence, 8,103 SNPs in introns, 752 SNPs in 5′untranslated regions and 706 SNPs in 3′UTR (Table S1). Within 2 kb upstream of genes there were a total of 7,707 SNPs and 6,723 SNPs were within 2 kb downstream of genes. Overall, more than 50% of the 60,800 SNPs resided within genes or within 2 kb upstream or downstream of genes.

Infinium Chip Performance

Of the 60,800 SNPs submitted to Illumina for inclusion in the SNP assay, 52,041 SNPs, or 86% of the 60,800 passed Illumina manufacturing quality control and were included in the final bead pool of the SoySNP50K Infinium BeadChip. Validation of the SoySNP50K Infinium BeadChip was performed via the analysis of diverse germplasm including 96 landraces, 96 elite cultivars and 96 wild soybean accessions. Of the 52,041 SNPs, 47,337 (91%) were polymorphic and yielded successful SNP allele calls based upon clear separation between the fluorescent signals from the two homozygous genotypic classes for a given SNP as assessed with the normalized theta analysis of the Illumina GenomeStudio SNP analysis software.

The rate of SNPs producing successful SNP allele calls was associated with the SNP selection index score, the Illumina Infinium SNP design score and the priority group (Table 4). The correlation coefficients of the SNP selection index score and SNP score vs. the rate of SNPs with successful allele calls were 0.93 and 0.94, respectively, and both were significant at the 1% level. Correlations with priority group were not calculated because priority group is not a continuous variable and the number of priority groups is small. However, it appeared that higher priority group resulted in an increased rate of SNPs with successful allele calls (Table 4).

Estimate of SNP Minor Allele Frequency in Diverse Germplasm

Of the 47,337 SNP assays that produced successful allele calls, 40,841 (86%) had a minor allele frequency ≥10% when tested on the 96 landraces, 96 elite cultivars and 96 wild soybeans. Within the individual groups, the landraces had the most SNPs with a minor allele frequency ≥10% while the elite cultivars had the fewest SNPs with a minor allele frequency ≥10% (Table 5). Only 3,294 of the 47,337 had a minor allele frequency ≤5% within the diverse germplasm (Table 5).

Estimate of Genetic Distance and Number of Polymorphic SNPs among Pairs of Genotypes within the 96 Landraces, the 96 Elite Cultivars and the 96 Wild Accessions

Genetic distances of 47.6% of the pairs of genotypes among the elite cultivars, 68.1% among the landraces and 48.1% among the wild accessions were greater than 0.3 based on the SNPs with minor allele frequency (MAF) >0.05 (Table 6). The mean genetic distance and the standard deviation were 0.294 and 0.0393 among pairs of elite cultivars, 0.322 and 0.0518 among pairs of landraces, and 0.292 and 0.0528 among pairs of wild accessions, respectively. On average, 9,974 and 18,151 SNPs were expected to be polymorphic between a random pair of elite cultivars or landraces, respectively. A total of 9,094 to15,565, 9,240 to 17,761 and 7,889 to 16,576 SNPs were projected to be polymorphic between any randomly selected pair of elite cultivars, landraces and wild soybean accessions at the 95% probability level, respectively.

Identification of Regions with Signatures of Selection

A total of 620 and 42 regions (100 kb windows), contained loci with high average fixation indices (F_{st}≥0.6) between *G. soja* and landrace and between landrace and elite populations, respectively (Table 7, Table S2). These regions may contain loci associated with domestication from the wild soybean or with selection imposed during 70 years of North American soybean breeding, respectively. The examination of these regions will provide the basis for much further investigation aimed at the genic content of regions that underwent selection during domestication and during the genetic improvement of soybean adapted to the diverse soybean producing environments of North America. At this time we thought it appropriate to briefly describe one of the genome regions that was apparently under selection during soybean domestication and one with the selection imposed by North American soybean breeding. In the case of a region associated with domestication, we observed that the region flanked by Sat_355 and Satt474 of soybean linkage group B2 [14], i.e., position13,784,007–33,076,661 of the Gm14 in Glyma1.01 [15], has been reported to contain major quantitative trait loci (QTL) controlling seed yield and plant height [16], oil [17], seed size [18–20], and stem strength [21]. Most of these traits were very likely affected during domestication, accordingly, we observed that most of the F_{st} values between *G. soja* and the landraces in this region were greater than 0.6. In contrast, most of the F_{st} values between the elite and landrace populations were less than 0.1 (Fig. 2a).

In the case of selection imposed during the years of North America soybean breeding, an example of a strong peak of F_{st} between the landrace and elite populations was observed between 2.7–5.0 mbp of Gm04 in Glyma1.01 (Fig. 2b). QTL for resistance to soybean cyst nematode (SCN) race 2 [22,23] seed size [24],

Table 3. Number and density of selected SNPs and the 146,161 SNPs in euchromatic and heterochromatic regions of each soybean chromosome.

Chromosome	Number of SNPs	Number of selected SNPs in the euchromatic regions	Number of selected SNPs in the heterochro-matic region	Sequence length of euchromatic regions(bp)	Sequence length of heterochroma-tic regions (bp)	SNP density in euchromatic regions (SNPs/Mb)	SNP density in the heterochro-matic regions (SNPs/Mb)	Number of SNPs of the 146,161 in the euchro-matic regions	Number of SNPs of the 146,161 in the heterochro-matic regions
Gm1	2489	1652	837	14841727	41073868	111.3	20.4	2688	5367
Gm2	3445	2929	516	26316426	25340287	111.3	20.4	4099	2852
Gm3	2691	2102	589	18879713	28901363	111.3	20.4	3245	4276
Gm4	2718	2099	619	18855914	30387938	111.3	20.4	3197	5888
Gm5	2927	2537	390	22797076	19139428	111.3	20.4	3062	1286
Gm6	3041	2458	583	22083366	28639455	111.3	20.4	2908	4851
Gm7	3421	3073	348	27609531	17073626	111.3	20.4	4586	1424
Gm8	3795	3473	322	31208512	15787020	111.3	20.4	4010	2062
Gm9	2556	1960	596	17602854	29240896	111.3	20.4	2828	5739
Gm10	3241	2696	545	24219274	26750361	111.3	20.4	3435	4328
Gm11	2610	2308	302	24367505	14805285	94.7	20.4	2441	1081
Gm12	2378	1910	468	17140105	22973035	111.3	20.4	2516	2340
Gm13	3591	3288	303	29558651	14850320	111.3	20.4	4286	2517
Gm14	2863	2265	598	20344958	29366246	111.3	20.4	3304	3014
Gm15	3164	2603	561	23378504	27560656	111.3	20.4	4398	5224
Gm16	2370	1969	401	17708632	19688753	111.3	20.4	3042	3155
Gm17	2694	2253	441	20240737	21666037	111.3	20.4	2957	4228
Gm18	4618	4099	519	36632197	25675943	111.3	20.4	8814	4680
Gm19	3520	3047	473	27373488	23215953	111.3	20.4	4371	4414
Gm20	2569	1980	589	17784173	28988994	111.3	20.4	2897	3546
Total	60701	50701	10000	4.59E+08	4.91E+08			73084	72272

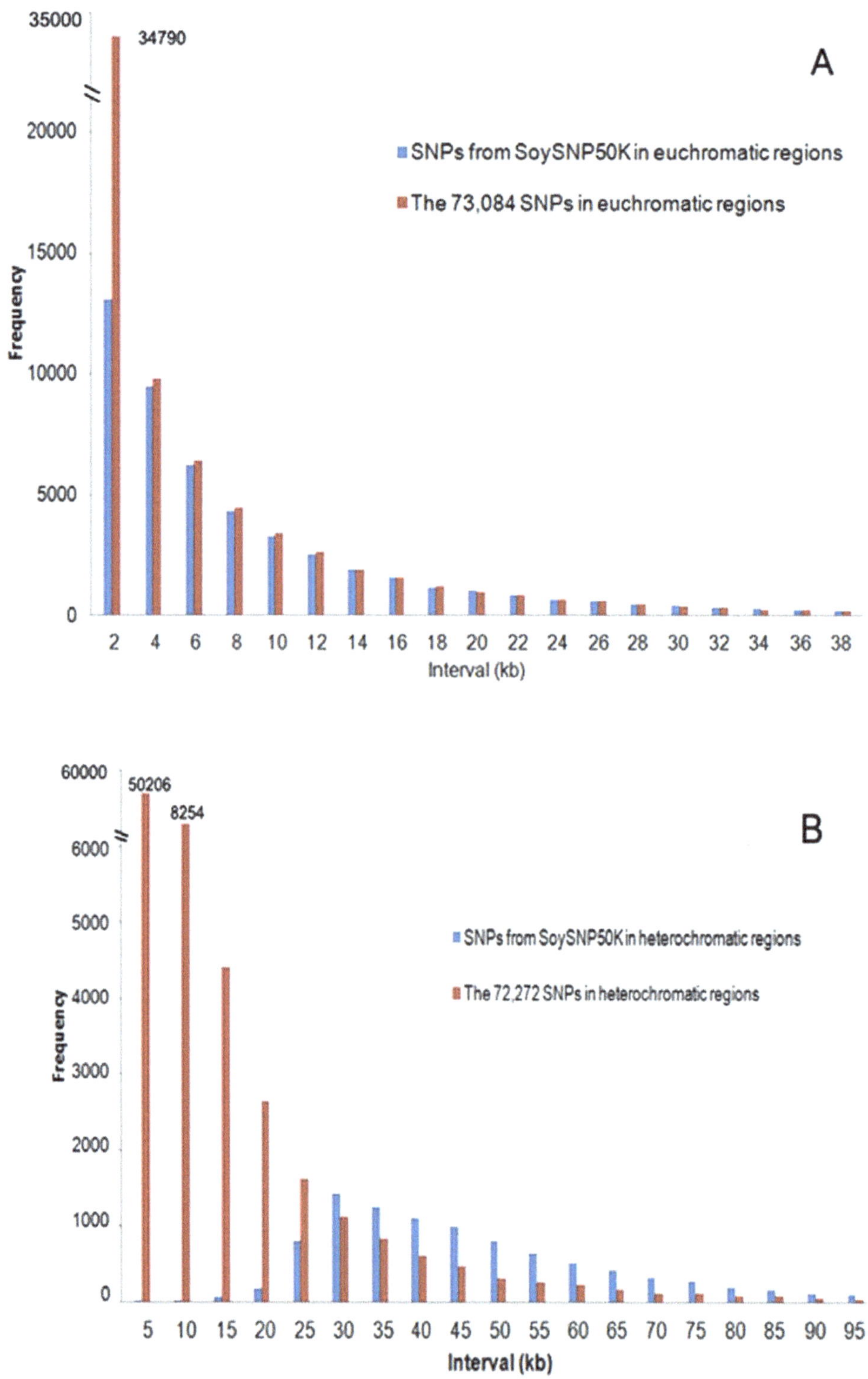

Figure 1. Distribution of distances between adjacent pre-selected and selected SNPs in euchromatic and heterochromatic regions. (A) Selected SNPs (50,701) in euchromatic regions of SoySNP50K and the pre-selected SNPs (73,084) in euchromatic regions. (B) Selected SNPs (10,000) in heterochromatic regions and pre-selected SNPs (72,272) in heterochromatic regions.

pods per nod [25] and seed protein concentration [20] have been reported in the region flanked by Satt396 and Sat_140 of soybean linkage group C1 [14], i.e., position 2,719,766–5,221,447 of Gm04 in Glyma1.01 [15].

Discussion

A number of options exist for high throughput SNP genotyping and the Illumina GoldenGate platform is a commonly used assay in plants. Examples of its application in plants include 1536-SNP assays in barley [4,26], maize [27], *Aegilops tauschii* [28], common

Table 4. Association of SNP selection index score, Illumina Infinium design score and priority group with the rate of SNPs with successful allele calls.

SNP selection index score (I_i)	Number of SNPs with success-ful allele calls	Number of SNPs not called	Rate of SNPs with success-ful allele calls	Illumina Infinium design score (D_i)	Number of SNPs with success-ful allele calls	Number of SNPs not called	Rate of SNPs with success-ful allele calls	Priority group	Number of SNPs with success-ful allele calls	Number of SNPs not called	Rate of SNPs with success-ful allele calls
≥0.9	5,974	210	0.97	≥0.9	22,359	1,326	0.94	A	10,943	582	0.95
≥0.8 & <0.9	7,630	410	0.95	≥0.8 & <0.9	10,316	800	0.93	B	16,163	1,536	0.91
≥0.7 & <0.8	10,995	749	0.94	≥0.7 & <0.8	5,755	605	0.90	C	16,505	2,003	0.89
≥0.6 & <0.7	11,102	1,000	0.92	≥0.6 & <0.7	3,649	460	0.89	D	3,835	474	0.89
≥0.5 & <0.6	5,507	663	0.89	≥0.5 & <0.6	2,414	458	0.84				
≥0.4 & <0.5	3,382	636	0.84	≥0.4 & <0.5	2,953	946	0.76				
<0.4	2,856	927	0.75								
Total	47,446	4,595			47,446	4,595			47,446	4,595	

bean [29], and a 96-SNP assay in wheat [30]. In soybean, a universal linkage panel containing 1536 SNPs for high-throughput soybean QTL mapping was developed [31]. The Illumina Infinium assay provides a significantly higher level of SNP genotyping capacity and recent Infinium assays with successful allele calls for nearly 50,000 SNPs were reported in maize (*Zea mays* L.) [32] and assays for nearly 8,000 SNPs have been reported in sunflower (*Helianthus annuus* L.) [33], apple (*Malus×domestica*) [34] and tomato (*Solanum lycopersicum*) [35]. The SoySNP50K is the first Infinium beadchip containing >50 k SNPs in soybean. This assay is being applied to genotype more than 19,000 cultivated and wild soybean accessions in the USDA Soybean Germplasm Collection and more than 960 recombinant inbred lines in each of two mapping populations at the USDA-ARS, Beltsville, MD. The resulting datasets will assist in the application of genome-wide association studies of important traits and to detect signatures of selection and evolution, pinpoint genes or regions controlling phenotypic variants and define whole-genome LD patterns and regions associated with soybean domestication. The SoySNP50K BeadChip will expedite the development of high resolution linkage maps to improve the previous whole genome sequence assembly and to position unanchored scaffolds to the current assembly (Glyma1.01).

The justification for the reduced representation library used in the SNP discovery was to reduce the complexity of the subset of the genome that was sequenced for SNP discovery and to generate large sets of common fragments for alignment. This approach may leave portions of the genome without sequence and thus without SNPs. However, due to the use of 10 different restriction enzymes and the resulting diversity of restriction sites, the identified SNPs were, for the most part, randomly distributed in the soybean genome.

The density of genes and the recombination rate in euchromatic and heterochromatic regions differ dramatically with much lower recombination and a lower gene density in the heterochromatic vs. the euchromatic regions [10]. Low recombination rate would be expected to result in more extensive LD in the heterochromatic regions, although the actual extent is unknown. Thus, we attempted to design a beadchip with higher SNP density in the euchromatic than in the heterochromatic regions i.e., 1 SNP every 9.1 kb in euchromatic regions and one SNP every 49.1 kb in heterochromatic regions. To avoid clustering, SNPs were selected such that the distance between SNPs was as near as possible to 9.1 kb in the euchromatic and 49.1 kb in the heterochromatic regions. Thus, the selected SNPs would evenly cover both the euchromatic and heterochromatic regions. The very different SNP densities in euchromatic and heterochromatic regions and their association with the anticipated difference in the magnitude of LD in the two regions were done in anticipation of the application of genome wide association studies in diverse populations of accessions selected from the USDA Soybean Germplasm Collection.

During the SoySNP50K BeadChip development, account was taken of the factors that were essential for a successful assay design i.e., high allele call success rate, high minor allele frequency and the even distribution of SNPs in heterochromatic and euchromatic regions across the soybean genome. The success rate of a beadchip is subject to the bead type success rate during manufacture and the rate of successful allele calls in genotyping. The bead type success rate is the proportion of the total number of attempted assays present in the final product. Illumina guarantees a bead type success rate of >80% but not 100% due to the imperfect efficiency of manufacturing steps such as synthesizing oligonucleotide probes, attaching probes to silica beads, pooling beads, loading

Table 5. Distribution of SNP minor allele frequency in elite cultivar, landrace and wild soybean populations.

Minor allele frequency	Elite population	Landrace population	Wild population	All 288 genotypes
<0.05	13,114(27.9%)	8,060(17.1%)	8,833(19.3%)	3,294(7.0%)
≥0.05 & <0.1	3,358 (7.1%)	4,432(9.4%)	5,186(11.3%)	3,202(6.8%)
≥0.1 & <0.2	6,377 (13.6%)	8,813(18.8%)	9,300(20.3%)	7,837(16.6%)
≥0.2 & <0.3	7,665(16.3%)	8,966(19.1%)	7,913(17.3%)	10,016(21.2%)
≥0.3 & <0.4	8,393(17.9%)	8,511(18.1%)	7,094(15.5%)	10,890(23.0%)
≥0.4 & ≤0.5	8,085(17.2%)	8,210(17.5%)	7,464(16.3%)	11,999(25.3%)
Total	46992	46992	45,790	47,337

bead pools and decoding the beads identified on the beadchip (http://www.illumina.com/Documents/products/technotes/technote_iselect_design.pdf). The 86% bead type success rate for the SoySNP50K beadchip is in the normal range defined by Illumina, Inc. The allele call success rate is the proportion of the total number of assays generating reliable allele calls among the total number of assays in the final product. Reliability of the SNP allele call is dependent on the intensity of fluorescent signals from both alleles in the SNP Graph displayed by the GenomeStudio software and the distinct separation of the clusters of the allelic classes i.e., the two homozygotes and the heterozygotes, as displayed in the normalized theta output of GenomeStudio. Because the fluorescent signal of SNPs may vary with DNA samples and assays, the incidence of poor allele clustering such as multiple or indistinguishable clusters of samples may occur and lead to inaccurate allele calls. Based on our standard to define SNPs with successful allele calls, the rate of successful allele calls using the SoySNP50K beadchip was 91%, which is comparable to the rate of 77–91% reported in human populations using the Illumina Infinium 550 K beadchip [36], 92% in cattle using Illumina Infinium assay BovineSNP50 [37] and 81% in horse using the Illumina commercial assay EquineSNP50 [38].

Rare alleles can produce spurious associations between SNP markers and phenotypes and are therefore often not used in certain genetic analyses. Therefore, SNPs with minor allele frequency below a certain threshold should be avoided. In order to detect SNPs with high minor allele frequency, we sequenced mixed DNAs from six diverse cultivated and two wild soybean genotypes and were able to identify SNPs by aligning reads from the mixed DNA to the reference sequence. Because the SNPs with one allele identical to the reference allele and another to the alternative allele that was present in at least one of the genotypes as determined using short reads of individual genotypes were given the highest priority score, these SNPs were most likely retained during subsequent selection. Statistically, in the mixed DNA, reads containing the minor allele were less likely to be present than reads containing the common allele. We observed that only 7% of the SNPs assayed with SoySNP50K had a minor allele frequency <0.05 in the 288 *G. max* and *G. soja* accessions analyzed. This proportion is generally lower than that reported in various populations of dog, cattle, horse, pig and sheep using Illumina BeadChips [37,39,40].

Numerous candidate regions associated with domestication and selection imposed by soybean breeding in North America were identified by examining the F_{st} between the populations. The example of the high fixation indices ($F_{st} \geq 0.6$) between *G. soja* and landrace in the 13–30 Mb region on Gm14 (Fig. 2a) provides one such region that is likely the result of a domestication sweep. While the high Fst between the elite cultivars and landraces on Gm04 (Fig. 2b) identifies a region that was likely under selection imposed by North American soybean breeders, it is not immediately clear why this region was the target of strong selection. As indicated earlier, QTL for resistance to soybean cyst nematode (SCN) race 2 was been reported in this region [22,23]. However, the QTL was present in PI 438498B and was not reported until 2001 and is thus unlikely to have been used in the breeding of the 96 elite cultivars included in this study that were released between 1990 and 2000. Relative to the other QTL reported in this region i.e., seed size [24], pods per node [25] and seed protein concentration [20] there is no suggestion that breeders have specifically targeted this region in any type of marker assisted selection program. However, it is possible that selection for improved agronomic traits and higher

Table 6. Distribution of genetic distance based upon the proportion of polymorphic SNPs between pairs of elite cultivar, landrace and wild soybean genotypes.

Genetic distance	Number of pairs among elite genotypes	Number of pairs among landrace genotypes	Number of pairs among wild genotypes
<0.1	1 (0.0%)	4 (0.1%)	51 (1.1%)
≥0.1 & <0.2	68 (1.5%)	71 (1.6%)	202(4.4%)
≥0.2 & <0.3	2,319 (50.9%)	1,377 (30.2%)	2,120 (46.5%)
≥0.3 & <0.4	2,172 (47.6%)	2,861 (62.7%)	2,167 (47.5%)
≥0.4 & ≤0.5	0 (0.0%)	247 (5.4%)	20 (0.4%)
Total	4,560	4,560	4,560

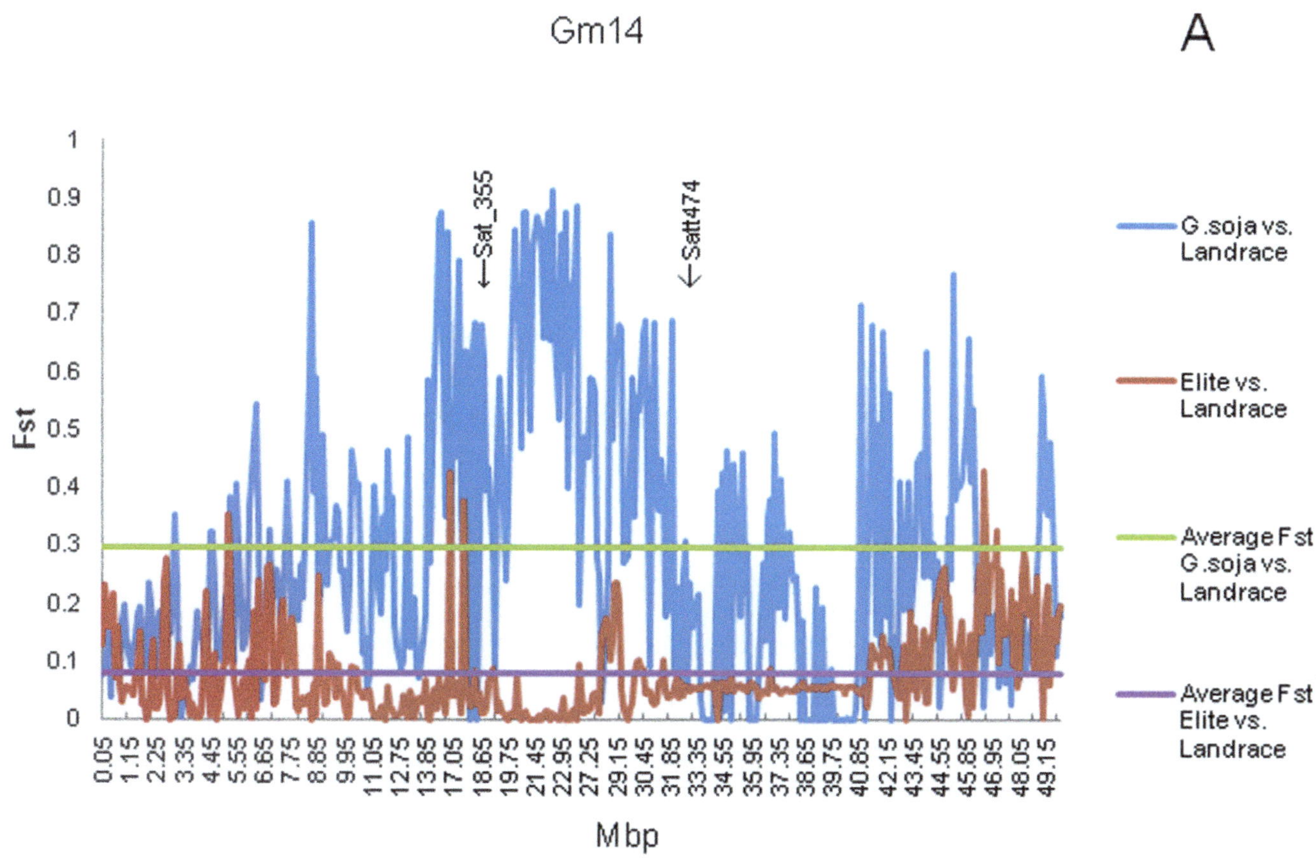
Gm14
A
Fst
Mbp
←Sat_355
←Satt474
G.soja vs. Landrace
Elite vs. Landrace
Average Fst G.soja vs. Landrace
Average Fst Elite vs. Landrace

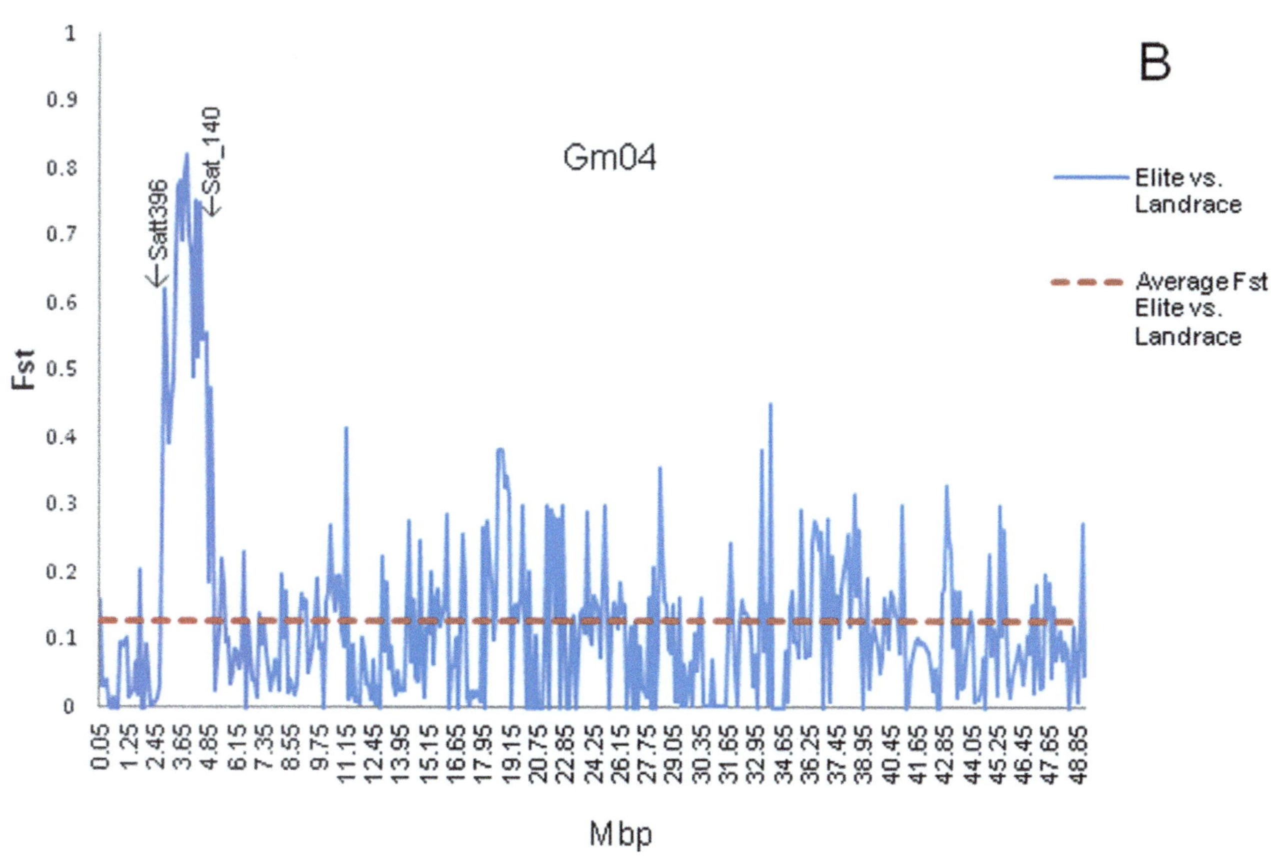
B
Gm04
Fst
Mbp
←Satt396
←Sat_140
Elite vs. Landrace
Average Fst Elite vs. Landrace

Figure 2. F_{st} values of *G. soja* vs. landraces and elite cultivars vs. landraces along chromosome Gm14 and Gm04. (A) Significant F_{st} values on Gm14 between SSR markers Sat_355 and Satt474 may contain loci associated with soybean domestication (B) Significant F_{st} values between SSR markers Satt396 and Sat_140 may contain loci that were under selection in N. American soybean breeding programs.

yields over many years of N. American soybean breeding has altered allele frequency in this region on Gm04. There are data available in the USDA Germplasm Resources Information Network (GRIN) database for seed size (g/100 seeds) for some of the 96 landrace and 96 elite cultivars we analyzed. Gizlice et al. [41] analyzed the pedigrees of modern N. American soybean cultivars and identified 25 landraces that were the ancestors of modern N. American cultivars that contributed 90% of the genetic base of modern cultivars. Seed size data for 23 of the 25 ancestors were available in GRIN and had a mean seed size of 16.4 g/100 seeds. GRIN also contained data for 39 of the 96 elite cultivars we analyzed. These had a mean seed size of 14.7 g/100 seeds. This suggests that selection over the many years of N. American soybean breeding has reduced seed size. The soybean decade's study that is comparing soybean ancestors along with cultivars developed over the years supports the contention of reduced seed size of modern cultivars versus ancestral cultivars (Personal communication – R. Nelson). Further analysis of this genome region is clearly of interest.

Knowledge of the positions of the regions associated with domestication and with gene frequency changes associated with N. American soybean breeding is critical for the identification of genes controlling important agronomic traits. However, due to limited sample size, the span of the regions identified in the current study are still quite large. As the analysis of all of the more than 19,000 annual *Glycine* accessions in the USDA Soybean Germplasm Collection with the SoySNP50K proceeds it is anticipated that regions under selection will be detected with high statistical power and resolution. We can anticipate that the identification of such regions with strong signatures of selection will be the targets of investigation with the goal of identifying the impacted phenotype and the gene or genes that underlie the phenotypic change(s).

Materials and Methods

Sequencing a Panel of Eight Soybean Genotypes with the Illumina Genome Analyzer

Six soybean genotypes, Essex, Evans, Archer, Minsoy, Noir 1, and Peking and two wild soybean accessions, PI 468916 and PI 479752, were used for re-sequencing. The eight accessions are parents of soybean mapping populations. DNA of each of the eight genotypes was extracted from leaf or root tissues and each was digested separately with two sets of five restriction enzymes: Set 1: HaeIII, PsiI, SspI, RsaI; and MslI and Set 2: Cvikl–1, DraI, SnaB1, PshA1, and Stu1. The resulting digests were combined and the restriction fragments were size selected on a 1.0% agarose gel and the 110–140 bp fraction was purified and prepared for Illumina GA sequencing (Illumina Inc, San Diego, CA). Protocols for adapter ligation, cluster generation and sequencing by

Table 7. Number of 100 kb regions across the 20 soybean chromosomes with Fst ≥0.6 in comparisons of *G. soja* vs. landraces and landraces vs. elite cultivars.

Chromosome	Number of regions with Fst ≥0.6 in *G. soja* vs. Landrace	Number of regions with Fst ≥0.6 landrace vs. elite and with 100 kb interval to the adjacent regions
Gm01	48	3
Gm02	21	
Gm03	26	
Gm04	11	11
Gm05	72	
Gm06	17	
Gm07	41	1
Gm08	26	3
Gm09	15	
Gm10	35	
Gm11	32	
Gm12	57	20
Gm13	28	
Gm14	51	
Gm15	2	
Gm16	8	1
Gm17	11	2
Gm18	14	
Gm19	44	
Gm20	61	1
Total	620	42

synthesis on the Illumina GAII were followed and read lengths of 35–45 bp were obtained. The Goat Pipeline (v 1.3) was used for base calling and quality assessment. A total of 56 lanes of sequence data were collected from the eight genotypes, in addition, 15 lanes of sequence data were collected from mixed DNA of Archer, Evans, Noir 1, Minsoy and Peking.

Sequence Alignment and SNP Identification

The whole genome sequence (WGS) of Williams 82 (Glyma1.01) [10] was used as the alignment reference and the short sequences from Essex, Evans, Archer, Minsoy, Noir 1, Peking, PI468916 and PI479752 were mapped to the reference sequence using MAQ (v0.6.8) [42] and Illumina ELAND software (Illumina Inc, San Diego, CA) of the Goat pipeline. SNPs were called using MAQ and Illumina CASAVA software (Illumina Inc, San Diego, CA). Default parameters of MAQ and CASAVA for SNP identification were applied except for the minimum read depth in MAQ, which was set to 1. Called SNPs were eliminated for Infinium design if any of the following conditions were met: both alleles were present in a genotype (indicative of paralogous sequence); the SNP allele from an individual genotype did not match the ambiguity code or the allele from the mixed DNA; 60 bases of SNP-flanking sequences contained N; there was another SNP residing within 25 bases of an identified SNP; the 25 nt SNP flanking sequences were not specific in the whole genome and the SNP type was A/T or G/C or an indel.

Initial Verification of SNP Calls from MAQ and CASAVA in Order to Set Parameters to Effectively Select SNPs for Infinium BeadChip Design

In order to optimize the selection criteria to effectively design the Infinium BeadChip, a total of 767 SNP-containing loci were randomly selected for preliminary validation of the accuracy of SNP calls. A PCR primer set targeting the SNP position was designed to each of the loci. Primers were designed using standalone Primer 3 (http://primer3.sourceforge.net/releases.php). The targeted PCR product length ranged from 400–600 bp, with the annealing temperature from 55–62°C and the primer length from 18–27 nucleotides. In order to reduce the likelihood of non-specific annealing of primers, all primer sequences were examined with e-PCR software [43] with the parameters of maximum number of mismatching nucleotides in one primer sequence of $N = 3$ and maximum number of gaps of one primer sequence of $G = 1$. Primer sets that were not uniquely aligned to the genome sequence under this stringency were discarded and reselected. The resulting amplification from genomic DNA of the genotypes containing the non-reference allele was sequenced with the ABI 3730 (Applied Biosystems, Foster City, CA) as described by Choi et al. [44], and the sequence was aligned to the soybean reference sequence. Variants were identified with PolyBayes SNP discovery software [45,46].

Identification and Screening of Pre-existing SNPs at dbSNP of NCBI

A total of 5300 SNPs in soybean had previously been genetically mapped and assayed in multiple genotypes using Sanger sequencing [9,31,44]. The flanking sequences of the SNPs were aligned to the WGS of Williams 82 using standalone Megablast software (http://www.ncbi.nlm.nih.gov/blast/megablast.shtml) and the specific positions of SNPs in the genome were determined with $W = 50$, cutoff percentage of alignment = 98 and low complexity filtered as described previously [15]. A total of 2739 of the 5300 SNPs passed filters as described above and were included for further selection.

Procedure and Iteration Algorithm for the Selection of a Core Set of SNPs

SNPs were assigned a priority group of A, B, C or D as follows: A. SNPs were called from at least one genotype with read depth greater than or equal to 1, and the two SNP alleles from the reference and non-reference sequences were consistent with the IUB code called from mixed DNA or the SNPs were already present in dbSNP at NCBI and had previously been successfully assayed in multiple genotypes. B. SNPs were identified in at least two genotypes. C. SNPs were detected in only one genotype but the read depth was greater than 3. D. SNPs were called from only one genotype with read depth less than three, and the non-reference SNP allele was also present in the mixed DNA. A priority score (R_i) was then calculated for each SNP as the ratio of verified to examined SNPs in each priority group among the 767 SNP-containing loci in the verification of SNP calls using ABI 3730 Sanger sequence analysis. In addition, the Illumina Infinium design score provided by Illumina, which is a predictor of assay success, was assigned for each SNP using Illumina's Assay Design Tool for Infinium (Illumina, Inc., San Diego). SNPs with a design score less than 0.4 were eliminated. For each SNP, a SNP selection index score (I_i) was calculated by multiplying the priority score (R_i) with the Illumina design score (D_i), i.e. $I_i = R_i \times D_i$.

Based on the SNP selection index scores, an iteration algorithm was developed to select the desired number of SNPs with desired physical distance between adjacent SNPs in euchromatic and heterochromatic regions and with elevated index scores. The total number of SNPs to be selected from the euchromatic and the heterochromatic regions was determined based upon the ratio of genetic map distance to physical distance in bp in the euchromatic versus the heterochromatic regions. The definition of the physical lengths of the euchromatic and heterochromatic regions was as follows: The physical positions of mapped SNPs and simple sequence repeats (SSRs) [9,14,44,47,48] in the soybean genome were determined by BLAST analysis of the SNP and SSR-containing source sequence to the soybean whole genome sequence using the standalone Megablast software as previously described [15]. The cumulative genetic distances (cM) [31] were plotted against their cumulative physical distance (Mbp) to determine the base pair/centiMorgan relationship via the common SSR and SNP loci positions on the genetic linkage map and their genome sequence along each chromosome. The region between the two inflection points of the cumulative genetic distance against cumulative physical distance on the plot was defined as the heterochromatic region. The reliability of defining heterochromatic regions using this method was examined in rice, and showed that the heterochromatic region defined by suppressed recombination rate was in agreement with that identified by the conventional DAPI staining method [48]. Since the ratio is about 5:1, the density of selected SNPs was determined to be 5 times greater in the euchromatic vs. the heterochromatic regions. The desired number of SNPs to be selected from euchromatic and heterochromatic regions of each chromosome was calculated as $L_{ij}/T_j \times W_j$, where L_{ij} is the sequence length (bp) of the euchromatic ($j = 1$) or heterochromatic region ($j = 2$) for the ith chromosome and T_j and W_j were the total sequence length (bp) and total number of SNPs to be selected in the euchromatic or heterochromatic regions across the 20 soybean chromosomes, respectively. For each iteration of the algorithm, gaps (sequence length) between adjacent SNPs were calculated along each chromosome. Subsequently, the smallest gap along each chromosome was identified and the SNP selection index scores of the SNPs flanking the gap were calculated. The SNP with the smallest SNP selection index score was removed. If the SNP selection index scores of the two SNPs were identical, the SNP with the smallest gap to another adjacent SNP was eliminated. If two or more gaps in a single iteration were identical, the SNP with the smallest SNP selection index score among the SNPs flanking the gaps was eliminated. The iteration continued until the desired number of SNPs in euchromatic and heterochromatic regions was reached. In addition, the whole genome sequence of Williams 82 (Glyma1.01) contains 236 unanchored scaffolds with lengths from 10 kb to 100 kb and 51 unanchored scaffolds with lengths greater than 100 kb. SNPs identified in these scaffolds were also included in order to facilitate anchoring of these scaffolds for the next genome assembly.

Validation of the SoySNP50K iSelect SNP Beadchip

A group of 96 diverse landraces, 96 elite cultivars and 96 *G. soja* accessions were used to determine the allele frequency of selected SNPs on the SoySNP50K iSelect SNP BeadChip. The 96 landraces [49] were from China, Japan and Korea and represent a wide range of geographic origin and maturity groups. The 96 elite cultivars were from North America and represent a diversity of publically developed cultivars released from 1990 to 2000 [31,50]. The 96 *G. soja* accessions were from China, Korea, Japan and Russia and were selected based upon the wide ranges of latitude and longitude from where they were collected (Table S3). SNP genotyping was conducted on the Illumina platform by following the Infinium® HD Assay Ultra Protocol (Illumina, Inc. San Diego, CA). The Infinium II assay protocol includes the procedures to make, incubate and fragment amplified DNA, prepare the bead assay, hybridize samples to the bead assay, extend and stain samples, and image the bead assay. The SNP alleles were called using the GenomeStudio Genotyping Module v1.8.4 (Illumina, Inc. San Diego, CA). A stringent standard to define SNPs with successful calls was adopted so that only those SNPs with two or three discrete clusters and both alleles with high signal intensity on the SNP Graph Alt based on the assessment populations were counted and included for the calculation of the success rate of genotyping even though some of the exclusions were likely real SNPs.

Statistical Analysis

A Pearson's correlation coefficient was used to investigate the correlation between the SNP selection index score, the Illumina Infinium SNP design score and the rate of SNPs with successful allele calls.

Genetic distance among pairs of genotypes within the 96 landraces, within the 96 elite cultivars and within the 96 wild accessions was calculated as the ratio of the number of polymorphic SNPs vs. total number of SNPs for which allele calls were made for the pair.

The fixation index (F_{st}) is a measure of population differentiation. F_{st} values range from 0 to 1, with zero representing no allele differentiation and 1 representing complete allele differentiation between the two populations. The F_{st} between *G. soja* and the landraces and between the landraces and elite cultivars were calculated for each locus using the software Arlequin v3.1 and the significance of the F_{st} was tested using a non-parametric permutation approach [51]. Subsequently, an average F_{st} of SNP loci in 100 kb windows along each chromosome was calculated.

The physical position of SSR markers flanking QTL controlling seed size, protein and oil content, seed hardness, seed color, and other traits that had been previously reported were obtained from BARCSOYSSR_1.0 [15].

Supporting Information

Table S1 SNP type, chromosome and SNP position, alternative alleles, SNP flanking sequences, priority group, scores, polymorphism, position in gene (if applicable), MAF in elite, landrace and wild soybean populations.

Table S2 Average F_{st} *G. soja* vs. Landraces and F_{st} Elite cultivars vs. Landraces in 100 kbp windows along each chromosome.

Table S3 Accession IDs, province and country of origin, and latitude and longitude at which the 96 *G. soja* accessions used in this study were collected.

Acknowledgments

We thank Rob Parry, Chris Pooley, Steven Schroder for their technical support in assembling the necessary hardware and software required for the Illumina sequence analysis.

Author Contributions

Designed the Illumina Infinium BeadChip: QS DLH PC. Conceived and designed the experiments: DLH PC. Performed the experiments: CQ EF GJ. Analyzed the data: QS DLH PC. Contributed reagents/materials/analysis tools: RLN. Wrote the paper: QS PC RLN.

References

1. Brookes AJ (1999) The essence of SNPs. Gene 234(2): 177–186.
2. Lin CH, Yeakley JM, McDaniel TK, Shen R (2009) Medium- to high-throughput SNP genotyping using VeraCode microbeads. Methods Mol Biol 496: 129–142.
3. Ding C, Jin S (2009) High-throughput methods for SNP genotyping. Methods Mol Biol 578: 245–254.
4. Close TJ, Bhat PR, Lonardi S, Wu Y, Rostoks N, Ramsay L (2009) Development and implementation of high-throughput SNP genotyping in barley. BMC Genomics 10: 582.
5. Barreiro LB, Henriques R, Mhlanga MM (2009) High-throughput SNP genotyping: combining tag SNPs and molecular beacons. Methods Mol Biol 578: 255–276.
6. Yoon MS, Song QJ, Choi IY, Specht JE, Hyten DL, et al. (2007) BARCSoySNP23: a panel of 23 selected SNPs for soybean cultivar identification. Theor Appl Genet 114: 885–899.
7. Nasu S, Suzuki J, Ohta R, Hasegawa K, Yui R, et al. (2002) Search for and analysis of single nucleotide polymorphisms (SNPs) in rice (*Oryza sativa, Oryza rufipogon*) and establishment of SNP markers. DNA Res 9: 163–171.
8. Somers DJ, Kirkpatrick R, Moniwa M, Walsh A (2003) Mining single-nucleotide polymorphisms from hexaploid wheat ESTs. Genome 46: 431–437.
9. Zhu YL, Song QJ, Hyten DL, Van Tassell CP, Matukumalli LK, et al. (2003) Single-nucleotide polymorphisms in soybean. Genetics 163: 1123–1134.
10. Schmutz J, Cannon SB, Schlueter J, Ma J, Mitros T, et al. (2010) Genome sequence of the palaeopolyploid soybean. Nature 463: 178–183.
11. Hyten DL, Cannon SB, Song QJ, Weeks N, Fickus EW, et al. (2010) High-throughput SNP discovery through deep resequencing of a reduced representation library to anchor and orient scaffolds in the soybean whole genome sequence. BMC Genomics 11: 38.
12. Kim MY, Lee S, Van K, Kim TH, Jeong SC, et al. (2010) Whole-genome sequencing and intensive analysis of the undomesticated soybean (*Glycine soja* Sieb. and Zucc.) genome. Proc Natl Acad Sci 107: 22032–22037.
13. Lam HM, Xu X, Liu X, Chen W, Yang G (2010) Resequencing of 31 wild and cultivated soybean genomes identifies patterns of genetic diversity and selection. Nat Genet 42: 1053–1059.

14. Song QJ, Marek LF, Shoemaker RC, Lark KG, Concibido VC, et al. (2004) A new integrated genetic linkage map of the soybean. Theor Appl Genet 109: 122–128.
15. Song QJ, Jia G, Zhu Y, Grant D, Nelson RT, et al. (2010) Abundance of SSR motifs and development of candidate polymorphic SSR markers (BARC-SOYSSR_1.0) in soybean. Crop Sci 50: 1950–1960.
16. Concibido VC, La Vallee B, McLaird P, Pineda N, Meyer J, et al. (2003) Introgression of a quantitative trait locus for yield from *Glycine soja* into commercial soybean cultivars. Theor Appl Genet 106: 575–582.
17. Qi ZM, Han X, Sun YN, Wu Q, Shan DP, et al. (2011) An integrated quantitative trait locus map of oil content in soybean, *Glycine max* (L.) Merr., generated using a meta-analysis method for mining genes. Agricultural Science in China 10: 1681–1692.
18. Hoeck JA, Fehr WR, Shoemaker RC, Welke GA, Johnson SL, et al. (2003) Molecular marker analysis of seed size in soybean. Crop Sci 43: 68–74.
19. Li D, Pfeiffer TW, Cornelius PL (2008) Soybean QTL for yield and yield components associated with *Glycine soja* alleles. Crop Sci 48: 571–581.
20. Specht JE, Chase K, Macaulay M, Graef GL, Chung J, et al. (2001) Soybean response to water: a QTL analysis of drought tolerance. Crop Sci 41: 493–509.
21. Chen H, Shan Z, Sha A, Wu B, Yang Z, et al. (2011) Quantitative trait loci analysis of stem strength and related traits in soybean. Euphytica 179: 485–497.
22. Yue P, Arelli PR, Sleper DA (2001) Molecular characterization of resistance to *Heterodera glycines* in soybean PI 438489B. Theor Appl Genet 102: 921–928.
23. Vuong TD, Sleper DA, Shannon JG, Nguyen HT (2011) Confirmation of quantitative trait loci for resistance to multiple-HG types of soybean cyst nematode (*Heterodera glycines* Ichinohe). Euphytica 181: 101–113.
24. Orf JH, Chase K, Jarvik T, Mansur LM, Cregan PB, et al. (1999) Genetics of soybean agronomic traits: I. comparison of three related recombinant inbred populations. Crop Sci 39: 1642–1651.
25. Zhang WK, Wang YJ, Luo GZ, Zhang CY, et al. (2004) QTL mapping of ten agronomic traits on the soybean (*Glycine max* L. Merr.) genetic map and their association with EST markers. Theor Appl Genet 108: 1131–1139.
26. Rostoks N, Ramsay L, MacKenzie K, Cardle L, Bhat PR, et al. (2006) Recent history of artificial outcrossing facilitates whole-genome association mapping in elite inbred crop varieties. Proc Natl Acad Sci 103: 18656–18661.
27. Hao Z, Li X, Xie C, Weng J, Li M, et al. (2011) Identification of functional genetic variations underlying drought tolerance in maize using SNP markers. J Integr Plant Biol 53: 641–652.
28. Luo MC, Xu K, Ma Y, Deal KR, Nicolet CM et al. (2009) A high-throughput strategy for screening of bacterial artificial chromosome libraries and anchoring of clones on a genetic map constructed with single nucleotide polymorphisms. BMC Genomics 10: 28.
29. Hyten DL, Song QJ, Fickus EW, Quigley CV, Lim JS, et al. (2010) High-throughput SNP discovery and assay development in common bean. BMC Genomics 11: 475.
30. Akhunov E, Nicolet C, Dvorak J (2009) Single nucleotide polymorphism genotyping in polyploid wheat with the Illumina GoldenGate assay. Theor Appl Genet 119: 507–517.
31. Hyten D, Choi IK, Song QJ, Specht JE, Carter TE, et al. (2010) A high density integrated genetic linkage map of soybean and the development of a 1536 universal soy linkage panel for quantitative trait locus mapping. Crop Sci 50: 960–968.
32. Ganal MW, Durstewitz G, Polley A, Bérard A, Buckler ES, et al. (2011) A large maize (*Zea mays* L.) SNP genotyping array: Development and germplasm genotyping, and genetic mapping to compare with the B73 reference genome. PLoS ONE 6: e28334.
33. Bachlava E, Taylor CA, Tang S, Bowers JE, Mandel JR, et al. (2012) SNP discovery and development of a high-density genotyping array for sunflower. PLoS ONE 7: e29814.
34. Chagné D, Crowhurst RN, Troggio M, Davey MW, Gilmore B, et al. (2012) Genome-wide SNP detection, validation, and development of an 8K SNP array for apple. PLoS ONE 7: e31745.
35. Sim SC, Durstewitz G, Plieske J, Wieseke R, Ganal MW, et al. (2012) Development of a large SNP genotyping array and generation of high-density genetic maps in tomato. PLoS ONE 7: e40563.
36. Klein RJ (2007) Power analysis for genome-wide association studies. BMC Genet 8: 58.
37. Matukumalli LK, Lawley CT, Schnabel RD, Taylor JF, Allan MF, et al. (2009) Development and characterization of a high density SNP genotyping assay for cattle. PLoS One 4: e5350.
38. Ricard A, Legarra A, Danvy S, Guyon C, Meriaux J, et al. (2011) Can we predict the quality of an equine breeding for the CSO from genomics? 37th Journée de la Recherche Équine, Le Pin au Haras, France, 24 February 227–230.
39. Fan B, Du ZQ, M GD, Rothschild MF (2010) Development and application of high-density SNP arrays in genomic studies of domestic animals. Asian-Aust J Anim Sci 23: 833–847.
40. Ramos AM, Crooijmans RP, Affara NA, Amaral AJ, Archibald AL, et al. (2009) Design of a high density SNP genotyping assay in the pig using SNPs identified and characterized by next generation sequencing technology. PLoS ONE 4: e6524.
41. Gizlice Z, Carter TE, and Burton JW (1994) Genetic base of North American public soybean cultivars released between 1947 and 1988. Crop Sci 34: 1143–1151.
42. Li H, Ruan J, Durbin R (2008) Mapping short DNA sequencing reads and calling variants using mapping quality scores. Genome Res 18: 1851–1858.
43. Schuler GD (1997) Sequence mapping by electronic PCR. Genome Res 7: 541–550.
44. Choi IY, Hyten DL, Matukumalli LK, Song QJ, Chaky JM, et al. (2007) A soybean transcript map: gene distribution, haplotype and single-nucleotide polymorphism analysis. Genetics 176: 685–696.
45. Marth GT, Korf I, Yandell MD, Yeh RT, Gu Z, et al. (1999) A general approach to single-nucleotide polymorphism discovery. Nat Genet 23: 452–456.
46. Matukumalli LK, Grefenstette JJ, Hyten DL, Choi IY, Cregan PB, et al. (2006) Application of machine learning in SNP discovery. BMC Bioinformatics 7: 4.
47. Cregan PB, Jarvik T, Bush AL, Shoemaker RC, Lark KG, et al. (1999) An integrated genetic linkage map of the soybean. Crop Sci 39: 1464–1490.
48. Zhao Q, Zhang Y, Cheng Z, Chen M, Wang S (2002) A fine physical map of the rice chromosome 4. Genome Res 12: 817–823.
49. Hyten DL, Song QJ, Choi IY, Yoon MS, Specht JE, et al. (2008) High-throughput genotyping with the GoldenGate assay in the complex genome of soybean. Theor Appl Genet 116: 945–952.
50. Hyten DL, Song QJ, Zhu Y, Choi IY, Nelson RL, et al. (2006) Impacts of genetic bottlenecks on soybean genome diversity. Proc Natl Acad Sci 103: 16666–16671.
51. Excoffier LGL, Schneider S (2005) Arlequin Ver. 3.0: An integrated software package for population genetics data analysis. Evolutionary Bioinformatics Online 1: 47–50.

6

Adjacent Habitat Influence on Stink Bug (Hemiptera: Pentatomidae) Densities and the Associated Damage at Field Corn and Soybean Edges

P. Dilip Venugopal*, Peter L. Coffey, Galen P. Dively, William O. Lamp

Department of Entomology, University of Maryland, College Park, Maryland, United States of America

Abstract

The local dispersal of polyphagous, mobile insects within agricultural systems impacts pest management. In the mid-Atlantic region of the United States, stink bugs, especially the invasive *Halyomorpha halys* (Stål 1855), contribute to economic losses across a range of cropping systems. Here, we characterized the density of stink bugs along the field edges of field corn and soybean at different study sites. Specifically, we examined the influence of adjacent managed and natural habitats on the density of stink bugs in corn and soybean fields at different distances along transects from the field edge. We also quantified damage to corn grain, and to soybean pods and seeds, and measured yield in relation to the observed stink bug densities at different distances from field edge. Highest density of stink bugs was limited to the edge of both corn and soybean fields. Fields adjacent to wooded, crop and building habitats harbored higher densities of stink bugs than those adjacent to open habitats. Damage to corn kernels and to soybean pods and seeds increased with stink bug density in plots and was highest at the field edges. Stink bug density was also negatively associated with yield per plant in soybean. The spatial pattern of stink bugs in both corn and soybeans, with significant edge effects, suggests the use of pest management strategies for crop placement in the landscape, as well as spatially targeted pest suppression within fields.

Editor: Youjun Zhang, Institute of Vegetables and Flowers, Chinese Academy of Agricultural Science, China

Funding: This work is supported by the United Soybean Board, http://unitedsoybean.org/ (GPD PDV); Maryland Soybean Board, http://mdsoy.com/ (GPD); Maryland Grain Producers Utilization Board, http://www.marylandgrain.com/ (GPD); Cosmos Club Foundation, http://www.cosmosclubfoundation.org/ (PDV); Northeast Sustainable Agriculture Research and Education - Graduate Student grant - GNE-12-047, http://www.nesare.org/ (PDV, WOL, GPD); Arthur B. Gahan Fellowship, Dept. of Entomology, Univ. of Maryland, http://entomology.umd.edu/academics/graduate/awardopportunities/gahanfellowshipfund (PDV); and United States Department of Agriculture - National Institute of Food & Agriculture Hatch Project #MD-ENTM-1016, http://www.csrees.usda.gov/ (WOL). Funding for Open Access provided by the UMD Libraries Open Access Publishing Fund. The funders had no role in study design, data collection and analysis, decision to publish, or preparation of the manuscript.

Competing Interests: The authors confirm that funding from the commercial source professional association (United Soybean Board Maryland Soybean Board, Maryland Grain Producers Utilization Board) does not alter the authors' adherence to PLOS ONE policies on sharing data and materials.

* Email: venugopal.dilip@gmail.com

Introduction

Agricultural fields are components within a heterogeneous landscape that strongly connect to and interact with surrounding habitats [1]. The movement of insects between natural and agricultural habitats has important implications for agricultural ecosystem function [2]. The movement of pest insects among seasonal crop resources is often non-random and directional as pest species disperse and colonize crops [3]. This movement may result in insect pests immigrating into an agricultural field in an aggregated manner in specific areas within the field [1]. In such cases, species-specific characteristics such as host range, vagility, chemical ecology, and host developmental status influence the spatial pattern of a pest population within crop fields, often resulting in the aggregation of pests along field edges as they disperse between habitats. The seasonal availability and suitability of source and recipient habitats in relation to the life stages of a mobile, polyphagous insect pest influence the dispersal dynamics of pests from sources to recipient habitats [2,4,5]. Knowledge about insect pest movement among habitats in the landscape, and the subsequent colonization of plants within crop fields, may inform risk of infestation by an insect pest prior to their subsequent population increase as well as provide opportunities for pest management [1].

Stink bugs in the family Pentatomidae are major pests of economically important crops globally [6], and are considered important pests in soybean *Glycine max* (L.) Merr. producing areas of the world [7]. While stink bugs caused economic losses in the southern United States, they were not considered serious pests of crops in the mid-Atlantic region until recently. The most common stink bugs in agricultural fields in the mid-Atlantic are *Chinavia hilaris* (Say 1832) and *Euschistus servus* (Say 1832), but these species have had little economic impact in the region [8,9]. The recent explosion in populations of the invasive brown marmorated stink bug, *Halyomorpha halys* (Stål 1855) however, has led to significant economic and ecological impacts.

Since its discovery near Allentown, Pennsylvania, USA, *H. halys* has been detected in 41 states within United States, and local populations and detections from Europe (Switzerland, France, Canada, Germany, Italy, and Liechtenstein) have also been reported [10]. This polyphagous stink bug has a wide range of host plants including tree fruits, vegetables, field crops, ornamental plants, and native vegetation in its native and invaded ranges. Since 2010, serious crop losses have been reported for apples, peaches, sweet corn, peppers, tomatoes, and row crops such as field corn and soybeans in the mid-Atlantic region [11]. *Halyomorpha halys* is also a nuisance pest in human dwellings. In this context, information on the movement of stink bugs into crops from adjacent managed and natural habitats, and the associated spatially-explicit crop damage, has direct implications for integrated pest management.

The dispersal and movement of various stink bug species between crops and other habitats has been addressed by many studies in the context of dispersal between habitats, adjacent habitat influences on densities at field edges, and their relationship to crop damage [12–23]. However, these studies mainly pertain to stink bug communities in crops in the southern U.S. Currently only anecdotal reports of high *H. halys* abundance in the edges of fields adjacent to woodlots [11] are available. These reports are consistent with the documented use of woody trees as a food source and as overwintering sites by *H. halys* [24]. This factor coupled with use of buildings as overwintering sites suggests that these habitats, which support high local stink bug populations that eventually invade soybean later in the season, are likely to play an important role in stink bug densities.

Many stink bug species cause significant seed quality and yield losses in field corn, *Zea mays* L., and soybean [19,23–29], and stink bugs are also associated with the transmission of plant diseases [32–34]. However, few studies quantify spatially-explicit field crop damage in relation to patterns of stink bug densities [35]. Soybean is one of the preferred hosts for *H. halys* [36], and both field corn and soybean constitute a very high proportion of overall crop area in the mid-Atlantic region and throughout the United States [37]. Research efforts aimed at determining the role of adjacent habitat in influencing stink bug dispersal, population density, and pattern of settlement into crop, contribute to the development of management strategies of *H. halys* in row crops.

In this study, our objectives were a) document the species composition and within-field distribution of stink bugs in field corn and soybean, b) examine the influence of adjacent managed and natural habitats on the density of stink bugs, and c) relate stink bug density to seed quality in field corn and soybean, and pod development and yield in soybean. We expected *H. halys* to be the most abundant stink bug in our study based on previous reports of stink species composition in mid-Atlantic row crops [8]. We predicted higher density of stink bugs along wooded habitats and buildings than open habitats as they provide host plants and over-wintering refuge [24]. We also predicted high density of stink bugs at the field edge, reducing with distance into the field interior as observed by anecdotal reports for *H. halys* [11].

Methods

Ethics statement

The field studies were conducted at facilities provided by the University of Maryland and the USDA Beltsville Agricultural Research Center where we had permission to conduct research. Stink bugs were enumerated *in situ* and were not collected, and no endangered or protected species were involved.

Field selection and stink bug sampling strategy

The study was conducted at the USDA Beltsville Agricultural Research Center at Beltsville, MD (39°01′50N; 76°53′28W) and University of Maryland Research and Education Center facilities at Beltsville (39°00′45N; 76°49′32W), Clarksville (39°15′18N; 76°55′51W) and Keedysville, MD (39°30′37N; 77°43′57W). At these sites, field corn (76.2 cm row spacing) and full season soybean (17.8 cm row spacing) fields with a portion of their perimeter directly adjacent to wooded areas (henceforth wooded habitats), buildings (buildings, houses and barns; henceforth building habitats), crops (alfalfa, sorghum, and vegetable crops; henceforth crop habitats) and open, non-crop areas that were grassy, untilled land (henceforth open habitats) were sampled in 2012 and 2013. Corn fields were chosen as one of the adjacent habitat types *in lieu* of crops for several soybean fields. In each field, the sampling layout included four transects spaced 20 m apart. Each transect was marked for sampling plots at distances 0, 1.5, 3, 4.5, 6, 9, 12, and 15 m from the edge to field interior. Stink bugs were enumerated at each plot by carefully examining 10 consecutive corn plants and later converted to densities based on planting density, or all plants within a semicircular area of 0.5 m radius (1.57 m^2) in soybean.

Visual counts of stink bug adults, nymphs, and egg masses of *H. halys*, *E. servus*, *C. hilaris*, *Murgantia histrionica* (Hahn 1834), and *Thyanta custator* (Fabricius 1803) were recorded and converted to densities. Fields and plots were sampled each week between mid-July and mid-August in field corn, and mid-August and late-September in soybean. Sampling coincided with the kernel development stages of corn (R2–R5; blister – dent; [38]) and the seed development stages of soybean (R4–R7; full pod to physiological maturity; [39]), which are associated with high *H. halys* and other stink bug species densities [8,11,40]. Details on the number of corn and soybean field edges with different adjacent habitats, and the sampling dates during 2012 and 2013 are provided in Table 1. Including the repeated sampling of fields, a total of 4,315 field corn plots in 31 fields, and 2,968 soybean plots in 26 fields across all sites were sampled for stink bugs during 2012 and 2013.

Assessing seed quality in field corn and soybean

To relate stink bug density to ear damage in corn, eight fields with the highest observed count of stink bugs in 2013 and with different adjacent habitats were selected. Of these fields, 3, 3, 1 and 1 were adjacent to wooded, building, crops and open habitats, respectively. In each field, 6–10 consecutive corn ears were collected at each sampling plot prior to harvest maturity and stored in cloth bags for drying. Planting details of the fields used for assessing corn damage are provided in Table 2. For each ear, the following data were recorded once at the end of the season: 1) number of kernels damaged by stink bugs (identified by a characteristic puncture scar typically surrounded by a discolored cloudy marking); 2) number of collapsed kernels due to stink bug damage (this type of damage was carefully examined to distinguish between kernels damaged by stink bugs and dusky sap beetles, *Carcophilus lugubris* Murray); 3) number of kernel rows around the ear; 4) length of one kernel row (mm); and 5) average width of individual kernels (mm). With the individual ear measurements, the total number of kernels was derived by dividing the kernel row length by the width of a kernel times the number of rows. Data were then summed across all ears of a plot and stink bug damage was expressed as the percentage of damaged and collapsed kernels in relation to total number of kernels per sample. A total of 2,326 ears of corn from 252 sampling plots across 8 fields were assessed for stink bug damage.

To relate stink bug density to soybean pod development prior to harvest, samples of 10 consecutive plants at each plot of seven soybean fields in 2012–2013 were examined once at the end of the season *in situ* to count the total numbers of pods with 3 or more

Table 1. Details on field corn and soybean field edges with different adjacent habitats sampled for stink bugs and the sampling occasions at each field in Maryland, USA during 2012–2013.

Crop	Year	Site	Adjacent habitats (number of field edges)	Sampling dates (frequency)
Field Corn	2012	Beltsville	woods (4), buildings (3), crops (1), open (4)	10 July–15 Aug (7–10 days)
		Clarksville	woods (1), buildings (1), crops (3)	10 July–15 Aug (7 days)
	2013	Clarksville	woods (3), buildings (2), crops (3), open (1)	18 July–22 Aug (7 days)
		Keedysville	woods (1), buildings (1), crops (1), open (2)	16 July–20 Aug (7 days)
		Overall	woods (9), buildings (7), crops (8), open (7)	10 July–22 Aug (7–10 days)
Soybean	2012	Beltsville	woods (2), buildings (3), corn (1), open (2)	23 Aug–20 Sept (7–10 days)
		Keedysville	woods (2), buildings (1), corn (2), open (1)	30 Aug–26 Sept (7–10 days)
	2013	Beltsville	woods (1), buildings (1), corn (1), open (1)	13 Aug–06 Sept (7 days)
		Clarksville	buildings (1), corn (1), open (1)	16 Aug–12 Sept (5–7 days)
		Keedysville	woods (2), buildings (1), corn (1), open (1)	15 Aug–18 Sept (5–7 days)
		Overall	woods (7), buildings (7), corn (6), open (6)	13 Aug–26 Sept (5–10 days)

fully developed seeds (full pods), pods with fewer than 3 seeds (half pods), and flat, immature pods (flat pods). For standardization, the proportions for each pod type were calculated for each sample. Pod quality data were collected from 64 plots in 2 fields adjacent to wooded habitat at Keedysville. Planting details for the fields used for assessing soybean seed damage are provided in Table 2. Of the 7 fields sampled for seed quality data, 2 were adjacent to buildings, and 5 fields were adjacent to wooded habitats, and all had the highest counts of stink bugs observed in each year. Just prior to harvest, 20 plants each from 154 plots in the fields were collected, stored in mesh bags, and allowed further drying for optimal thrashing. Seeds were removed from pods for each sample by a stationary motor-driven thrasher (RL Brownfield-Swanson Machine Company; Model B1). Dirt, chaff, or un-thrashed pods were removed, and the remaining seed samples were then weighed to measure yield (measured as grams/20 plants).

To assess seed quality, subsamples of 200–300 seeds were removed from each sample, counted, and weighed to calculate test weight (expressed as weight per 100 seeds). Seed samples in 2012 were sieved to separate smaller, immature seeds (<0.3 cm), whereas these smaller seeds were not removed from subsamples in 2013. In both years, seeds were individually examined and categorized into six groups as follows: 1) stink bug damaged seed, distinguished by a puncture scar and often surrounded by a discolored cloudy area; 2) moldy seed, characterized by having milky white or grayish crusty growth on surface, sometimes with cracks and fissures; 3) shriveled seed that appeared wrinkled and often undersized; 4) purple seed recognized as purple or pink areas on the seed coat due to the fungus *Cercospora kikuchii* Matsumoto & Tomoy 1925 [41]; 5) green seed showing discolored green tissue in cross section, rather than the normal yellow; and 6) normal, undamaged seed. Seeds with characteristics of more than one

Table 2. Details on the field corn and soybean fields used for analyzing grain and seed damage in Maryland, USA during 2012–2013.

Crop	Year	Site	Field ID	Variety	Planting Date	Density/acre
Field Corn	2013	Clarksville	Corn1	P1319HR (113)	2May13	26,000
			Corn2	DKC61-21 (111)	15May13	26,000
			Corn3	P1319HR (113)	2May13	26,000
			Corn4	P1319HR (113)	2May13	26,000
			Corn5	DKC61-21 (111)	16May13	26,000
			Corn6	NK74R3000GT (114)	16May13	26,000
		Keedysville	Corn7	Doebler's 633HXR (110)	23Apr13	26,000
			Corn8	Doebler's 633HXR (110)	23Apr13	26,000
Soybean	2012	Beltsville	Soy1	AG3030 (3.0)	11May12	155,555
			Soy2	AG3030 (3.0)	11May12	155,555
		Keedysville	Soy3	Doebler 3809RR (3.8)	26May12	180,000
			Soy4	Doebler 3809RR (3.8)	4Jun12	180,000
			Soy5	Doebler 3809RR (3.8)	4Jun12	180,000
	2013		Soy6	SCS9360RR (3.5)	22May13	180,000
			Soy7	Doebler 3809RR (3.8)	27May13	180,000

For each of the corn and soybean varieties, the corn relative maturity in days and the soybean maturity groups respectively, have been provided in parenthesis.

damage category were assigned to whichever category seemed dominant, except for stink bug damaged seeds which were all assigned to that category. To standardize across samples, the percentage of seeds in each category was calculated. Soybean seed quality data were collected for a total of 154 plots from 4 fields in 2012 and 2 fields in 2013.

Statistical analyses

Adjacent habitat and distance from edge influences. The influence of adjacent habitat and distance from field edge on the density and distribution of stink bugs was analyzed by Generalized Linear Mixed Models (GLMMs) based on Laplace approximation, with a Poisson-lognormal error distribution and log link function [42]. All analyses were performed with fields as replicates and the sampling plots along transects within the field as subsamples. For corn and soybean data, analyses were performed on –data pooled over all stink bug life stages. For each of these datasets, GLMMs were performed on the data pooled across species, years, and study sites, and on data from each study site pooled across years. Sampling points within each field edge was treated as a random factor to control for repeated measurement [43]; adjacent habitat, distance from edge, and their interaction were the fixed effects, and stink bug density was the response variable. For the overall data models, study site and year were also treated as random effects.

Model building and selection procedures for the mixed effects modeling followed the procedures used by Zuur and others [44]. First, several candidate models, each with different random effects but identical fixed effects, were tested to choose the optimal random effect model using a combination of AIC and BIC values for selection criteria [36]. For all optimum fixed effect models, an initial full model analysis including individual and interactive effects of adjacent habitat (4 levels - woods, buildings, mix crop/ corn, and open) and distance from edge was performed. Specifically we designed a model matrix that directly estimated the intercepts and the slopes for each level of adjacent habitat (extension of means parameterization; see pages 61–64 in [45]) in relation to distance from field edge. The significance of the fixed effects was determined by Wald χ^2 tests. If a significant interaction was found these estimates, directly interpreted as the intercept and slope of the regression of stink bug density on distance for each of the adjacent habitat types, were then simultaneously compared through post-hoc comparisons with a Bonferroni correction. Models were evaluated for assumption appropriateness by testing for over-dispersion and correlations among random effect terms, and by visualizing variances in a location-scale plot with superimposed loess fit [42].

Relating stink bug density and seed damage. Influence of stink bug density on damage to corn kernels was assessed using generalized linear models (GLMs) with Poisson or quasi-Poisson error distribution and log link function [46]. Percentage of collapsed and stink bug damaged seeds were used as the response variables and mean stink bug density, the explanatory variable. For significant results, the coefficient of determination was calculated by Nagalkerke's pseudo r^2 statistic [47].

Linear regression was used to assess the influence of stink bug density on soybean pod development. To meet normality assumptions, response variables (% flat or full pods) was square root transformed prior to analysis. Influence of stink bug density on soybean seed quality was assessed by linear mixed models (LMMs) with year as a random effect to account for minor differences in grading seed size protocols. LMMs related stink bug density to the percentage of seeds in each category of seed quality. Influence of stink bug density on soybean yield per plot was assessed by LMMs with field as a random effect to account for differences in soybean variety and other field conditions among sites. Response variables (% seeds in various damage categories, or yield) were log or square root transformed to meet normality requirements and the significance of the fixed effect was determined by Wald t-tests. Diagnostic plots of the models visualizing within-group residuals (standardized residuals vs fitted values, normal Q-Q plots, and histograms of residuals) and estimated random effects (normal Q-Q plots and pairs-scatter plot matrix) were used to assess model appropriateness. The coefficients of determination for the LMMs, based on the likelihood-ratio test, were calculated using Nagalkerke's pseudo r^2 statistic [47]. Patterns in damage to corn kernels, soybean pods, and seeds at different distances from the edge, in relation to stink bug density, were visualized by plotting average values of damage and stink bug density aggregated by distance.

All statistical analyses were performed in R program [48] and associated statistical packages. GLMMs were performed with package 'lme4' [49] and LMMs with package 'nlme' [50]. Multiple comparisons of means for GLMMs were computed with R packages 'contrast' [51] and 'multcomp' [52]. GLMMs and LMMs estimated coefficients were extracted through package 'effects' [53], converted to densities/m^2 and plotted using 'ggplot2' [54]. Coefficient of determination (pseudo r^2) for the GLM was calculated with package 'rms' [55], and with package 'MuMin' [56] for LMMs.

Results

Species composition and density

A total of 9440 individuals (66% nymphs; 34% adults) of four phytophagous stink bug species (*E servus*, *H. halys*, *C. hilaris*, and *M. histrionica*) were recorded in field corn, of which *H. halys* accounted for 97% of the total. Species composition varied among study sites and crop systems. *Halyomorpha halys* comprised 57% of the sampled populations in corn at Beltsville, followed by *E. servus* (35%), whereas *H. halys* accounted for ~97% of all stink bugs at Clarksville and Keedysville. In soybean, a total of 9867 individuals (68% nymphs; 32% adults) of five phytophagous stink bug species (*E. servus, H. halys, C. hilaris, M. histrionic, T. custator*) were recorded, of which *H. halys* accounted for 93% of the total. *Halyomorpha halys* comprised 83–85% of the stink bug numbers in soybean at Beltsville, while greater than 92% were *H. halys* at Clarksville and Keedysville. Results obtained from the statistical analyses hence pertain mainly to patterns of *H. halys* density, since this species constituted ~95% of all observed stink bugs in both field corn and soybean.

Adjacent habitat and distance from field edge influences

The trends for density among adjacent habitats at different distances from the edge observed in adult and nymph data sets were broadly similar and were hence pooled for the analyses, and we report only these results.

Field corn. For the analysis of overall stink bug data from field corn edges, the random effects used for the GLMM included the field, study site and year. Results showed significant interactive influences of adjacent habitat and distance from edge on stink bug density (Wald $\chi^2 = 492.6$, DF = 8, $P<0.001$). Simultaneous comparison of intercepts and slopes among adjacent habitats accounting for the influence of distance from edge revealed significant differences. Overall, the density of stink bugs at different distance from the edge was significantly higher in fields adjacent to wooded habitats compared to density in fields adjacent to buildings and open habitats (Fig. 1A). Density between wooded and crop habitats did not significantly differ, although mean densities of stink bugs were consistently higher adjacent to wooded habitats (see Fig. S1 for variance around estimated means). Similarly stink bug density did not differ significantly between crop habitats and buildings although higher mean density was observed along buildings.

For site level GLMMs, significant interactive influences of adjacent habitat and distance from edge on stink bug density were observed for all the sites - Clarksville (Wald $\chi^2 = 426.9$, DF = 8, $P<0.001$), Keedysville (Wald $\chi^2 = 367.9$, DF = 8, $P<0.001$) and Beltsville (Wald $\chi^2 = 43.1$, DF = 3, $P<0.001$). Multiple comparisons of means of stink bug density at Clarksville showed similar trends to the results of analyses of data pooled over all study sites, with density in fields adjacent to wooded habitats higher than density adjacent to buildings and open habitats (Fig. 1C). Similar to overall data, at Clarksville, stink bug density did not differ significantly between wooded and crop habitats, although mean densities of stink bugs were consistently higher adjacent to wooded habitats. Similarly stink bug density did not differ significantly between crop habitats and buildings although higher mean density was observed along buildings. At Keedysville however, stink bug density was highest along fields adjacent to crops, and the slopes and intercepts were significantly different among adjacent habitats (Fig. 1B). Additionally, corn fields at Keedysville had higher stink bug density along the outside rows (0 m) adjacent to crops than adjacent to woods. At Beltsville, with the least stink bug density, multiple comparisons did not reveal significant differences in stink bug density among adjacent habitats, after accounting for the distance from edge influence (Fig. 1D; also see Fig. S1 for variance around estimated means for each site). The raw means for each site and year (Fig. S2) are provided as supplementary information.

Soybean. The GLMM analysis of overall stink bug data treated field as the only random effect and showed significant interactive influences of adjacent habitat and distance on stink bug density (Wald $\chi^2 = 717.2$, DF = 8, $P<0.001$). Multiple comparisons of means showed that density of stink bugs was significantly higher at all distances in fields adjacent to wooded habitats compared to density in fields adjacent to buildings and open habitats (Fig. 2A). Stink bug density adjacent to wooded habitats was also consistently higher than density in fields adjacent to corn

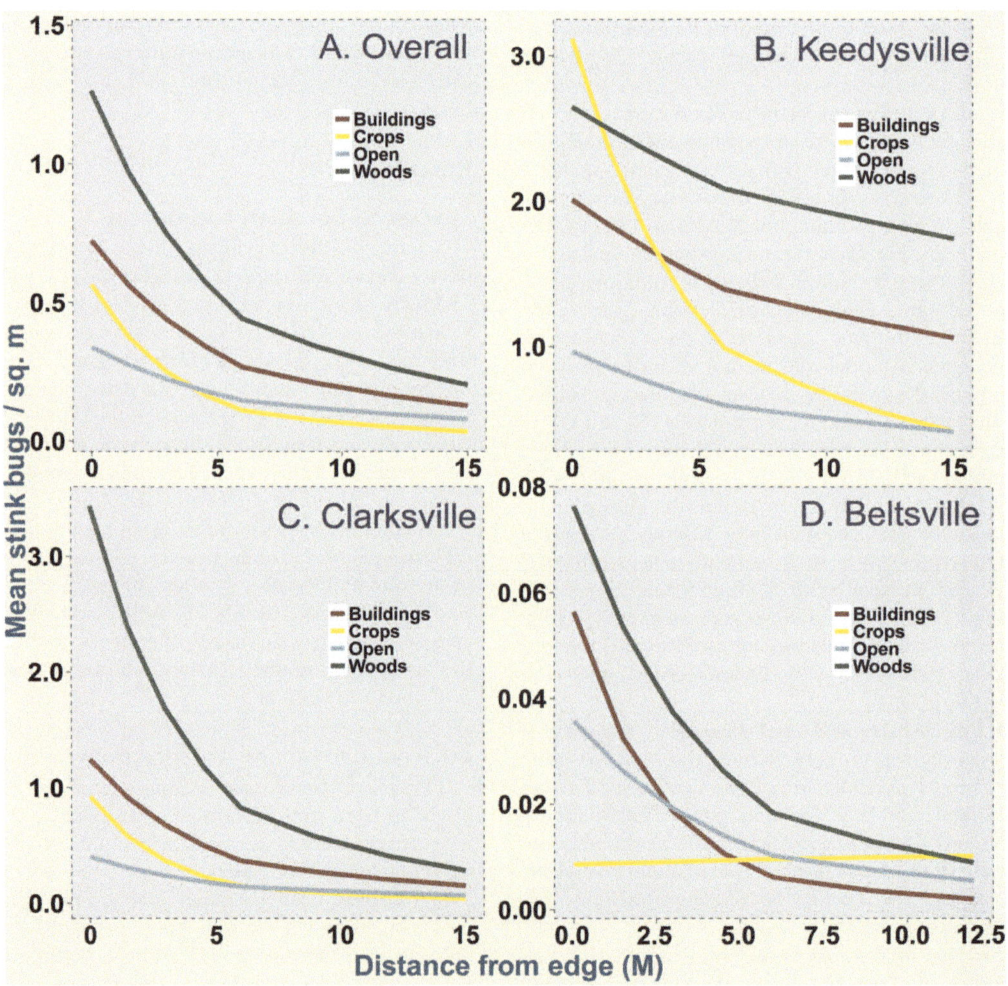

Figure 1. Mean stink bug density in field corn in relation to different adjacent habitats and distance from the field edge. Estimates derived from Poisson-lognormal GLMMs are plotted for overall stink bug data pooled over all study sites (A), Keedysville (B), Clarksville (C) and Beltsville (D). Values presented here have been back transformed from their original link function estimated model coefficients. Multiple comparison of means with a bonferroni correction ($\alpha = 0.05$) showed significant differences in: overall (A) - wooded habitats and buildings and, wooded and open habitats; Keedysville (B) – all adjacent habitats significantly different from each other; and Clarksville (C) - wooded habitats and buildings and, wooded and open habitats. For Beltsville (D), all multiple means comparisons were non-significant.

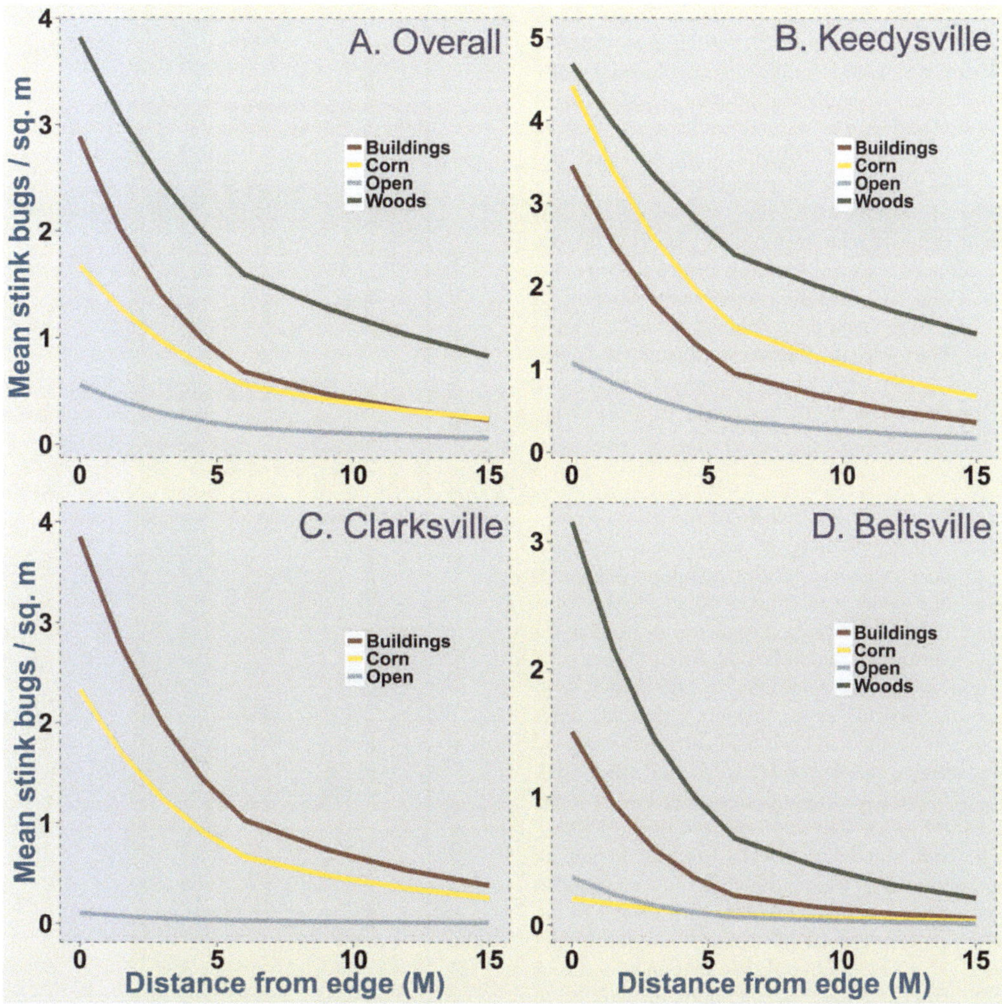

Figure 2. Mean stink bug density in soybean field edges in relation to different adjacent habitats and distance from field edge. Estimates derived from Poisson-lognormal GLMMs are plotted for overall stink bug data pooled over all study sites (A), Keedysville (B), Clarksville (C) and Beltsville (D). Values presented here have been back transformed from its original link function estimated model coefficients. Multiple comparison of means with a bonferroni correction ($\alpha = 0.05$) showed significant differences in: overall (A) - wooded habitats and buildings, wooded and open habitats, buildings and corn habitats, buildings and open habitats; Keedysville (B) – wooded habitats and buildings, wooded and open habitats, buildings and corn habitats, buildings and open habitat; Clarksville (C) - buildings and open habitats, and corn and open habitats; Beltsville (D) – buildings and open habitats, buildings and wooded habitats, and wooded and corn habitats.

field habitats, but differences were not significant (see Fig. S3 for variances around estimated means). Across all the adjacent habitat types the highest density of stink bugs was recorded at the immediate field edge (0 m), and was lowest at 15 m from edge (< $1/m^2$).

GLMMs performed by study site on overall stink bug data showed significant interactive influence of adjacent habitat and distance from edge on density at all the sites - Keedysville (Wald $\chi^2 = 485.4$, DF = 8, $P<0.001$), Clarksville (Wald $\chi^2 = 216.63$, DF = 5, $P<0.001$) and Beltsville (Wald $\chi^2 = 279.2$, DF = 8, $P< 0.001$). Multiple means comparisons for Keedysville data showed significantly higher densities adjacent to wooded habitats buildings and open habitats (Fig. 2B). The intercept and slopes for buildings and corn, and buildings and open areas were also significantly different. At Clarksville, buildings and corn habitats had significantly higher density of stink bugs than open areas (Fig. 2C). Model estimated stink bug densities for buildings and corn habitats were not significantly different. At Beltsville, wooded habitats and buildings had significantly higher stink bug density than corn as adjacent habitat (Fig. 2D). The slope and intercept were significantly different for wooded habitats and buildings. Unlike the other sites where stink bugs were fewest adjacent to open habitats, at Beltsville, corn as adjacent habitat harbored the lowest density of stink bugs. The variance around the estimated density of stink bugs in soybean fields for each site (Fig. S3), and the raw means for each site and year (Fig. S4) are provided as supplementary information.

Corn and soybean seed damage

For field corn, results from the quasi-Poisson GLM showed a significant positive association between percentage of stink bug damaged kernels and mean stink bug density ($y = 0.57 + 0.15x$, n = 252, $P<0.001$, pseudo $r^2 = 0.47$). A Poisson GLM showed that the percentage of collapsed kernels was not significantly associated with mean stink bug density ($y = -6.75 + 0.14x$, n = 252, $P = 0.50$, pseudo $r^2 = 0.17$). For soybean pod development data, regression analysis revealed that the percentage of full pods was negatively influenced by mean stink bug density ($y = 5.9 - 0.17x$, n = 63, $P<$

0.001, $r^2 = 0.51$), while percentage of flat pods (square root transformed) was positively influenced ($y = 2.18+0.26x$, $n = 63$, $P < 0.001$, $r^2 = 0.63$). Results of LMMs analyzing each seed quality category (Table 3) showed a significant positive association of mean stink bug density and purple stained seeds, percentage of stink bug damaged seeds, percentage of immature, shriveled and moldy seeds, and overall percentage of damaged seeds.

Overall percentage of normal, undamaged soybean seeds and yield had a significant negative relationship with stink bug density (Table 3). The overall seed damage by stink bugs in both corn (Fig. 3A) and soybean (Fig. 3C), and their impact on soybean pod development (Fig. 3D), were highest at field edge and declined gradually towards the field interior. Furthermore, soybean yields (grams/20 plants) were lowest at field edge, gradually increasing from field edge to highest yields at 12 and 15 m from the edge (Fig. 3B).

Discussion

Seasonal availability and suitability of source and recipient habitats influence dispersal dynamics of pest populations [2,4,5]. Aggregation of stink bugs at the field edge of target crops results from their directional movement among a sequence of hosts in an area in response to deteriorating suitability of the host crops or native vegetation and increasing suitability of the target crop plants [57]. Our results provide strong evidence that *H. halys* density exhibits a clear edge effect in both field corn and soybean. Across all adjacent habitats, density of *H. halys* and other stink bugs was highest within the first 3 meters, reaching lowest levels between 9–15 m from the edge, least at 15 m from the edge ($<1/m^2$). In addition to host suitability, species specific behaviors contribute to the resulting aggregations of stink bugs along crop edges. The strong edge effect exhibited by *H. halys* is similar to the within-field infestation pattern reported for other native stink bug species in U.S. crops [12,15,17–19,23,58]. We demonstrated that the aggregations are associated with specific kinds of damage to both corn grain and soybean seeds, thus suggesting that sitenspecific management at the field and farmscape levels may improve crop protection from stink bugs.

The timing of infestations during mid to late July in corn, and then later colonization of soybean fields in August, suggests that the majority of *H. halys* adults were offspring of the generation which developed on host plants earlier in the season (June/July). *Halyomorpha halys* is known to feed on a wide range of cultivated and wild hosts (up to 170 species) [59], of which many tree and shrub species are likely present in wooded habitats. Particularly high density of *H. halys* was observed (personal field observations, Dilip Venugopal) in soybean fields bordering wooded habitats with tree of heaven (*Ailanthus altissima* Swingle), princess tree (*Paulownia tomentosa* Baill.), and black cherry (*Prunus serotina* Ehrhart), all which support high population densities of reproducing *H. halys* [59,60]. Similarly, earlier reports show that *C. hilaris* populations in soybean are greater adjacent to wooded borders with black cherry and elderberry (*Sambucus canadensis* L.) [13]. Additionally, *E. servus* populations are greater in cotton fields with adjacent woods containing many oak species (*Quercus* spp.) and black cherry [18].

This study also addressed the influence of adjacent habitats on stink bug density and quantified differences in density at various distances from the field edges in soybean and corn crops. We found that adjacent habitats, particularly wooded habitats, influenced the densities of *H. halys* and other stink bugs. In both corn and soybeans, fields adjacent to wooded habitats, across all study sites and distances from field edge, consistently harbored

Table 3. Statistical results of LMMs for analyzing the relationship between stink bug density and various soybean seed damage categories and yield.

Response variable	Data Transformation	Intercept	Intercept SE	Estimate	SE	DF	Wald t	Pval	psuedo r^2
% normal seeds	None	75.8	8.04	−2.11	0.25	145	8.28	<0.001	0.30
% stink bug damaged seeds	Square Root	3.41	0.5	0.07	0.01	145	4.58	<0.001	0.12
% purple damaged seeds	log	1.39	0.14	0.09	0.01	145	9.99	<0.001	0.44
% moldy + shriveled + immature seeds	Square Root	2.59	0.47	0.09	0.02	148	5.87	<0.001	0.19
% all damaged seeds	Square Root	4.78	0.63	0.18	0.02	145	9.03	<0.001	0.35
Total Yield (grams/20 plants)	Square Root	17.1	1.04	−0.20	0.04	140	4.67	<0.001	0.13

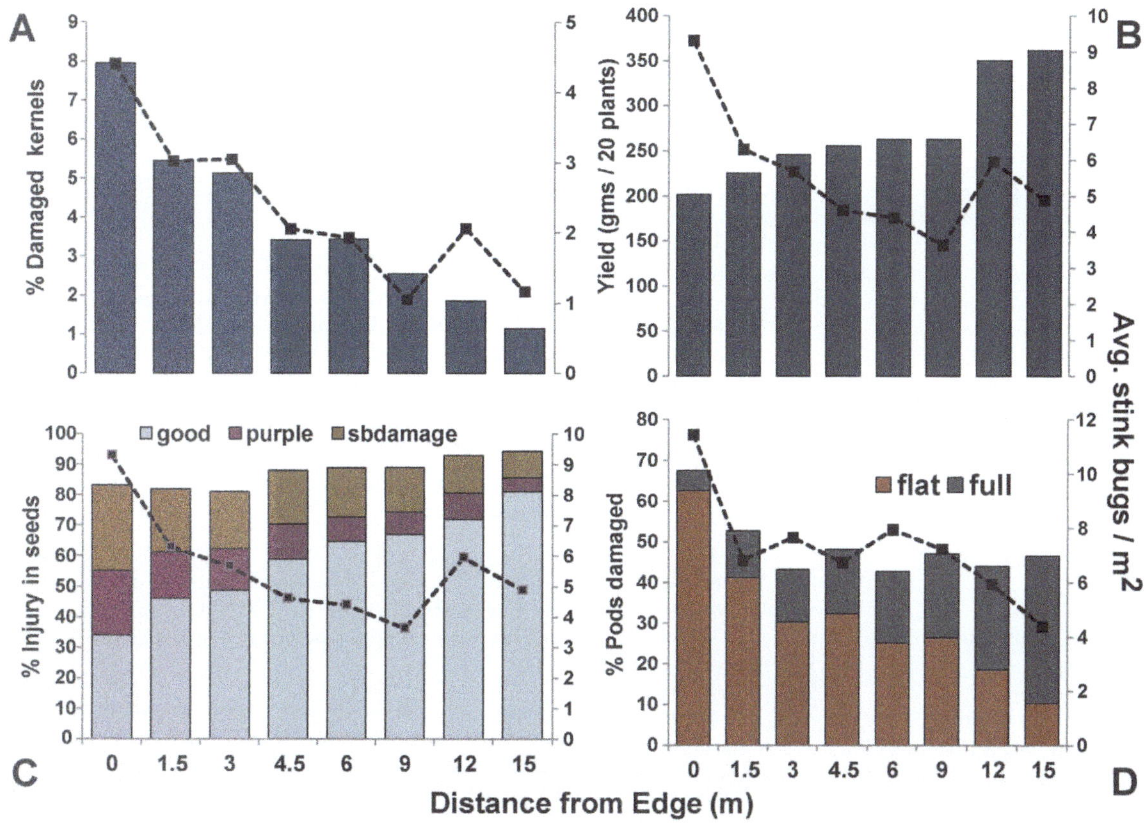

Figure 3. Patterns of kernel damage in field corn (A), soybean yield (B), soybean seed damage by category (C), and soybean pod development (D) in relation to mean stink bug density at different distance from field edge. The proportions of soybean seeds in each seed quality category (stink bug damaged, purple damaged, and normal seeds) and pod types (flat and full) are also provided. Mean stink bug abundance are denoted by the dashed lines which represent the second y axis.

significantly higher densities of stink bugs than in fields adjacent to open habitats. Also, stink bug density in fields adjacent to wooded habitats was consistently higher than in fields adjacent to buildings and corn habitats. These results suggest that wooded habitats play an important role in serving as sources of stink bug populations that colonize crops.

Patterns of infestation along field edges of corn and soybean differed among study sites, partly influenced by overall stink bug density and other adjacent habitats. Stink bug density in both row crops at Keedysville was consistently (3–5 times) greater than the mean density at the other two study sites, and this could be attributed to the higher overall densities of *H. halys* observed in the Western Maryland region over the past 5 years. In addition to wooded habitats, building and other crop habitats served as sources of colonizing adults in row crops. *H. halys* often utilize buildings as overwintering sites and thus these structures would likely influence stink bug populations earlier in the spring when post-diapause adults are moving to host plants, thereby supporting high local populations that eventually invade soybean later in the season. At Keedysville, stink bug density was higher in corn fields adjacent to alfalfa and in soybean fields adjacent to corn than in corn or soybean fields adjacent to building habitats. However, at Clarksville and Beltsville adjacent building habitats showed a higher density of stink bugs than crop habitats.

As with other studies, our results highlight the role of other adjacent cultivated crops in influencing stink bug populations. For example, adjacent fields of alfalfa, field corn, and other cultivated borders have been reported as a sources contributing to higher densities of stink bugs in tomato, cotton, sorghum, and peanut fields [12,16,19,23]. However, differences in the relative influence of adjacent habitats in our study could be related to overall stink bug density at each of the study sites. For example, stink bug density did not differ significantly among adjacent habitats in field corn at Beltsville where *H. halys* densities were lowest. Moreover, the influence of the landscape heterogeneity on stink bug density could extend to larger spatial scales beyond habitats adjacent to a crop. Since insect population dynamics and spatial patterns are affected by regional landscape context and species traits such as dispersal ability [61], distribution, and density of *H. halys* may depend on habitat and other environmental characteristics at spatial scales greater than the local agricultural field [62,63]. Differences in landscape structure among our study sites may explain the higher density of stink bugs at Keedysville, and specifically the role of adjacent crop habitats as sources of stink bugs in field corn and soybean.

We related the various corn and soybean damage measurements to stink bug density within plots. As expected, stink bug damage to corn kernels increased with stink bug density. The percentage of damaged kernels reached levels up to 8% at the field edge to less than 3% between 9–15 m from field edge, and was positively correlated with the stink bug density within plots. The percentage of collapsed kernels was negligible and not significantly influenced by stink bug density. Based on findings by earlier studies [25,64], neither kernel damage, ear weight or grain weight are affected beyond tasseling stage (VT) from feeding damage by *E. servus* or *N. viridula*. Although *H. halys* density can be high along edges of corn fields, our result suggests that *H. halys* kernel quality loss is restricted to about 10 m from the field edge.

Halyomorpha halys populations in soybean had a significant impact on pod development, with the percentage of flat pods significantly increasing with increasing stink bug density. Concomitantly, the proportion of fully developed pods significantly decreased with increasing stink bug density. Changes in the development and maturation of soybean pods due to *H. halys* feeding have been recently documented [35], showing that most severe pod loss occurred at the R4 (full pod) growth stage. Observed effects on pod and seed development with higher stink bug density were similar to damage caused by other stink bug species [26–31]. Our finding that higher stink bug density is related to increased proportions of moldy and purple stained seeds suggest a potential role of *H. halys* in transmitting or facilitating various pathogens; however, this needs to be further investigated experimentally. Our study found a significant, yet weak negative association between soybean yield and stink bug density per plot. In contrast, recent field research using cages addressed the effects of *H. halys* feeding on soybean growth, and did not detect a significant relationship between H. *halys* stink bug densities and yield loss [35]. Soybean field studies to compare yields of insecticide-treated and untreated plots are needed to establish the relationship between soybean yield losses and stink bug density.

Knowledge of how adjacent habitats influence *H. halys* populations and the within-field distribution has several implications in stink bug management. First, our results indicate that scouting corn and soybean fields for decision-making is more efficient if efforts initially concentrated on field edges bordering wooded habitats where there is a greater likelihood of colonization and higher risk of infestation. Secondly, the infestation patterns of stink bug communities dominated by *H. halys* are primarily edge-centric, and population densities beyond 12 m are generally low ($<1/m^2$). Based on our results, if insecticides are to be applied, edge-only applications, particularly along wooded and building habitats, could reduce the cost of control while preventing damage caused by stink bugs in field corn and soybean. Preliminary studies show that treating just 12 m into soybean field prevented further invasion by *H. halys* and other stink bugs (Ames Herbert, personal communication). The edge-only treatment prevented reinvasion and also resulted in an 85–95% reduction in insecticide used compared with whole-field treatments [11]. Results presented here showing highest stink bug density and associated damage limited to the immediate field edge provide validity for the edge-only treatment. While experimental research on effective insecticide application strategies are currently underway, based on our findings, we suggest that integrated pest management programs for the stink bug complex in field crops should include farmscape-level planning - crop location with regards to adjacent habitats, and targeted interventions in the form of edge-only treatments to prevent seed quality and yield losses.

Supporting Information

Figure S1 GLMM estimated mean stink bug densities in field corn (bold lines) and 95% CI (shaded region) among adjacent habitats and distance from the field edge.

Figure S2 Site and year wise raw stink bug averages in field corn among adjacent habitats and distance from the field edge.

Figure S3 GLMM estimated mean stink bug densities in soybean (bold lines) and 95% CI (shaded region) among adjacent habitats and distance from the field edge.

Figure S4 Site and year wise raw stink bug averages in soybean among adjacent habitats and distance from the field edge.

Acknowledgments

The farm managers and staff at each of the research facilities - Kevin Conover (Beltsville), Tim Ellis (Keedysville), David Justice and Mike Dwyer (Clarksville), and David Swain (USDA-BARC) - all extended excellent support and assistance for conducting the study. We thank Terry Patton for his help to identify sampling fields and for data collection. We thank the summer technicians, Jesse Ditillo, Jessie Saunders, Jake Bodart, Taylor Schulden, Kyle Runion, and Emily Zobel for data collection.

Author Contributions

Conceived and designed the experiments: PDV GPD WOL. Performed the experiments: PDV GPD PLC. Analyzed the data: PDV. Contributed reagents/materials/analysis tools: PDV. Contributed to the writing of the manuscript: PDV PLC GPD WOL.

References

1. Nestel D, Carvalho J, Nemny-Lavy E (2004) The spatial dimension in the ecology of insect pests and its relevance to pest management. In: Horowitz PDAR, Ishaaya PDI, editors. Insect pest management. Springer Berlin Heidelberg. 45–63.
2. Ekbom BS, Erwin ME, Robert Y (2000) Interchanges of insects between agricultural and surrounding landscapes. Kluwer Netherlands. 266 p.
3. Stinner RE, Barfield CS, Stimac JL, Dohse L (1983) Dispersal and movement of insect pests. Annu Rev Entomol 28: 319–335. doi:10.1146/annurev.en.28.010183.001535.
4. Kennedy GG, Margolies DC (1985) Mobile arthropod pests: management in diversified agroecosystems. Bull Entomol Soc Am 31: 21–35.
5. Kennedy GG, Storer NP (2000) Life systems of polyphagous arthropod pests in temporally unstable cropping systems. Annu Rev Entomol 45: 467–493. doi:10.1146/annurev.ento.45.1.467.
6. Panizzi AR (1997) Wild hosts of Pentatomids: ecological significance and role in their pest status on crops. Annu Rev Entomol 42: 99–122. doi:10.1146/annurev.ento.42.1.99.
7. Panizzi AR, Slansky F (1985) Review of phytophagous Pentatomids (Hemiptera: Pentatomidae) associated with soybean in the Americas. Fla Entomol 68: 184–214.
8. Nielsen AL, Hamilton GC, Shearer PW (2011) Seasonal phenology and monitoring of the non-native *Halyomorpha halys* (Hemiptera: Pentatomidae) in soybean. Environ Entomol 40: 231–238. doi:10.1603/EN10187.
9. Hooks CRR (2011) Stink bugs and their soybean obsession. Agron News 2: 1–4.
10. CABI (2014) *Halyomorpha halys* [original text by Leskey TC, Hamilton GC, Biddinger DC, Buffington ML, Dicekhoff C, et al.]. In: Invasive Species Compendium. Wallingford, UK: CAB International. www.cabi.org/isc. Available: http://www.cabi.org/isc/?compid=5&dsid=27377&loadmodule=datasheet&page=481&site=144. Accessed 19 March 2014.
11. Leskey TC, Hamilton GC, Nielsen AL, Polk DF, Rodriguez-Saona C, et al. (2012) Pest status of the brown marmorated stink bug, *Halyomorpha halys* in the USA. Outlooks Pest Manag 23: 218–226. doi:10.1564/23oct07.
12. Toscano NC, Stern VM (1976) Dispersal of *Euschistus conspersus* from alfalfa grown for seed to adjacent crops. J Econ Entomol 69: 96–98.
13. Jones WA, Sullivan MJ (1982) Role of host plants in population dynamics of stink bug pests of soybean in South Carolina. Environ Entomol 11: 867–875.
14. Outward R, Sorenson CE, Bradley JR (2008) Effects of vegetated field borders on arthropods in cotton fields in eastern North Carolina. J Insect Sci 8. Available: http://www.ncbi.nlm.nih.gov/pmc/articles/PMC3061576/.
15. Tillman PG, Northfield TD, Mizell RF, Riddle TC (2009) Spatiotemporal patterns and dispersal of stink bugs (Heteroptera: Pentatomidae) in peanut-

cotton farmscapes. Environ Entomol 38: 1038–1052. doi:10.1603/022.038.0411.

16. Toews MD, Shurley WD (2009) Crop juxtaposition affects cotton fiber quality in Georgia farmscapes. J Econ Entomol 102: 1515–1522. doi:10.1603/029.102.0416.
17. Pease CG, Zalom FG (2010) Influence of non-crop plants on stink bug (Hemiptera: Pentatomidae) and natural enemy abundance in tomatoes. J Appl Entomol 134: 626–636. doi:10.1111/j.1439-0418.2009.01452.x.
18. Reay-Jones FPF (2010) Spatial and temporal patterns of stink bugs (Hemiptera: Pentatomidae) in wheat. Environ Entomol 39: 944–955. doi:10.1603/EN09274.
19. Reeves RB, Greene JK, Reay-Jones FPF, Toews MD, Gerard PD (2010) Effects of adjacent habitat on populations of stink bugs (Heteroptera: Pentatomidae) in cotton as part of a variable agricultural landscape in South Carolina. Environ Entomol 39: 1420–1427. doi:10.1603/EN09194.
20. Reisig DD (2011) Insecticidal management and movement of the brown stink bug, Euschistus servus, in corn. J Insect Sci 11: Article 168. doi:10.1673/031.011.16801.
21. Herbert JJ, Toews MD (2011) Seasonal abundance and population structure of brown stink bug (Hemiptera: Pentatomidae) in farmscapes containing corn, cotton, peanut, and soybean. Ann Entomol Soc Am 104: 909–918. doi:10.1603/AN11060.
22. Olson DM, Ruberson JR, Zeilinger AR, Andow DA (2011) Colonization preference of *Euschistus servus* and *Nezara viridula* in transgenic cotton varieties, peanut, and soybean. Entomol Exp Appl 139: 161–169. doi:10.1111/j.1570-7458.2011.01116.x.
23. Tillman PG (2011) Influence of corn on stink bugs (Heteroptera: Pentatomidae) in subsequent crops. Environ Entomol 40: 1159–1176. doi:10.1603/EN10243.
24. Lee D-H, Short BD, Joseph SV, Bergh JC, Leskey TC (2013) Review of the biology, ecology, and management of *Halyomorpha halys* (Hemiptera: Pentatomidae) in China, Japan, and the Republic of Korea. Environ Entomol 42: 627–641. doi:10.1603/EN13006.
25. Ni X, Da K, Buntin GD, Cottrell TE, Tillman PG, et al. (2010) Impact of brown stink bug (Heteroptera: Pentatomidae) feeding on corn grain yield components and quality. J Econ Entomol 103: 2072–2079. doi:10.1603/EC09301.
26. Daugherty DM, Neustadt MH, Gehrke CW, Cavanah LE, Williams LF, et al. (1964) An evaluation of damage to soybeans by brown and green stink bugs. J Econ Entomol 57: 719–722.
27. Todd JW, Turnipseed SG (1974) Effects of southern green stink bug damage on yield and quality of soybeans. J Econ Entomol: 421–426.
28. McPherson RM, Newsom LD, Farthing BF (1979) Evaluation of four stink bug species from three Genera affecting soybean yield and quality in Louisiana. J Econ Entomol 72: 188–194.
29. Brier HB, Rogers DJ (1991) Susceptibility of soybeans to damage by *Nezara viridula* (l.) (Hemiptera: Pentatomidae) and *Riptortus serripes* (f.) (Hemiptera: Alydidae) during three sages of pod development. Aust J Entomol 30: 123–128. doi:10.1111/j.1440-6055.1991.tb00403.x.
30. McPherson RM, Douce GK, Hudson RD (1993) Annual variation in stink bug (Heteroptera: Pentatomidae) seasonal abundance and species composition in Georgia soybean and its impact on yield and quality. J Entomol Sci 28: 61–72.
31. Corrêa-Ferreira BS, De Azevedo J (2002) Soybean seed damage by different species of stink bugs. Agric For Entomol 4: 145–150. doi:10.1046/j.1461-9563.2002.00136.x.
32. Clarke RG, Wilde GE (1971) Association of the green stink bug and the yeast-spot disease organism of soybeans. III. Effect on soybean quality. J Econ Entomol: 222–223.
33. Mitchell PL (2004) Heteroptera as vectors of plant pathogens. Neotrop Entomol 33: 519–545. doi:10.1590/S1519-566X2004000500001.
34. Medrano EG, Esquivel J, Bell A, Greene J, Roberts P, et al. (2009) Potential for *Nezara viridula* (Hemiptera: Pentatomidae) to transmit bacterial and fungal pathogens into cotton bolls. Curr Microbiol 59: 405–412. doi:10.1007/s00284-009-9452-5.
35. Owens DR, Herbert Jr DA, Dively GP, Reisig DD, Kuhar TP (2013) Does feeding by *Halyomorpha halys* (Hemiptera: Pentatomidae) reduce soybean seed quality and yield? J Econ Entomol 106: 1317–1323. doi:10.1603/EC12488.
36. Hoebeke ER, Carter ME (2003) *Halyomorpha halys* (Stål) (Heteroptera: Pentatomidae): a polyphagous plant pest from Asia newly detected in North America. Proc Entomol Soc Wash 105: 225–237.
37. USDA National Agricultural Statistics Service (2014) Crop production. Available: http://www.usda.gov/nass/PUBS/TODAYRPT/crop0614.pdf. Accessed 14 June 2014.
38. Hanway JJ (1963) Growth stages of corn (*Zea mays,* L.). Agron J 55: 487. doi:10.2134/agronj1963.00021962005500050024x.
39. Fehr WR, Caviness CE, Burmood DT, Pennington JS (1971) Stage of development descriptions for soybeans, *Glycine Max* (L.) Merrill. Crop Sci 11: 929. doi:10.2135/cropsci1971.0011183X001100060051x.
40. Schumann FW, Todd JW (1982) Population dynamics of the southern green stink bug (Heteroptera: Pentatomidae) in relation to soybean phenology. J Econ Entomol 75: 748–753.
41. Walters HJ (1980) Soybean leaf blight caused by *Cercospora kikuchii*. Plant Dis 64: 961–962. doi:10.1094/PD-64-961.
42. Bolker BM, Brooks ME, Clark CJ, Geange SW, Poulsen JR, et al. (2009) Generalized linear mixed models: a practical guide for ecology and evolution. Trends Ecol Evol 24: 127–135. doi:10.1016/j.tree.2008.10.008.
43. Pinheiro JC, Bates DM (2000) Mixed effects models in S and S-Plus. New York: Springer. 560 p.
44. Zuur AF, Ieno EN, Walker N, Saveliev AA, Smith GM (2009) Mixed effects models and extensions in ecology with R. New York: Springer. 580 p.
45. Kéry M (2010) Introduction to WinBUGS for ecologists: bayesian approach to regression, ANOVA, mixed models and related analyses. Academic Press. 322 p.
46. Ver Hoef JM, Boveng PL (2007) Quasi-Poisson Vs. negative binomial regression: How should we model overdispersed count data? Ecology 88: 2766–2772. doi:10.1890/07-0043.1.
47. Nagelkerke NJD (1991) A note on a general definition of the coefficient of determination. Biometrika 78: 691–692. doi:10.1093/biomet/78.3.691.
48. R Development Core Team (2011) R: A language and environment for statistical computing. The R foundation for statistical computing, Vienna, Austria. Available: http://www.R-project.org/.
49. Bates D, Maechler M, Bolker B, Walker S (2013) lme4: Linear mixed-effects models using Eigen and S4. Available: http://cran.r-project.org/web/packages/lme4/index.html. Accessed 25 December 2013.
50. Pinheiro J, Bates DM, Debroy S, Sarkar D, R Core Team (2013) nlme: Linear and nonlinear mixed effects models R package version 3.1-117. Available: http://cran.r-project.org/web/packages/nlme/index.html. Accessed 25 August 2014.
51. Kuhn M, Weston S, Wing J, James JF, Thaler T (2013) contrast: A collection of contrast methods. Available: http://cran.r-project.org/web/packages/contrast/index.html. Accessed 25 December 2013.
52. Hothorn T, Bretz F, Westfall P (2008) Simultaneous inference in general parametric models. Biom J 50: 346–363.
53. Fox J, Weisberg S, Friendly M, Hong J, Andersen R, et al. (2013) effects: effect displays for linear, generalized linear, multinomial-logit, proportional-odds logit models and mixed-effects models. Available: http://cran.r-project.org/web/packages/effects/index.html. Accessed 25 December 2013.
54. Wickham H (2009) ggplot2: elegant graphics for data analysis. New York: Springer. Available: http://had.co.nz/ggplot2/book.
55. Harrell FE (2013) rms: Regression modeling strategies. Available: http://cran.r-project.org/web/packages/rms/index.html. Accessed 25 December 2013.
56. Bartoń K (2013) MuMIn: Multi-model inference. Available: http://cran.r-project.org/web/packages/MuMIn/index.html. Accessed 25 December 2013.
57. Todd JW (1989) Ecology and behavior of *Nezara viridula*. Annu Rev Entomol 34: 273–292.
58. Tillman PG (2010) Composition and abundance of stink bugs (Heteroptera: Pentatomidae) in corn. Environ Entomol 39: 1765–1774. doi:10.1603/EN09281.
59. BMSB IPM Working Group & Northeastern IPM Center (2013) Host plants of the brown marmorated stink bug in the U.S. Available: http://www.stopbmsb.org/where-is-bmsb/host-plants/. Accessed 2 January 2014.
60. Nielsen AL, Hamilton GC (2009) Life history of the invasive species *Halyomorpha halys* (Hemiptera: Pentatomidae) in northeastern United States. Ann Entomol Soc Am 102: 608–616. doi:10.1603/008.102.0405.
61. Tscharntke T, Brandl R (2004) Plant-insect interactions in fragmented landscapes. Annu Rev Entomol 49: 405–430. doi:10.1146/annurev.ento.49.061802.123339.
62. Thies C, Steffan-Dewenter I, Tscharntke T (2003) Effects of landscape context on herbivory and parasitism at different spatial scales. Oikos 101: 18–25. doi:10.1034/j.1600-0706.2003.12567.x.
63. Tscharntke T, Rand TA, Bianchi F (2005) The landscape context of trophic interactions: insect spillover across the crop-noncrop interface. Ann Zool Fenn 42: 421–432.
64. Negrón JF, Riley TJ (1987) Southern green stink bug, *Nezara viridula* (Heteroptera: Pentatomidae), feeding in corn. J Econ Entomol 80: 666–669.

7

Biochemical Aspects of the Soybean Response to Herbivory Injury by the Brown Stink Bug *Euschistus heros* (Hemiptera: Pentatomidae)

Renata Velozo Timbó[1,2], Marcelo Hermes-Lima[1], Luciano Paulino Silva[2], Angela Mehta[2], Maria Carolina Blassioli Moraes[2], Débora Pires Paula[2]*

1 Department of Cell Biology, University of Brasília, Campus Universitário Darcy Ribeiro, Brasília, Distrito Federal, Brazil, **2** Department of Biological Control, Embrapa Genetic Resources and Biotechnology, Brasília, Distrito Federal, Brazil

Abstract

Plant defense response is an elaborate biochemical process shown to depend on the plant genetic background and on the biological stressor. This work evaluated the soybean biochemical foliar response to brown stink bug herbivory injury through an analysis of redox metabolism and proteomic 2DE profiles of susceptible (BRS Silvania RR) and resistant (IAC-100) varieties. The activity of lipoxygenase-3, guaiacol peroxidase, catalase and ascorbate peroxidase was monitored every 24 h up to 96 h. In the susceptible variety, injury caused an increase in the activities of lipoxygenase 3 and guaiacol peroxidase, no change in ascorbate peroxidase, and a decrease in catalase. In the resistant variety, injury did not cause an alteration of any of these enzymes. The proteomic profiles were evaluated after 24 h of injury and revealed to have a similar proportion (4–5%) of differential protein expression in both varieties. The differential proteins, identified by mass spectrometry, in the susceptible variety were related to general stress responses, to plant defense, and to fungal infections. However, in the resistant variety, the identified change in protein profile was related to Calvin cycle enzymes. While the susceptible variety showed adaptive changes in redox metabolism and expression of stress-responsive proteins, the resistant showed a defense response to circumvent the biological stressor.

Editor: Joseph Clifton Dickens, United States Department of Agriculture, Beltsville Agricultural Research Center, United States of America

Funding: This work was supported by Embrapa (www.embrapa.br) (DPP AM LPS), CNPq (www.cnpq.br) (LPS), and CAPES (www.capes.gov.br) (RVT). The funders had no role in study design, data collection and analysis, decision to publish, or preparation of the manuscript.

Competing Interests: The authors have declared that no competing interests exist.

* Email: debora.pires@embrapa.br

Introduction

Stink bugs (Hemiptera: Pentatomidae) are one of the most important pests of soybean in Brazil [1]. Pentatomids feed on various structures of the host plant, mainly fruits and immature seeds, affecting their quality, development and maturation [2]. The brown stink bug *Euschistus heros* is considered one of the main soybean pests [3,4]. Information about the soybean defense mechanisms to this herbivore when it colonizes in the late vegetative stages is important to support pest management.

Several varieties of soybean have been studied regarding their defense response to herbivores in order to identify herbivory resistant varieties [5,6,7,8]. These studies have mostly focused on primary and secondary metabolite production as parts of the plant defense strategy aiming to support environmentally friendly soybean production. Plant defense to herbivory injury involves two major systems: (i) the direct defense, involving structural components (e.g. thorns, trichomes), and production of primary metabolites (e.g. proteinase inhibitors, antioxidant enzymes) and non-volatile secondary metabolites (e.g. phenolic acids and isoflavonoids) [9,10]; and (ii) the indirect defense, which is related to the production of volatile organic compounds (VOCs), such as (*E*,*E*)-α-farnesene, methyl salicylate and *cis*-jasmone, which attract natural enemies (predators and parasitoids) of the herbivores that feed on the plants, or in some cases act as herbivore repellents [6,11].

Plant defense response varies according to the plant genetic background, and the type of stressor [10]. Therefore, it is valuable to study the response elicited by the key herbivores in different varieties. Biochemical studies on plant defense have focused on evaluating the activity of enzymes related to the redox metabolism (including antioxidant enzymes) [9,12–14], such as lipoxygenase (LOX), peroxidase (POD), glutathione reductase (GR), ascorbate peroxidase (APX), and catalase (CAT), as well as on the modification of protein profiles [15]. However, despite the significant economic impact of the damage caused by stink bugs to soybean [16,17], there are no reports regarding the soybean redox response and the regulation of protein expression induced by stink bug herbivory.

Plants differ in resistance to their key herbivorous pests according to tolerance, non-preference and antibiosis [18]. In this work we used two soybean varieties with large differences in antibiosis resistance to brown stinkbug: resistant IAC-100 and susceptible BRS Silvania RR. IAC-100 was developed by crossing the parental varieties IAC-12 and IAC 78-2318, showing resistance to chewing and sucking insects [19]. BRS Silvania RR was developed for glyphosate tolerance [20]. The goal was to characterize the biochemical effects of the brown stink bug herbivory injury on the foliar redox response, through the analysis of the activity of lipoxygenase-3 (LOX3), guaiacol peroxidase (GOPx), catalase (CAT) and ascorbate peroxidase (APX), as well as on the foliar protein expression, through two-dimensional electrophoresis and mass spectrometry.

Materials and Methods

Soybean cultivation and stink bug rearing

The seeds of IAC-100 and BRS Silvania RR were germinated in a neutral pH germination paper. Seedlings with equivalent development were transplanted to plastic pots of 350 mL containing mixture of soil and organic substrate (in a proportion of 3:1), maintained in a germination chamber at 25±2°C and 13 h photoperiod, and irrigated every two days. The bioassays were conducted with plants in vegetative stage 3 (V3, first three trifoliolate leaves fully expanded). Stink bugs colonize soybeans during the vegetative stage, during which soybean plants produce higher amount of volatiles compared to the reproductive stage, and are known to have their indirect defense elicited by injury of the brown stink bug [10,17,21]. Understanding the plant defense mechanisms during this stage could help control stink bug pest populations that result in the reproductive stage.

The brown stink bugs were obtained from a laboratory colony at Embrapa Genetic Resources and Biotechnology (Brasília, Brazil), reared according to Moraes *et al.* [18] in plastic cages (26×22 cm) containing cotton fiber soaked in water and fed on peanut seeds [*Arachis hypogaea* (L.)], soybean grains [*Glycine max* (L.) Merrill.], sunflower seeds [*Helianthus annus* (L.)] and green beans [*Phaseolus vulgaris* (L.)], at 26±2°C, 60±10% RH and 14 h photoperiod.

Herbivory bioassays

The herbivory bioassays consisted of no injury (control) or injury to individual V3 plants of the two soybean varieties by three male adult brown stink bugs. Plants were exposed to stink bugs for 24, 48, 72, or 96 h after which the plants were harvested. Simultaneously, plants were harvested from the control. There were six replicate plants per harvest per treatment in a completely randomized design. Only males were used to prevent possible interference of oviposition on the plants, which could trigger other plant responses [6,21]. The stink bugs were deprived of food for 24 h before the bioassays. The plants were covered with transparent plastic bags (with micro holes for aeration) to guarantee the constant presence of the stink bugs during the bioassay. All assays were performed in separate germination chambers (FANEM) for each treatment, so that the VOC released could not interfere with the other treatments. All leaves of the six plants per treatment were collected and transferred to 50 mL tubes and immediately frozen in liquid nitrogen and stored at −80°C for further biochemical analysis.

Quantification of enzyme activities

The enzymatic activity was measured for LOX3, GOPx, APX and CAT in the soybean foliar extracts of three plants from each treatment in the four 24 h sampling intervals. Foliar homogenates were prepared in a mortar containing liquid nitrogen, where the leaves were macerated until the formation of a fine powder. For every 100 mg of macerated leaf, 400 μL of 50 mM potassium phosphate buffer (KPi) containing 10 mM phenylmethanesulfonyl fluoride (PMSF), 5 mM ethylenediaminetetraacetic acid (EDTA) was added. After homogenization, the samples were centrifuged at 14,000×*g* for 15 min at 4°C. The supernatant was quantified by the Bradford method [22] and used in triplicate for each enzyme assay.

The conditions for the enzymatic activity assays are detailed in Table 1. The assays for GOPx, APX and CAT were performed in quartz cuvettes (10×10 mm) by adding 900 μL of KPi pH 7.0 and 1 to 20 μL of foliar supernatant. For LOX3, 2,975 μL Kpi pH 6.5 and 1 to 20 μL of foliar supernatant were used. The reactions were initiated by adding the substrate for each enzyme (100 μL for GOPx, APX and CAT and 25 μL for LOX3) and the absorbance was measured for 150 s at intervals of 10 s with a Smart Plus spectrophotometer (BioRad).

The enzymatic activities ($\mu mol.min^{-1}.mg^{-1}$) were compared by one-way ANOVA, and the paired comparisons among the treatments were performed by Tukey's HSD ($p<0.05$).

Total protein extraction for the proteomic analysis

The total foliar protein profiles of all treatments (three plants/each) were analyzed after 24 h of bioassay by 2-DE. Total protein extraction was performed adapting the TCA/acetone methodology [27]. According to Wildgruber *et al.* [28], this method inhibits the degradation of proteins by minimizing protease activity (due to low pH), and also improves cell lysis and subsequent precipitation of proteins at −20°C. The leaves from the same treatment were pooled and macerated in liquid nitrogen until the formation of a fine powder and chilled cold acetone solution, containing 10% (w/v) TCA and 0.07% (v/v) β-mercaptoethanol, was added in a proportion of 1:2 v/v. After vortex homogenization for 5 min and incubation for 24 h at −20°C, the samples were centrifuged at 10,000×*g* for 20 min at 4°C. The pellets were washed three times in cold acetone containing 0.07% (v/v) β-mercaptoethanol and centrifuged at 10,000×*g* for 10 min at 4°C. To each 12 mg of pellets 1 mL of solubilization buffer [8.5 M urea, 2.5 M thiourea, 2% (w/v) CHAPS, 1% (w/v) DTT, 1% (w/v) Serdolit, 2% (v/v) IPG buffer nonlinear 3–11, 1% (w/v) bromophenol blue] [29] was added and incubated at room temperature for 1 h under gentle stirring. After centrifugation at 14,000×*g* for 1 h at 25°C, the supernatants were collected and the proteins quantified by the Bradford method [22].

Two-dimensional electrophoresis (2-DE) and gel image analysis

The isoelectric focusing (IEF) was carried out in 13 cm Immobiline DryStrips pH 3–11 NL (GE Healthcare). The strips were hydrated overnight with 250 μL of solubilization buffer containing 500 μg of total protein extract (three strips for each treatment). The IEF was conducted at 20°C using the Multiphor II Electrophoresis System (GE Healthcare) coupled to a cooling system bath in one step of 300 V for 1 min and one step of 3,500 V for 5.5 h (all at 2 mA and 5 W). After IEF, the strips were reduced by the addition of 8 mL of equilibration buffer [50 mM Tris-HCl pH 8.8, 6 M urea, 30% (v/v) glycerol, 2% (w/v) SDS] containing 1% (w/v) DTT under gentle agitation for 15 min at room temperature; and alkylated by adding 8 mL of equilibration buffer containing 2.5% (w/v) iodoacetamide under gentle agitation for 15 min. The equilibrated strips were transferred to a 12.5% denaturaing polyacrylamide gel and electrophoresis was performed at 25°C in Tris/glycine/SDS buffer on a multiple Ettan DALTsix (GE Healthcare), coupled to a cooling bath system, according to manufacturer's recommendation. For molecular mass reference, the BenchMark Protein ladder (Invitrogen) was applied on a side of the gel using filter paper. The electrophoresis was carried out in two steps: first at 10 mA/80 V for 1 h and 40 mA/500 V for 4.3–6 h. The 2-DE gels were

Table 1. Enzymatic activity assay conditions in leaves of the soybean varieties BRS Silvania RR and IAC-100 in the V3 stage, with and without (control group) brown stink bug herbivory injury.

Enzyme	Substrate	ε ($mM^{-1}cm^{-1}$)	Wavelength (nm)	Final Volume (μL)	References
LOX3	80 mM sodium linoleate	22.00	234	3,000	[23]
GOPx	10 mM hydrogen peroxide +20 mM guaiacol	26.60	470	1,000	[24]
APX	10 mM hydrogen peroxide +40 mM ascorbic acid	14.50	265	1,000	[25]
CAT	10 mM hydrogen peroxide	0.04	240	1,000	[26]

stained for two days in a solution of 0.8% (w/v) Coomassie Brilliant Blue G-250, 20% (v/v) ethanol, 1.6% (v/v) phosphoric acid, 8% (w/v) ammonium sulfate, and rinsed five times in water to remove excess dye.

The 2-DE gel images were obtained by ImageScanner III (GE Healthcare). Spot detection and measurements, as well as spot matching in the three gel images per treatment, were performed using the ImageMaster 2D Platinum 7.0 software (GE Healthcare). A spot was considered differential when its intensity and area were different in the injured and uninjured gels. The comparison of the total average number of spots between the injured and uninjured treatments was performed by one-away ANOVA ($p<$ 0.05).

Protein identification

For the soybean variety BRS Silvania RR the differential spots were excised and sliced into small fragments. Spot destaining was conducted by successive washes in 30% ethanol and incubation in 50% (v/v) acetonitrile solution containing 25 mM ammonium bicarbonate under agitation for 15 min. The gel fragments were dehydrated with 100% acetonitrile for 10 min under agitation and dried in a centrifugal evaporator for 20 min. The spots were rehydrated with 50 μL of hydrolysis solution consisting of 50 mM Tris-HCl pH 7.6 added to 1 mM calcium chloride and 0.5 μL of trypsin at 1 μg/μL (Sequencing Grade Modified Trypsin, Promega), which was previously heated to 30°C for 15 min, and to 50 μL of 50 mM ammonium bicarbonate. The samples were incubated at 37°C for 14 to 18 h. For the soybean variety IAC-100, trypsin hydrolysis was carried out using the Trypsin Profile IGD Kit (Sigma), according to the manufacturer instructions.

The trypsin-hydrolyzed peptides were analyzed by MALDI-TOF mass spectrometry using an UltraFlex III MALDI TOF/TOF (Bruker Daltonics) controlled by the software FlexControl 3.3.108.0 using the following parameters: mass range 600-3,500 Da; RP_PEPMIX method; 200 automatic shooting at a frequency of 200 Hz, laser intensity between 20–45%; deflection mass of 600 Da; reflective mode; and positive external calibration by peptide calibration standard I. The method to obtain the MS/MS data was the LIFT; AUTO_LIFT method; using the following parameters: 2,000 shots for which, 400 and 1,600 shots for the precursor and the fragments, respectively; laser intensity between 22–55%; and the other parameters were the same as mentioned above. The peaks list files containing the m/z ratios of precursor ions and MS/MS fragmented ions were used to search similar oligopeptides in the National Center for Biotechnology Information (NCBI) database sequences using BLASTp and Mascot software. The parameters used in the Mascot program for MS/MS were: (i) NCBI-nr database, (ii) Viridiplantae taxonomy, (iii) fixed modifications: carbamidomethylation of cysteine, (iv) variable modification: oxidation of methionine, (v) monoisotopic mass spectra data, (vi) mass tolerance of 150 ppm, (vii) fragment tolerance of ±0.05 Da. Only significant hits, as defined by Mascot ($p<0.05$), were accepted.

Results and Discussion

Redox response

The herbivory injury by the stink bug affected the foliar enzymatic response ($p<0.05$) in the susceptible BRS Silvania RR variety (Figure 1). The LOX3 activity was significantly higher at 72 h of herbivory, with a decline at 96 h, but still significantly higher than in the first 48 h of injury and in the uninjured control. The GOPx activity did not differ over the time of injury, but was higher along the time evaluated compared to the uninjured plants. On the other hand, APX activity was not affected during the herbivory period, and CAT activity decreased after 48 h of injury compared to uninjured plants. In contrast, the herbivory injury by the brown stink bug did not affect the foliar activity of any of the enzymes evaluated even after 96 h of injury (Fig. 2) for the resistant IAC-100 variety, suggesting that the redox metabolism was not affected in this variety under *E. heros* feeding injury.

From the enzymes studied, LOX3 and GOPx have a clear relation to the plant defense response and an enhanced activity under herbivory injury has been reported [30–32]. LOX3 catalyzes the production of jasmonic acid (JA) from linolenic acid, which stimulates the expression of genes responsible for producing defense compounds and is related to plant defense against herbivorous insects, and GOPx activates defense genes [30–32]. Chen *et al.* [24] observed that herbivory by *Melanocallis caryaefoliae* (Hemiptera: Aphididae) induced higher activity of LOX3 and GOPx in susceptible pecans, although in moderate and high resistance pecans the GOPx activity did not change. Similarly, Ni *et al.* [33] reported increased activity of GOPx, CAT and polyphenol oxidase in wheat leaves after herbivory by *Diuraphis noxia* (Hemiptera: Aphididae), and that peroxidase activity in barley leaves was significantly higher in susceptible than in resistant plants after herbivory by the same aphid species.

Although uninjured BRS Silvania RR generally had higher activity of three enzymes related to the redox metabolism than IAC-100, LOX3 activity was higher in uninjured IAC-100 than in BRS Silvania RR. This result may be directly related to resistance, because one of the functions of the lipoxygenase pathways the production of JA. The constitutive high activity of lipoxygenase in plants is related to their resistance against insects [34], which may help to explain the resistance of IAC-100 against stink bugs [6,16] and why it produces high level of induced plant volatiles when injured by them, which were not detected in BRS Silvania RR [6].

APX catalyzes the redox reaction between hydrogen peroxide and ascorbate [35]. Bi & Felton [36] reported that APX activity in soybean leaves increased over time after herbivory by *Helicoverpa zea* (Lepidoptera: Noctuidae), with maximum activity at 96 h. In our results, the activity of APX was not affected by stink bug

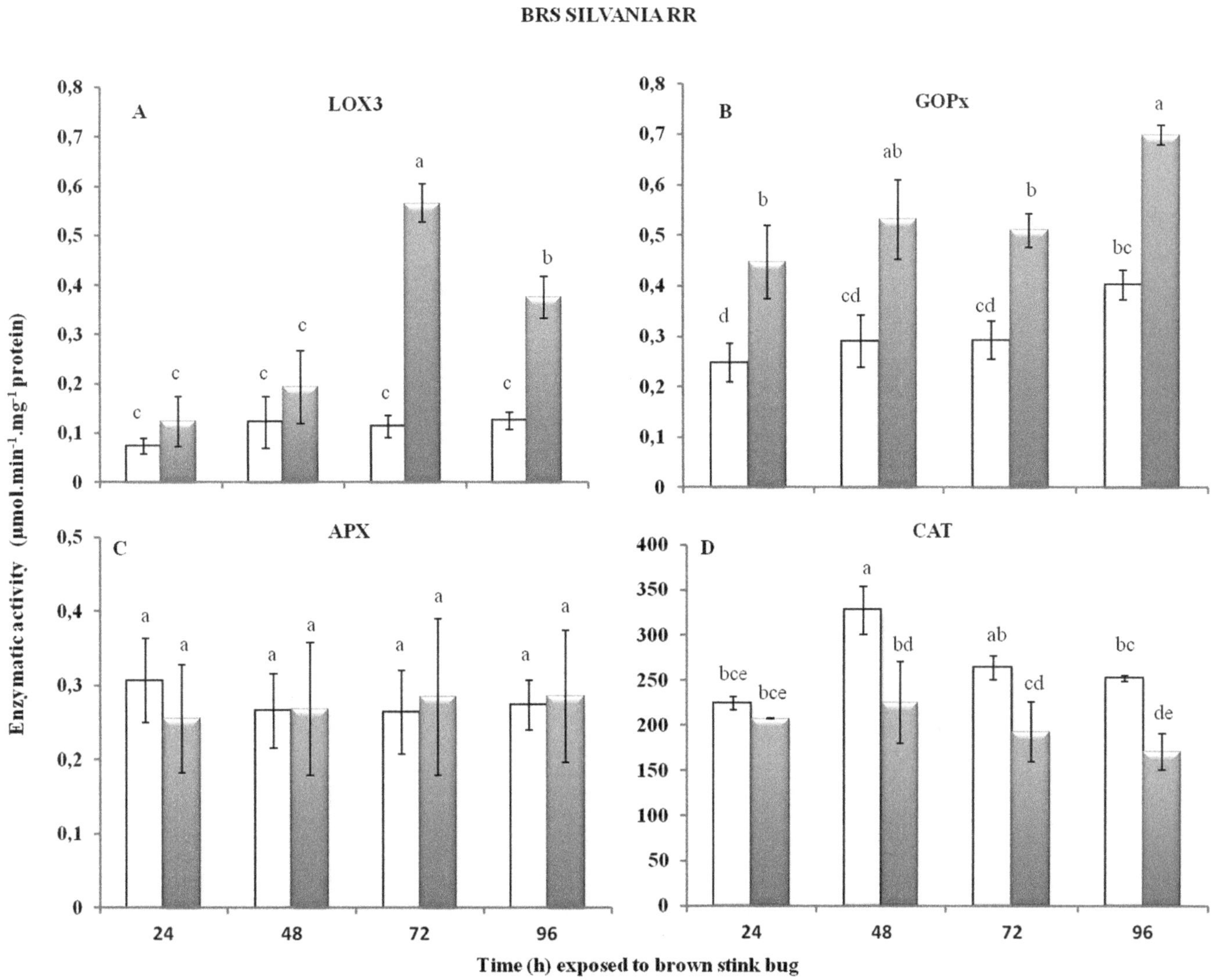

Figure 1. Enzymatic activity for (A) LOX3 ($F_{7,16}=2.32$; $p=0.0779$), (B) GOPx ($F_{7,16}=1.01$; $p=0.4629$), (C) APX ($F_{7,16}=3.35$; $p=0.0213$), and (D) CAT ($F_{7,16}=1.78$; $p=0.1609$) in leaves of BRS Silvania RR soybean variety (V3) with or without brown stink bug herbivory injury along different sampling periods. Injured is represented in grey and not injured in white. Bars with the same letter are not significantly different by Tukey's HSD ($p<0.05$).

herbivory on the susceptible soybean variety. Stink bugs are piercing-sucking insect, therefore provoking less damage to plants compared to chewing insects [37], and possibly inducing lower levels of hydrogen peroxide in the injury process.

CAT performs two major functions: (i) decomposition of hydrogen peroxide (a product of photorespiration and other plant metabolic activities) and (ii) oxidation of hydrogen donors such as methanol, ethanol, formic acid, phenols, with consumption of hydrogen peroxide [25,38]. An increase in CAT activity may be related to the control of hydrogen peroxide levels (which can be stressful at high concentrations) formed in response to herbivory injury [39]. In herbivory studies on various plant species, CAT activity remained unchanged [12,24,40], increased [33] or decreased [36,37]. In our study, a decrease in CAT activity was observed in response to stink bug injury only in the susceptible variety, showing that the varieties are managing the intracellular hydrogen peroxide levels differently. The observed increase in GOPx activity could be compensating the reduction in CAT levels in BRS Silvania RR leaves.

Overall, the susceptible soybean variety (BRS Silvania RR) showed redox gene expression changes, while the resistant variety (IAC-100) did not. The fluctuation in the activity of the redox metabolism enzymes might be associated with redox stress, which could be an indication of herbivory susceptibility. In the resistant variety, the redox genes might be constitutively expressed, so the stink bug injury might not produce enough reactive oxygen species (ROS) to elicit changed expression of these genes. Additional assessment of the redox alterations generated by herbivory across time would include measures such as protein carbonyls, lipid peroxidation and oxidized glutathione (GSSG), which are also classic indicators of redox oxidative stress, as well as other antioxidant enzymes, such as superoxide dismutases and peroxyredoxins [26].

Expression of foliar proteins

In both varieties, the foliar protein profile changed after 24 h of herbivore injury by the brown stink bug. Not all differentially expressed proteins could be identified, even when using the

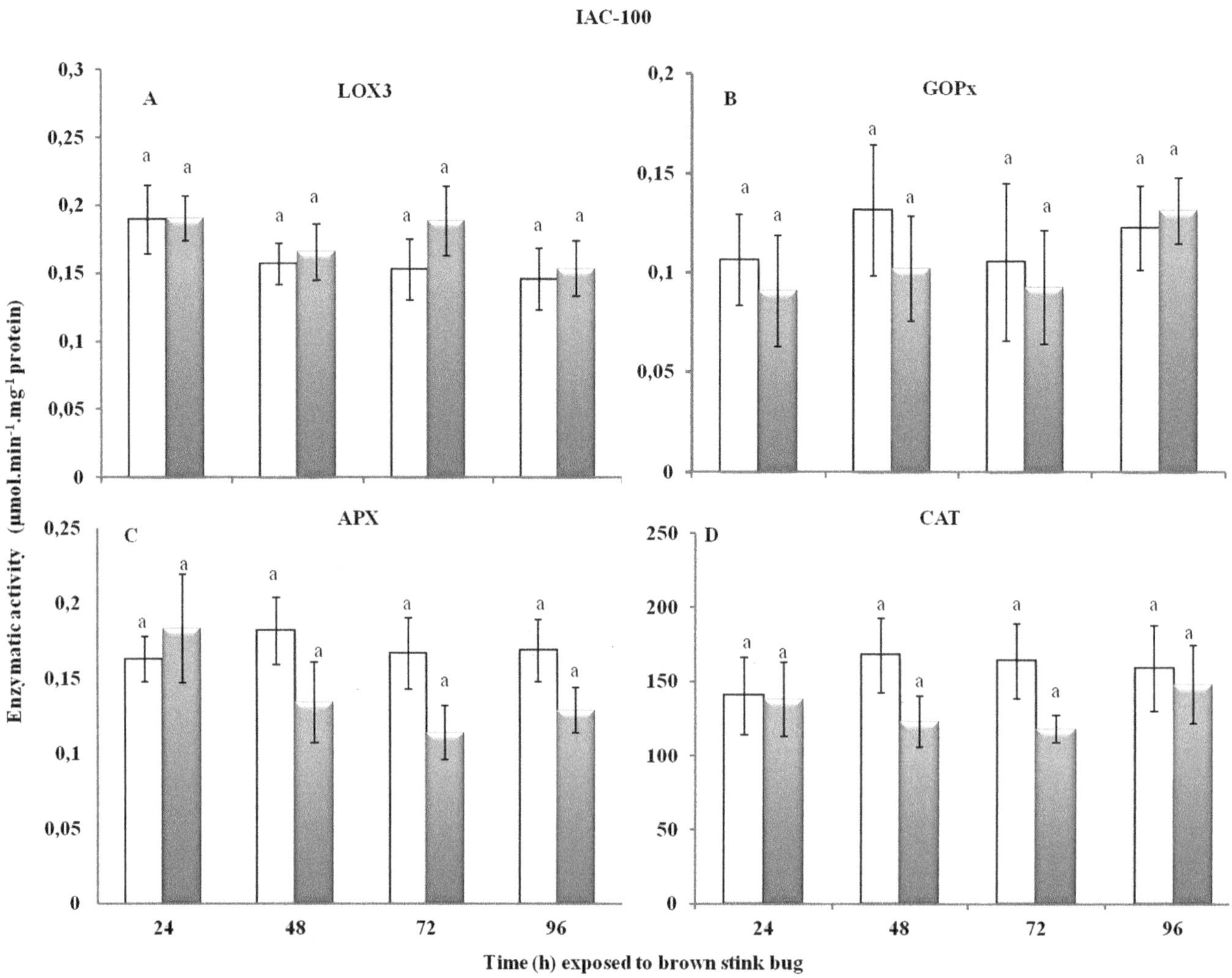

Figure 2. Enzymatic activity for (A) LOX3 ($F_{7,16}=34.52$; $p<0.0001$), (B) GOPx ($F_{7,16}=29.6$; $p<0.0001$), (C) APX ($F_{7,16}=0.16$; $p=0.9907$), and (D) CAT ($F_{7,16}=12.59$; $p<0.0001$) in leaves of IAC-100 soybean variety (V3) with and without brown stink bug herbivory injury along different sampling periods. Injured is represented in grey and not injured in white. Bars with the same letter are not significantly different by Tukey's HSD ($p<0.05$).

soybean genome reference database, possibly due to their low concentration or even absence in the database.

The 2-DE gel analyses of the BRS Silvania RR foliar protein profiles had an average of 386 protein spots per gel, while IAC-100 showed an average of 191 protein spots per gel, and most proteins were observed in the pH range from 4 to 8 and in the molecular mass range from 15 to 100 kDa (Fig. 3). The gel replicates of uninjured and injured plants had a high reproducibility: $r^2=0.95$ and 0.96, respectively for the susceptible; and $r^2=0.93$ and 0.97, respectively for the resistant.

When the foliar protein profile of the injured susceptible plants was compared to the uninjured profile, 17 proteins were differentially expressed, of which eight were up-regulated (Table 2). The identified up-regulated proteins were the chaperonin HSP20, glycine-rich 2b-like protein, both related to biotic and abiotic stress responses [41,42], and one hypothetical protein (gi|147800453).

The identified down-regulated proteins were CP4-EPSPS enzyme, a protease inhibitor, related to plant defense to herbivory, a hypothetical hydrolase function, related to plant defense against fungal infections [43], a heat-shock chaperonin 70 kDa protein (HSP), a glycine-rich RNA binding protein (GRP), related to general plant stress responses, a type 1 glutamine amidotransferase-like (GATase1) and three unknown hypothetical proteins (gi|40538952, gi|242054087, gi|147800453).

CP4-EPSPS participates in the shikimic acid pathway synthetizing the aromatic amino acids phenylalanine, tyrosine and tryptophan, which act as precursors for a wide range of products such as alkaloids, the plant hormone auxin, phytoalexins [44], flavonoids, and phenolic compounds such as lignin [45], and secondary volatile compounds such as methyl salicylate and indole, which are involved in indirect plant defense [6,20,21]. Michereff *et al.* [6] showed that the soybean variety BRS Silvania RR did not release methyl salicylate or indole when injured by the brown stink bug. Therefore the decrease of CP4-EPSPS in BRS Silvania RR might be related to its susceptibility to stink bugs [6,16]. HSPs are not exclusively related to heat stress, and some are induced by stresses such as injury, drought, and salinity [46]. In plants, GRPs are responsible for regulating cellular processes of genic regulation post-transcription, and translation [42,47], as well as cell elongation [48], protoxylem development [49,50], signal transduction [51], RNA chaperone activity [52], plant defense

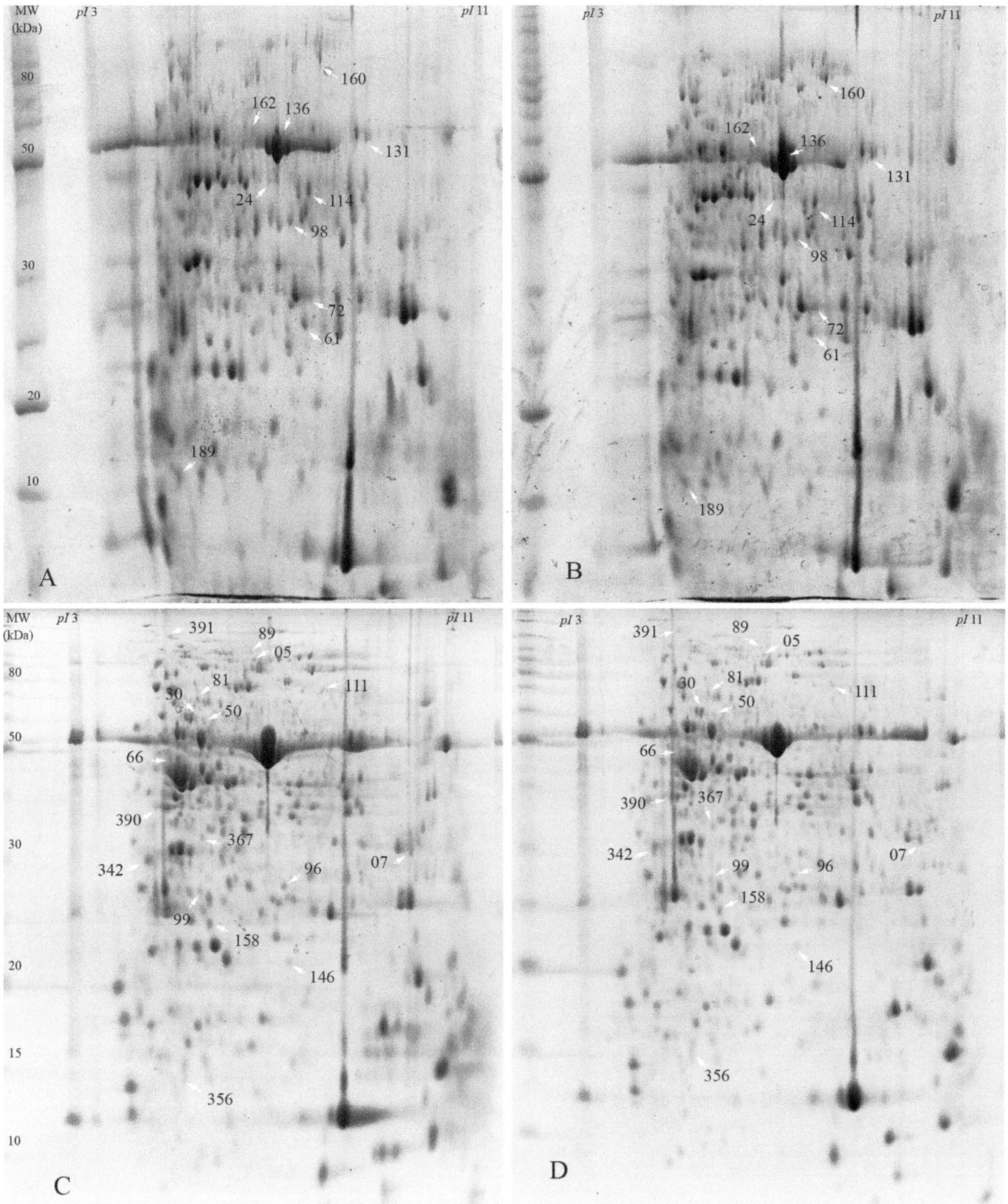

Figure 3. Foliar protein profiles (2-DE, 500 μg) of soybean varieties (V3). Variety IAC-100: (A) injured and (B) not injured, from the total average of 191 (±2) protein spots, 10 were differentially expressed and two identified (Table 2). Variety BRS Silvania RR: (C) injured and (D) not injured, from the total average of 386 (±4) protein spots, 17 were differentially expressed and 11 identified (Table 3). The differentially expressed proteins are indicated with white arrows and numbers, which correspond to identification numbers listed in the respective table.

Table 2. Identification of differentially expressed proteins by MS/MS in BRS Silvania RR soybean leaves (V3) ($p<0.05$).

Spot number	Protein identification	Accession (NCBI)	Molecular mass (kDa) theoretical/experimental	*p*I theoretical/experimental	Peptide sequence	Mascot/Blast Score	Search Type	RMS Error (ppm)	E-value	Expression level (%)
07	GATase1-like domain	gi|255639997	27.0/31.5	5.43/9	LSDFFEILATSEDR	117	MS/MS	44.40	4e-09	↓ 16.7
35*	Unknown	gi|40538952	-/57.8	-/5.6	GFIMKGATNATVDSMV	-	MS/MS	-	6e-13	↑ 30.4
66	CP4-EPSPS	gi|18266432	47.6/47.7	5.13/5	SAVLLAGLNTPGITTVIEPIMTR	57	MS/MS	12.80	2.2e-3	↓ 54.6
81*	70 kDa chaperonin-like	gi|473930207	-/64.6	-/5.6	AVITVPAYFNDAQR	-	MS/MS		1e-06	↓ 56.0
99*	STI-Soybean trypsin inhibitor	gi|351726331	19.8/26.8	-/5.5	GFPITISSPAR	-	MS/MS	-	0.003	↓ 24.0
111*	Unknown	gi|242054087	-/74.6	-/7.7	LSGIIPVHIGKLTK	-	MS/MS	-	5e-06	↓ 15.5
146	Glycine-rich protein 2b-like	gi|356538622	19.3/21.6	6.29/7.09	SLAEGESVEFAIESESDGR	132	MS/MS	11.60	1.6e-10	↑ 250.0
158	20 kDa chaperonin-like	gi|356556406	26.6/21.6	7.79/5.76	DKPSIGTVIAVGPGHLDEEGNR	146	MS/MS	0.86	3.2e-12	↑ 88.0
342	Unknown	gi|1478004532	50.0/31.0	5.43/4.7	SLLYVETADRPGLLVDLVK	90	MS/MS	4.51	1.6e-06	↓ 68.1
356	Glycine-rich RNA-binding	gi|351725469	15.8/15.0	6.58/5.3	AFSQYGEIVETK	77	MS/MS	20.90	2.9e-05	↓ 35.0
367	SGNH plant lipase-like	gi|255646252	40.0/33.4	5.83/5.6	AFFVFGDSLVDNGNNNFLATTAR	85	MS/MS	50.90	4.2e-06	↑ 58.9

* Identification by similarity search with BLASTp using peptide sequence obtained by manual *de novo* sequencing.

against pathogens [53], RNA binding activity [54], among others. Sachetto-Martins *et al.* [55] reported that the expression of GRP is regulated by a number of external stimuli, including stress by cold, water, high salinity, injury, and viral infection. The GATase1 belongs to a large group of enzymes capable of removing ammonia from glutamine and to transfer this group to other substrates to form amino acids, purine, pyrimidine nucleotides, and coenzymes [56]. Further studies are necessary to understand why these proteins were increased or decreased when the plants suffer injury by brown stink bug.

When the protein profiles of the injured and uninjured resistant plants were compared, 10 proteins had quantitative differential expression levels, of which four were increased in the injured plants. Two differentially expressed proteins were identified as ribulose bisphosphate carboxylase oxygenase (RuBisCO), which increased, and glyceraldehyde-3-phosphate dehydrogenase (GAPDH), which decreased (Table 3).

RuBisCO and GAPDH are enzymes in the Calvin cycle [57]. RuBisCO catalyzes the reaction of one carbon dioxide with ribulose-1,5-bisphosphate to form 3-phosphoglycerate (3-PGA), which in turn is catalyzed by phosphoglycerate kinase to 1,3-diphosphoglycerate. This compound, after catalysis by GAPDH produces glyceraldehyde-3-phosphate [43]. The increased expression of RuBisCO, and decreased expression of GAPDH, increases the level of the intermediate 3-phosphoglycerate. 3-Phosphoglycerate is metabolized during glycolysis into phosphoenolpyruvate, which in turn can be used in the shikimic acid pathway. Similar to what occurred in BRS Silvania RR, a protein involved in the shikimic pathway was affected by a piercing-sucking insect on IAC-100 variety. In general, sucking insects act on the shikimic acid pathway, and chewing insects act on the jasmonic acid pathway. Our results corroborate this hypothesis, because the main proteins identified here are involved in the shikimic acid pathway. However in contrast to BRS Silvania RR, the shikimic pathway was up-regulated in IAC-100. Michereff *et al.* [6] showed that IAC 100 injured by brown stink bugs produced higher amounts of methyl salicylate, and this volatile appears to be involved in the indirect defense of soybean.

Overall, the resistant variety (IAC-100) did not exhibit physiological stress to feeding injury by the brown stink bug, and seemed to shift its metabolism to produce secondary compounds for insect defense. In contrast, the susceptible variety (BRS Silvania RR) exhibited signs of physiological stress, both in its enzymes related to redox metabolism and in the protein profile response, which could affect its subsequent development in long term exposure to herbivory injury.

Conclusions

This work assessed the foliar redox response and protein expression profile of two soybean varieties (IAC-100, resistant, and BRS Silvania RR, susceptible, at stage V3) under herbivory injury by the brown stink bug. As could be expected from previous reports, the two soybean varieties responded differently at the biochemical level to herbivory by the brown stink bug. The susceptible variety BRS Silvania RR showed adaptive changes in the activity of enzymes related to redox metabolism and in the expression of stress-responsive proteins, probably to minimize biochemical damage caused by herbivory injury. On the other hand, in the resistant variety, the redox genes could be constitutively expressed, and the stink bug injury might not produce enough ROS to elicit changed expression of these genes. Overall, the susceptible variety responded to herbivory in ways that could compromise its development in the long term, while the

Table 3. Identification of differentially expressed proteins (by MS/MS) in IAC-100 soybean leaves (V3) ($p<0.05$).

Spot number	Protein identification	Accession (NCBI')	Molecular mass (kDa) theoretical/experimental	pI theoretical/experimental	Peptide sequence	Mascot Score	Search Type	RMS Error (ppm)	E-value	Expression level (%)
114	GAPDH	gi|66026	36.0/42.7	6.66/7.2	K.VIAWYDNEWGYSQR.V	64	MS/MS	61.7	8.4e-4	↓ 32.0
136	RuBisCO large subunit	gi|131898	52.0/53.9	6.23/6.24	K.TFQGPPHGIQVER.D	107	MS/MS	11.0	9.1e-08	↑ 33.0

resistant variety IAC-100 showed a defense response oriented to circumvent the biological stressor. Both varieties were injured by a piercing-sucking insect and the proteins regulated by this stress were linked to the shikimic acid pathway corroborating the hypothesis that sucking insects induce the shikimic acid route of defense more than the jasmonic acid route defense.

Acknowledgments

We would like to thank: Dra. Clara Beatriz Hoffmann-Campo for providing the soybean seeds; Dr. Miguel Borges and Dr. Raul A. Laumann for providing the brown stink bugs; Msc. Démerson André Polli for statistics consulting; MSc. Gabriela Villeth and Dr. Lilian Travassos for supporting the proteomic experiments and analysis; Dr. David A. Andow for English revisions.

Author Contributions

Conceived and designed the experiments: DPP RVT MCBM. Performed the experiments: RVT LPS. Analyzed the data: DPP MHL LPS RVT MCBM AM. Contributed reagents/materials/analysis tools: MHL DPP. Wrote the paper: DPP MHL LPS RVT MCBM AM.

References

1. Corrêa-Ferreira BS, Panizzi AR (1999) Percevejos da soja e seu manejo. Londrina: Embrapa - CNPSo, Circular Técnica 24.
2. Fernandes AP (2000) Análise comparativa da espermogênese normal e diapáusica em percevejos fitófagos (Hemiptera, Pentatomidae) – Revisão Universitas. Biociências (UNICEUB). Brasília-Brasil. 1(1): 87–95.
3. Todd JW, Herzog DC (1980) Sampling phytophagous Pentatomidae on soybean. In: Kogan M, Herzog D C (Ed) Sampling.
4. Companhia Nacional de Abastecimento (CONAB), Acompanhamento da Safra Brasileira de Grãos (2012) http://www.conab.gov.br/OlalaCMS/uploads/arquivos/12_10_17_16_09_58_boletim_graos_-_julho_2012.pdf.
5. Vieira CR, Moraes MCB, Borges M, Sujii ER, Laumann RA (2013) *cis*-Jasmone indirect action on egg parasitoids (Hymenoptera: Scelionidae) and its application in biological control of soybean stink bugs (Hemiptera: Pentatomidae). Biol. Control 64: 75–82.
6. Michereff MFF, Laumann RA, Borges RAM, Michereff-Filho M, Diniz IR, et al. (2011) Volatiles mediating a plant-herbivore-natural enemy interaction in resistant and susceptible soybean cultivars. J. Chem. Ecol. 37: 273–285.
7. Moraes MCB, Laumann RA, Pareja M, Sereno FTPS, Michereff MFF, et al. (2009) Attraction of the stink bug egg parasitoid *Telenomus podisi* to defence signals from soybean activated by treatment with *cis*-jasmone. Entomol. Exp. Appl. 131: 178–188.
8. Vieira SS, Bueno AF, Boff MIC, Bueno RC, Hoffman-Campo CB (2011) Resistance of soybean genotypes to *Bemisia tabaci* (Genn.) biotype B. Neo. Entomol. (Impress) 40: 117–122.
9. Dicke M, Hilker M (2003) Induced plants defence: from molecular biology to evolutionary ecology. Basic Appl. Ecol. 4: 3–14.
10. Courtois EA, Paine CE, Blandinieres PA, Stien D, Bessiere JM, et al. (2009) Diversity of the volatile organic compounds emitted by 55 species of tropical trees: a survey in French Guiana. J. Chem. Ecol. 35: 1349–1362.
11. Colazza S, Wajnberg E (2013) Chemical ecology of insect parasitoids: towards a new era. In: Wajnberg E, Colazza S, Chemical Ecology of Insect Parasitoids, 1st Edition, John Wiley & Sons (2013) 1–8.
12. McPherson RM, Buss GR (2007) Evaluation lepidopteran defoliation resistance in soybean breeding lines containing the stink bug (Hemiptera: Pentatomidae) resistance IAC-100 cultivar in their pedigrees. J. Econ. Entomol. 100: 962–968.
13. Panizzi AR, Mcpherson JE, James JG, Javabery M, McPherson RM (2000) Stink bugs (Pentatomidae). In: Schaefer CW, Panizzi AR (Eds.), Heteroptera of economic importance. Boca Raton: Florida, USA: CRC 432–434 pp.
14. Corrêa-Ferreira BS, Azevedo J (2002) Soybean seed damage by different species of stink bugs. Agr. Forest Entomol. 4: 145–150.
15. Bi JL, Murphy JB, Felton GW (1997) Antinutritive and oxidative components as mechanisms of induced resistance in cotton to *Helicoverpa zea*. J. Chem. Ecol. 23: 97–117.
16. Laumann RA, Farias Neto AL, Moraes MCB, Silva AP, Vieira CR, et al. (2008) Dinâmica populacional de percevejos (Hemiptera: Pentatomidae) em diferentes genótipos de soja. In: IX Simpósio Nacional Cerrado; II Simpósio Internacional Savanas Tropicais, Brasília, DF. Desafios e estratégias para o equilíbrio entre sociedade, agronegócio e recursos naturais: Anais. Planaltina, DF: Embrapa Cerrados, HTTP://www.cpac.embrapa.br/download/747t.
17. Rostas M, Eggert K (2008) Ontogenetic and spatio-temporal patterns of induced volatiles in *Glycine max* in the light of the optimal defense hypothesis. Chemoecol. 18: 29–38.
18. Moraes MCB, Laumann RA, Sujii ES, Pires CSS, Borges M (2005) Induced volatiles in soybean and pigeon pea plants artificially infested with the Neotropical brown stink bug, *Euschistus heros*, and their effect on the egg parasitoid, *Telenomus podisi*. Entomol. Exp. Appl. 115: 227–237.
19. Veiga RFA, Rossetto CJ, Razera LF, Galo PB, Bortoleto M (1999) Caracterização morfológica e agronômica da cultivar de soja 'IAC-100'. Instituto Agronômico de Campinas. Boletim Técnico, 175. 25p.
20. Gris CF, Pinho EVRV, Carvalho MLM, Diniz RP, Andrade T (2013) Lignificação da planta e qualidade de sementes de soja RR pulverizadas com herbicida glifosato. Rev. Ceres. Viçosa 60(2): 286–292.
21. Moraes MCB, Pareja M, Laumann RA, Hoffmann-Campo CB, Borges M (2008) Response of the parasitoid to induced volatiles from soybean damaged by stink bug herbivory and oviposition. J. Plant Interact. 3: 111–118.
22. Bradford MM (1976) A rapid and sensitive method for quantification of microgram quantities of protein utilizing the principle of protein-dye binding. Anal. Biochem. 72: 248–254.
23. Axelrod BT, Cheesbrough TM, Laakso S (1981) Lipoxygenase from soybeans. Methods Enzymol. 71: 441–445.
24. Chen Y, Ni X, Cottrell E, Wood BW, Buntin GD (2009) Changes of oxidase and hydrolase activities in pecan leaves elicited by black pecan aphid (Hemiptera: Aphididae) feeding, J. Econ. Entomol. 102 (3): 1262–1269.
25. Anderson JV, Chevone BI, Hess JL (1992) Seasonal variation in the antioxidant system of Eastern white pine needles. Plant Physiol. 98: 501–508.
26. Hermes-Lima M, Carreiro C, Moreira DC, Polcheira C, Machado DP, et al. (2012) Glutathione status and antioxidant enzymes in a crocodilian species from the swamps of the Brazilian Pantanal. Comp. Biochem. Physiol. Part A. 163: 189–198.
27. Méchin V, Damerval C, Zivy M (2007) Total protein extraction with TCA/acetone. In: Thiellement H, Zivy M, Damerval M, Méchin V (2007) Plant proteomics methods and protocols. Totowa, New Jersey, John M. Walker, 1–8.
28. Wildgruber R, Reil G, Drews O, Parlar H, Görg A (2002) Web-based two dimensional database of *Saccharomyces cerevisae* proteins using immobilized pH gradients from pH 6 to pH 12 and matrix-assisted laser desorption/ionization-time of flight mass spectrometry. Proteomics 2: 727–732.
29. Weiss W, Gorg A (2007) Two-dimensional electrophoresis for plant proteomics. Methods Mol. Biol. 355: 1421–143.
30. Farmer EF, Johnson RR, Ryan CA (1992) Regulation of expression of proteinase inhibitor genes by methyl jasmonate and jasmonic acid. Plant Physiol. 98: 995–1002.
31. Prost I, Dhondt S, Rothe G, Vicente J, Rodriguez MJ, et al. (2005) Evaluation of the antimicrobial activities of plant oxylipins supports their involvement in defense pathogens. Plant Physiol. 139: 1902–1913.
32. Ishige F, Mori H, Yamazaki K, Imaseki H (1993) Identification of a basic glycoprotein induced by ethylene in primary leaves of azuki bean as cationic peroxidase. Plant Physiol. 101: 193–199.
33. Ni X, Quisenberry SS, Heng-Moss T, Markwell J, Sarath G, et al. (2001) Oxidative responses of resistant and susceptible cereal leaves to symptomatic and nonsymptomatic cereal aphid (Hemiptera: Aphididae) feeding. Faculty Publications: Department of Entomology. Paper 113. http://digitalcommons.unl.edu/entomologyfacpub/113.
34. Devi PUM, Reddy PS, Rani NRU, Reddy KJ, Reddy MN, et al. (2000) Lipoxygenase metabolites of alpha-linolenic acid in the development of resistance in pigeonpea, *Cajanus cajan* (L.) Millsp, seedlings against *Fusarium udum* infection. European J. Plant Pathol. 106: 857–865.
35. Pérez FJ, Villegas D, Mejia N (2002) Ascorbic acid and flavonoid-peroxidase reactions as a detoxifying system of H_2O_2 in grapevine leaves. Phytochemistry 60: 573–580.
36. Bi L, Felton GW (1995) Foliar oxidative stress and insect herbivory: primary compounds, secondary metabolites, and reactive oxygen species as components of induced resistance. J. Chem. Ecol. 21: 1511–1530.
37. Ruuhola T, Yang S (2005) Wound-induced oxidative responses in mountain birch leaves. Ann. Botany 97: 29–37.
38. Aebi H (1984) Catalase *in vitro*. B. Chance. Act Chem. Scand. 1: 121–126.
39. Karuppanapandian T, Moon J, Kim C, Manoharan K, Kim W (2011) Reactive oxygen species in plants: their generation, signal transduction, and scavenging mechanisms. Australian J. Crop Sci. 5 (6): 709–725.
40. Gomez SK, Oosterhuis DM, Rajguru SN, Johnson DR (2004) Foliar antioxidant enzyme responses in cotton after aphid herbivory. J. Cotton Sci. 8: 99–104.
41. Li D-C, Yang F, Lu B, Chen DF, Yang WJ (2012) Thermotolerance and molecular chaperone function of the small heat shock protein HSP20 from hyperthermophilic aschaeon, *Sulfolobus solfataricus* P2. Cell Stress and Chaperon. 17: 103–108.
42. Mangeon A, Junqueira RM, Sachetto-Martins G (2010) Functional diversity of the plant glycine-rich proteins superfamily. Plant Signaling Behav. 5 (2): 99–104.
43. Taiz L, Zeiger E (2006) Secondary metabolites and plant defense. In: Taiz L, Zeiger E (Eds.) Plant Physiology, Sinauer Associates Inc., Sunderland, MA, 4th edition. 283–306 pp.
44. Kutchan TM (1995) Alkaloid biosynthesis: the basis for metabolic engineering of medicinal plants. Plant Cell 7: 1059–1070.
45. Vogt T (2010) Phenylpropanoid biosynthesis. Mol. Plant. 3: 2–20.

46. Lindquist S, Craig EA (1998) The heat-shock proteins. Annu. Rev. Genet. 22: 631–677.
47. Burd CG, Dreyfuss G (1994) Conserved structures and diversity of functions of RNA-binding proteins. Science 265: 615–21.
48. Mangeon A, Magioli C, Menezes-Salgueiro AD, Cardeal V, de Oliveira C, et al. (2009) AtGRP5, a vacuole-located glycine-rich protein involved in cell elongation. Planta 230: 253–265.
49. Ringle C, Hauf G, Keller B (2001) Hydrophobic interactions of the structural protein GRP1.8 in the cell wall of protoxylem elements. Plant Physiol. 125: 673–82.
50. Ryser U, Schorderet M, Guyot R, Keller B (2004) A new structural element containing glycine-rich proteins and rhamnogalacturonan I in the protoxylem of seed plants. J. Cell Sci. 117: 1179–90.
51. Park AR, Cho SK, Yun SJ, Jin MY, Lee SH, et al. (2001) Interaction of the arabidopsis receptor protein kinase Wak1 with a glycine-rich protein, AtGRP-3. J. Biol. Chem. 276: 26688–26693.
52. Kim JY, Park SJ, Jang B, Jung CH, Ahn SJ, et al. (2007) Functional characterization of a glycine-rich RNA-binding protein 2 in *Arabidopsis thaliana* under abiotic stress conditions. Plant J. 50: 439–451.
53. Fu ZQ, Guo M, Jeong BR, Tian F, Elthon TE, et al. (2007) A type III effector ADP-ribosylates RNA-binding proteins and quells plant immunity. Nature 447: 284–288.
54. Kim YO, Kim JS, Kang H (2005) Cold-inducible zinc finger-containing glycine-rich RNA-binding protein contributes to the enhancement of freezing tolerance in *Arabidopsis thaliana*. Plant J. 42: 890–900.
55. Sachetto-Martins G, Franco LO, de Oliveira DE (2000) Plant glycine-rich proteins: a family or just proteins with a common motif? Biochim Biophys Acta. 1492: 1–14.
56. Mäntsälä P, Zlakin H (1984) Replacement of the active-site cysteine in glutamine phosphoribosylpyrophosphate amidotranferase by site-directed mutagenesis. J. Soc. Biol. Chem. 259: 14230–14236.
57. Brynat DA, Friggard NU (2006) Porkaryotic photosynthesis and phototrophy illuminate. Trends Microbiol. 14: 488–496.

8

Phenotypic Plasticity of HSP70s Gene Expression during Diapause: Signs of Evolutionary Responses to Cold Stress among Soybean Pod Borer Populations (*Leguminivora glycinivorella*) in Northeast of China

Ling Wang[¶], **Shuai Yang**[¶], **Lanlan Han, Dong Fan, Kuijun Zhao***

College of Agriculture, Northeast Agricultural University, Harbin, China

Abstract

The soybean pod borer (*Leguminivora glycinivorella* Matsumura) successfully survives the winter because of its high expression of 70-kDa heat shock proteins (HSP70s) during its overwintering diapause. The amount of HSP70s is different under different environmental stresses. In this study, inducible heat shock protein 70 and its constitutive heat shock cognate 70 were cloned by RT-PCR and RACE. These genes were named *Lg-hsp70* and *Lg-hsc70*, respectively. Gene transcription and protein expression after cold stress treatment (5°C to −5°C) were analyzed by western blotting and by qRT-PCR for four populations that were sampled in the northeast region of China, including Shenyang, Gongzhuling, Harbin and Heihe, when the soybean pod borer was in diapause. As the cold shock temperature decreased, the levels of Lg-HSP70s were significantly up-regulated. The amount of cold-induced Lg-HSP70s was highest in the southernmost population (Shenyang, 41°50′N) and lowest in the northernmost population (Heihe, 50°22′N). These results support the hypothesis that the soybean pod borer in the northeast region of China displays phenotypic plasticity, and the accumulation of Lg-HSP70s is a strategy for overcoming environmental stress. These results also suggest that the induction of HSP70 synthesis, which is a complex physiological adaptation, can evolve quickly and inherit stability.

Editor: Sue Cotterill, St. Georges University of London, United Kingdom

Funding: This work is supported by the Special Fund for Modern Agro-industry Technology Research System (CARS-04) and the Fund of Common Wealth Industry (agriculture) Special Research (201103002). The funders had no role in study design, data collection and analysis, decision to publish, or preparation of the manuscript.

Competing Interests: The authors have declared that no competing interests exist.

* Email: kjzhao@neau.edu.cn

¶ These authors contributed equally to this work.

Introduction

Diapause is a strategy by which insects can escape death and gradually adapt to the environment in response to adverse environmental conditions [1]. Studying the molecular mechanisms of diapause is of great significance because of the important role of diapause in insect survival. Many studies have shown that environmental stress is harmful to insects, even those exposed to sub-zero temperatures over a short period, and that a strong relation between insect death and protein denaturation exists [2]. The expression of HSP70s (70-kDa Heat Shock Proteins) is a normal strategy used by insects to avoid damage arising from environmental stress [3,4].

As molecular chaperones, HSP70s can recover the normal structure of denatured proteins to repair and protect the cell as well as increase the tolerance and resistance of insects [4,5,6]. HSP70s have a conserved structure and prevail in living organisms [7]. Depending on the expression style, these proteins are classified as inducible HSP70 (70-kDa Heat shock protein,HSP70) or constitutive HSC70 (Heat shock cognate protein, HSC70) [8,9]. In general, the levels of HSP70 are low in normal conditions and increase upon environmental stress, such as high or low temperatures and exposure to heavy metals or microbial infection. HSC70 is stably expressed and only slightly increases upon environmental stress [10].

The expression level of HSP70s can be up-regulated by low temperature during insect diapause. This up-regulation has been observed in *Chilo suppressalis* and *Sesamia nonagrioides*, which belong to Lepidoptera [11,12], in *Drosophila triauraria* and *Drosophila melanogaster*, which belong to Diptera [13,14], and in *Leptinotarsa decemlineata*, which belongs to Coleoptera [15]. RNAi was used to study the effect of HSP70 expression on cold-stress resistance. Similar to conditions when HSP70 expression

Figure 1. Collection sites of the soybean pod borer Northeast region map of China.

was inhibited, *Pyrrhocoris apterus* has a less intense response to cold resistance in diapause, and the pupae of *Sarcophaga crassipalpis* have feeble cold resistance [4,16]. Therefore, we hypothesize that high HSP70 expression plays an important role in increasing insect resistance to cold stress.

HSP70s are an example of adaptive phenotypic plasticity providing inducible tolerance to the environment [17,18,19]. Phenotypic plasticity allows insects to react to different environments [20,21]. This characteristic prevails in different organisms and represents an important aspect of adaptive evolution [22,23,24]. Phenotypic plasticity does not only increase cold stress resistance but also allows it to be stably inherited, playing an important role in insect evolution [25,26,27]. Because *hsp70* is a plasticity gene, understanding the function of *hsp70s* could supply a good theoretical foundation for understanding insect adaptability to the environment [28,29,30]. To date, the homologous genes of *hsp70s* in relation to phenotypic plasticity have been studied in *Rhagoletis cerasi*, *Calliphora vicina*, *Lymantria dispar* and *Melitaea cinxia* [19,31,32,33]. However, the variation between conspecifics of different geographical populations has not been widely assessed [34].

The soybean pod borer (*Leguminivora glycinivorella* Matsumura) belongs to Lepidoptera: Olethreutidae and is one of the most important soybean pests in the world. Only one generation of *L. glycinivorella* lives in the northeast region of China, which is the main soybean production region, and its diapause stage lasts from October to July [35,36]. *L. glycinivorella* can cause severe losses in soybean yield and quality; the loss rate is from 10% to 30% in a general year and up to 50% for more severe situations in a typical year [37,38]. To date, research concerning *L. glycinivorella* has primarily focused on environmental ecology and physiology and has not focused on the genetic regulation of the diapause stage. Because a significant difference in climatic conditions exists in the vast, northeast region of China, we presumed that phenotypic plasticity occurred in different populations of *L. glycinivorella* in response to cold stress. In this study, we cloned *Lg-hsp70s* and analyzed their expression levels under cold stress induction (5°C to −5°C) to establish a foundation for understanding the molecular regulation of cold resistance and of the ecological adaptability of different populations of *L. glycinivorella*.

Materials and Methods

Materials

Insect collection sites: The soybean pod borer originated from four different populations: Harbin (Heilongjiang Province, 45°45′N, 126°41′E), Heihe (Heilongjiang Province, 50°22′N, 127°53′E), Gongzhuling (Jilin Province, 43°11′N, 124°02′E) and Shenyang (Liaoning Province, 41°50′N, 123°24′E). The total distance between the south and north populations is approxi-

Table 1. Minimum and maximum mean air temperatures and their average from the nearest weather stations to the collection location for the soybean pod borer.

Collection site	Coordinates	October-June (°C)*		
		Min	Max	Mean
Heihe	50°22′N, 127°53′E	−22.6	16.2	−5.2
Harbin	45°45′N, 126°41′E	−19.4	18.9	−1.7
Gongzhuling	43°11′N, 124°02′E	−15.3	20.4	1.6
Shenyang	41°50′N, 123°24′E	−12.0	21.5	3.1

Note: * represents the air temperatures. Overwinter temperatures below ground typically correlate with air temperatures but are slightly higher (Xu *et al*, 1965).

mately 1182 km, and the linear distance from the north to south was approximately 700 km (Figure 1). For our study, we assumed that there was geographical isolation between the different collection sites. The average monthly temperature and the temperature range during diapause are shown in Table 1.

Insect feeding: The adult insects were harvested in soybean fields in late July and reared in a glass on fresh soybean pods, with 20 adults in one glass. The female:male ratio was 1:1. The glass was placed in a RXZ climate chamber (Ningbo, Jiangsu, China) at 25°C, with a light:dark cycle of 14 h:10 h. The adults were removed after 12 h, and the eggs were harvested and transferred into a new glass with a soybean pod. After hatching, the larvae were reared to the fourth instar stage in the soybean pod and then harvested in the soybean pod. To induce diapause, the samples were placed in a sterilized soil box in a climate chamber with a mean temperature of 17°C. The temperature was 20°C during the 14-h photophase and 13°C during the 6 h of darkness. The lights were dimmed gradually over 2 h before and after the photophase, imitating sunrise and sunset. After the larvae became cocoons, we transferred the cocoons into a 5°C chamber, where the cocoons overwintered [39]. To diminish the impact of maternal effects, the fourth instar larvae used in the experiment were the second laboratory generation that was collected during the second consecutive year. To test the levels of the Lg-HSP70s, one set consisted of individuals that hatched from eggs collected during the same year that the experiment was performed.

Methods

Cold shock treatment. The experiment began when the fourth instar larvae of the Heihe population were in the diapause

Table 2. Primers used for the experiments in this study.

Primer type	Sequences of primers (5′-3′)	Use of primers
P1	GGCARGCCCHACVAARGATG	PCR for *Lg-hsp70*
P2	GTTGTCYTTBGTCATVGCTC	
P3	GAGGGMRTCGACTTCTAYACBTCCATCAC	PCR for Lg-hsc70
P4	CTTGTACTTCTCTGCNTCYTGBACCAT	
P5	GCGCACCTTTGCTGAGTTACTCTAC	For full-length *Lg-hsp70*
P6	CTGTCTTGTCTGAACGGAATGTTC	
P7	CTAAGCTGTTCATCTACTAGTTAAAACTACGACTGAAG	For full-length *Lg-hsc70*
P8	GATGGTGATTGCGTATGGAATGTTTAG	
5′PGSP1	CTCCTGCGGCGGACTTCACTTCG	5′ RACE for *Lg-hsp70*
5′PGSP2	TCTTTTGTGGCCTGTCTTTGTGAGTCGTTGAAGT	
5′CGSP1	GTTCACCATACGGTTGTCGAAGTC	5′ RACE for *Lg-hsc70*
5′CGSP2	GACACATCGAAGGTACCGCCGCCGAGATCG	
3′PGSP1	AGGTGCAGGACCTGCTGCTGCTCGA	3′ RACE for *Lghsp70*
3′PGSP2	ATCAACCCCGACGAGGCCGTCGCTTA	
3′CGSP1	CGGCAAGTTCGAGCTCACCGGCATC	3′ RACE for *Lg-hsc70*
3′CGSP2	GAAGGAGACCATCCAGGCCAAG	
hsp70F	CGTGTCCATCCTGACCATCG	qRT-PCR for *Lg-hsp70*
*hsp70*R	CAGTGCGCAGACGACGGAG	
hsc70F	GTCCAGGAGTTCAAGAGGAAGTACAAG	qRT-PCR for *Lg-hsc70*
*hsc70*R	GAAGAGAGAGTCGATTTCGATGCTAG	
*β-actin*F	GGCGACATAGCACAGCTTCTC	qRT-PCR for *β-actin*
*β-actin*R	ATCCTCCGTCTGGACTTGGC	

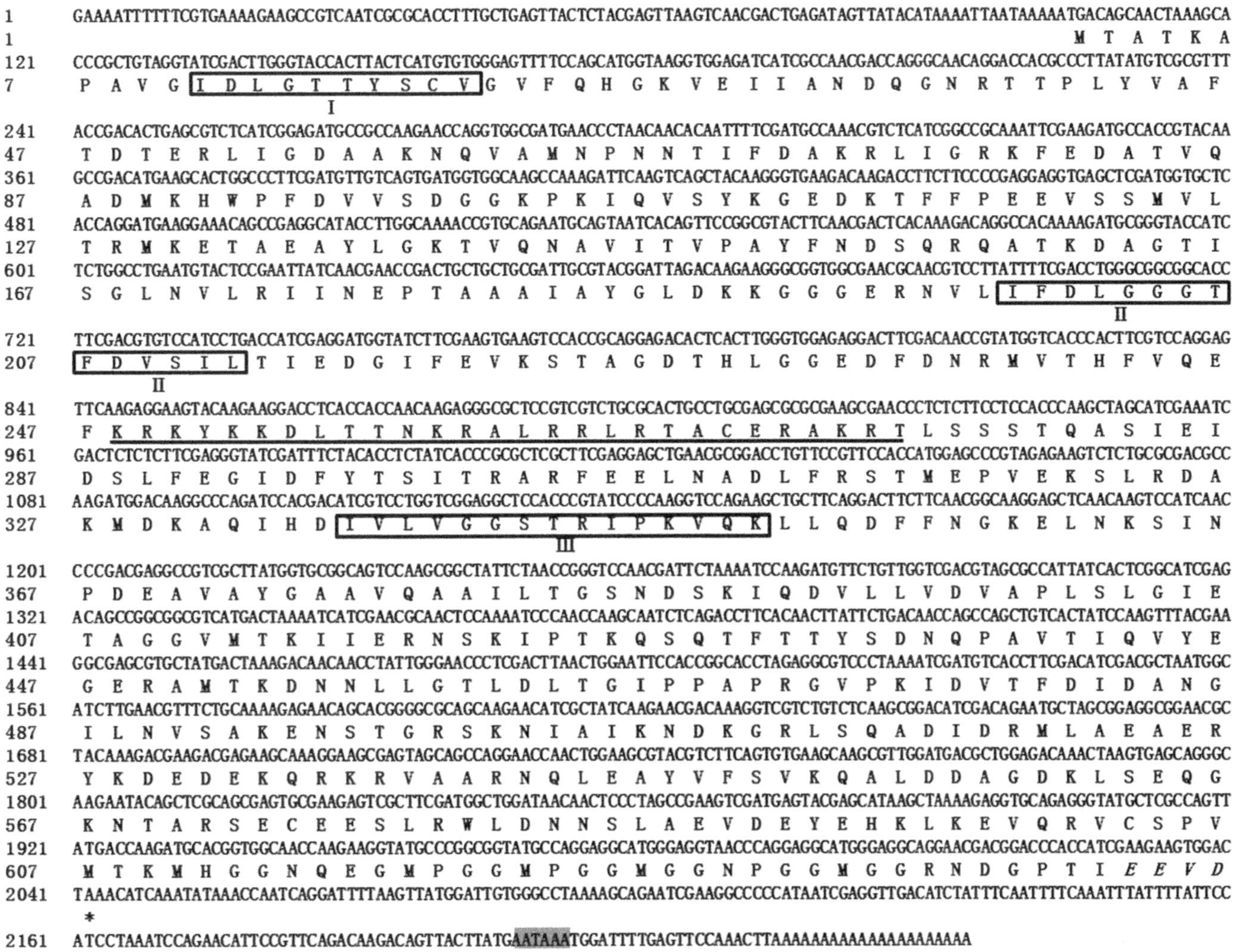

Figure 2. Nucleotide and deduced amino acid sequences of *Lg-hsp70*. The upper sequences indicate the nucleotides, and the lower sequences show the inferred amino acids. Start and stop codons are in bold. The signature sequences of the HSP70s family are shown in boxes, the nuclear localization signal sequence is underlined, the consensus sequence EEVD at the C-terminus is indicated in italics and the consensus polyadenylation signal (AATAAA) is shown in a gray box.

midpoint (150 days). For the cold shock treatment, all fourth instar larvae were exposed to 5°C to −5°C for 1.5 h and then recovered at room temperature (25°C) for 1.5 h (none of the individuals died). Then, the larvae were frozen in liquid nitrogen and stored at −80°C. To determine the constitutive level of HSP70s, the control was a fourth instar larvae maintained at 25°C.

Cloning of *Lg-hsp70* and of *Lg-hsc70*. Total RNA isolation and first cDNA strand synthesis: RNA was extracted using an RNA isolation kit (Omega, USA), and the concentration and purity were determined using a NanoDrop spectrophotometer (NanoDrop 8000, Parallels, USA) and agarose gel electrophoresis. Then, first-strand cDNA synthesis was performed using a reverse transcription kit (ToYoBo, Japan).

Primer design and gene cloning: Based on the sequences of *hsp70* and *hsc70* from other insects in the GenBank database, degenerate primers for the conserved region of the gene family were designed using Primer Premier 5.0 software. The primer sequences are listed in Table 2. The PCR reaction for the conserved region was as follows: 3 min at 94°C; 35 cycles of 30 s at 94°C, 30 s at 57.8°C for *Lg-hsp70* or 30 s at 58.0°C for *Lg-hsc70*, 1 min at 72°C; and 10 min at 72°C. Then, the amplified fragment was isolated on a 1.0% agarose gel and ligated into the pMD18-T vector for sequencing. Based on the sequenced fragments of *Lg-hsp70* and *Lg-hsc70*, 5′ RACE and 3′ RACE primers were designed (see Table 2) and used following the manufacturer's protocol for the RACE kits. Full-length *Lg-hsp70* and *Lg-hsc70* were amplified using the same procedure and used for the following experiment.

Bioinformatics software: The full-length sequences of *Lg-hsp70* and *Lg-hsc70* were analyzed using DNAStar software. The identity analysis was performed using the BLAST search program, which is available on the NCBI website (http://www.ncbi.nlm.gov/BLAST/). Multiple sequence alignment and phylogenetic analysis were performed using the ClustalW program (http://www.ebiac.uk/clustalw/) and the MEGA 4.1 software with the neighbor-joining method. The PSORT II program (http://psort.hgc.jp/form2.html) was used to calculate the nuclear localization signal fragment. The analyses of protein domains were performed using ProtParam (http://www.cn.expasy.org/tolls/ProtParam), ScanProsite (http://prosite.expasy.org/scanprosite/) and Swiss-Model (http://www.swissmodel.expasy.org/SWISS-MODEL.html) software tools.

```
1     GAAAATTATTTCAAGAAAAGAAACCGTCAATCGCGCACGTTTGCTAAGCTGTTCATCTACTAGTTAAAACTACGACTGAAGTATTAGTTAAATTTATACAAAATGGCCAAGGCACCAGCG
1                                                                                                           M  A  K  A  P  A
121   GTAGGAATCGATTTGGGCACCACGTACTCTTGCGTGGGTGTGTTCCAGCATGGAAAGGTGGAAATCATCGCCAACGACCAGGGCAACAGGACTACGCCCTCGTACGTGGCCTTCACCGAC
7     V  G [I  D  L  G  T  T  Y  S  C  V] G  V  F  Q  H  G  K  V  E  I  I  A  N  D  Q  G  N  R  T  T  P  S  Y  V  A  F  T  D
                       I
241   ACCGAGCGTCTCATCGGAGATGCCGCCAAGAACCAGGTGGCGATGAACCCCAACAACACTATTTTCGATGCCAAACGACTCATCGGCCGCAAGTTCGAAGATGCCACAGTACAGTCCGAC
47    T  E  R  L  I  G  D  A  A  K  N  Q  V  A  M  N  P  N  N  T  I  F  D  A  K  R  L  I  G  R  K  F  E  D  A  T  V  Q  S  D
361   ATGAAGCACTGGCCCTTCGATGTTGTCAGTGATGGTGGCAAGCCAAAGACTCAAGTCAGCTACAAGGGTGAAAACAAGACCTTCTTCCCCGAGGAGGTGAGCTCGATGGTGCTCACCAAG
87    M  K  H  W  P  F  D  V  V  S  D  G  G  K  P  K  T  Q  V  S  Y  K  G  E  N  K  T  F  F  P  E  E  V  S  S  M  V  L  T  K
481   ATGAAGGAAACAGCCGAGGCATACCTTGGCAAAACCGTGCAGAATGCAGTAATCACAGTTCCGGCGTACTTCAACGACTCACAAAGACAGGCCACAAAAGATGCGGGTACCATCTCTGGC
127   M  K  E  T  A  E  A  Y  L  G  K  T  V  Q  N  A  V  I  T  V  P  A  Y  F  N  D  S  Q  R  Q  A  T  K  D  A  G  T  I  S  G
601   CTGAATGTACTCCGAATTATCAACGAACCGACTGCTGCTGCGATTGCGTACGGATTAGACAAGAAGGGCGGTGGCGAACGCAACGTCCTTATTTTCGATCTCGGCGGCGGTACCTTCGAT
167   L  N  V  L  R  I  I  N  E  P  T  A  A  A  I  A  Y  G  L  D  K  K  G  G  G  E  R  N  V  L  I  F  D  L  G  G  G  T  F  D
721   GTGTCGATTGCGTACGGCCTTGACAAGAAGGGCGGTGGCGAACGCAACGTCCTTATTTTCGACCTGGGCGGCGGCACCTTCGACGTGTCCATCCTGACCATCGAGGATGGTATCTTCGAA
207   V  S  I  A  Y  G  L  D  K  K  G  G  G  E  R  N  V  L [I  F  D  L  G  G  G  T  F  D  V  S  I  L] T  I  E  D  G  I  F  E
                                                                       II
841   GTGAAGTCCACCGCAGGAGACACTCACTTGGGTGGAGAGGACTTCGACAACCGTATGGTCACCCACTTCGTCCAGGAGTTCAAGAGGAAGTACAAGAAGGACCTCACCACCAACAAGAGG
247   V  K  S  T  A  G  D  T  H  L  G  G  E  D  F  D  N  R  M  V  T  H  F  V  Q  E  F  K  R  K  Y  K  K  D  L  T  T  N  K  R
961   GCGCTCCGTCGTCTGCGCACCGCCTGCGAGCGCGCGAAGCGAACCCTCTCTTCCTCCACCCAAGCTAGCATCGAAATCGACTCTCTCTTCGAGGGTATCGATTTCTACACCTCTATCACC
287   A  L  R  R  L  R  T  A  C  E  R  A  K  R  T  L  S  S  S  T  Q  A  S  I  E  I  D  S  L  F  E  G  I  D  F  Y  T  S  I  T
1081  CGCGCTCGCTTCGAGGAGCTGAATGCGGACCTGTTCCGTTCCACCATGGAGCCCGTAGAGAAGTCTCTGCGCGACGCCAAGATGGACAAGGCCCAGATCCACGACATCGTCCTGGTCGGA
327   R  A  R  F  E  E  L  N  A  D  L  F  R  S  T  M  E  P  V  E  K  S  L  R  D  A  K  M  D  K  A  Q  I  H  D [I  V  L  V  G]
                                                                                                                     III
1201  GGCTCCACCCGTATCCCCAAGGTCCAGAAGCTGCTTCAGGACTTCTTCAACGGCAAGGAGCTCAACAAGTCCATCAACCCCGACGAGGCCGTCGCTTACGGCGCCGCCGTGCAGGCCGCC
367  [G  S  T  R  I  P  K  V  Q  K] L  L  Q  D  F  F  N  G  K  E  L  N  K  S  I  N  P  D  E  A  V  A  Y  G  A  A  V  Q  A  A
                III
1321  ATCCTCCACGGAGACAAGTCCGAGGAGGTGCAGGACCTGCTGCTGCTCGACGTCACCCCGCTGTCGCTCGGTATCGAGACCGCCGGCGGCGTCATGACCACGCTCATCAAGCGCAACACC
407   I  L  H  G  D  K  S  E  E  V  Q  D  L  L  L  L  D  V  T  P  L  S  L  G  I  E  T  A  G  G  V  M  T  T  L  I  K  R  N  T
1441  ACCATCCCCACCAAGCAGACCCAGACCTTCACCACCTACTCTGATAACCAGCCCGGTGTATTGATCCAAGTGTTCGAGGGCGAGCGCGCCATGACCAAGGACAATAACCTGCTCGGCAAG
447   T  I  P  T  K  Q  T  Q  T  F  T  T  Y  S  D  N  Q  P  G  V  L  I  Q  V  F  E  G  E  R  A  M  T  K  D  N  N  L  L  G  K
1561  TTCGAGCTCACCGGCATCCCGCCCGCGCCGCGTGGCGTGCCCCAGATCGAGGTCACCTTCGACATCGACGCCAACGGCATCCTCAACGTGTCCGCCGTCGAGAAGTCCACCAACAAGGAG
487   F  E  L  T  G  I  P  P  A  P  R  G  V  P  Q  I  E  V  T  F  D  I  D  A  N  G  I  L  N  V  S  A  V  E  K  S  T  N  K  E
1681  AACAAGATCACCATCACCAACGACAAGGGCCGCCTGTCCAAGGAGGAGATCGAGCGCATGGTGAACGAGGCCGAGCGCTACCGCAGCGAGGACGACAAGCAGAAGGAGACCATCCAGGCC
527   N  K  I  T  I  T  N  D  K  G  R  L  S  K  E  E  I  E  R  M  V  N  E  A  E  R  Y  R  S  E  D  D  K  Q  K  E  T  I  Q  A
1801  AAGAACGCGCTCGAGTCGTACTGCTTCAACATGAAGTCCACCATGGAGAAGGAGACCATCCAGGCCAAGAACGCGCTCGAGTCGTACTGCTTCAACATGAAGTCCACCATGGAGGACGAG
567   K  N  A  L  E  S  Y  C  F  N  M  K  S  T  M  E  K  E  T  I  Q  A  K  N  A  L  E  S  Y  C  F  N  M  K  S  T  M  E  D  E
1921  AAGCTCAAGGGGAAGCTCGCCGACTCGGACAAGCAGACCATCCTCGACAAGTGCAACGACACCATCAAGTGGCTGGATTCGACCCAGCTCGCCGACAAGGAGGAGTACGAGCACAAGCAG
607   K  L  K  G  K  L  A  D  S  D  K  Q  T  I  L  D  K  C  N  D  T  I  K  W  L  D  S  T  Q  L  A  D  K  E  E  Y  E  H  K  Q
2041  AAGGAGCTGGAGGGAGTGTGCAACCCCACCATCGAGGAGGTCGACTAAACATTCCATACGCAATCACCATCCGTTGCTAATTGAAGTCGATTAATAAATTGAAACATAAAAAAAAAAAAA
647   K  E  L  E  G  V  C  N  P  T  I  E  E  V  D  *
2161  AAAAA
```

Figure 3. Nucleotide and deduced amino acid sequences of *Lg-hsc70*. The upper sequences indicate the nucleotides, and the lower sequences show the inferred amino acids. The start and stop codons are in bold. The signature sequences of the HSP70s family are shown in boxes, the nuclear localization signal sequence is underlined, the consensus sequence EEVD at the C-terminus is indicated in italics and the consensus polyadenylation signal (AATAAA) is shown in a gray box.

Determining the total protein concentration. The fourth instar larvae were weighed after cold shock. Then, the total protein was isolated using a CytoBuster kit (Novagen, Germany), and the concentration of the protein was determined using a BCA kit (Thermo, USA). The isolated protein was diluted to 2.5 ng/µl in PBS and stored at −80°C. Then, 25 ng of each sample was separated by 12% SDS-PAGE (Mini-PROTEAN Tetra Cell, Bio-Rad). Each treatment was performed in 3 replicates, and β-actin (approximately 42 kDa) was used as a reference.

After SDS-PAGE, the sample was electroblotted onto a PVDF membrane and blocked in 5% dried skim milk.The membrane was incubated with a mouse monoclonal HSP70 antibody that could identify both HSP70 and HSC70 (H5147, Sigma, USA) (1:5000, V:V) and with a mouse monoclonal β-actin antibody (1:5000, V:V) (A5441, Sigma, USA) at 4°C overnight. After the membrane was washed three times in TBST for 10 min each, goat anti-mouse lgG (H+L) was used as the secondary antibody (Sigma, USA) (1:1000, V:V) at room temperature for 2 h and then washed three times in TBST for 10 min each. The blots were developed in DAB and photographed. The bandwidth and gray level were analyzed using ImageJ 4.0 software. Then, a one-way ANOVA was computed using SPSS 16.0 software, with $P<0.05$.

Quantitative real-time PCR. qRT-PCR primers were designed using the *Lg-hsp70* and *Lg-hsc70* sequences (see Table 2). The products were 179 bp and 141 bp, respectively. Using *β-actin* as the reference gene, the primers were designed according to Li, and the amplified fragment was 132 bp [40]. Total RNA was extracted using an RNA isolation kit (Omega, USA), and the concentrations were brought to 100 ng/µl using DEPC-treated water in all samples. Then, cDNA synthesis was performed using a reverse transcription kit (ToYoBo, Japan).

Standard curve: Using the first-strand cDNA as a template and the above-mentioned primers for PCR amplification, the PCR products were purified using a Gel Extraction Kit (Omega, USA), ligated into the pMD18-T vector (TaKaRa, Japan) and transformed into *Escherichia coli* competent cells with blue-white color selection. Positive clones were harvested and cultured overnight. The plasmid was extracted and identified by PCR. Then, the positive clones were sequenced at Shanghai Sangon Biological Engineering Technology & Services Co., Ltd. (Shanghai, China).

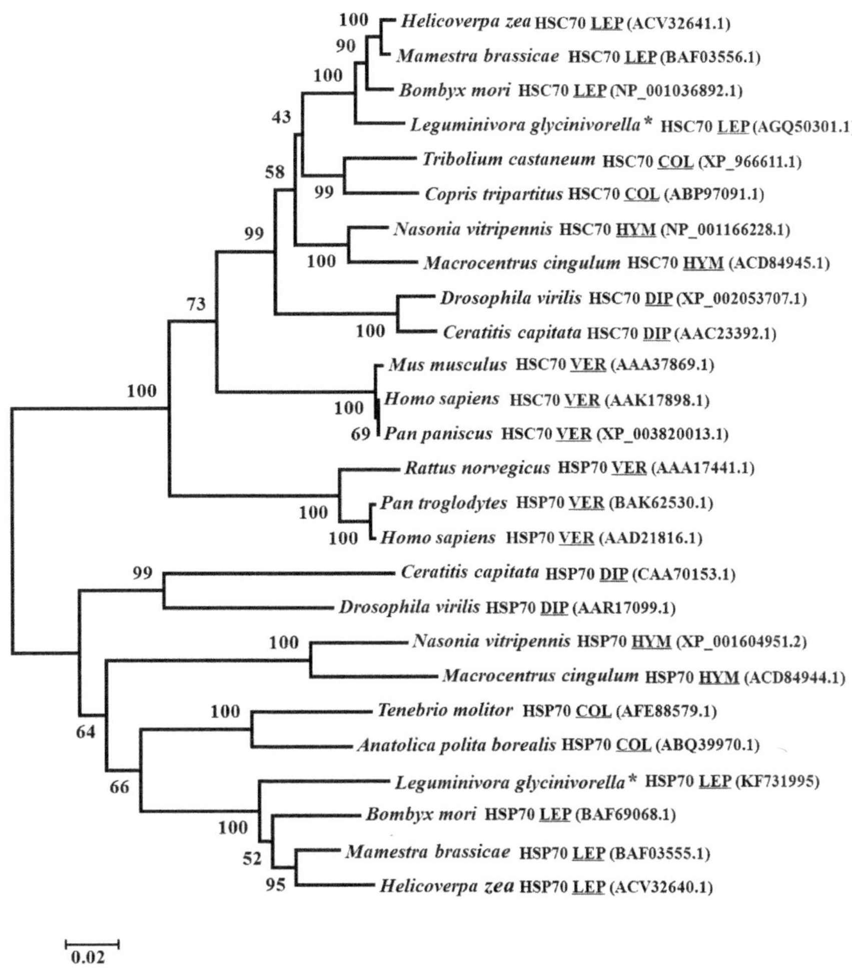

Figure 4. Phylogenetic tree of the HSP70 and HSC70 sequences, constructed using the neighbor-joining method. A three-letter code has been included to indicate the order name of the corresponding insects and vertebrates (COL = Coleoptera, LEP = Lepidoptera, DIP = Diptera, HYM = Hymenoptera and VER = Vertebra). The values indicated on the branches correspond to bootstrap percentages (BP).

The correctly identified plasmid was quantified using a NanoDrop (NanoDrop 8000, Parallels, USA) and used at a 10x dilution as the standard for the qRT-PCR reaction. Each sample had 3 replicates. The standard equation was derived based on the critical number of cycles (threshold cycle, Ct) and on the logarithmic values of the initial template copy number. The qRT-PCR reaction used a fluorescent dye, SYBR Green I. The qRT-PCR reaction conditions were as follows: 95°C denaturation 30 s, 95°C denaturation 10 s, and 40 cycles of 60°C annealing and extension for 30 s. The Ct value of the reaction was read by the Opticon 3 software.

Data processing: The data were calculated using the $2^{-\Delta\Delta Ct}$ method after standard genetic homogenization treatment. For the control, the value of ΔΔCt was 0 ($2^0 = 1$). Statistical significance was determined using a one-way ANOVA and post-hoc Duncan multiple range tests with the SPSS 16.0 software. The significance was established at $P<0.05$.

Results

Sequence analysis of *Lg-hsp70* and *Lg-hsc70*

Lg-hsp70 and *Lg-hsc70* were cloned and deposited into the GenBank database under accession numbers KF731995 and KC844150, respectively. The coding amino acids are shown in Figures 2 and 3. The full-length *Lg-hsp70* sequence was 2253 bp [excluding poly(A)] and included a 102-bp 5′-UTR, a 189-bp 3′-UTR, and a 1941-bp ORF, which encoded 646 aa. The molecular weight was 70.73 kDa, and the PI was 5.54. The positive charge of disability (Arg + Lys) was 83, and the negative charge of disability (Asp + Glu) was 93. In addition, the full-length *Lg-hsc70* sequence was 2147 bp [excluding poly(A)] and included a 102-bp 5′-UTR, a 59-bp 3′-UTR and a 1986-bp ORF, which encoded 661 aa. The molecular weight was 73.23 kDa, and the PI was 5.30. The positive charge of disability (Arg + Lys) was 76, and the negative charge of disability (Asp + Glu) was 101. The protein sequences showed that both Lg-HSP70 and Lg-HSC70 have 3 conserved HSP70 family domains, EEVD at the end of the C-terminus [41] and two putative bipartite nuclear localization signal (NLS) sequences (Figures 1 & 2).

Phylogenetic analysis of Lg-HSP70 and Lg-HSC70

Using homology analysis, we found that Lg-HSP70 was more than 73.6% similarity with other known HSP70; in particular, it had 89.5% identity with *Bombyx mori* (BAF69068) and *Mamestra brassicae* (BAF03555) and 70.6% identity with vertebrates. Lg-HSC70 had more than 78.1% similarity with other known HSC70s, and it had 86% identity with *Helicoverpa zea* (ACV32641) and 72.6% identity with vertebrates. A phylogenetic

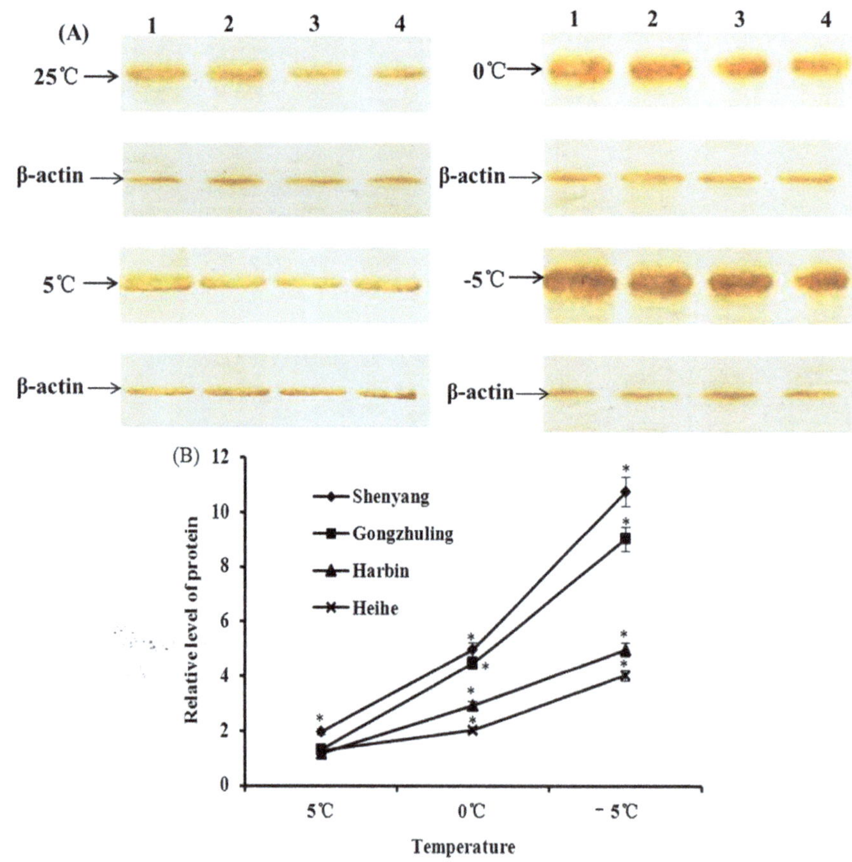

Figure 5. Relative analysis of Lg-HSP70s levels in different geographical populations after cold treatment. (A) Western blotting of Lg-HSP70s. 1: Shenyang, 2: Gongzhuling, 3: Harbin, 4: Heihe. (B) Relative gray levels of Lg-HSP70s (*: significant difference compared with control, $P< 0.05$).

tree was constructed, revealing that the HSP70 family could be divided into two main branches (Figure 4). One branch is HSC70, which also contains the other 11 HSC70 genes, and the other branch is HSP70, which contains all other HSP70 genes. Therefore, we could confirm that *Lg-hsc70* and *Lg-hsp70* are constitutive and inducible genes, respectively.

Expression of the Lg-HSP70 genes

The transcription and expression of Lg-HSP70s were determined using qRT-PCR and western blotting, respectively. The results showed that the levels of the Lg-HSP70s were not significantly different ($P>0.05$) between field- and laboratory-derived populations in the four cities. Three factors were considered to analyze the relationships among temperature (cold stress and control), sex (female and male) and population (50°22′N, 45°45′N, 43°11′N and 41°50′N). However, only the treatment between temperature and population had a significant effect on the Lg-HSP70 levels ($P<0.05$). Western blotting showed that the expression of Lg-HSP70s that originated from the same geographical population was significantly up-regulated as the temperature decreased, and it peaked at -5°C (Figure 5). Compared with the amount of Lg-HSP70s at the same temperature, there was a significant increasing trend as the latitude decreased. Shenyang (41°50′N) expressed the most Lg-HSP70s, and Heihe (50°22′N) expressed the least Lg-HSP70s. Because no differences in the HSP70 level were observed, we concluded that the laboratory-rearing conditions had little effect on the HSP70 level. These results revealed that cold stress has a significant inducible effect on the expression of Lg-HSP70s, indicating that phenotypic plasticity variation for Lg-HSP70s occurs in different geographical populations.

qRT-PCR analysis: The positive clones were identified by PCR. The soybean pod borer cDNA sequences for *Lg-hsp70*, *Lg-hsc70* and *β-actin* had 100% homology, indicating that the recombinant plasmid contained the target fragment. Using a UV spectrophotometer, the standard concentrations of *Lg-hsp70*, *Lg-hsc70* and *β-actin* were found to be 22 ng/μl, 18 ng/μl and 15 ng/μl, respectively. The copy numbers of the *Lg-hsp70*, *Lg-hsc70* and *β-actin* standards were 7.01×10^9, 5.79×10^9 and 4.87×10^9 copies/μl, respectively. The equations of the standard curves of *Lg-hsp70*, *Lg-hsc70* and *β-actin* were $Y=-3.124X+40.601$, $Y=-3.124X+46.327$ and $Y=-3.297X+39.876$, respectively, where Y represents the value of Ct, and X represents the logarithmic value of the template. The correlation coefficients were $R^2=0.992$, $R^2=0.998$ and $R^2=0.997$, respectively. When the template cDNA was diluted, the Ct value was up-regulated 3 to 4 cycles, and the concentration decreased with each 10-fold dilution. The slopes of three standard curves were near the expected value of -3.322, with a correlation coefficient close to 1.000, suggesting that the results of the qRT-PCR could be used for the quantitative detection of *Lg-hsp70* and *Lg-hsc70* over a wide range.

Expression of *Lg-hsp70* induced by cold shock: Compared with the control (25°C), the expression level of *Lg-hsp70* could be induced in the four populations at 5°C. *Lg-hsp70* increased 7.08

Table 3. Differences in *Lg-hsp70* and *Lg-hsc70* levels induced by cold stress temperatures in the soybean pod borer.

Temperature (°C)	Induced level of *Lg-hsp70*				Induced level of *Lg-hsc70*			
	Shenyang	Gongzhuling	Harbin	Heihe	Shenyang	Gongzhuling	Harbin	Heihe
−5°C	80.18±0.54a	69.32±0.11b	41.37±0.92c	25.01±0.45d	7.91±0.07a	5.09±0.14b	4.67±0.82c	2.07±0.06d
0°C	26.74±0.04a	20.02±0.79b	13.02±0.03c	9.57±0.25d	3.51±0.53a	2.44±0.06b	2.21±0.46b	1.62±0.47c
5°C	7.08±0.12a	5.58±0.17b	5.02±0.01b	5.34±0.37b	1.25±0.31a	1.21±0.52a	1.06±0.73a	0.93±0.46a

Note: The different letters above the bars (a–d) were the result of a multi-comparison, which indicated that the means are significantly different ($P<0.05$).

fold in Shenyang, which was significantly higher than the levels in the other populations, and there was no significant difference among the other three populations. At 0°C, the *Lg-hsp70* transcripts were present at a high level in each population, and there was a significant difference in their levels compared with the level of the control ($P<0.05$). The expression of *Lg-hsp70* in each population reached its peak at −5°C, increasing by 80.18 fold, 69.32 fold, 41.37 fold and 25.01 fold in Shenyang, Gongzhuling, Harbin and Heihe, respectively, compared with the control. A significant difference existed among the different geographical populations ($P<0.05$) (see Table 3). Thus, cold treatment had a significant effect on inducing *Lg-hsp70* at diapause.

Expression of *Lg-hsc70* induced by cold shock: There was a small, inducible effect on the transcription of *Lg-hsc70* by cold treatment from −5°C to 5°C, and the level was significantly lower than that of *Lg-hsp70*. At 5°C, the expression level of *Lg-hsc70* did not differ from the control. At 0°C, the expression level of *Lg-hsc70* significantly increased ($P<0.05$); however, there was no significant difference between Gongzhuling and Harbin ($P>0.05$). The highest transcription of *Lg-hsc70* was observed at −5°C, increasing by 7.91 fold, 5.09 fold, 4.67 fold and 2.07 fold in Shenyang, Gongzhuling, Harbin and Heihe, respectively, compared with the control. A significant difference existed among the four geographical populations (see Table 3).

Discussion

The Lg-HSP70 and Lg-HSC70 sequences both contain a conserved amino acid site, and the consensus sequence EEVD at the end of the C-terminus revealed that the two proteins belong to the DanK cytoplasmic-type protein of the HSP70 family. In addition, two nuclear localization sites (NLS) that play important roles in the inhibitory effect of HSP70s against nucleolar impairment induced by oxidative stress were identified [42].

To determine whether *Lg-hsp70* and *Lg-hsc70* were inducible or constitutive, a cluster analysis was performed. The results showed that *Lg-hsp70* and *Lg-hsc70* belong to two different branches of *hsp70* and *hsc70*, respectively. Some recent studies have reported that the expression level of *hsp70* increases significantly after induction by environmental stress but that *hsc70* is expressed with a time lag and with a low range compared with *hsp70* [43]. In our experiment, the expression time and rate of increase of *Lg-hsp70* significantly differed from that of *Lg-hsc70*, which supports the conclusion that *Lg-hsp70* is an inducible gene and that *Lg-hsc70* is a constitutive gene. The difference in expression of the two genes may be because the *Lg-hsp70* gene has no intron and *Lg-hsc70* does, which affects the split joint of the mRNA [44,45].

Selection for low-temperature stress: Although *L. glycinivorella* overwinters in buried soil, this insect can still be exposed to harmful environmental temperatures. In northeastern China, the depth of the frozen soil varies with year and location [46]. Consequently, the optimum overwinter depth for *L. glycinivorella* is unpredictable, and exposure to the cold is possible. According to the temperature data for frozen soil from the northeastern region and the report that −5°C was the lethal temperature for *L. glycinivorella* to overwinter [47], −5°C and −10°C were designed as the pretreatment temperatures in this study. The results showed that all larvae were dead at −10°C; therefore, −5°C was selected as the treatment temperature.

Lg-HSP70 protein expression and mRNA transcription were analyzed after cold stress in the four geographical populations. The results revealed that HSP70s had a strong relation with insect resistance to cold stress [11,34]. Thus, this mechanism is most likely a strategy that *L. glycinivorella* uses to react to cold stress. Additionally, we found that the up-regulation of Lg-HSP70s varies with latitude. A southern population may have less chance of experiencing cold stress during their overwintering diapause compared with a northern population [48]. Some studies have shown that the production of HSP70s as a response to temperature shock is less intense in organisms that are more frequently exposed to unfavorable temperatures in their habitat compared with those organisms exposed to benign conditions [49]. Research concerning *Leptinotarsa decemlineata* showed the same trend when comparing a southern population (Bonin, Poland, 54°N) to a northern population (Lodeynoye Pole, Russia, 60°N) [50].

Because HSP70 production requires more energy, this production may incur significant costs to the synthesis of other proteins [51,52]. Therefore, the northern population might have evolved other more reasonable, less costly mechanisms to cope with cold stress [48,53]. In recent years, possible mechanisms have been studied and include accumulating low-molecular-weight cryoprotectants [54], synthesizing antifreeze proteins [55] and remodeling the structure of the cell membrane [56]. In contrast, with the development of molecular biology, the function, reactions and location of HSP70s also could be identified by bioinformatics tools and proteomics techniques, such as 2-dimensional electrophoresis and mass spectra which have become powerful tools to discover and validate protein-coding genes. Some proteomic studies showed that different isoforms of the HSP70 family, including HSP68, HSP70Ba, and HSP70Bb, exist and that several of them displayed strong increases in protein abundance during environmental variations [57,58]. Furthermore, the gene annotation and localization might play a pivotal role to facilitate identification and analysis of HSP70s in the integrated genome of *L. glycinivorella*. Thus, on the basis of this research, application of proteomic and genomic strategies could supply good support for the relevant mechanism of *L. glycinivorella* from different populations.

L. glycinivorella from different populations showed a differential response to cold stress. The significant differences in the expression levels of LgHSP70s in the southern (Shenyang, 41°50′N) and northern populations (Heihe, 50°22′N) showed strong evidence for phenotypic plasticity among the four different populations [59]. The phenotypic plasticity in these populations contributes to the soybean pod borer response to different selective pressures under different environmental conditions and has great potential for improving their ability to resist environmental stress [60,61]. Because relative reports of the damage of *L. glycinivorella* date to before the 1950s [38], we assume that there must be a long history of *L. glycinivorella* damaging the soybean crop in northeastern China. Although no reports of genetic variability can be found, we believe that phenotypic plasticity variation based on epigenetic mechanisms has occurred for a long period in different geographic populations of *L. glycinivorella* so that they could adapt to the different ecological conditions. Thus, because this variation can be stably inherited, it will undoubtedly have a profound effect on population adaptability [62,63,64].

Acknowledgments

We gratefully acknowledge Dr. Xin Dawei (Northeast Agriculture University, China) for his constructive reviews of the manuscript.

Author Contributions

Conceived and designed the experiments: LW SY LH DF KZ. Performed the experiments: LW SY. Analyzed the data: LW SY. Contributed reagents/materials/analysis tools: LW SY LH DF KZ. Wrote the paper: LW SY KZ.

References

1. Danks HV (2007) The elements of seasonal adaptations in insects. The Canada Entomologist 139(1): 1–44.
2. Watson P, Morris G (1987) Cold shock injury in animals. In: Bowler K, Fuller B, eds. Temperature and animal cells. Cambridge: The Company of Biologist Limited. pp 311–340.
3. Bettencourt BR, Hogan CC, Nimali M, Drohan BW (2008) Inducible and constitutive heat shock gene expression responds to modification of Hsp70 copy number in *Drosophila melanogaster* but does not compensate for loss of thermotolerance in Hsp70 null flies. BMC Biology 6(5): 1–15.
4. Rinehart JP, Li A, Yocum GD, Robich RM, Hayward SA, et al. (2007) Up-regulation of heat shock proteins is essential for cold survival during insect diapause. Proceedings of the National Academy of Sciences of the United States of America 104: 11130–11137.
5. Mads D (2007) The heat shock protein 70 family: Highly homologous proteins with overlapping and distinct functions. FEBS Letters 581: 3702–3710.
6. Doucet D, Walker VK, Qin W (2009) The bugs that came in form the cold: molecular adaptations to low temperatures in insects. Cellular and Molecular life Sciences 66(8): 1404–1418.
7. Schlesinger MJ (1990) Heat shock proteins. The Journal of Biological Chemistry 265: 12111–12114.
8. Qin W, Tyshenko MG, Wu BS, Walker VK, Robertson RM (2003) Cloning and characterization of a member of the Hsp70 gene family from *Locusta migratoria*, a highly thermotolerant insect. Cell Stress and Chaperones 8, 144–152.
9. Franzellitti S, Fabbri E (2005) Differential HSP70 gene expression in the *Mediterranean mussel* exposed to various stressors. Biochemical and Biophysical Research Communications 336(4): 1157–163.
10. Daugaard M, Rohde M, Jäättela M (2007) The heat shock protein 70 family: highly homologous proteins with overlapping and diatinct functions. FEBS Letters 581(19): 3702–3710.
11. Sonoda S, Fukumoto K, Izumi Y, Yoshida H, Tsumuki H (2006) Cloning of heat shock protein genes (hsp90 and hsc70) and their expression during larval diapause and cold tolerance acquisition in the rice stem borer, *Chilo suppressalis* Walker. Archives of Insect Biochemistry and Physiology 63(1): 36–47.
12. Gkouvitsas T, Kontogiannatos D, Kourti A (2009) Cognate Hsp70 gene is induced during deep larval diapause in the moth *Sesamia nonagrioides*. Insect Molecular Biology 18(2): 253–264.
13. Goto SG, Yoshida KM, Kimura MT (1998) Accumulation of Hsp70 mRNA under environmental stresses in diapausing and nondiapausing adults of *Drosophila triauraria*. Journal of Insect Physiology 44: 1009–1015.
14. Sinclair BJ, Gibbs AG, Roberts SP (2007) Gene transcription during exposure to, and recovery from, cold and desiccation stress in *Drosophila melanogaster*. Insect Molecular Biology 16: 435–443.
15. Yocum GD (2001) Differential expression of two HSP70 transcripts in response to cold shock, thermoperiod, and adult diapause in the Colorado potato beetle. Journal of Insect Physiology 47: 1139–1145.
16. Kostál V, Tollarová-Borovanská M (2009) The 70 kDa heat shock protein assists during the repair of chilling injury in the insect, *Pyrrhocoris apterus*. PLoS One 4(2): e4546.
17. Callahan HS, Maughan H, Steiner UK. (2008) Phenotypic Plasticity, Costs of Phenotypes, and Costs of Plasticity. Annals of the New York Academy of Sciences 1133: 44–46.
18. Schulte PM, Healy TM, Fangue NA (2011) Thermal performance curves, phenotypic plasticity, and the time scales of temperature exposure. Integrative and Comparative Biology 51(5): 691–702.
19. Luo SQ, Wong SC, Xu CR, Hanskic I, Wang RJ, et al. (2014) Phenotypic plasticity in thermal tolerance in the Glanville fritillary butterfly. Journal of Thermal Biology 42: 33–39.
20. Bradshaw AD (1965) Evolutionary significance of phenotypic plasticity in plants. Advances in Genetics 13: 115–155.
21. Van Kleunen M, Fischer M (2005) Constrains on the evolution of adaptive phenotypic plasticity in plants. New Phytologist 166: 49–60.
22. Sultan SE (2004) Promising directions in plant phenotypic plasticity. Perspectives in Plant Ecology 6(4): 227–233.
23. Pigliucci M, Murren CJ, Schlichting CD (2006) Phenotypic plasticity and evolution by genetic assimilation. Journal of Experimental Biology 209: 2362–2367.
24. Bourgeois M, Jacquin F, Savois V, Sommerer N, Labas V, et al. (2009) Dissecting the proteome of pea mature seeds reveals the phenotypic plasticity of seed protein composition. Proteomics 9(2): 254–271.
25. Whitman DW, Ananthakrishnan TN (2009) Phenotypic plasticity of insects: mechanisms and consequences. pp. x+ 894 pp.
26. Stillwell RC, Blanckenhorn WU, Teder T, Davidowitz G, Fox CW (2010) Sex differences in phenotypic plasticity affect variation in sexual size dimorphism in insects: from physiology to evolution. Annual Review of Entomology 55: 227–245.
27. Schlichting CD, Wund MA (2014) Phenotypic plasticity and epigenetic marking: an assessment of evidence for genetic accommodation. Evolution 68(3): 656–672.

28. Zhou XZ, Tarver MR, Scharf ME (2007) Hexamerin-based regulation of juvenile hormone-dependent gene expression underlies phenotypic plasticity in a social insect. Development 134: 601–610.
29. Moczek AP (2010) Phenotypic plasticity and diversity in insects. Philosophical Transactions of the Royal Society B: Biological Sciences 365(1540): 593–603.
30. Morris M, Rogers SM (2014) Integrating phenotypic plasticity within an ecological genomics framework: recent insights from the genomics, evolution, ecology, and fitness of plasticity. Advances in Experimental Medicine and Biology Volume 781: 73–105.
31. Moraiti CA, Nakas CT, Papadopoulos NT (2014) Diapause termination of *Rhagoletis cerasi* pupae is regulated by local adaptation and phenotypic plasticity: escape in time through bet-hedging strategies. Journal of Evolutionary Biology 27(1): 43–54.
32. Coleman PC, Bale JS, Hayward SA (2014) Cross generation plasticity in cold hardiness is associated with diapause, but not the non-diapause developmental pathway, in the blowfly, *Calliphora vicina*. The Journal of Experimental Biology http://dx.doi: 10.1242/jeb.098053 .
33. Mrdaković M, Stojković B, Perić-Mataruga V, Ilijin L, Vlahović M, et al. (2014) Adaptive phenotypic plasticity of gypsy moth digestive enzymes. Central European Journal of Biology 9(3): 309–319.
34. Zhang Q, Denlinger DL (2010) Molecular characterization of heat shock protein 90, 70 and 70 cognate cDNAs and their expression patterns during thermal stress and pupal diapause in the corn earworm. Journal of Insect Physiology 56(2): 138–150.
35. Zhao GY, Wang J, Han YP, Teng WL, Sun GL, et al. (2008) Identification of QTL underlying the resistance of soybean to pod borer, *Leguminivora glycinivorella* (Mats.) obraztsov, and correlations with plant, pod and seed traits. Euphytica 164 (1): 275–282.
36. Zhang T, Feng J, Cai C, Zhang X (2011) Synthesis and field test of three candidates for soybean pod borer's sex pheromone. Natural Product Communications 6(9): 1323–1326.
37. Hu DH, He J, Zhou YW, Feng JT, Zhang X (2012) Synthesis and Field Evaluation of the Sex Pheromone Analogues to Soybean Pod Borer *Leguminivora glycinivorella*. Molecules, 17(10): 12140–12150.
38. Wang KQ, Mei XR, Liu XL, Shao TY, Li XM, et al. (2013) Field trapping effect of two types of sex pheromone lures and traps of *Leguminivora glycinivorella*. Heilongjiang Agricultural Sciences 5: 34–36. (in Chinese)
39. Zang LS, Song LV, Yuan XH, Ruan CQ, Shi SS (2012) Parasitism activity of four indigenous Trichogramma species on eggs of different ages of the soybean pod borer, *Leguminivora glycinivorella*. Acta Phytophylacica Sinica 39(4): 381–382. (in Chinese)
40. Li DM, Meng FL, Wang ZK, Zang ZY, Sui J, et al. (2012) Cloning and characterization of 28SrDNA gene form soybean pod borer (*Leguminivora glycinivorella* Matsumurav). Journal of Northeast Agricultural University 43(7): 13–17. (in Chinese)
41. Demand J, Lüders J, Höhfeld J (1998) The carboxy-terminal domain of Hsc70 provides binding sites for a distinct set of chaperone cofactors. Molecular Cell Biology 18(4): 2023–2028.
42. Wang KK, E SM, Jiang L, Zhang HL, Liu K, et al. (2005) Roles of nuclear localization signal (NLS) in inhibitory effect of HSP70 on nucleolar segregation induced by oxidative stress. Progress Biochemistry Biophysics 32(5): 456–462. (in Chinese)
43. Deane EE, Woo NYS (2005) Cloning and characterization of the hsp70 Multigene family from silver sea bream: modulated gene expression between warm and cold temperature acclimation. Biochemical and Biophysical Research Communications 330(3): 776–783.
44. Yost HJ, Lindquist S (1991) Heat shock proteins affect RNA processing during the heat shock response of *Saccharomyces cerevisiae*. Molecular and Cellular Biology 11: 1062–1068.
45. Chuang K, Ho S, Song Y (2007) Cloning and expression analysis of heat shock cognate 70 gene promoter in tiger shrimp (*Penaeus monodon*). Gene 405: 10–18.
46. Zhang W, Song SY, Jing QM, Li H, Sha HL (2005) Identification of resistance of new soybean varieties (Lines) to soybean pod borer in Jilin Province. Journal of Jilin Agricultural Sciences 30(6): 40–41. (in Chinese)
47. Xu QF, Guo SG, Han YM, Feng Z, Zhang R, et al. (1965) A study on the soybean pod borer (*Leguminivora glycinivorella* (Mats.) Obraztsov). Acta Entomologlca sinica 14(5): 461–475. (in Chinese)
48. Lansing E, Justesen J, Loeschcke V (2000) Variation in the expression of Hsp70, the major heat-shock protein, and thermotolerance in larval and adult selection lines of *Drosophila melanogaster*. Journal of Thermal Biology 25: 443–450.
49. Sørensen J, Dahlgraad J, Loeschcke V (2001) Genetic variation in thermal tolerance among natural populations of *Drosophila buzzatii*: down regulation of Hsp70 expression and variation in heat stress resistance traits. Functional Ecology 15: 289–296.
50. Lyytinen A, Mappes J, Lindström L (2012) Variation in Hsp70 Levels after Cold Shock: Signs of Evolutionary Responses to Thermal Selection among *Leptinotarsa decemlineata* Populations. PLoS One 7(2): e31446.
51. Krebs RA, Loeschcke V (1994) Costs and benefits of activation of the heat-shock response in *Drosophila melanogaster*. Functional Ecology 8: 730–737.
52. Feder JH, Rossi JM, Solomon J, Solomon N, Lindquist S (1992) The consequences of expressing hsp70 in Drosophila cells at normal temperatures. Genes and Development 6: 1402–1413.
53. Sørensen J, Kristensen TN, Loeschcke V (2003) The evolutionary and ecological role of heat shock proteins. Ecology Letters 6: 1025–1037.
54. Crowe JH, Crowe LM, Carpenter JF, Rudolph AS, Wistrom CA, et al. (1988) Interactions of sugars with membranes. Biochimicaet Biophysica Acta 947: 367–384.
55. Duman JG (2001) Antifreeze and ice nucleator proteins in terrestrial arthropods. Annual Review of Physiology 63: 327–357.
56. Tomcala A, Tollarová M, Overgaard J, Simek P, Kostál V (2006) Seasonal acquisition of chill-tolerance and restructuring of membrane glycerophospholipids in an overwintering insect: triggering by low temperature, desiccation and diapause progression. Journal of Experimental Biology 209: 4102–4114.
57. Aparna HS, Rajesh RK, Manjunatha HB (2010) Molecular characterization of heat shock proteins 90 (HSP83?) and 70 in tropical strains of *Bombyx mori*. Proteomics 10: 2734–2745.
58. Hervé C, Johannes O, Emmanuelle C, Jesper GS (2013) Proteomic profiling of thermal acclimation in *Drosophila melanogaster*. Insect Biochemistry and Molecular Biology 43: 352–365.
59. Schlichting CD, Smith H (2002) Phenotypic plasticity: linking molecular mechanisms with evolutionary outcomes. Evolutionary Ecology 16: 189–211.
60. Ghalambor CK, Mckay JK, Carroll SP, Reznick DN (2007) Adaptive versus non-adaptive phenotypic plasticity and the potential for contemporary adaptation in new environments. Functional Ecology 21: 394–407.
61. Sexton JP, McKay JK, Sala A (2002) Plasticity and genetic diversity may allow saltcedar to invade cold climate in North America. Ecological Applications 12: 1652–1660.
62. King EG, Roff DK (2010) Modeling the Evolution of Phenotypic Plasticity in Resource Allocation in Wing-Dimorphic Insects. The American Naturalist 175(6): 702–716.
63. Moczek AP, Kijimoto T, Snell-Rood E, Rocha G, Pespeni M, et al. (2014) Evolutionary and ecological genomics of developmental plasticity: novel approaches and first insights from the study of horned beetles. Ecological Genomics 781: 127–148.
64. Tobin PC, Parry D, Aukema BH (2014) The influence of climate change on insect invasions in temperate forest ecosystems. Forestry Sciences 81: 267–293.

The IQD Gene Family in Soybean: Structure, Phylogeny, Evolution and Expression

Lin Feng[1]◔, Zhu Chen[1]◔, Hui Ma[1], Xue Chen[1], Yuan Li[1], Yiyi Wang[1], Yan Xiang[1,2]*

1 Laboratory of Modern Biotechnology, School of Forestry and Landscape Architecture, Anhui Agricultural University, Hefei, China, **2** Key Laboratory of Crop Biology of Anhui Agriculture University, Hefei, China

Abstract

Members of the plant-specific IQ67-domain (IQD) protein family are involved in plant development and the basal defense response. Although systematic characterization of this family has been carried out in *Arabidopsis*, tomato (*Solanum lycopersicum*), *Brachypodium distachyon* and rice (*Oryza sativa*), systematic analysis and expression profiling of this gene family in soybean (*Glycine max*) have not previously been reported. In this study, we identified and structurally characterized IQD genes in the soybean genome. A complete set of 67 soybean IQD genes (*GmIQD1-67*) was identified using Blast search tools, and the genes were clustered into four subfamilies (IQD I–IV) based on phylogeny. These soybean IQD genes are distributed unevenly across all 20 chromosomes, with 30 segmental duplication events, suggesting that segmental duplication has played a major role in the expansion of the soybean IQD gene family. Analysis of the Ka/Ks ratios showed that the duplicated genes of the GmIQD family primarily underwent purifying selection. Microsynteny was detected in most pairs: genes in clade 1–3 might be present in genome regions that were inverted, expanded or contracted after the divergence; most gene pairs in clade 4 showed high conservation with little rearrangement among these gene-residing regions. Of the soybean IQD genes examined, six were most highly expressed in young leaves, six in flowers, one in roots and two in nodules. Our qRT-PCR analysis of 24 soybean IQD III genes confirmed that these genes are regulated by MeJA stress. Our findings present a comprehensive overview of the soybean IQD gene family and provide insights into the evolution of this family. In addition, this work lays a solid foundation for further experiments aimed at determining the biological functions of soybean IQD genes in growth and development.

Editor: Marc Robinson-Rechavi, University of Lausanne, Switzerland

Funding: This work was supported by grants from the National Natural Science Foundation of China (No. 31370561),Specialized research Fund for the Doctoral Program of Higher Education(No.20133418110005), Anhui Provincial Natural Science Foundation (No. 1308085MC36)and Anhui Agricultural University disciplinary construction Foundation(No.XKTS2013001) The funders had no role in study design, data collection and analysis, decision to publish, or preparation of the manuscript.

Competing Interests: The authors have declared that no competing interests exist.

* Email: xiangyanahau@sina.com

◔ These authors contributed equally to this work.

Introduction

Ca^{2+} is a pivotal cytosolic second messenger involved in many physiological processes such as plant growth [1], plant-pathogen interactions [2], photosynthetic electron transport and photophosphorylation [3], regulation of stomatal aperture [4], hormonal regulation [5] and so on. Plants produce calcium signals by adjusting cytoplasm Ca^{2+} levels at specific times, places and concentrations [6], responding to numerous extracellular stimuli including physical signals (light, temperature, gravity, etc.) and chemical signals (plant hormones, pathogenic bacteria inducing factors, etc.) [7].

The transmission of these intracellular calcium signals relies on the oscillation signal generated by voltage- and ligand-gated Ca^{2+}-permeable channels (influx) and by Ca^{2+}-ATPases and antiporters (efflux) to return to resting Ca^{2+} levels [8,9]. The conduction of calcium signals is also dependent on downstream Ca^{2+} sensors. These Ca^{2+} sensors detect changes in Ca^{2+} levels by binding to Ca^{2+} via domains such as EF hands, which undergo conformational changes [10]. Consequently, calcium signature information is decoded and relayed by these Ca^{2+} sensors [6,11–13].

To date, approximately four major classes of Ca^{2+} sensors have been identified in plants. Most of these sensors contain the Ca^{2+}-binding EF-hand motif, a conserved helix-loop-helix structure that can bind to a single Ca^{2+} ion [7]. The four major classes of Ca^{2+} sensors are as follows: class A: calmodulin (CaM), containing four EF-hand motifs; class B: calcineurin B-like (CBL) proteins, possessing three EF-hand motifs; class C: Ca^{2+}-dependent protein kinases (CDPK), containing four EF-hand motifs and a Ca^{2+}-dependent Ser/Thr protein kinase domain and class D: lacking EF-hand motifs [7,14–19].

Calmodulin (CaM) and calcineurin B-like (CBL) proteins, which lack catalytic activity, are sometimes referred to as "Ca^{2+} sensor relays" [15,19,20]. In contrast, CDPK proteins, which function as catalytic effectors, are referred to as "Ca^{2+} sensor responders" [18]. Among these Ca^{2+}-binding proteins, calmodulin is the most extensively studied Ca^{2+} sensor. Calmodulin is small, acid resistant, heat resistant and highly stable. This multifunctional protein is widespread in eukaryotic cells, highly conserved and has at least 30 multiple target proteins or enzymes [21–23].

To mediate intracellular calcium signaling pathways, Ca^{2+} sensor relays expose their negative hydrophobic surfaces after they undergo conformational changes induced by Ca^{2+} binding. As a result, the affinity between Ca^{2+} sensor relays and their effectors are enhanced, and the biochemical activities of target proteins are modulated by Ca^{2+} sensor relays [6,12,14,19].

In the final phase of the calcium signal transduction process, the target effectors respond to specific extracellular signals by regulating various cellular activities. Calmodulin interacts with numerous target proteins termed calmodulin-binding proteins (CaMBPs), mainly by recognizing and targeting calmodulin-binding domains (CaMBD; basic amphiphilic helices usually composed of 16–35 amino acid residues) in the CaMBPs via its negative hydrophobic pockets [12,22,24].

CaMBD amino acid sequences contain three CaMBD motifs that are grouped into two categories, including a Ca^{2+}-independent motif termed the IQ motif and two Ca^{2+}-dependent motifs referred to as the l-5-10 motif and the l-8-14 motif. The number and positions of these motifs in different CaMBPs are variable [25–27]. Due to the diversity of the motif arrangement, there are a variety of diverse CaMBPs with disparate functions, which are implicated in plant development, metabolic regulation, stress reactions, defense reactions, transcriptional regulation and so on [28,29].

Plant-specific IQ67 domain (IQD) protein families were first identified in *Arabidopsis* and rice by Abel et al. (2005) [30]. These proteins have two common features in their IQ67 domains (67 conserved amino acid residues) [31]. One feature is 1–3 copies IQ motifs are separated by 11 and 15 residues and overlapped certain regions with 1–4 copies 1-5-10 motif as well as 1-8-14 motif. The other hallmark is a highly conserved exon-intron boundary that interrupts codons 16 and 17 with a 0 phase intron [31–33]. To date, IQD gene families have been identified in four genomes (*Arabidopsis*, rice, tomato and *Brachypodium distachyon*), including approximately 30 IQD genes per species (33 in *Arabidopsis*, 29 in rice, 34 in tomato and 23 in *Brachypodium distachyon*), and the functions of several members of the IQD family have been reported [30,34,35]. Overexpression of *Arabidopsis IQD1* can mediate the accumulation of glucosinolate in response to insect herbivory [31]. *Arabidopsis IQD22* contributes to the negative feedback regulation of gibberellin (GA) [36]. The tomato *SUN* gene plays a role in elongating tomato fruit by increasing the vertical division of cells and reducing horizontal cell divisions [37–39].

Soybean serves as a major source of vegetable proteins and edible oil and own the ability to fix atmospheric nitrogen via its intimate symbiosis with microorganisms. This crucial leguminous seed crop has high economic and nutritional value [40,41], serving as a main food crop for humans and animals in many parts of the world. Nevertheless, soybean production is limited by many biotic stresses. For example, Asian soybean rust (ASR, caused by the fungus *Phakopsora pachyrhizi*) results in soybean yield losses ranging from 10 to 80% in various countries [42,43].

In this study, we identified and characterized 67 soybean IQD genes. Among these, we subjected 24 IQD III genes to qRT-PCR analysis to investigate their response to MeJA stress. We determined that all 24 soybean IQD genes are stress-responsive. Our findings lay the foundation of further investigations of the functions of these calmodulin target proteins in soybean.

Results

Identification and annotation of IQD genes in soybean

As described in previous studies, IQD proteins, which are plant-specific calcium-dependent calmodulin-binding proteins, contain 67 amino acid residues in their central regions referred to as the IQ67 domain, including three CaM recruitment motifs exhibiting unique repetitive patterns. The Ca^{2+}-independent IQ motif (IQxxxRGxxxR or its more relaxed version [ILV]QxxxRxxxx [R, K]) is present in 1–3 copies and overlaps with 1–4 copies of the Ca^{2+}-dependent 1-5-10 motif ([FILVW]x_3 [FILV]x_4 [FILVW]) and the 1-8-14 motif ([FILVW]x_6 [FAILVW]x_5 [FILVW]) by several conserved basic and hydrophobic amino acid residues flanking these motifs [30,34]. These features allow the IQ67 domain to fold into a basic amphiphilic helix structure, which enables these proteins to perform specific roles.

To conducted genome-wide identification of IQD gene families in soybean, we performed *Glycine max* genome BLASTP analysis. Through removing redundant sequences and pattern identificating, a total of 67 IQD genes were identified in the soybean genome, which is twice that of *Arabidopsis* (Table 1 and 2). We named these 67 IQD genes *GmIQD1* to *GmIQD67* according to their physical locations (from top to bottom) on chromosomes 1–20 (Table 1).

The physicochemical parameters of each gene were calculated using ExPASy. Although all of the GmIQD genes encode the conserved IQ67 domains (Figure S1), their sequences are highly diverse with respect to size (141–904 aa) and molecular mass (16.3–99.2 kDa; Table 1). Almost all soybean IQD proteins (97%) have relatively high isoelectric points (pI>7.0 with an average of 10.1), except for *GmIQD9* (pI 5.4) and *GmIQD17* (pI 5.7; Table 1). All soybean IQD proteins were submitted to TargetP and Wolf PSORT to predict their subcellular localizations. Wolf PSORT revealed that fifty-seven soybean IQD proteins are localized to the nucleus, nine to the chloroplast and one to the endoplasmic reticulum. TargetP analysis revealed that fifteen soybean IQD proteins are located in the mitochondria, five in the chloroplast, one in the secretory pathway and forty-six in other compartments (Table 1). The detailed parameters are provided in Table 1.

Phylogenetic and structural analyses of the soybean IQD genes

To infer the similarity and evolutionary ancestry of soybean IQD proteins, we constructed an unrooted phylogenetic tree of the 67 soybean protein sequences. The soybean IQD gene family was categorized into four major subfamilies (subfamily I, II, III and IV; Figure 1a) according to phylogenetic analysis of IQD genes in *Arabidopsis*, rice, tomato and *Brachypodium distachyon* [30,34,35]. Subfamily I was further divided into four subclasses (clade Ia, Ib, Ic and Id), and subfamily II and III were divided into two subclasses (clade IIa and IIb; clade IIIa and IIIb) based on bootstrap values, the existence and positions of introns and the presence of protein motifs flanking the IQ67 domain (Figure 1 and 2). Subfamily I (containing 27 members) is the largest group, followed by subfamily III (24) and subfamily IV (10). Subfamily II has the fewest IQD gene members (6). This distribution pattern is similar to that observed for IQD genes in *Arabidopsis* and rice (Table 2). The phylogenetic tree reveals that 62 of the 67 soybean IQD genes form 31 gene pairs with strong bootstrap values (Figure 1a).

To further examine the structural diversity of the IQD genes in soybean, we deduced the exon/intron organization of individual GmIQD genes (Figure 1b). A comparison of the 67 genomic loci

Table 1. List of 67 IQD genes identified in soybean, their sequence characteristics and subcellular localization.

Name	Gene Identifier	Chr.	Location coordinates (5′-3′)	ORF length (bp)	Protein			Exons	subcellular localization	
					Length (aa.)	pI	Mol.Wt. (kD)		WoLF PSORT	TargetP
GmIQD1	Glyma01g01030	1	681417–683646	1263	420	10.3	46.6	3	N	M0.65/4
GmIQD2	Glyma01g05100	1	4750065–4755456	1692	563	9.7	61.7	6	N	C0.68/3
GmIQD3	Glyma01g42620	1	53843322–53846963	1191	396	10.4	44.4	4	N	M0.54/5
GmIQD4	Glyma02g00710	2	502944–506146	1254	417	9.6	46.8	3	N	?
GmIQD5	Glyma02g02370	2	1778568–1785636	1692	563	9.8	61.7	6	N	M0.43/5
GmIQD6	Glyma02g15590	2	14083370–14089609	1608	535	10.8	60.1	5	N	?
GmIQD7	Glyma03g33560	3	41092523–41096935	1434	477	10.0	53.3	5	N	?
GmIQD8	Glyma03g40630	3	46330165–46332185	1125	374	10.5	42.4	3	N	?
GmIQD9	Glyma04g02830	4	2030287–2036251	2715	904	5.4	99.2	6	N	?
GmIQD10	Glyma04g05520	4	4187757–4190317	1353	450	10.5	49.9	5	N	?
GmIQD11	Glyma04g23760	4	27192306–27195532	1353	450	9.8	50.8	5	N	?
GmIQD12	Glyma04g34150	4	40144241–40151603	1752	583	9.4	64.6	6	N	?
GmIQD13	Glyma04g41380	4	47220698–47225472	1392	463	9.6	51.7	4	N	?
GmIQD14	Glyma05g01240	5	785189–792757	1761	586	9.7	64.9	6	N	?
GmIQD15	Glyma05g03450	5	2638386–2641896	1338	445	10.0	48.9	4	N	?
GmIQD16	Glyma05g35920	5	39871246–39873985	1128	375	10.0	41.4	4	N	M0.81/3
GmIQD17	Glyma06g02841	6	1950849–1956820	2532	843	5.7	92.8	6	N	?
GmIQD18	Glyma06g05530	6	3957759–3960421	1353	450	10.7	49.8	5	E.R.	?
GmIQD19	Glyma06g13470	6	10606168–10611219	1341	446	9.7	50.1	4	N	?
GmIQD20	Glyma06g20341	6	16752231–16759304	1755	584	9.5	64.9	6	N	?
GmIQD21	Glyma07g01040	7	607467–610485	1302	433	10.0	47.9	5	N	?
GmIQD22	Glyma07g01760	7	1164144–1167157	1191	396	10.2	44.6	3	N	?
GmIQD23	Glyma07g05680	7	4335391–4339373	1641	546	10.3	61.2	5	N	?
GmIQD24	Glyma07g14910	7	14801071–14803234	1398	465	10.0	51.8	3	C	S0.90/1
GmIQD25	Glyma07g32531	7	37416802–37421879	873	290	10.6	32.8	5	N	?
GmIQD26	Glyma07g32860	7	37753882–37759623	1602	533	10.9	59.7	5	N	?
GmIQD27	Glyma08g03710	8	2630927–2633769	1311	436	10.2	48.2	3	N	M0.77/4
GmIQD28	Glyma08g20430	8	15453660–15456579	1266	421	10.4	46.4	5	N	M0.56/5
GmIQD29	Glyma08g21430	8	16271106–16273575	1209	402	10.3	45.2	3	N	?
GmIQD30	Glyma08g40880	8	40742659–40748073	1644	547	9.8	60.6	6	C	C0.75/3
GmIQD31	Glyma09g26630	9	33163730–33169453	1449	482	10.0	53.3	4	C	M0.51/4
GmIQD32	Glyma09g30780	9	37552192–37557238	1305	434	10.1	48.0	6	N	?
GmIQD33	Glyma09g35920	9	41794962–41798738	1407	468	9.9	52.6	5	N	M0.60/4

Table 1. Cont.

Name	Gene Identifier	Chr.	Location coordinates (5′-3′)	ORF length (bp)	Protein			Exons	subcellular localization	
					Length (aa.)	pI	Mol.Wt. (kD)		WoLF PSORT	TargetP
GmIQD34	Glyma10g00630	10	386683–389158	1272	423	9.5	47.5	3	N	M0.33/5
GmIQD35	Glyma10g05720	10	4477640–4481520	1425	474	10.0	52.8	5	N	?
GmIQD36	Glyma10g35721	10	43974896–43978361	1452	483	10.6	53.0	5	N	?
GmIQD37	Glyma10g38310	10	46118444–46123432	1395	464	10.4	51.0	4	C	M0.54/5
GmIQD38	Glyma10g39030	10	46764292–46767407	1410	469	9.7	52.0	4	N	?
GmIQD39	Glyma11g20880	11	17714458–17717939	1374	457	10.0	51.7	5	N	M0.67/3
GmIQD40	Glyma12g01410	12	842971–846738	1383	460	10.0	51.8	5	N	M0.56/4
GmIQD41	Glyma12g31610	12	35181013–35188577	1269	422	9.9	46.5	6	N	?
GmIQD42	Glyma12g35711	12	38833825–38837834	885	294	9.8	34.2	5	N	?
GmIQD43	Glyma13g20070	13	23539750–23543840	1413	470	10.1	52.3	5	N	C0.34/5
GmIQD44	Glyma13g24070	13	27399608–27404534	774	257	10.5	29.3	4	N	?
GmIQD45	Glyma13g30590	13	33154582–33158861	900	299	10.4	33.6	5	N	?
GmIQD46	Glyma13g34700	13	36237460–36241896	1173	390	9.8	45.5	6	N	?
GmIQD47	Glyma13g38800	13	39521853–39528595	1278	425	9.9	47.1	6	N	?
GmIQD48	Glyma13g42440	13	42441870–42445047	1239	412	10.3	45.8	5	N	?
GmIQD49	Glyma13g43031	13	42796469–42804226	1143	380	10.2	43.4	3	N	?
GmIQD50	Glyma14g11050	14	9335703–9339095	1254	417	10.3	47.3	5	N	?
GmIQD51	Glyma14g25860	14	31470493–31475301	1377	458	10.0	51.3	4	N	?
GmIQD52	Glyma15g02370	15	1595640–1598698	1137	378	10.2	43.3	3	N	?
GmIQD53	Glyma15g02940	15	2051157–2053854	1251	416	10.3	45.9	5	C	?
GmIQD54	Glyma15g08660	15	6125483–6129362	927	308	10.3	34.7	5	N	?
GmIQD55	Glyma16g02240	16	1759053–1762330	1653	550	10.2	61.6	5	N	M0.37/5
GmIQD56	Glyma16g22935	16	26564269–26565120	426	141	11.1	16.3	2	C	M0.82/4
GmIQD57	Glyma16g32161	16	35337880–35343544	1434	477	10.0	52.8	4	C	M0.56/4
GmIQD58	Glyma17g10660	17	8002515–8009332	1767	588	9.5	65.0	6	N	?
GmIQD59	Glyma17g14000	17	10763173–10767584	1344	447	10.0	48.9	4	N	?
GmIQD60	Glyma17g23770	17	23932487–23938307	1386	461	10.4	50.7	5	N	?
GmIQD61	Glyma17g34520	17	38500561–38503843	1242	413	10.4	46.7	5	N	?
GmIQD62	Glyma18g16130	18	16440695–16446996	1644	547	9.7	60.3	6	N	C0.65/4
GmIQD63	Glyma19g36270	19	43610551–43615073	1434	477	10.0	53.3	5	N	?
GmIQD64	Glyma19g43300	19	48995941–48998264	1113	370	10.6	42.2	3	N	?
GmIQD65	Glyma20g28800	20	37708013–37709907	1434	477	9.8	52.7	3	N	C0.67/5
GmIQD66	Glyma20g29550	20	38392614–38397440	1371	456	10.5	50.3	4	C	?
GmIQD67	Glyma20g31810	20	40423269–40426995	1470	489	10.4	53.7	5	C	?

bp, base pair; aa, amino acids; kD, kilo Dalton.
WoLF PSORT predictions: N (nucleus), C (chloroplast), ER (endoplasmic reticulum).
TargetP predictions: C (chloroplast), M (mitochondrion), S (secretory pathway),? (any other location); values indicate score (0.00–1.00) and reliability class (1–5; best class is 1).

Table 2. Number of IQD genes in the soybean, rice, *Arabidopsis*, tomato and *Brachypodium distachyon* genomes.

Subfamily	Soybean	Arabidopsis	Tomato	Rice	Brachypodium distachyon
I	27	13	15	11	9
II	6	4	6	1	2
III	24	10	10	10	9
IV	10	5	3	3	2
Outgroup		1		4	1
Total number	67	33	34	29	23

with corresponding cDNA sequences revealed that most of the gene models predicted by GSDS are correct, except for one pair of genes (*GmIQD9/-17*). Both *GmIQD9* and *GmIQD17* encode six exons, but GSDS predicted that these genes contain only five exons. This unconformity is caused by the missing annotation of the fifth intron by GSDS. The schematic structures reveal that the coding sequence of each IQD gene contains 2–6 translated exons (Figure 1b), which is similar to that reported in *Arabidopsis*, rice and *Brachypodium distachyon* [30,34]. More than three-fifths of the soybean IQD gene family (41 members) contain five or six protein-coding exons, and one gene (*GmIQD56*, encoding the smallest protein, comprising 141 aa) contains two exons (Figure 1b). Most closely related soybean IQD members in the same subfamilies share similar intron numbers and exon lengths. Soybean IQD genes in subfamily II and IV possess five and six exons, respectively. Most members in subfamily III contain five exons, except for *GmIQD44* (four exons) and *GmIQD32*, *GmIQD41*, *GmIQD46* and *GmIQD47* (six exons). Subfamily I genes harbor 2–5 exons. All introns of most IQD genes are in phase-0 (interrupting two triplet codons exactly); a phase-1 intron (separating the first and second nucleotides of a codon) was found in 15 remaining IQD genes, and no phase-2 intron (splitting the second and third nucleotides of a codon) was found (Figure 1b).

The exon/intron organization of 31 paralogous pairs that clustered together at the terminal branch of the phylogenetic tree was further examined to obtain traceable intron gain/loss information. Although twenty-seven paralogous pairs exhibited conserved exon/intron structures, four pairs (*GmIQD16/-27*, *GmIQD38/-65*, *GmIQD25/-44* and *GmIQD42/-46*) showed certain variations (Figure 1b). These differences may have been derived from single intron loss or gain events during the long evolutionary period. Based on analysis of the exon/intron organization of IQD genes from soybean, *Arabidopsis* [30], rice [30], and *Brachypodium distachyon* [34], we infered that both *GmIQD16* and *GmIQD38* gained the third intron; *GmIQD46* gained the first intron while *GmIQD44* lost the first intron. The second or third exons in the central regions of most members encode amino acids 17–67 of the IQ67 domain, except for *GmIQD46* (the fourth exon) and *GmIQD56* (the C-terminal exon), with a conserved phase-0 intron separating codons 16 and 17 (Figure 1b and Figure S1).

A total of 67 IQD genes from soybean were subjected to analysis with MEME to reveal conserved domains or motifs shared among related proteins. We identified 10 conserved motifs (Figure 2 and Table S1). Each of the putative motifs was annotated by searching Pfam and SMART. Motif 1 was found to encode the IQ domain. Motif 2 and motif 7 were found to encode proteins of unknown function (DUF4005) and (DUF3982). While the other subfamily-specific motifs have not functional annotation. As expected, most of the closely related members had common motif compositions, suggesting functional similarities among IQD proteins within the same subfamily (Figure 2). The most common motif is motif 1, found in all sixty-seven soybean IQD genes (Figure 2). Motif 8 is mainly present in subfamily I besides one of *GmIQD14* exists in subfamily IV. Subfamily III members contain motif 1, motif 10, motif 4 and motif 3 in order, except for *GmIQD28* lacking motif 10. Motif 7 is peculiar to subfamily IV. To some extent, these subfamily-specific motifs may contribute to the functional divergence of IQD genes in soybean. The detailed information is shown in table S1. To predict calmodulin-binding sites, we searched the Calmodulin Target Database, which provides various structural and biophysical parameters for the 67 soybean IQD protein sequences. This analysis predicted that all soybean IQD proteins contain multiple IQ motifs and 1–3 strings of high-scoring amino acid residues (Table 3). These IQ motifs and amino acid residues indicate the locations of putative calmodulin interaction sites. Among the 67 IQD protein sequences, the predicted calmodulin binding sites of 50 sequences overlap with the IQ67 domain (Figure 2).

Chromosomal location and gene duplication

The 67 soybean IQD genes were mapped to all 20 soybean chromosomes. The distribution of soybean IQD genes varies depending on the chromosome and appears to be unequal. Both chromosomes 11 and 18 contain only one soybean IQD gene, while chromosomes 13, which possesses seven IQD genes, has the highest number of IQD genes per chromosome. Although high densities of IQD genes were found on some chromosomal regions, for instance, the bottom of chromosome 13, these is no substantial clustering of soybean IQD genes on the map (Figure 3).

We investigated gene duplication events to further understand the expansion mechanism of the soybean IQD family. Except for three genes (*GmIQD11*, *-39* and *-51*) located outside of a duplicated block, 64 genes were mapped onto 48 related duplicated blocks (Figure 3 and Table S2). Among these, twenty-two block pairs retained thirty GmIQD gene pairs, whereas the remaining four duplication blocks harbor *GmIQD3*, *-32*, *-56* and *-60* respectively, but lack IQD sisters in their corresponding synteny blocks (Figure 3 and Table S2). Analysis of GmIQD paralogous pairs showed that one pair (*GmIQD11/-39*) appear to be closely related paralogs, sharing 91.2% identity (Table S3) as well as similar exon–intron organization. However, both of them exist outside of any duplicated blocks. Except for *GmIQD11/-39*, 30 out of 31 gene pairs have remained in conserved positions on segmental duplicated blocks, indicating that these genes were generated by segmental duplication. Furthermore, we analyzed the adjacent genes to determined whether tandem duplication has taken place. A pair of genes separated by three or fewer genes within a 100-kb region on a

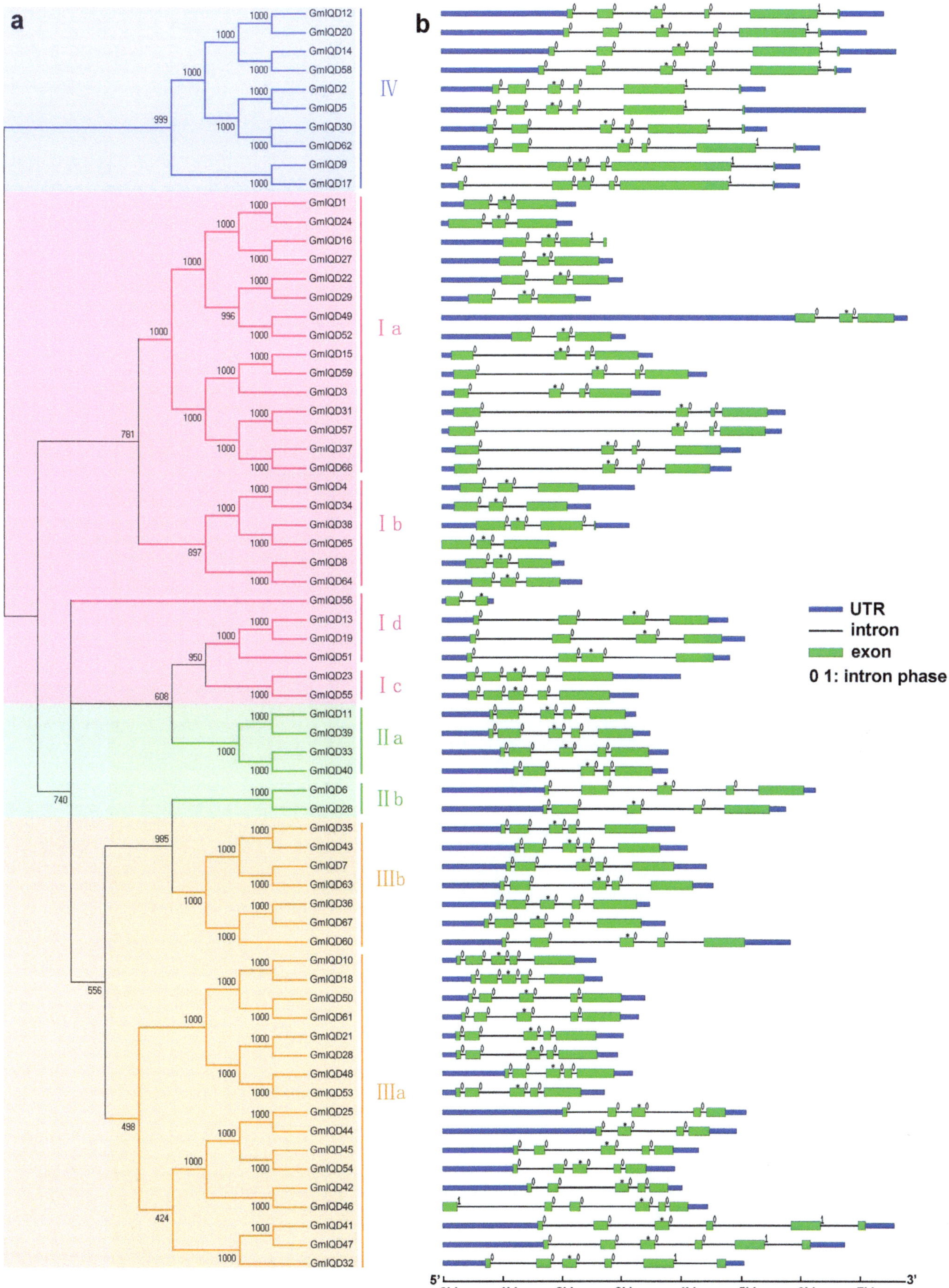
a
b
GmIQD12
GmIQD20
GmIQD14
GmIQD58
GmIQD2
GmIQD5
GmIQD30
GmIQD62
GmIQD9
GmIQD17
GmIQD1
GmIQD24
GmIQD16
GmIQD27
GmIQD22
GmIQD29
GmIQD49
GmIQD52
GmIQD15
GmIQD59
GmIQD3
GmIQD31
GmIQD57
GmIQD37
GmIQD66
GmIQD4
GmIQD34
GmIQD38
GmIQD65
GmIQD8
GmIQD64
GmIQD56
GmIQD13
GmIQD19
GmIQD51
GmIQD23
GmIQD55
GmIQD11
GmIQD39
GmIQD33
GmIQD40
GmIQD6
GmIQD26
GmIQD35
GmIQD43
GmIQD7
GmIQD63
GmIQD36
GmIQD67
GmIQD60
GmIQD10
GmIQD18
GmIQD50
GmIQD61
GmIQD21
GmIQD28
GmIQD48
GmIQD53
GmIQD25
GmIQD44
GmIQD45
GmIQD54
GmIQD42
GmIQD46
GmIQD41
GmIQD47
GmIQD32
IV
I a
I b
I d
I c
II a
II b
IIIb
IIIa
UTR
intron
exon
0 1: intron phase
5'
3'
0kb
1kb
2kb
3kb
4kb
5kb
6kb
7kb

Figure 1. Phylogenetic relationships and exon/intron organization of soybean IQD genes. a: Unrooted tree generated with Clustal X2.0 using the full-length amino acid sequences of the 67 soybean IQD proteins by the Neighbor-Joining (NJ) method with 1,000 bootstrap replicates. Subfamilies and subclasses of IQD genes (I–IV) are highlighted with different colored backgrounds and vertical bars next to the gene names of the tree. **b:** Exon/intron organization of soybean IQD genes. Green boxes represent exons and black lines represent introns. Untranslated regions (UTRs) are indicated by blue boxes. Numbers 0 and 1 represent the splicing phases. The sizes of exons and introns can be estimated using the scale at the bottom. The exon encoding amino acids 17–67 of the IQ67 domain in each soybean gene is indicated with an asterisk.

chromosome may have resulted from tandem duplication. According to this criterion, no pair was found to be generated by tandem duplication. Therefore, segmental duplication appears to have played a crucial role in the expansion of the IQD gene family in soybean (Figure 3 and Table S2).

To explore the selective constraints on duplicated soybean IQD genes, we calculated the ratio of nonsynonymous versus synonymous substitutions (Ka/Ks) for each pair of duplicated IQD genes. In general, a ratio of 1 indicates that both genes are drifting neutrally; a Ka/Ks ratio >1 indicates accelerated evolution with positive selection, while a ratio <1 indicates functional constraint, with negative or purifying selection of the genes[44]. The Ka/Ks ratios from 31 soybean IQD paralogous pairs (Table 4) were less than 0.6. This result suggests that the soybean IQD gene family has evolved mainly under the influence of strong purifying selection pressure, with limited functional divergence occurring after segmental duplication. Duplication of these 31 paralogous pairs was estimated to have occurred between 6.39 to 17.94 Mya (Table 4), according to the divergence rate of 6.1×10^{-9} synonymous mutations per synonymous site per year, as previously proposed for soybean [45,46].

Comparative analysis of the IQD genes in soybean, *Arabidopsis*, rice, tomato and *Brachypodium distachyon*

The development of comparative genomics has enabled the analysis of the same protein families among different species. We constructed an NJ phylogenetic tree using 184 full-length protein sequence to reveal the evolutionary relationships among soybean, *Arabidopsis*, rice, tomato and *Brachypodium distachyon* IQD proteins [34]. In *Arabidopsis*, the IQD gene family is divided into four subfamilies, with *AtIQD33* (containing a C-terminally truncated IQ67 domain) as the outgroup. Therefore, based on their phylogenetic relationships, the combined phylogenetic tree can be divided into five distinct subfamilies (I to V; Figure 4) [30]. In general, IQD I genes comprise the largest subfamily in these plant species, except for *Brachypodium distachyon*, where both IQD I and III comprise the largest subfamilies. By contrast, IQD V genes comprise the smallest IQD subfamily (Figure 4, Table 2).

To illustrate the paralogous and orthologous relationships among IQD family members, the subfamilies were further divided into subgroups using previously defined clades from studies of *Arabidopsis*, rice and tomato IQDs, as shown in Figure 4. IQD subfamily I was divided into four subclasses, i.e., a, b, c and d, and clade b was further divided into two clades, b_1 and b_2. Because one of the IQD Ib clades only contains four IQD genes (*BdIQD11*, *BdIQD19*, *OsIQD19* and *OsIQD20*) from monocots, we assigned these four genes to the rice- and *Brachypodium distachyon*-specific Ib_2 clade. The clade containing the genes encoding C-terminal IQ67 domains was defined as Id. Notably, no members of *Brachypodium distachyon* were detected in this clade, suggesting that *Brachypodium distachyon* IQD family lost its members of this subgroup during the long period of evolution. Both IQD II and IQD III subfamilies were divided into two subclasses, a and b, which were designated as described by Zejun et al.(2013) and Abel et al. (2005) [35]. The C-terminally truncated IQ67 domain-containing genes (*At IQD33*, *OsIQD28* and *BdIQD14*) comprise IQD V subfamily (Figure 4) [30,34].

The combined phylogenetic tree reveals that most genes in the IQD family, especially the duplicated genes, are contained in paralogous pairs in each species, which supports the occurrence of species-specific IQD gene duplication events. By contrast, we identified 20 pairs of orthologous genes from monocotyledons (rice and *Brachypodium distachyon*) distributed among all of the subfamilies. In addition, two pairs of orthologous genes from dicotyledons (soybean and tomato) stemming from subfamily I (*GmIQD56* and *SlSUN9*) and subfamily III (*GmIQD60* and *SlSUN13*) were found. And *AtIQD20* and *OsIQD26*, members of subfamily I, formed a pair of orthologous genes.

Conserved microsynteny of IQD III genes from soybean, *Arabidopsis* and tomato

The analysis of microsynteny provides valuable information for identifying gene expansion patterns and inferring gene orthology or paralogy. We combined genetic and phylogenetic analyses to perform microsynteny analysis of three dicotyledons, i.e., soybean, tomato and *Arabidopsis*.

To provide a basic framework for the identification of IQD III orthologous or paralogous genes, 44 IQD III genes, including 24 predicted soybean IQDs, 10 *Arabidopsis* IQDs and 10 tomato IQDs, were classified into four distinct clades, clade 1 (thirteen genes), clade 2 (five genes), clade 3 (eleven genes) and clade 4 (fifteen genes), based on phylogenetic analysis (Figure S2). Clade 1, 2 and 3 correspond to IQD IIIa and clade 4 corresponds to IQD IIIb (Figure 4 and S2). Each clade contains at least one gene from soybean, tomato and *Arabidopsis*, indicating that members from different species may be derived from a common ancestor.

Subsequently, to produce a comparative genetic map, 44 IQD III genes from the three dicot genomes were used as anchor genes. Conserved microsynteny was identified through reciprocal pairwise comparisons of the chromosomal regions containing IQD III genes. Microsynteny relationships among *AtIQD3*, *AtIQD4*, *AtIQD5*, *GmIQD32*, *GmIQD56* or *SlSUN13* with other IQD III members in these three dicot genomes were not observed. The map reveals that the 38 conserved syntenic segments diverged into four groups (Figure 5), which were anastomosed with the classification revealed by phylogenetic tree analysis.

In clade 1 (Figure 5a), Map a shows a higher level of microsynteny, with both the same and opposite directions. *SlSUN11/GmIQD21* and *GmIQD53/GmIQD48* exhibit remarkable opposite-direction microsynteny, while *GmIQD10/GmIQD18*, *GmIQD21/GmIQD28*, *AtIQD7/AtIQD8* and *AtIQD8/SlSUN11* are aligned with flanking gene pairs in the same order but discordant transcriptional orientation. In addition, genes in map a were divided into two groups (Figure 5a), i.e., one group with higher levels of microsynteny (*GmIQD21*, *GmIQD28*, *GmIQD53*, *GmIQD48*, *AtIQD7*, *AtIQD8* and *AtIQD8* and *SlSUN11*) and the other group with lower levels of microsynteny (*GmIQD61*, *GmIQD50*, *GmIQD10*, *GmIQD18*, *AtIQD6* and *SlSUN22*). These two groups were also detected in the phylogenetic tree of IQD III genes (Figure S2). In clade 2 (Figure 5b), two pairs from soybean and tomato, *GmIQD41/*

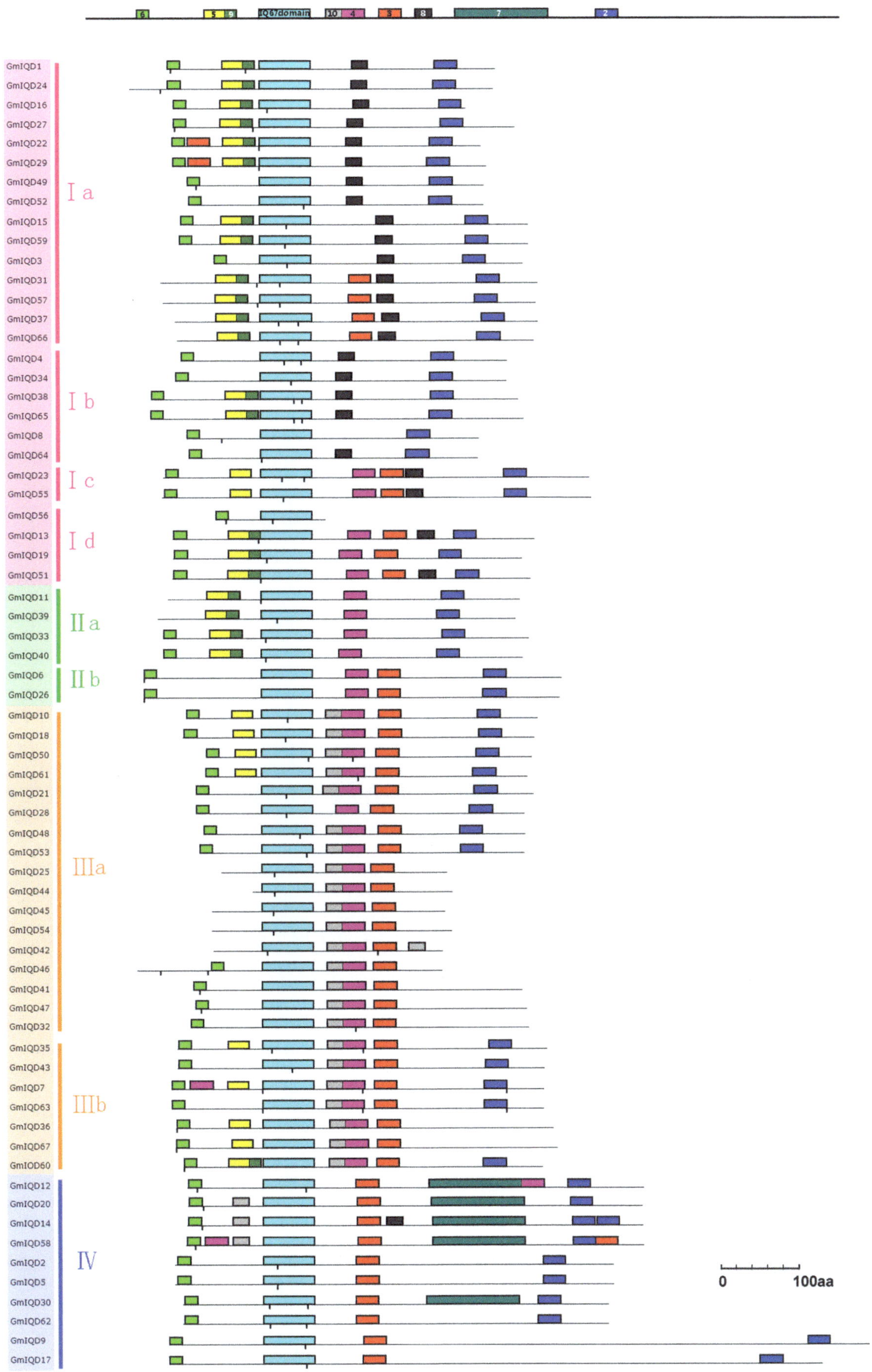
6
5
9
IQ67domain
10
4
3
8
7
2
GmIQD1
GmIQD24
GmIQD16
GmIQD27
GmIQD22
GmIQD29
GmIQD49
GmIQD52
GmIQD15
GmIQD59
GmIQD3
GmIQD31
GmIQD57
GmIQD37
GmIQD66
Ⅰa
GmIQD4
GmIQD34
GmIQD38
GmIQD65
GmIQD8
GmIQD64
Ⅰb
GmIQD23
GmIQD55
Ⅰc
GmIQD56
GmIQD13
GmIQD19
GmIQD51
Ⅰd
GmIQD11
GmIQD39
GmIQD33
GmIQD40
Ⅱa
GmIQD6
GmIQD26
Ⅱb
GmIQD10
GmIQD18
GmIQD50
GmIQD61
GmIQD21
GmIQD28
GmIQD48
GmIQD53
GmIQD25
GmIQD44
GmIQD45
GmIQD54
GmIQD42
GmIQD46
GmIQD41
GmIQD47
GmIQD32
Ⅲa
GmIQD35
GmIQD43
GmIQD7
GmIQD63
GmIQD36
GmIQD67
GmIQD60
Ⅲb
GmIQD12
GmIQD20
GmIQD14
GmIQD58
GmIQD2
GmIQD5
GmIQD30
GmIQD62
GmIQD9
GmIQD17
Ⅳ
0
100aa

Figure 2. Motif patterns in 67 IQD proteins of soybean. The schematic soybean IQD proteins were aligned relative to the IQ67 domain (motif 1, light blue box). The lengths of the proteins and motifs can be estimated using the scale. Motifs shared by at least four soybean IQD proteins are depicted at the reference bar on top of each alignment. The positions of putative calmodulin-binding sites predicted by the Calmodulin Target Database are indicated by vertical tick marks below each protein model. Subfamilies and subclasses (I–IV) of IQD proteins are highlighted with colored backgrounds and colored vertical bars as in Figure 1a to the right of the gene names.

SlSUN31 and *GmIQD47/SlSUN31*, exhibited microsynteny. However, the predicted duplicated pair *GmIQD41/GmIQD47* had no detectable linkage with each other. High level of microsynteny exists in Clade 3, with most pairs in reverse order (Figure 5c), especially *GmIQD42/GmIQD46*, *GmIQD45/GmIQD54* and *GmIQD25/GmIQD44*. *GmIQD45/GmIQD44* and *GmIQD44/SlSUN33* were identified as having same-direction microsynteny. In clade 4 (Figure 5d), we also observed a higher level of microsynteny. Except for *AtIQD2/SlSUN29*, *SlSUN30/SlSUN29* and *GmIQD67/GmIQD36*, which are aligned in the opposite direction, most gene pairs in this clade have successive collinearity in order and the same orientation.

Two regions are considered to have originated from a large-scale duplication event when five or more protein-coding gene pairs flanking the anchor point are ligatured with the best non-self match (E-value <1e−10) [47,48]. Applying this standard, except for the pair *GmIQD41/GmIQD47*, all soybean IQD III paralogous gene pairs were generated from a large-scale duplication event, which further supports the results of soybean gene duplication analysis (Figure 5, Table S4).

To estimate the degree of conserved gene content and order, the synteny quality was calculated [47]. The average synteny quality of IQD III genes from the three dicotyledons genomes was 18.41% (Table S5d). Due to the large number of syntenic genes between tomato and soybean, the synteny quality between these genomes is 26.39%; this value is higher than that observed in the Sl/At synteny blocks (16.68%). The lowest synteny quality (12.15%) was found between soybean and *Arabidopsis* (Table S5). Details of this comparative analysis are shown in Table S5.

Expression patterns of soybean IQD genes in various tissues

To gain insight into the expression patterns of soybean IQD genes in various tissues, we searched the RNA-Seq Atlas of *Glycine max*; this atlas provides high-resolution gene expression data from 14 diverse tissues, including aerial tissues (young leaf, flower, one-cm pod, pod-shell 10-DAF and pod-shell 14-DAF), underground tissues (root and nodule) and seed tissue at various stages of development (seed 10-DAF, seed 14-DAF, seed 21-DAF, seed 25-DAF, seed 28-DAF, seed 35-DAF and seed 42-DAF). Because the expression profiles of eight IQD genes (*GmIQD17*, *-20*, *-25*, *-36*, *-42*, *-49*, *-56*, *-57*) weren't obtained in the database, we only examined the expression patterns of fifty-nine IQD genes (Figure 6 and Table S6).

Most soybean IQD genes exhibit broad expression patterns (Figure 6). Forty-four soybean IQDs are expressed in all of the seven tissues (young leaves, flowers, one-cm pod, pod-shell, roots, nodules and seed). The heat map also revealed that the majority of GmIQDs showed preferential expression. Based on a hierarchical clustering analysis, fifty-nine IQD genes were mainly clustered into four groups (A–D) (Figure 6). Group A showed partial expression in young leaves, group B in roots, group C in nodules and group D in flowers. Eight GmIQDs (*GmIQD8*, *-13*, *-19*, *-29*, *-36*, *-52*, *-64* and *-67*) showed marked high transcript abundance profiles in only a single tissue. Among the fifty-nine soybean IQD genes examined, six showed the highest transcript accumulation in young leaves (*GmIQD13*, *-18*, *-19*, *-29*, *-50* and *-61*), six in flowers (*GmIQD11*, *-15*, *-38*, *-52* *-65* and *-67*), one in roots (*GmIQD26*) and two in nodules (*GmIQD8* and *-64*; Figure 6). Genes in different subfamilies have their primary abundant transcripts, for instance, GmIQD I in leaves, flowers and nodules, GmIQD II in flowers and roots, GmIQD III in young leaves, flowers and roots and GmIQD IV in roots and young leaves (Figure 6). These subfamily-specific tissue expression patterns may be closely related to gene functions. The expression patterns of the paralogous pairs were also revealed by heat maps; paralogous pairs with high sequence similarity have similar expression patterns. The best examples of this include *GmIQD8/-64* and *GmIQD6/-26*, which are strongly expressed in nodules and root respectively, with little or no expression in other tissues. Expression divergence was also found in paralogous pairs. For example, *GmIQD15* is highly expressed in flowers, while its paralog, *GmIQD59*, is highly expressed in nodules.

Examination of soybean IQD gene expression by qRT-PCR

Since soybean production is limited by stress, it is important to identify the master regulators of stress responses in soybean, as well as their regulatory pathways. According to microsynteny analysis, the high level of microsynteny indicates that IQD III genes existed before the divergence of the three dicotyledon genomes examined (soybean, tomato and *Arabidopsis*). In addition, IQD III genes in the same clade may share common ancestors and play similar roles in these species. *AtIQD1*, which belongs to the IQD III subfamily, plays a major role in the response to biotic stress, as it mediates the accumulation of glucosinolate in response to phytophagous insect attack. Jasmonic acid methyl ester (MeJA), the plant hormones and the signal molecules, widely exists in plants, which triggers expression of plants defense genes by exogenous applications and has similar effects with mechanical damage and insect herbivory [49,50]. Based on these, we subjected 24 members of the soybean IQD III subfamily to real-time quantitative PCR (qRT-PCR) analysis to examine their regulation by MeJA.

The qRT-PCR results show that all 24 genes are MeJA-responsive, but some differences were observed among these genes (Figure 7). Although 23 genes were upregulated by MeJA treatment, *GmIQD21* was obviously downregulated (<0.5 folds) at all time points. Eleven of the twenty-three upregulated GmIQD III genes exhibited minor changes in expression (relative expression scale from 0 to 5 and lower), including *GmIQD10*, *-18*, *-21*, *-25*, *-42*, *-47*, *-53*, *-61*, *-7*, *-36* and *-63*. By contrast, 12 genes (*GmIQD2*, *-32*, *-41*, *-44*, *-45*, *-46*, *-48*, *-50*, *-53*, *-54*, *-35*, *-43*, *-60* and *-67*) exhibited major changes in expression (relative expression scales from 0 to 5 up 0 to 80). The expression of six genes (*GmIQD35*, *-36*, *-47*, *-54*, *-63* and *-67*) peaked relatively early (at 1 h of treatment); *GmIQD54* and *-67* were strongly upregulated (more than 26-fold and 34-fold, respectively). Eight genes (*GmIQD10*, *-18*, *-28*, *-41*, *-42*, *-43*, *-53* and *-60*) were highly expressed at 4 h; *GmIQD28* and *-60* had the highest expression level more than 12-fold and *GmIQD41* had the highest expression level more than 28-fold. While seven genes (*GmIQD7*, *-25*, *-32*, *-44*, *-46*, -48 and *-61*) exhibited the highest expression levels at 8 h; *GmIQD48* were strongly induced by more than 35-fold. Only one gene (*GmIQD45*) had the highest expression level

Table 3. Predicted calmodulin binding sites in soybean IQD proteins.

Group	Name	Gene Identifier	Predicted calmodulin binding sequence	
Ia	GmIQD1	Glyma01g01030	7-WVKSLFGIRREKEKKLN	100-V**AVVRLTSQGRGRTMFG**
	GmIQD3	Glyma01g42620	94-VRGH**IERK**RTAEW	
	GmIQD15	Glyma05g03450	136-LVRG**HIERKR**TAEWL	
	GmIQD16	Glyma05g35920	120-G**QERLAVVKIQT**FFR	
	GmIQD22	Glyma07g01760	109-FSGS**REKWA**AVKI	
	GmIQD24	Glyma07g14910	39-MGRATRW**VKSLFGIRKE**	
	GmIQD27	Glyma08g03710	2-GRAIRWLKGLFGIRTDRER	102-RDTTFGGAGQERL**AVVKI**
			164-LIRAQATVRSKKSRNEAHR	
	GmIQD29	Glyma08g21430	108-FSGS**REKWA**AVKI	
	GmIQD31	Glyma09g26630	123-RRVAEETTA**AAVKIQSAFR**	153-K**ALVKLQALVRGHIVRKQT**
	GmIQD37	Glyma10g38310	136-ALVKLQALVRGHIVRKQS	158-**RRMQTLVRLQAQARASRA**
	GmIQD49	Glyma13g43031	8-**LKG LLGKKKEKDYCGY**	
	GmIQD52	Glyma15g02370	148-**AQAVARSVRARRSM**	
	GmIQD57	Glyma16g32161	121-RVANETTA**AAVKIQSAFRG**	150-K **ALVKLQALVRGHIVRKQT**
	GmIQD59	Glyma17g14000	137-LVRG**HIERKR**TAEW L	
	GmIQD66	Glyma20g29550	133-LKALVKLQALVRGHIVRKQS	155-**RRMQTLVRLQAQARASRA**
Ib	GmIQD4	Glyma02g00710	133-LQALVRGHLVRKQ**A**RETL	155-AL**VIAQ**SRARAQRA
	GmIQD8	Glyma03g40630	46-RR**WSFGKLTGAGH**KF	
	GmIQD34	Glyma10g00630	148-LVRK**QARETLRCIQ**ALVIA	
	GmIQD38	Glyma10g39030	181-RKQ**AKATL**RC	193-ALVTAQ
	GmIQD64	Glyma19g43300	92-KDK**NKAATKIQA**SF	
	GmIQD65	Glyma20g28800	182-RKQ**AKATL**RC	194-AL**VTAQ**AR
Ic	GmIQD23	Glyma07g05680	152-LV**KLQALVRGHNVR**KQA	180-RVQARVLDQRIRSSL
	GmIQD55	Glyma16g02240	154-LV**KLQALVRGHN**VR	
Id	GmIQD13	Glyma04g41380	109-YGRQ**SKEERAAILIQ**SYYR	
	GmIQD19	Glyma06g13470	119-IL**IQSYYRGYL**ARRALRALKG	
	GmIQD51	Glyma14g25860	111-RQ**SKEERAATLIQ**SYYRGYLARRALRAL	
	GmIQD56	Glyma16g22935	13-RGRFLRSS	73-GHLARR**AYKALKSLVKLQA**LVR
IIa	GmIQD11	Glyma04g23760	119-**KIQESSAIKIQIAFRGY**L	
	GmIQD33	Glyma09g35920	125-IKESA**AAIKIQTAFRG**Y	
	GmIQD39	Glyma11g20880	132-KIQES**SAIKIQTAYRGYL**A	
	GmIQD40	Glyma12g01410	125-IKESA**AAIKIQTAFRG**Y	
IIb	GmIQD6	Glyma02g15590	1-**MGKKGSWFSAI**	
	GmIQD26	Glyma07g32860	1-**MGKKGSWFSAI**	
IIIa	GmIQD10	Glyma04g05520	131-VRG**RQVRKQAAVTLRCMQ**ALVRVQA	
	GmIQD18	Glyma06g05530	136-VRG**RQVRKQAAVTLRCMQ**ALVRVQAR	
	GmIQD21	Glyma07g01040	117-AIFR**GWQVRKQAAVTLR**CMQ	
	GmIQD25	Glyma07g32531	67-AYK**ARKYL**HRLR	
	GmIQD28	Glyma08g20430	117-AIFR**GWQVRKQAAVTLR**CMQ	
	GmIQD32	Glyma09g30780	205-RQEAAA**KRGRAMAY**AL	
	GmIQD41	Glyma12g31610	3-V**SGKWIKALVGLKKSEKP**G	90-R EELAAIRIQTAFRGFLA
			207-AKRERAMAYALSHQWQAG	
	GmIQD42	Glyma12g35711	68-AATRIQNAFRSFMARRTL	210-**LGKESWGWSWTERWVAAR**
	GmIQD44	Glyma13g24070	27-AYK**ARKYLH**RLRG	
	GmIQD45	Glyma13g30590	78-RAY**KARKAL**RRMKGFTK**LKI**LTEG	
	GmIQD46	Glyma13g34700	25-EIKHLIQRGWVV	90-LKR**NKRMGAK**KWF
	GmIQD47	Glyma13g38800	3-**VSGKWIKALVGLKKSEKP**	204-**AKRERAMAYALSHQWQAG**
	GmIQD48	Glyma13g42440	123-LRCMQA**LVRVQARVRA**R	
	GmIQD50	Glyma14g11050	126-VRVQARVRAR	187-GAF**K RERAIAYS**LA
	GmIQD53	Glyma15g02940	138-LRCMQA**LVRVQARVRA**R	

Table 3. Cont.

Group	Name	Gene Identifier	Predicted calmodulin binding sequence	
	GmIQD54	Glyma15g08660	78-RAY**KARKAL**RRMKGFTK**LKI**LTEG	
	GmIQD61	Glyma17g34520	197-EGA**F KR**ERAIAYSL	
IIIb	GmIQD7	Glyma03g33560	116-**PKDEVAAIKIQTAFRGY**L	227-LSKYEATTRRERALAYA
			427-NG**KAEKGSFGSAKKRL**SF	
	GmIQD35	Glyma10g05720	111-EEM**AAIRIQKAFRGYLA**	218-KLLSKYEASMRRERAMAYS
	GmIQD36	Glyma10g35721	1-**MGRKGGWFSAV**	292-**HASAKSVASQTMSV**
	GmIQD43	Glyma13g20070	126-LARR**ELR**ALRGLV	
	GmIQD60	Glyma17g23770	1-**MGKKGSWFSAV**	
	GmIQD63	Glyma19g36270	116-**PKDEVAAIKIQTAFRGY**L	227-LSKYEATMRRERALAYA
			427-NA**KAEKGSFGSAKKRL**SF	
	GmIQD67	Glyma20g31810	1-**MGRKGGWFSAV**	293-**HASAKSVASQTMSV**
IV	GmIQD2	Glyma01g05100	130-LARQ**TFK**KLEGV	175-RGYNVRRS
	GmIQD5	Glyma02g02370	130-LARR**TLQ**KLKGV	
	GmIQD9	Glyma04g02830	173-QA**IIKMQILVRARR**AR	
	GmIQD12	Glyma04g34150	13-LFGKKSS**K SNI**SK	153-KLQALVRGGRIRQS
	GmIQD14	Glyma05g01240	19-SKS**NISKGRE**KLV	
	GmIQD17	Glyma06g02841	175-II**KMQILVRARRA**WQ	
	GmIQD20	Glyma06g20341	20-KS**NIS**KGRE	
	GmIQD30	Glyma08g40880	113-QAA**IRGYQ**ARG	163-LARGYKVRHS
	GmIQD58	Glyma17g10660	12-VL**FGKKSSKSNI**SK	
	GmIQD62	Glyma18g16130	115-IRGYQARGTFKTL	161-QA**LARGYKVRHS**DV

Predicted calmodulin binding sites obtained from the Calmodulin Target Database are shown for strings of amino acid residues with a score of at least 7. Residues with the highest score (9) are highlighted in bold. Numbers before strings indicate the location of the first amino acid residues of the strings in soybean IQD protein sequences.

at 12 h, with a relative expression level approaching 70-fold. We also compared the expression profiles of paralogous pairs. Most paralogs in a pair had different expression profiles. For example, the expression of *GmIQD28* peaked at 4 h while its sister gene, *GmIQD21*, was downregulated at all time points, suggesting that these genes may play diverse roles in the response to MeJA stress.

Discussion

Structural characteristics of IQD proteins

The plant-specific IQD gene family has previously been comprehensively analyzed in *Arabidopsis*, rice, tomato and *Brachypodium distachyon*, this gene family has not been previously identified and annotated in soybean. We identified and characterized 67 IQ67 domain-encoding genes in soybean using genome-wide analysis. The IQD gene family in soybean is by far the largest one compared to that in other plant species (33 in *Arabidopsis*, 29 in rice, 34 in tomato and 23 in *Brachypodium distachyon*). At ~1,150 Mbp, with ~46,400 predicted coding genes, soybean possesses 9.2-fold larger genome size and 1.75-fold higher gene count than *Arabidopsis*, which has a genome of 125 Mbp and ~26,500 coding genes [51]. Given the obvious difference in genome size and estimated gene count between soybean and *Arabidopsis*, the IQD genes in soybean seems to be highly expanded. The presence of twice as many of these genes in soybean versus *Arabidopsis* may be mainly due to the recent polyploidy event and segmental duplication events in soybean's evolutionary history. It can be speculated that the presence of more IQD genes in soybean genome may reflect the great needs for these genes coding for calcium signal regulatory components with functions in plant development, defense response or others.

The common feature of IQ67 domain proteins is the arrangement of three IQ motifs separated by 11 and 15 intervening amino acid residues (Figure S1). To date, at least five protein families containing IQ motifs, which play a role in the calcium signaling pathway, have been identified in *Arabidopsis*. These protein families include the cyclic nucleotide gated channels family (CNGC), the IQ-Motif family (IQM), the CaM-binding transcriptional activator family (CAMTA), the myosin family and the IQD family, which contain one, one, two, five and up to three IQ motifs, respectively [52–54]. The unique spacing of IQ motifs and exon/intron organization of each family suggest that these IQD protein families represent separate classes of putative calmodulin targets. The calmodulin-interacting peptides in AtIQD20 and CNGC proteins, which were experimentally verified, were previously predicted using the algorithm provided by the Calmodulin Target Database successfully [30]. In the current study, using the Calmodulin Target Database, we detected calmodulin-binding sites in all soybean IQD proteins, which strongly suggests that all IQD proteins have the potential to interact with calmodulin (Figure 2 and Table 3). Three aspects of IQD proteins appear to underlie the mechanism of interaction between IQD proteins and calmodulin: the number and specific composition of the IQ, 1-5-10 and 1-8-14 motifs, the predicted calmodulin binding site and the overall tertiary structure of the IQD protein.

Of the 31 soybean IQD paralogs examined, 27 exhibit highly conserved exon-intron structures, which is consistent with the high

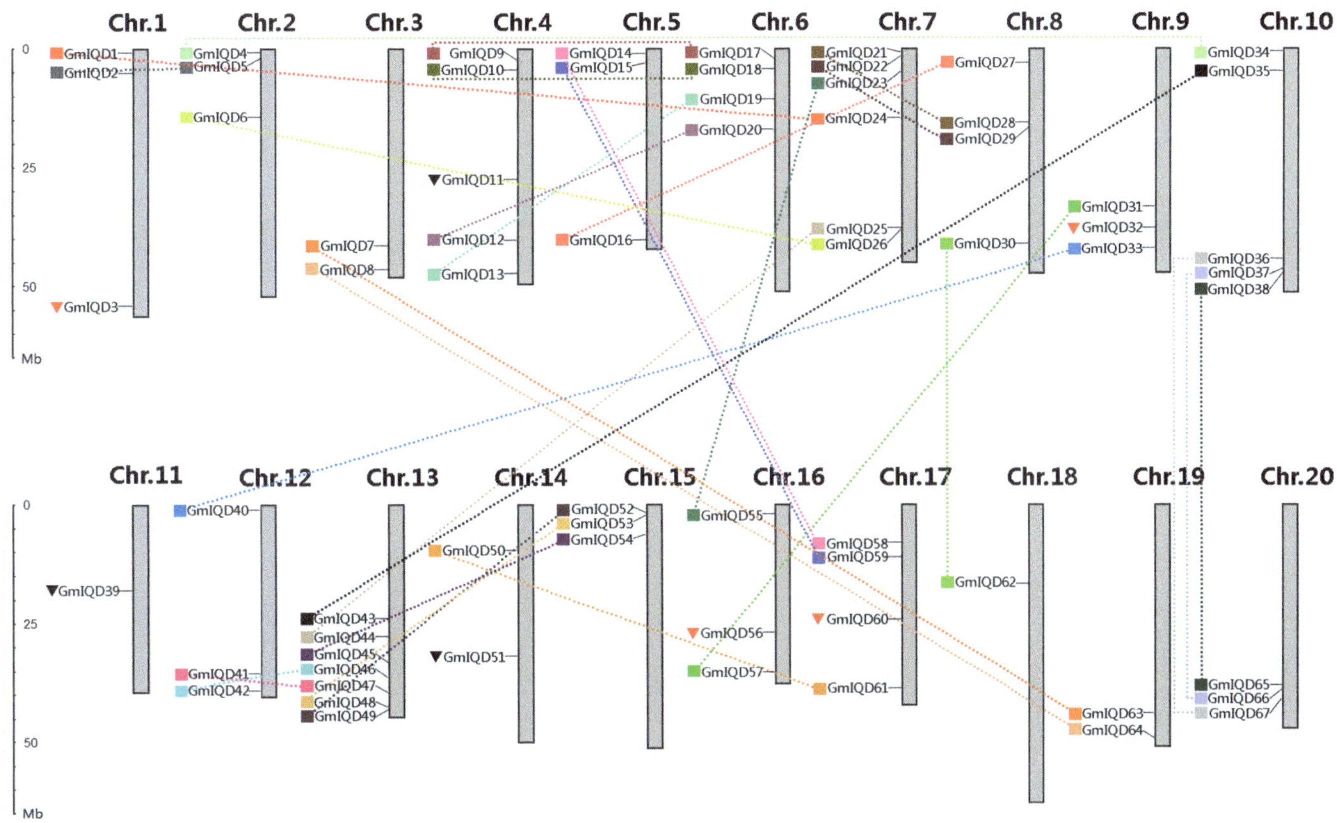

Figure 3. Chromosomal distribution and segmental duplication events for soybean IQD genes. The 67 IQD genes were mapped to the 20 soybean chromosomes. The duplicated paralogous pairs of IQD genes in the segmental duplicated blocks are indicated with small boxes of the same color and connected by dashed lines of the same color. Red triangles represent soybean IQD genes located on duplicated segments with the corresponding members lost. Black circles indicate soybean IQD genes not located in duplicated regions. Scale represents the length of the chromosome.

degree of position and phase conservation broadly found across angiosperms [41]. In addition, the sizes of related introns between paralogs are also highly conserved, indicating that few insertions and deletions have accumulated within introns over the past 13 million years [41]. Most introns in GmIQD genes are in phase-0. This strong bias for phase-0 introns in soybean IQD genes is also found in IQD genes of *Arabidopsis*, rice and *Brachypodium distachyon*. The strong bias for one intron phase class, along with the variation in the number of exons (two to six) and the sizes of encoded proteins, suggests that exon shuffling has played a prominent role during the evolution and diversification of IQD genes [30].

65 of 67 soybean IQD proteins have relatively high isoelectric points with an average of 10.1. It is very similar to *Arabidopsis* (10.3), rice (10.4) and *Brachypodium distachyon* (10.3) [30,34]. The extensive presence of the basic isoelectric point and high frequency of serine residues (*Arabidopsis*: ~11%, rice: ~11%, *Brachypodium distachyon*: ~11.5% and soybean: ~12%; Table S7) in IQDs suggest that the basic nature of IQDs is crucial to their biochemical functions [30,34]. The high isoelectric points are evocative of RNA-binding proteins although IQD proteins don't comprise currently known RNA-binding motifs. Fifty-seven soybean IQD proteins are localized to the nucleus, because of their high content of basic residues revealed by Wolf PSORT. TargetP analysis revealed that fifteen soybean IQD proteins are located in the mitochondria by identifying the presence of mitochondrial targeting peptide (mTP). The contradicting subcellular localization predictions is due to the different algorithm used by Wolf PSORT and TargetP. Most soybean IQD protein members are likely to function in the nucleus, as nucleus specific Ca^{2+}-signatures are reported to generate in plant cells [55–57] and calmodulin and related Ca^{2+}sensor proteins may play a regulatory role in nuclear processes such as transcription [58,59]. Observably, *Arabidopsis IQD1* was rereraled to target to microtubules as well as the cell nucleus and nucleolus [32]. In vitro binding to single-stranded nucleic acids suggests *AtIQD1* and other IQD family members may control and fine-tune gene expression and protein sorting by facilitating cellular RNA localization [32].

Phylogenetic analysis and evolution of IQD family genes

IQD proteins are an ancient family of CaM/CML binding proteins that originated during the early evolution of land plants, as IQD genes are present in *Physcomitrella patens*. ESTs corresponding to IQD proteins for angiosperm species (*Arabidopsis*, rice, etc.) and at least nine homologous sequences in the gymnosperm pine (*Pinus* ssp.) corresponding to IQD proteins were identified suggesting that the IQD gene family originated not later than the split of gymnosperms and angiosperms about 300 Myr ago [30]. We performed a genome-wide comparison of plant IQD members from monocots (rice and *Brachypodium distachyon*) and eudicots (soybean, *Arabidopsis* and tomato) to explore how the IQD gene family has evolved. The plant IQD members from monocots (rice and *Brachypodium distachyon*) and eudicots (soybean, Arabidopsis and tomato) appear to be more closely

Table 4. Divergence between paralogous IQD gene pairs in soybean.

Group	No.	Paralogous pairs	Ka	Ks	Ka/Ks	Duplication date (MY)	Duplicate type
Ia	1	GmIQD1-GmIQD24	0.0474	0.0802	0.5914	6.57	S
	2	GmIQD16-GmIQD27	0.045	0.195	0.228	15.99	S
	3	GmIQD22-GmIQD29	0.029	0.108	0.267	8.83	S
	4	GmIQD49-GmIQD52	0.041	0.106	0.388	8.66	S
	5	GmIQD15-GmIQD59	0.041	0.158	0.258	12.96	S
	6	GmIQD31-GmIQD57	0.029	0.147	0.194	12.03	S
	7	GmIQD37-GmIQD66	0.030	0.116	0.260	9.48	S
Ib	8	GmIQD4-GmIQD34	0.044	0.124	0.356	10.18	S
	9	GmIQD38-GmIQD65	0.054	0.111	0.485	9.07	S
	10	GmIQD8-GmIQD64	0.039	0.134	0.293	11.00	S
Ic	11	GmIQD23-GmIQD55	0.017	0.086	0.193	7.08	S
Id	12	GmIQD13-GmIQD19	0.057	0.164	0.346	13.47	S
IIa	13	GmIQD11-GmIQD39	0.043	0.093	0.460	7.60	O
	14	GmIQD33-GmIQD40	0.022	0.094	0.238	7.70	S
IIb	15	GmIQD6-GmIQD26	0.022	0.091	0.245	7.49	S
IIIa	16	GmIQD10-GmIQD18	0.030	0.152	0.197	12.43	S
	17	GmIQD50-GmIQD61	0.031	0.162	0.189	13.30	S
	18	GmIQD21-GmIQD28	0.041	0.125	0.325	10.22	S
	19	GmIQD48-GmIQD53	0.029	0.111	0.262	9.08	S
	20	GmIQD25-GmIQD44	0.052	0.157	0.335	12.84	S
	21	GmIQD45-GmIQD54	0.034	0.105	0.325	8.57	S
	22	GmIQD42-GmIQD46	0.058	0.219	0.263	17.94	S
	23	GmIQD41-GmIQD47	0.037	0.095	0.387	7.78	S
IIIb	24	GmIQD35-GmIQD43	0.033	0.114	0.293	9.34	S
	25	GmIQD7-GmIQD63	0.024	0.093	0.253	7.61	S
	26	GmIQD36-GmIQD67	0.045	0.127	0.349	10.43	S
IV	27	GmIQD12-GmIQD20	0.054	0.134	0.400	10.99	S
	28	GmIQD14-GmIQD58	0.035	0.118	0.297	9.66	S
	29	GmIQD2-GmIQD5	0.059	0.109	0.537	8.93	S
	30	GmIQD30-GmIQD62	0.067	0.151	0.443	12.35	S
	31	GmIQD9-GmIQD17	0.028	0.078	0.363	6.39	S

S: segmental duplication, O: other events.

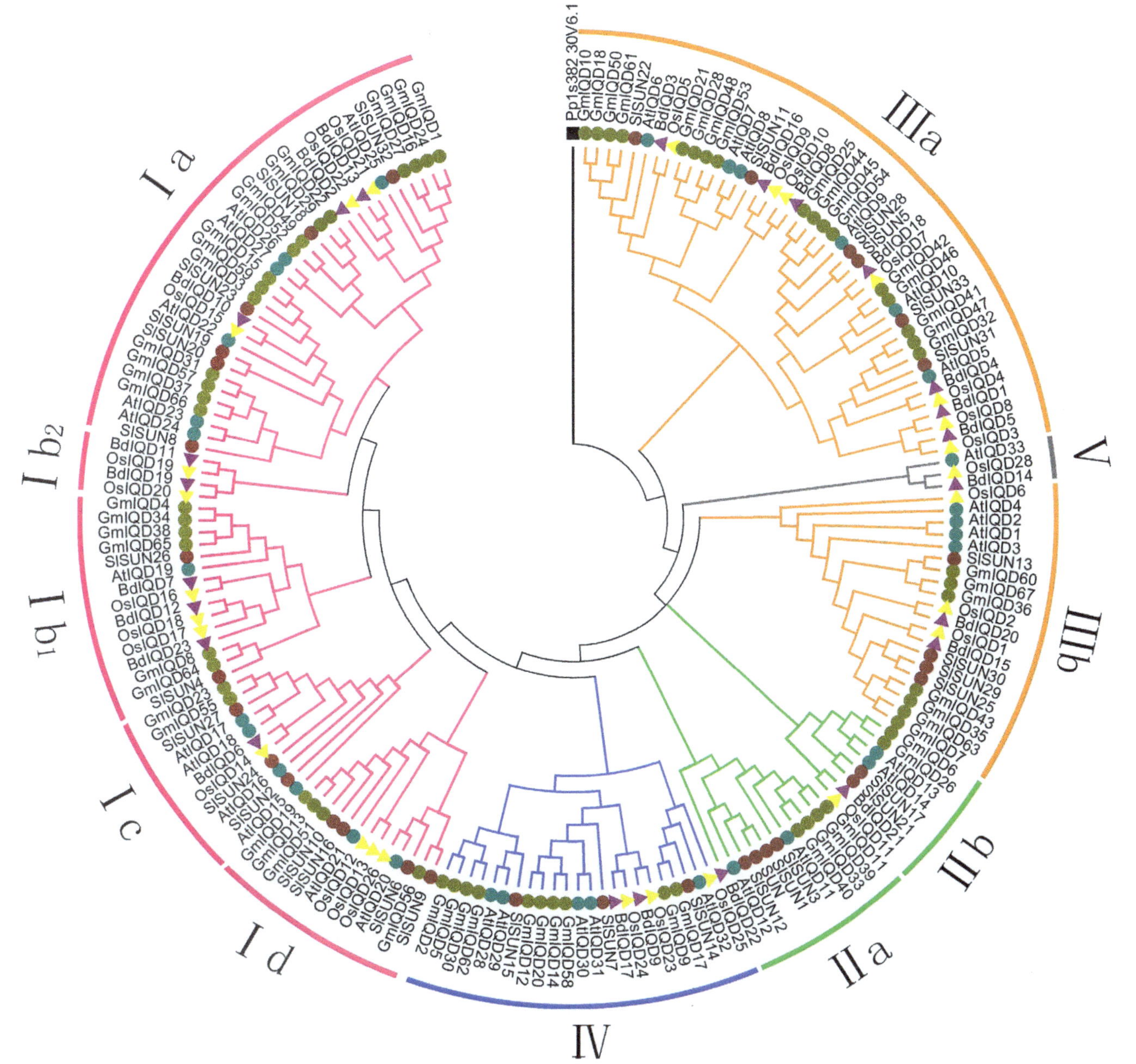

Figure 4. Phylogenetic tree of full-length IQD proteins from soybean, *Arabidopsis*, rice, tomato and *Brachypodium distachyon*. The tree was generated with Clustal X2.0 using the NJ method. Dicotyledons (soybean, tomato and *Arabidopsis*) IQD proteins are marked with colored dots. Monocotyledons (rice and *Brachypodium distachyon*) are marked with colored triangles. A moss IQD protein (Pp1s38230v6), used as the outgroup, is marked with a black box. Each IQD subfamily is indicated by a specific color.

related to each other than to IQD genes of the same species in a different subfamily. This alternating distribution of monocots and eudicots in all subfamilies suggests that an ancestral set of IQD genes have existed before the dicot–monocot split (Figure 4, Table 4). The presence of five distinct subfamilies of IQD genes and the presence of both monocots and eudicots containing members in all five subfamilies indicate IQD genes have diversified before the monocot–eudicot split (Figure 4). These subfamilies include 23 pairs of orthologous genes, suggesting that orthologous genes may have originated from a common ancestor (Figure 4). About half of the orthologous genes (10 pairs; *BdIQD1/OsIQD8*:N, *BdIQD5/OsIQD3*:N, *BdIQD8/OsIQD10*:C, *BdIQD9/OsIQD23*:C, *BdIQD11/OsIQD19*:N, *BdIQD14/OsIQD28*:N, *BdIQD17/OsIQD24*:N *BdIQD18/OsIQD7*:N, *BdIQD20/OsIQD2*:N, *SlSUN13/GmIQD60*:N) have the same predicted subcellular localization suggesting that the encoded proteins may play similar roles in both species [30,34]. A total of 87% (20 pairs) of orthologous gene pairs from rice and *Brachypodium distachyon* are distributed in all subfamilies. However, only two pairs of orthologous genes from dicotyledons (soybean and tomato) are from subfamily I and III. This difference may be due to the fact that both rice and *Brachypodium distachyon* are in the grass family and are therefore more closely related than *Arabidopsis*, soybean and tomato, which belong to Cruciferae, Solanaceae and Leguminosae, respectively. The number of soybean genes in each subfamily is greater than that of the other four species examined

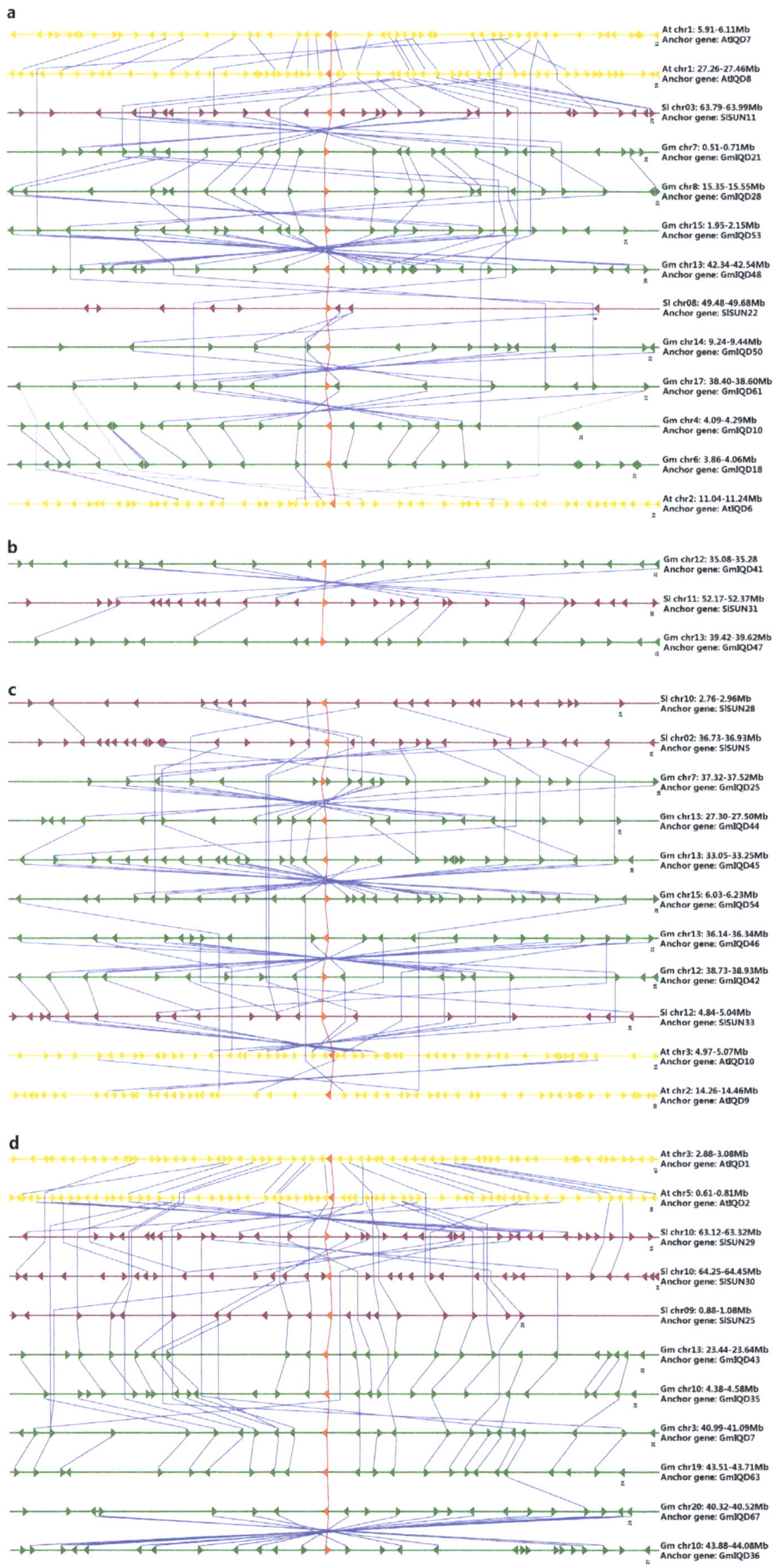
a
At chr1: 5.91-6.11Mb
Anchor gene: AtIQD7
At chr1: 27.26-27.46Mb
Anchor gene: AtIQD8
Sl chr03: 63.79-63.99Mb
Anchor gene: SlSUN11
Gm chr7: 0.51-0.71Mb
Anchor gene: GmIQD21
Gm chr8: 15.35-15.55Mb
Anchor gene: GmIQD28
Gm chr15: 1.95-2.15Mb
Anchor gene: GmIQD53
Gm chr13: 42.34-42.54Mb
Anchor gene: GmIQD48
Sl chr08: 49.48-49.68Mb
Anchor gene: SlSUN22
Gm chr14: 9.24-9.44Mb
Anchor gene: GmIQD50
Gm chr17: 38.40-38.60Mb
Anchor gene: GmIQD61
Gm chr4: 4.09-4.29Mb
Anchor gene: GmIQD10
Gm chr6: 3.86-4.06Mb
Anchor gene: GmIQD18
At chr2: 11.04-11.24Mb
Anchor gene: AtIQD6
b
Gm chr12: 35.08-35.28
Anchor gene: GmIQD41
Sl chr11: 52.17-52.37Mb
Anchor gene: SlSUN31
Gm chr13: 39.42-39.62Mb
Anchor gene: GmIQD47
c
Sl chr10: 2.76-2.96Mb
Anchor gene: SlSUN28
Sl chr02: 36.73-36.93Mb
Anchor gene: SlSUN5
Gm chr7: 37.32-37.52Mb
Anchor gene: GmIQD25
Gm chr13: 27.30-27.50Mb
Anchor gene: GmIQD44
Gm chr13: 33.05-33.25Mb
Anchor gene: GmIQD45
Gm chr15: 6.03-6.23Mb
Anchor gene: GmIQD54
Gm chr13: 36.14-36.34Mb
Anchor gene: GmIQD46
Gm chr12: 38.73-38.93Mb
Anchor gene: GmIQD42
Sl chr12: 4.84-5.04Mb
Anchor gene: SlSUN33
At chr3: 4.97-5.07Mb
Anchor gene: AtIQD10
At chr2: 14.26-14.46Mb
Anchor gene: AtIQD9
d
At chr3: 2.88-3.08Mb
Anchor gene: AtIQD1
At chr5: 0.61-0.81Mb
Anchor gene: AtIQD2
Sl chr10: 63.12-63.32Mb
Anchor gene: SlSUN29
Sl chr10: 64.25-64.45Mb
Anchor gene: SlSUN30
Sl chr09: 0.88-1.08Mb
Anchor gene: SlSUN25
Gm chr13: 23.44-23.64Mb
Anchor gene: GmIQD43
Gm chr10: 4.38-4.58Mb
Anchor gene: GmIQD35
Gm chr3: 40.99-41.09Mb
Anchor gene: GmIQD7
Gm chr19: 43.51-43.71Mb
Anchor gene: GmIQD63
Gm chr20: 40.32-40.52Mb
Anchor gene: GmIQD67
Gm chr10: 43.88-44.08Mb
Anchor gene: GmIQD36

Figure 5. Microsynteny map of IQD III genes in soybean, *Arabidopsis* and tomato. a, b, c and **d:** four groups of syntenic segments. The relative positions of all flanking protein-coding genes were defined by the anchored IQD III genes (red). Colored horizontal lines represent chromosome segments of soybean (green), *Arabidopsis* (yellow) and tomato (purple). Triangles of the same color represent individual genes and their transcriptional orientations. The total number of genes on each segment is indicated to the right below the segment. Colored lines connect the conserved gene pairs among the segments (anchor gene pairs, red; others, blue).

suggesting that IQD counterparts in soybean may have undergone gene expansion.

The duplication of individual genes, chromosomal segments or entire genomes has been a major force in the evolution of plant genome structure and content during the process of genome evolution [60,61]. The soybean genome has undergone at least two round of duplication, resulting in the presence of significant features of remnants of a glycine-specific genome duplication that occurred ~13 Mya and fainter remnants of older polyploidies prior to the divergence of the papilionoids (58–60 Mya) that occurred ~58 Mya [41,51]. Thus, 75% of soybean genes are present in multiple copies [41]. Among 67 soybean IQD genes, *GmIQD3*, *-32*, *-56* and *-60* were found as single copies on duplication blocks. These results suggest that segmental duplication has occurred as a continuous process and dynamic changes may have occurred in a chromosomal segment that contained two ancestral IQD genes, leading to corresponding sister gene loss [62]. One paralogs (*GmIQD11/-39*) shares 91.2% identity and similar exon/intron organization, but exists outside of any duplicated blocks. this pair might have been produced by retrotransposition. A high proportion (approximately 96%) of soybean IQD genes reside preferentially in duplicated segments, suggesting that segmental duplications have played a prominent role in the expansion of the soybean IQD gene family. The duplicated IQD genes in soybean have been preferentially retained at the high rate of 92.5% (62/67), which is distinctly higher than the retention rate (67.3%) of duplicated paralogs in the 1.1-gigabase sequence of the soybean (cv. Williams 82) genome, in which 31,264 genes exist as 15,632 paralog pairs (out of the 46,430 predicted high-confidence genes that were duplicated and retained after the 13-Mya tetraploidy event) [63]. The higher retention rate corroborates previous findings that genes involved in signal transduction are preferentially retained following duplications[64]. Our calculation of the duplication dates of the 31 paralogous pairs revealed that all of the segmental duplication events in the soybean IQD family occurred during the recent whole genome duplication event.

During evolution, eukaryotic genomes have retained genes on corresponding chromosomes (synteny) and in corresponding orders (collinearity) to various degrees. Synteny broadly refers to parallels in gene arrangement in dissimilar genomes. Collinearity, a specific form of synteny, requires genes to occur in largely corresponding orders along the chromosomes of respective genomes. According to the microsynteny analysis, microsynteny relationships among *AtIQD3*, *AtIQD4* or *SlSUN13* with other IQD III members in these three dicot genomes were not observed indicating that either these genes are ancient genes without detectable linkage to other IQD genes or that they were formed through complete transposition and loss of their primogenitors. In addition, three different duplicated chromosomal segments (harboring *AtIQD5*, *GmIQD32* and *GmIQD6*) that lost their sister IQD genes lack detectable microsynteny relationships to all other IQD III genes in the soybean and *Arabidopsis* genome, respectively. In the four IQD III gene clades, genes from soybean, tomato and *Arabidopsis* exhibit high levels of microsynteny, which indicates the IQD III genes existed before the divergence of the three dicotyledons genomes (soybean, tomato and *Arabidopsis*). Microsynteny was detected in most pairs, and alignment in clade 1–3 was discordant, suggesting that these genes may all be present in genome regions that were inverted, expanded or contracted after the divergence. Notably, most gene pairs in clade 4 have successive collinearity in order and the same orientation, which indicates high conservation among these IQD III gene-residing regions, with little rearrangement. The low (18.41%) synteny quality of IQD III genes from the three dicotyledon genomes (soybean, tomato and *Arabidopsis*) may have been due to the fact that these plants are not closely related; moreover, the gene density differs between *Arabidopsis* and the two other species. Significantly, the number of synteny blocks (31) within the soybean genome is much more than the number (3 or 4) of synteny blocks between tomato or *Arabidopsis* genomes, which suggests that soybean IQD III genes may have undergone large-scale duplication events and less rearrangement was followed (Figure 5 and Table S5b). The gene expansion pattern analysis of soybean paralogs indicates that most pairs were generated from a large-scale duplication, which supports the results of soybean gene duplication analysis, with the exception of *GmIQD41/-47*.

Organ- or tissue-specific expression of IQD genes and expression of GmIQD III genes under MeJA stress treatment

Organ- or tissue-specific expression patterns have been observed for quite a few members of the IQD family. However, the functions of soybean IQD genes remain unclear. We therefore performed a thorough analysis of the RNA-Seq Atlas to investigate organ- or tissue-specific expression of IQD genes and qRT-PCR to examine the expression of GmIQD III members under MeJA stress treatment.

The tissue expression data deficiency of eight soybean IQD genes potentially indicated that these are pseudogenes or express only at specific developmental stages or under special conditions. 65.7% soybean IQD genes constitutively express in all of the seven tissues suggesting that GmIQDs may play roles at multiple developmental stages. Eight GmIQD proteins peak in only one tissue indicating that these tissue-specific calmodulin target proteins may be limited to discrete cells or organs to regulate various cellular activities.

Except for group C comprised of genes from GmIQD I, group A, B and D comprise genes from four subfamilies indicating these soybean IQD genes exhibit similar transcript abundance profiles but are relatively phylogenetically distinct. The analysis indicated that only some members within the same phylogenetic subgroup share a similar expression profile in soybean organs/tissues during development, excluding *GmIQD6* and *GmIQD26* belong to GmIQD IIb. For instance, *GmIQD4*, *-34*, *-38* and *-65* belong to GmIQD Ib clustered in group with high expression in flowers suggesting their potential roles in flower formation. While the other two GmIQD Ib members (*GmIQD8* and *-64*) were detected in nodules indicating they may involve in fixing atmospheric nitrogen.

Members possessing similar sequences are clustered in the same subfamilies, which may have similar expression patterns or functions. In IQD subgroup Ia, *Arabidopsis IQD22* is involved in the negative feedback regulation of GA-responsive DELLA

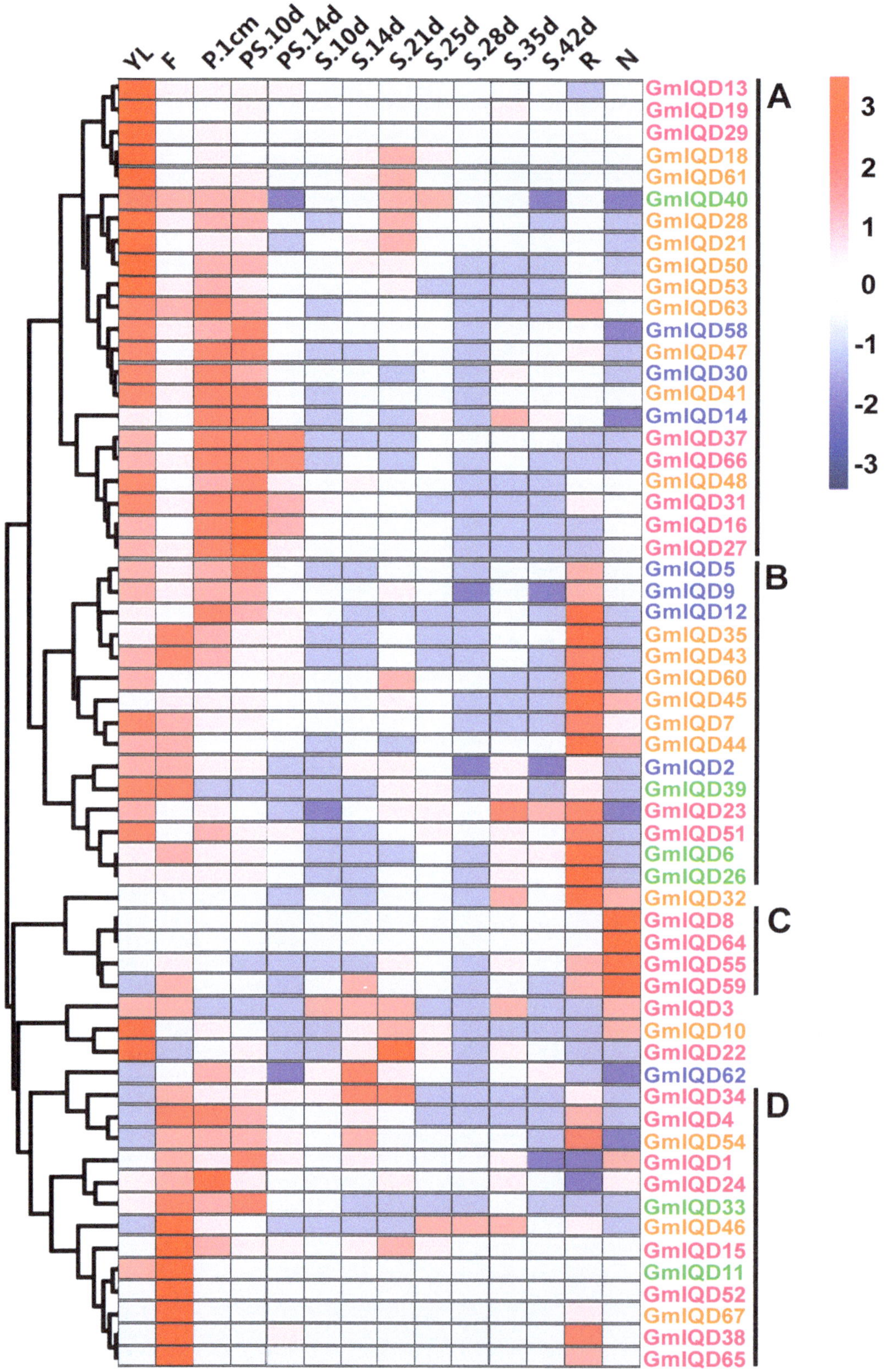

Figure 6. Hierarchical clustering of the expression profiles of soybean IQD genes in 14 tissues. RNA-seq relative expression data from 14 tissues were used to reconstruct the expression patterns of soybean genes. The raw data was normalized and retrieved from the online database http://soybase.org/soyseq/. The normal relative expression levels of 67 IQD genes are shown in Table S6. YL, young leaf; F, flower; P.1cm, one cm pod; PS.10d, pod shell 10 DAF; PS.14d, pod shell 14 DAF; S.10d, seed 10 DAF; S.14d, seed 14 DAF; S.21d, seed 21 DAF; S.25d, seed 25 DAF; S.28d, seed 28 DAF; S.35d, seed 35 DAF; S.42d, seed 42 DAF; R, root; N, nodule. Gene names in different subfamilies are highlighted with various colors. Genes clustered into four groups (A–D) are indicated by the black vertical bars.

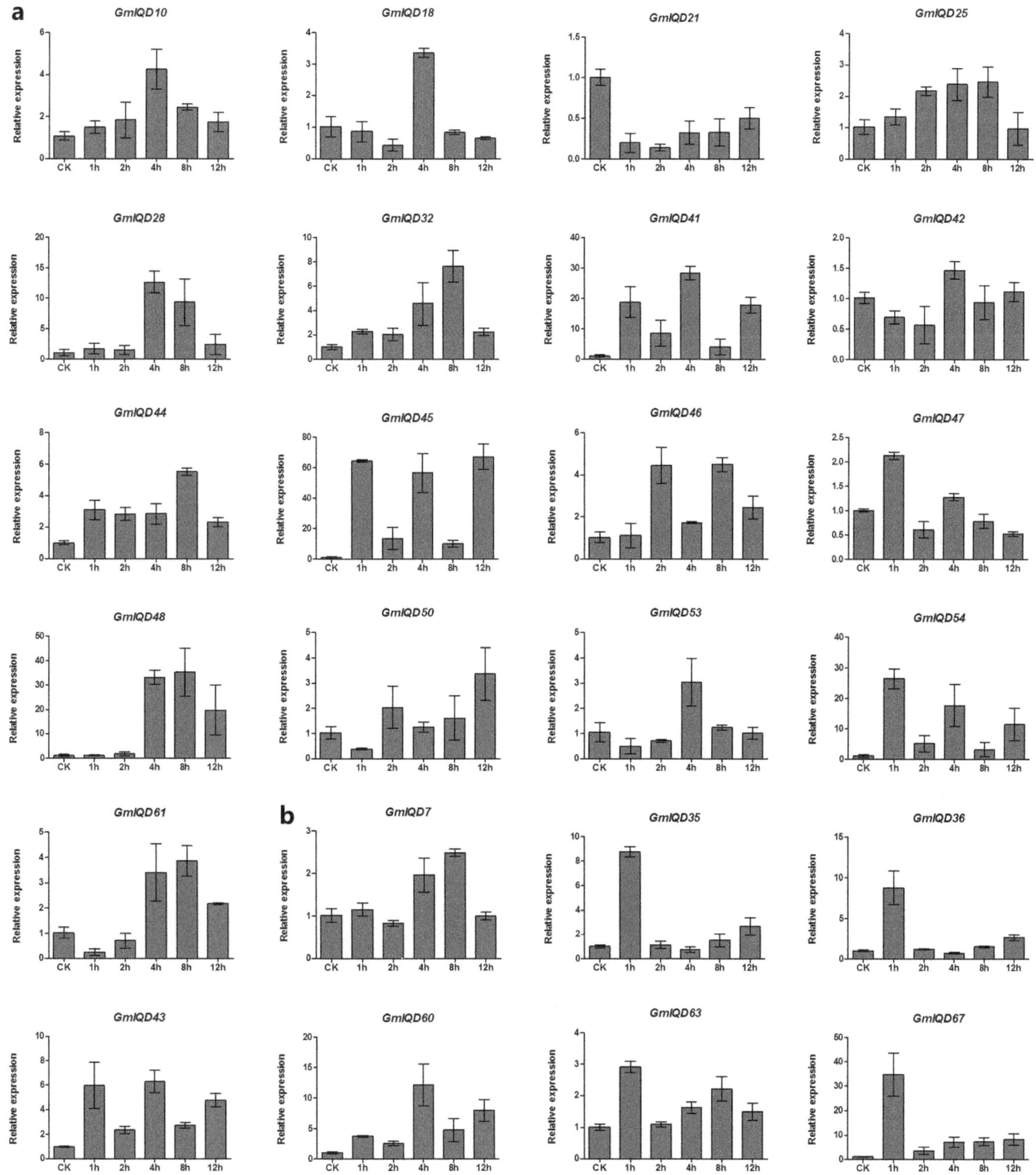

Figure 7. Expression patterns of 24 selected IQD III genes under MeJA stress using qRT-PCR. Relative expression levels of 24 IQD genes were examined by qRT-PCR and normalized to the expression of *CYP2*. Bars represent standard deviations (SD) of three biological replicates. Y-axes indicate the scale of the relative expression levels. X-axes show time courses of MeJA stress treatments for each gene. **a** and **b**: data for genes from IIIa and IIIb, respectively.

genes [36]. Subgroup Ia members of *Arabidopsis IQD26* has higher expression level in parts with divided vigorous growth and microtubule organization of leaves, root and flowers [65]. Eight of the of the thirteen soybean IQD Ia members (*GmIQD3*, *-16*, *-22*, *-27*, *-29*, *-31 -37* and *-66*) have high expression in young fleaves. And six of them (*GmIQD15*, *-16*, *-27*, *-31 -37* and *-66*) have high expression in one-cm pod or pod-shell 10-DAF. *GmIQD22* have obviously higher expression at seed 21-DAF, the period of seed cell division (3, 4 weeks after flowering) [66]. These founding suggested that soybean IQD Ia members may function in transport of signaling molecules, nutrient transport and cell division.

Mapping and positional cloning of the *SUN* locus revealed that this member of the IQD II subfamily was generated by duplication of a 24.7-kb region carrying the tomato *IQD12* gene, a major gene involved in the control of fruit shape, particularly length, in tomato [35]. *SUN* is expressed the highest in hypocotyl and shoot apex. Overexpression of *SUN* causes root reduction when applied auxin and prostrate growth and twisted stems indicating that *SUN* can affect auxin transport or response [38]. *SlSUN1* shows slightly higher expression in the hypocotyl, flower at anthesis and fruit at 10 and 20 DPA [35]. *SlSUN12* and *SlSUN21* highly are expressed in the hypocotyls and root respectively [35]. *SlSUN17* evenly expressed in almost all tissues [35]. Soybean IQD II members showed the similar expression profile (*GmIQD11*: highest in flowers; *GmIQD33*, *-39* and *-40*: slight high in flowers and *GmIQD6* and *-26*: highest in roots) indicating that GmIQD II members may play similar role in plant development.

Arabidopsis IQD1 from the IQD III subfamily modulates the expression of several glucosinolate (GS) pathway genes, resulting in the alteration of glucosinolate content and composition to promote resistance to herbivory. *Arabidopsis IQD1*, the first functionally characterized IQD gene, is expressed in vascular tissues of hypocotyls, leaves, stems, flowers and roots, as revealed by histochemical analysis. Expression pattern analysis of soybean IQD genes revealed that genes from the soybean IQD III subfamily were mainly expressed in young leaves, flowers and roots (Figure 6). Jasmonic acid(JA) treatment leads to elevating levels of specific glucosinolate in *Arabidopsis* [67,68]. And overexpression of *AtIQD1* causes the accumulation of glucosinolates. However, *AtIQD1* expression is independent of JA, as steady-state *AtIQD1* mRNA expression levels are not appreciably altered when externally applied JA and are also unaffected in mutants defective in JA synthesis or signaling (JA *-jar1* and *fad3-2 fad7-2 fad8*) [31]. Indeed, *AtIQD1* increases resistance against herbivory by augmenting and fine-tuning glucosinolate accumulation [31]. Glucosinolates with important roles in plant defense and human nutrition are a small but diverse class of defense related secondary metabolites in cruciferous species such as *Brassica* crops and the *Arabidopsis thaliana* [69,70]. Obviously, soybean doesn't synthesize glucosinolates. Base on the microsynteny analysis of IQDIII members, we auspiciously found there exsited highly conserved microsynteny relationship between *AtIQD1* and soybean IQDIII members. And combined with phylogenetic analysis of IQDIII members, we speculated soybean IQDIII members might have the similar biological function with *AtIQD1* in defenses to insect herbivory.

Therefore, we performed qRT-PCR of 24 soybean IQD III with MeJA treatment to detect whether soybean IQD III genes defense to insect herbivory. Compared to *AtIQD1*, soybean IQD III genes exhibited different responses with the MeJA treatment. The qRT-PCR results showed that 23 of the 24 soybean IQD III subfamily genes were upregulated by MeJA treatment, except *GmIQD21*, seven genes (*GmIQD28*, *-41*, *-45*, *-48*, *-54*, *-60* and *-67)* were strongly induced by MeJA with relative expression more than 10-fold. *GmIQD45* even accumulated the highest transcripts approaching 70-fold at 12 h (Figure 7). Based on these, we speculate that IQD III genes in soybean may involve in defense to insect herbivory by JA pathway.

Orthologs may have equivalent functions, as they originated from a single ancestral gene in the last common ancestor of the species. Two pairs of orthologous genes (*SlSUN22/AtIQD6* and *SlSUN31/AtIQD5*) were found between tomato and *Arabidopsis* (Figure S2). Similar expression patterns of these two pairs in tomato and *Arabidopsis* have been reported; *SlSUN31* and *AtIQD5* are almost ubiquitously expressed, whereas *SlSUN22* and *AtIQD6* are highly expressed in young flower buds [30,34,35].

Duplicated genes may face three different fates: nonfunctionalization (one copy becomes silenced); neofunctionalization (one copy acquires a novel, beneficial function while the other copy retains the original function) or subfunctionalization (both copies become partially compromised by the accumulation of mutations) [45,71]. Paralogs originating from duplication within one organism may have more divergent functions. In the current study, several pairs of paralogs showed similar expression patterns, which suggests that they may share a common or similar function. For example, *GmIQD10/GmIQD18* were highly expressed in young leaves, and their expression peaked at 4 h in response to MeJA (Figure 7). Several pairs of paralogs have different expression patterns, suggesting that they play diverse roles in soybean development. For example, *GmIQD21/GmIQD28* are mainly expressed in young leaves. Upon MeJA treatment, *GmIQD28* was most highly expressed at 4 h while its sister gene *GmIQD21* was downregulated at all time points.

In conclusion, IQD proteins play fundamental roles in various plant developmental processes. Therefore, the systematic analysis of the soybean IQD gene family performed in the current study provides an important reference for further characterization of the biological functions of these proteins.

Materials and Methods

Identification of IQD family genes in soybean

To identify IQD proteins in soybean, the *Glycine max* genome database (release 1.0, http://www.phytozome.net/soybean.php) was searched using Basic Local Alignment Search Tool algorithms (BLASTP), with the published *Arabidopsis* IQD protein sequences and their IQ67 domain used as initial query sequences. Redundant sequences were then removed manually, and the Hidden Markov Model of the Pfam (http://pfam.sanger.ac.uk/search) [72] and SMART (http://smart.embl-heidelberg.de/) [73] databases were used to confirm each candidate sequence as a member of the IQD family [74]. A total of 33 *Arabidopsis*, one moss (*Physcomitrella patens*) and 23 *Brachypodium* IQD protein sequences were downloaded from Phytozome v9.0 (http://www.phytozome.net/), and 34 tomato IQD protein sequences were retrieved from the tomato WGS chromosomes (2.40; SL2.40) (SGNhttp://solgenomics.net). Finally, 27 rice IQD protein sequences were obtained from the TAIR database (http://rice.plantbiology.msu.edu). Accession numbers of published IQD proteins for *Arabidopsis*, rice, tomato, *Brachypodium* and moss were listed in Table S8.

Soybean IQD gene information, including the number of amino acids, ORF lengths and chromosome locations, was obtained from the Phytozome database. Physicochemical parameters including the molecular weight (kDa) and isoelectric point (pI) of each gene product were calculated using compute pI/Mw tool from ExPASy (http://www.expasy.org/tools/) and parameter (resolution) was set to average [75]. Subcellular localization was predicted using the TargetP 1.1 (http://www.cbs.dtu.dk/services/TargetP/) server and WoLF PSORT (http://wolfpsort.org/).

Multiple alignment and phylogenetic analysis of IQD family genes

Multiple sequence alignment of all predicted soybean IQD protein sequences was performed with Clustal X2.0 software using default parameters. Then, based on this alignment, phylogenetic trees were constructed using Clustal X2.0 with the Neighbor-Joining (NJ) method, and bootstrap analysis was conducted using 1,000 replicates [76]. An unrooted NJ tree of 184 the full-length IQD protein sequences from soybean, rice, *Arabidopsis*, tomato, *Brachypodium* was constructed using Clustal X2.0 with one moss IQD protein (Pp1s38230v6.1) as the outgroup. The GmIQD genes were classified according to their phylogenetic relationships with the corresponding *Arabidopsis* and rice IQD genes. For microsynteny analysis of IQD III genes across soybean, tomato and *Arabidopsis*, a phylogenetic tree was constructed using MEGA 5.0 with the NJ method and bootstrap analysis was conducted using 1,000 replicates.

Genomic structure

Exon and intron structures of individual soybean IQD genes were deduced using GSDS (Gene structure display server; http://gsds.cbi.pku.edu.cn/) via alignment of the cDNAs with their corresponding genomic DNA sequences [77].

Identification of conserved motifs and putative calmodulin-binding sites

Online MEME (Multiple Expectation Maximization for Motif Elicitation) (http://meme.sdsc.edu/meme4_4_0/intro.html) was performed to identify the conserved motif structures encoded by GmIQD genes. The parameters were as followings: number of repetitions - any, maximum number of motifs -10, and the optimum motif width was constrained to between 6 and 200 residues. In addition, each structural motif was annotated using the Pfam (http://pfam.sanger.ac.uk/search) and SMART (http://smart.embl-heidelberg.de/) tools. All IQD protein sequences were examined against the Calmodulin Target Database (http://calcium.uhnres.utoronto.ca/ctdb/ctdb/home.html) to predict putative calmodulin-binding sites.

Chromosomal location and gene duplication

The chromosomal location image of GmIQD genes was generated by MapInspect (http://www.plantbreeding.wur.nl/uk/software_mapinspect.html) according to chromosomal position information provided in the Phytozome database. To identify tandem and segmental duplications, two genes in the same species located in the same clade of the phylogenetic tree were defined as coparalogs. The SoyBase browser (http://soybase.org/gb2/gbrowse/gmax1.01/) [78] was queried to detect the segmental duplication coordinates of the target genes. Coparalogs were deemed to result from segmental duplication if they were located on duplicated chromosomal blocks [79]. Paralogs were deemed to be tandem duplicated genes if two genes were separated by five or fewer genes in a 100-kb region [80]. The local alignment of two protein sequences was calculated using the Smith-Waterman algorithm (http://www.ebi.ac.uk/Tools/psa/).

Calculation of Ka/Ks Values

Amino acid sequences of each paralog pair were first aligned using Clustal X2.0. Then, the multiple sequence alignments of proteins and the corresponding cDNA sequences were converted to codon alignments using PAL2NAL (http://www.bork.embl.de/pal2nal/) [81]. Finally, the resulting codon alignment was subjected to calculation of synonymous (Ks) and non-synonymous (Ka) substitution rates using the CODEML program of PAML [82].

Based on a rate of 6.1×10^{-9} substitutions per site per year, divergence time (T) was calculated using the Ks value with the formula: $T = Ks/(2\times6.1\times10^{-9})\times10^{-6}$ Mya [45].

Microsynteny analysis

For microsynteny analysis, IQD III genes from soybean, tomato and *Arabidopsis* as the anchors were localized to specific target genomic regions. Then, all protein-coding sequences of 100 kb flanking each anchor point were compared by pairwise BLASTP analysis. The syntenic blocks used to construct synteny analysis maps of the IQD III genes were obtained from the Plant Genome Duplication Database (http://chibba.agtec.uga.edu/duplication/), a web service providing synteny information in terms of collinearity between chromosomes [83–85]. A synteny block is defined as a region where three or more conserved homologs are located within a 100-kb DNA stretch in both genomes. Two regions were considered to have originated from a large-scale duplication event when five or more protein-coding gene pairs flanking the anchor point were ligatured with the best non-self match (E-value$<$1e$-$10) [48]. The relative syntenic quality in a region was calculated from the sum of the total number of genes in both conserved gene regions (excluding retroelements and transposons and collapsing tandem duplications) [47].

RNA-Seq atlas analysis

To acquire the tissue-specific transcript data, a list of 67 GmIQD gene names was entered to the RNA-Seq Atlas of *Glycine max* (http://soybase.org/soyseq/) [86]. The raw digital gene expression counts of the uniquely mappable reads were normalized using a slight variation of the reads/Kb/Million (RPKM) method and the normalized data was download from this database [86]. Hierarchical clustering analysis was conducted using clustering distance "correlation" (Pearson correlation) and the clustering method used "complete" (complete linkage method) in R [87]. A heat map was generated in R using the pheatmap function [87].

Plant growth and treatments

Soybean (*Glycine max* L.) Williams 82 was used in this study. Seedlings were grown in a growth chamber under the following conditions: temperature, 30°C; photoperiod, 12 h/12 h; photon flux density, 80 μmolm^{-2} s^{-1} and relative humidity, 50% [88]. For expression pattern analysis of soybean IQD genes under stress, four-week-old seedlings were treated with 100 µm MeJA in the growth chamber [89]. Jasmonic acid methyl ester (MeJA) (Sigma, 95%) was diluted 1:10 with 95% ethanol, followed by a further dilution with MilliQ water containing 0.1% Triton X-100, resulting in a final concentration of 100 µmol/L MeJA. Untreated seedlings were used as a control. Leaves of MeJA-treated plants were collected at 0, 1, 2, 4, 8 and 12 h. After collection, the samples were immediately frozen in liquid N_2 and stored at $-$80°C for RNA extraction. Three biological replicates were conducted per sample.

RNA extraction and qRT-PCR analysis

An RNAprep Pure Plant Kit (Tiangen) was used to isolate total RNA from each frozen sample. Possible contaminating genomic DNA was removed using DNaseI supplied in the kit. First-strand cDNA was synthesized from the RNA using a PrimeScript™ RT Master Mix Kit (TaKaRa) according to the manufacturer's instructions. Gene-specific primers for the 24 GmIQD genes were designed using Primer5.0 (Table S9). Primer specificity was first checked using the primer-BLAST tool available on the NCBI website. Subsequently, by performing analysis of melting curves and analysis of visualization of amplicon fragments, we found primers were gene-specific only when corresponding melting curves generated a single sharp peak and the primers demonstrated an electrophoresis pattern of a single amplicon with the correct predicted length. A housekeeping gene constitutively expressed in soybean, *CYP2* (cyclophilin) [46,90–93], was used as a reference for normalization. The qRT-PCR analysis was conducted on an ABI 7500 Real-Time PCR system (Applied Biosystems). The reactions were performed in a 20 μl volume containing 10 μl 2×SYBR® *Premix Ex Taq*™ II (TaKaRa), 6.0 μl ddH$_2$O, 0. 4 μl ROX Reference Dye II, 2.0 μl diluted cDNA and 0.8 μl of each gene-specific primer. The PCR conditions were as follows: Stage 1: 95°C for 30 s; stage 2: 40 cycles of 5 s at 95°C and 34 s at 60°C; stage 3: 95°C for 15 s, 60°C for 1 min, 95°C for 15 s. At stage 3, a melting curve was generated to estimate the specificity of the reactions. Three biological replicates were used per sample.

The relative expression levels were calculated as $2^{-\Delta\Delta CT}$ [$\Delta C_T = C_{T,\ Target} - C_{T,\ CYP2}$. $\Delta\Delta CT = \Delta C_{T,\ treatment} - \Delta C_{T,\ CK\ (0\ h)}$]. The relative expression levels ($2^{-\Delta\Delta CT,\ CK\ (0\ h)}$) in the untreated control plants were normalized to 1 as described previously [46,94,95]. If an efficiency of amplification was less than 2, the result was proofread. Statistical analyses were conducted using GraphPad Prism 5.01 software [96].

Supporting Information

Figure S1 Amino acid sequence alignments of IQ67 domains in soybean IQD protein sequences. The multiple alignment results indicate the highly conserved IQD domains among the 67 identified soybean IQD protein sequences. The positions of the conserved IQ calmodulin binding motifs are shown. Identical residues of proteins are marked with an asterisk. The consensus sequence at the bottom was constructed with greater than 50% conservation among the 67 soybean IQD proteins. Red arrow indicates the position of the conserved phase-0 intron, which divides codons 16 and 17 of the IQ67 domain.

Figure S2 Phylogenetic tree of full-length IQD III proteins from soybean, *Arabidopsis* and tomato. The tree was generated with MEGA 5.0 using the NJ method with 1,000 bootstrap replicates. Dicotyledon (soybean, tomato and *Arabidopsis*) IQD proteins are marked with colored dots. IQD III proteins from soybean, *Arabidopsis* and tomato were divided into four clades (1–4) presented by different color.

Table S1 Detailed information about the 10 motifs in soybean IQD proteins.

Table S2 Recent synteny blocks of soybean and soybean (13 Mya) genomes containing IQD genes.

Table S3 Pairwise identities between paralogous pairs of IQD genes from soybean.

Table S4 The synteny blocks used to construct micro-synteny map.

Table S5 Number of conserved gene pairs and synteny blocks and relative syntenic quality.

Table S6 Transcription of soybean IQD genes, as determined by RNA-seq analysis.

Table S7 Animo acid content of 67 soybean IQD proteins.

Table S8 Accession numbers of IQDs for *Arabidopsis thaliana*, rice, tomato, *Brachypodium distachyon* and moss.

Table S9 List of primer sequences used for qRT-PCR analysis of the 24 soybean IQD III genes.

Author Contributions

Conceived and designed the experiments: LF YX. Performed the experiments: LF XC. Analyzed the data: LF ZC. Contributed reagents/materials/analysis tools: LF XC YW. Wrote the paper: LF ZC YL.

References

1. Hepler PK, Vidali L, Cheung AY (2001) Polarized cell growth in higher plants. Annual review of cell and developmental biology 17: 159–187.
2. Du L, Ali GS, Simons KA, Hou J, Yang T, et al. (2009) Ca2+/calmodulin regulates salicylic-acid-mediated plant immunity. Nature 457: 1154–1158.
3. Harada A, Shimazaki K-i (2009) Measurement of changes in cytosolic Ca2+ in Arabidopsis guard cells and mesophyll cells in response to blue light. Plant and cell physiology 50: 360–373.
4. Ng C, Mcainsh MR, Gray JE, Hunt L, Leckie CP, et al. (2001) Calcium-based signalling systems in guard cells. New Phytologist 151: 109–120.
5. Reddy AS (2001) Calcium: silver bullet in signaling. Plant Science 160: 381–404.
6. Dodd AN, Kudla J, Sanders D (2010) The language of calcium signaling. Annu Rev Plant Biol 61: 593–620.
7. Day IS, Reddy VS, Shad Ali G, Reddy AS (2002) Analysis of EF-hand-containing proteins in Arabidopsis. Genome Biol 3: RESEARCH0056.
8. Evans NH, McAinsh MR, Hetherington AM (2001) Calcium oscillations in higher plants. Current opinion in plant biology 4: 415–420.
9. Harper JF (2001) Dissecting calcium oscillators in plant cells. Trends in plant science 6: 395–397.
10. Ali GS, Reddy VS, Lindgren PB, Jakobek JL, Reddy A (2003) Differential expression of genes encoding calmodulin-binding proteins in response to bacterial pathogens and inducers of defense responses. Plant Molecular Biology 51: 803–815.
11. Snedden WA, Fromm H (2001) Calmodulin as a versatile calcium signal transducer in plants. New Phytologist 151: 35–66.
12. Bouché N, Yellin A, Snedden WA, Fromm H (2005) Plant-specific calmodulin-binding proteins. Annu Rev Plant Biol 56: 435–466.
13. Luan S, Kudla J, Rodriguez-Concepcion M, Yalovsky S, Gruissem W (2002) Calmodulins and calcineurin B–like proteins calcium sensors for specific signal response coupling in plants. The Plant Cell Online 14: S389–S400.
14. Ranty B, Aldon D, Galaud J-P (2006) Plant calmodulins and calmodulin-related proteins: multifaceted relays to decode calcium signals. Plant Signal Behav 1: 96.
15. Zhang H, Yin W, Xia X (2008) Calcineurin B-Like family in Populus: comparative genome analysis and expression pattern under cold, drought and salt stress treatment. Plant Growth Regulation 56: 129–140.
16. Zhang C, Bian M, Yu H, Liu Q, Yang Z (2011) Identification of alkaline stress-responsive genes of CBL family in sweet sorghum (Sorghum bicolor L.). Plant Physiol Biochem 49: 1306–1312.
17. Yu Y, Xia X, Yin W, Zhang H (2007) Comparative genomic analysis of CIPK gene family in Arabidopsis and Populus. Plant Growth Regulation 52: 101–110.
18. Zuo R, Hu R, Chai G, Xu M, Qi G, et al. (2013) Genome-wide identification, classification, and expression analysis of CDPK and its closely related gene families in poplar (Populus trichocarpa). Mol Biol Rep 40: 2645–2662.
19. Reddy AS, Ali GS, Celesnik H, Day IS (2011) Coping with stresses: roles of calcium- and calcium/calmodulin-regulated gene expression. Plant Cell 23: 2010–2032.

20. Batistic O, Kudla J (2004) Integration and channeling of calcium signaling through the CBL calcium sensor/CIPK protein kinase network. Planta 219: 915–924.
21. Boonburapong B, Buaboocha T (2007) Genome-wide identification and analyses of the rice calmodulin and related potential calcium sensor proteins. BMC Plant Biol 7: 4.
22. Perochon A, Aldon D, Galaud J-P, Ranty B (2011) Calmodulin and calmodulin-like proteins in plant calcium signaling. Biochimie 93: 2048–2053.
23. DeFalco T, Bender K, Snedden W (2010) Breaking the code: Ca2+ sensors in plant signalling. Biochem J 425: 27–40.
24. Chang'en T, Yuping Z (2014) Research Progress in Plant IQ Motif-containing Calmodulin-binding Proteins. Chinese Bulletin of Botany 48: 447–460.
25. Fischer C, Kugler A, Hoth S, Dietrich P (2013) An IQ domain mediates the interaction with calmodulin in a plant cyclic nucleotide-gated channel. Plant and cell physiology 54: 573–584.
26. Bähler M, Rhoads A (2002) Calmodulin signaling via the IQ motif. FEBS letters 513: 107–113.
27. Hoeflich KP, Ikura M (2002) Calmodulin in action: diversity in target recognition and activation mechanisms. Cell 108: 739–742.
28. Bhattacharya S, Bunick CG, Chazin WJ (2004) Target selectivity in EF-hand calcium binding proteins. Biochimica et Biophysica Acta (BBA)-Molecular Cell Research 1742: 69–79.
29. Clapperton JA, Martin SR, Smerdon SJ, Gamblin SJ, Bayley PM (2002) Structure of the complex of calmodulin with the target sequence of calmodulin-dependent protein kinase I: studies of the kinase activation mechanism. Biochemistry 41: 14669–14679.
30. Abel S, Savchenko T, Levy M (2005) Genome-wide comparative analysis of the IQD gene families in Arabidopsis thaliana and Oryza sativa. BMC Evol Biol 5: 72.
31. Levy M, Wang Q, Kaspi R, Parrella MP, Abel S (2005) Arabidopsis IQD1, a novel calmodulin-binding nuclear protein, stimulates glucosinolate accumulation and plant defense. Plant J 43: 79–96.
32. Burstenbinder K, Savchenko T, Muller J, Adamson AW, Stamm G, et al. (2012) Arabidopsis Calmodulin-binding Protein IQ67-Domain 1 Localizes to Microtubules and Interacts with Kinesin Light Chain-related Protein-1. Journal of Biological Chemistry 288: 1871–1882.
33. Abel S, Burstenbinder K, Muller J (2013) The emerging function of IQD proteins as scaffolds in cellular signaling and trafficking. Plant Signal Behav 8: e24369.
34. Filiz E, Tombuloglu H, Ozyigit II (2013) Genome-wide analysis of IQ67 domain (IQD) gene families in Brachypodium distachyon. Plant Omics 6: 425–432.
35. Huang Z, Van Houten J, Gonzalez G, Xiao H, van der Knaap E (2013) Genome-wide identification, phylogeny and expression analysis of SUN, OFP and YABBY gene family in tomato. Mol Genet Genomics 288: 111–129.
36. Zentella R, Zhang ZL, Park M, Thomas SG, Endo A, et al. (2007) Global analysis of della direct targets in early gibberellin signaling in Arabidopsis. Plant Cell 19: 3037–3057.
37. Xiao H, Jiang N, Schaffner E, Stockinger EJ, van der Knaap E (2008) A retrotransposon-mediated gene duplication underlies morphological variation of tomato fruit. Science 319: 1527–1530.
38. Wu S, Xiao H, Cabrera A, Meulia T, van der Knaap E (2011) SUN regulates vegetative and reproductive organ shape by changing cell division patterns. Plant Physiol 157: 1175–1186.
39. Kamenetzky L, Asis R, Bassi S, de Godoy F, Bermudez L, et al. (2010) Genomic analysis of wild tomato introgressions determining metabolism- and yield-associated traits. Plant Physiol 152: 1772–1786.
40. Soto-Valdez H, Colin-Chavez C, Peralta E. Fabrication and Properties of Antioxidant Polyethylene-based Films Containing Marigold (Tagetes erecta) Extract and Application on Soybean Oil Stability; 2012. DEStech Publications, Inc. pp. 206.
41. Schmutz J, Cannon SB, Schlueter J, Ma J, Mitros T, et al. (2010) Genome sequence of the palaeopolyploid soybean. Nature 463: 178–183.
42. Silva DC, Yamanaka N, Brogin RL, Arias CA, Nepomuceno AL, et al. (2008) Molecular mapping of two loci that confer resistance to Asian rust in soybean. Theor Appl Genet 117: 57–63.
43. Helfer S (2014) Rust fungi and global change. New Phytologist 201: 770–780.
44. Nekrutenko A, Makova KD, Li W-H (2002) The KA/KS ratio test for assessing the protein-coding potential of genomic regions: an empirical and simulation study. Genome research 12: 198–202.
45. Lynch M, Conery JS (2000) The evolutionary fate and consequences of duplicate genes. Science 290: 1151–1155.
46. Chen X, Chen Z, Zhao H, Zhao Y, Cheng B, et al. (2014) Genome-Wide Analysis of Soybean HD-Zip Gene Family and Expression Profiling under Salinity and Drought Treatments. PloS one 9: e87156.
47. Cannon SB, McCombie WR, Sato S, Tabata S, Denny R, et al. (2003) Evolution and microsynteny of the apyrase gene family in three legume genomes. Molecular Genetics and Genomics 270: 347–361.
48. Zhang X, Feng Y, Cheng H, Tian D, Yang S, et al. (2011) Relative evolutionary rates of NBS-encoding genes revealed by soybean segmental duplication. Molecular Genetics and Genomics 285: 79–90.
49. Creelman RA, Tierney ML, Mullet JE (1992) Jasmonic acid/methyl jasmonate accumulate in wounded soybean hypocotyls and modulate wound gene expression. Proceedings of the National Academy of Sciences 89: 4938–4941.
50. HU W-z, JIANG A-l, YANG H, LIU C-h, HE Y-b (2012) Effect of Jasmonic acid methyl ester treatment on the physiological and biochemical reactions of fresh-cut apple. Science and Technology of Food Industry 16: 086.
51. Cannon SB, Shoemaker RC (2012) Evolutionary and comparative analyses of the soybean genome. Breeding science 61.
52. Reddy A, Day IS (2001) Analysis of the myosins encoded in the recently completed Arabidopsis thaliana genome sequence. Genome Biol 2: 1–17.
53. Gao F, Han X, Wu J, Zheng S, Shang Z, et al. (2012) A heat-activated calcium-permeable channel–Arabidopsis cyclic nucleotide-gated ion channel 6–is involved in heat shock responses. The Plant Journal 70: 1056–1069.
54. Bürstenbinder K, Savchenko T, Müller J, Adamson AW, Stamm G, et al. (2013) Arabidopsis Calmodulin-binding Protein IQ67-Domain 1 Localizes to Microtubules and Interacts with Kinesin Light Chain-related Protein-1. Journal of Biological Chemistry 288: 1871–1882.
55. Pauly N, Knight MR, Thuleau P, van der Luit AH, Moreau M, et al. (2000) Cell signalling: control of free calcium in plant cell nuclei. Nature 405: 754–755.
56. Xiong TC, Jauneau A, Ranjeva R, Mazars C (2004) Isolated plant nuclei as mechanical and thermal sensors involved in calcium signalling. The Plant Journal 40: 12–21.
57. Lecourieux D, Ranjeva R, Pugin A (2006) Calcium in plant defence-signalling pathways. New Phytologist 171: 249–269.
58. Anandalakshmi R, Marathe R, Ge X, Herr J, Mau C, et al. (2000) A calmodulin-related protein that suppresses posttranscriptional gene silencing in plants. Science 290: 142–144.
59. Yoo JH, Park CY, Kim JC, Do Heo W, Cheong MS, et al. (2005) Direct interaction of a divergent CaM isoform and the transcription factor, MYB2, enhances salt tolerance in Arabidopsis. Journal of Biological Chemistry 280: 3697–3706.
60. Du J, Tian Z, Sui Y, Zhao M, Song Q, et al. (2012) Pericentromeric effects shape the patterns of divergence, retention, and expression of duplicated genes in the paleopolyploid soybean. The Plant Cell Online 24: 21–32.
61. Ohno S (1970) Evolution by gene duplication: London: George Alien & Unwin Ltd. Berlin, Heidelberg and New York: Springer-Verlag.
62. Schlueter JA, Dixon P, Granger C, Grant D, Clark L, et al. (2004) Mining EST databases to resolve evolutionary events in major crop species. Genome 47: 868–876.
63. Shoemaker RC, Schlueter J, Doyle JJ (2006) Paleopolyploidy and gene duplication in soybean and other legumes. Current opinion in plant biology 9: 104–109.
64. Dreze M, Carvunis A-R, Charloteaux B, Galli M, Pevzner SJ, et al. (2011) Evidence for network evolution in an Arabidopsis interactome map. Science 333: 601–607.
65. Su-Juan WH-YGZ-QWZ-JLZ-WWZ-JC (2008) Isolation and Characterization of Calmodulin-binding Protein *AtIQD26* in *Arabidopsis thaliana*. Progress in Biochemistry and Biophysics 35: 703–711.
66. Hajduch M, Ganapathy A, Stein JW, Thelen JJ (2005) A systematic proteomic study of seed filling in soybean. Establishment of high-resolution two-dimensional reference maps, expression profiles, and an interactive proteome database. Plant Physiol 137: 1397–1419.
67. Brader G, Tas É, Palva ET (2001) Jasmonate-dependent induction of indole glucosinolates in arabidopsis by culture filtrates of the nonspecific pathogen-erwinia carotovora. Plant Physiology 126: 849–860.
68. Cipollini D, Enright S, Traw M, Bergelson J (2004) Salicylic acid inhibits jasmonic acid-induced resistance of Arabidopsis thaliana to Spodoptera exigua. Molecular Ecology 13: 1643–1653.
69. Fahey JW, Zalcmann AT, Talalay P (2001) The chemical diversity and distribution of glucosinolates and isothiocyanates among plants. Phytochemistry 56: 5–51.
70. Wittstock U, Halkier BA (2002) Glucosinolate research in the *Arabidopsis* era. Trends in plant science 7: 263–270.
71. Dittmar K, Liberles D (2011) Evolution after gene duplication: John Wiley & Sons.
72. Finn RD, Mistry J, Schuster-Böckler B, Griffiths-Jones S, Hollich V, et al. (2006) Pfam: clans, web tools and services. Nucleic acids research 34: D247–D251.
73. Letunic I, Copley RR, Schmidt S, Ciccarelli FD, Doerks T, et al. (2004) SMART 4.0: towards genomic data integration. Nucleic Acids Research 32: D142–D144.
74. Bateman A, Coin L, Durbin R, Finn RD, Hollich V, et al. (2004) The Pfam protein families database. Nucleic acids research 32: D138–D141.
75. Gasteiger E, Gattiker A, Hoogland C, Ivanyi I, Appel RD, et al. (2003) ExPASy: the proteomics server for in-depth protein knowledge and analysis. Nucleic acids research 31: 3784–3788.
76. Hu R, Qi G, Kong Y, Kong D, Gao Q, et al. (2010) Comprehensive analysis of NAC domain transcription factor gene family in Populus trichocarpa. BMC plant biology 10: 145.
77. Guo A-Y, Zhu Q-H, Chen X, Luo J-C (2007) GSDS: a gene structure display server]. Yi chuan = Hereditas/Zhongguo yi chuan xue hui bian ji 29: 1023.
78. Grant D, Nelson RT, Cannon SB, Shoemaker RC (2010) SoyBase, the USDA-ARS soybean genetics and genomics database. Nucleic acids research 38: D843–D846.
79. Wei F, Coe E, Nelson W, Bharti AK, Engler F, et al. (2007) Physical and genetic structure of the maize genome reflects its complex evolutionary history. PLoS Genetics 3: e123.

80. Wang L, Guo K, Li Y, Tu Y, Hu H, et al. (2010) Expression profiling and integrative analysis of the CESA/CSL superfamily in rice. BMC plant biology 10: 282.
81. Suyama M, Torrents D, Bork P (2006) PAL2NAL: robust conversion of protein sequence alignments into the corresponding codon alignments. Nucleic acids research 34: W609–W612.
82. Yang Z (2007) PAML 4: phylogenetic analysis by maximum likelihood. Molecular biology and evolution 24: 1586–1591.
83. Lee T-H, Tang H, Wang X, Paterson AH (2013) PGDD: a database of gene and genome duplication in plants. Nucleic acids research 41: D1152–D1158.
84. Tang H, Bowers JE, Wang X, Ming R, Alam M, et al. (2008) Synteny and collinearity in plant genomes. Science 320: 486–488.
85. Tang H, Wang X, Bowers JE, Ming R, Alam M, et al. (2008) Unraveling ancient hexaploidy through multiply-aligned angiosperm gene maps. Genome research 18: 1944–1954.
86. Severin AJ, Woody JL, Bolon Y-T, Joseph B, Diers BW, et al. (2010) RNA-Seq Atlas of Glycine max: a guide to the soybean transcriptome. BMC plant biology 10: 160.
87. Team RC (2012) R: A language and environment for statistical computing.
88. Hyun TK, Eom SH, Jeun YC, Han SH, Kim J-S (2013) Identification of glutamate decarboxylases as a γ-aminobutyric acid (GABA) biosynthetic enzyme in soybean. Industrial Crops and Products 49: 864–870.
89. Cheng Q, Zhang B, Zhuge Q, Zeng Y, Wang M, et al. (2006) Expression profiles of two novel lipoxygenase genes in *Populus deltoides*. Plant Science 170: 1027–1035.
90. Gutierrez N GMJ, Palomino C, et al. (2011) Assessment of candidate reference genes for expression studies in Vicia faba L. by real-time quantitative PCR. Molecular Breeding: 13–24.
91. Vívian de Jesus Miranda RRC, Antônio Américo Barbosa Viana , Osmundo Brilhante de Oliveira Neto , Regina Maria Dechechi Gomes Carneiro , Thales Lima Rocha , Maria Fatima Grossi de Sa , Rodrigo Rocha Fragoso (2013) Validation of reference genes aiming accurate normalization of qPCR data in soybean upon nematode parasitism and insect attack. BMC Research Notes 6.
92. Le DT, Nishiyama R, Watanabe Y, Mochida K, Yamaguchi-Shinozaki K, et al. (2011) Genome-wide expression profiling of soybean two-component system genes in soybean root and shoot tissues under dehydration stress. DNA Res 18: 17–29.
93. Jian B, Liu B, Bi Y, Hou W, Wu C, et al. (2008) Validation of internal control for gene expression study in soybean by quantitative real-time PCR. BMC Mol Biol 9: 59.
94. Livak KJ, Schmittgen TD (2001) Analysis of Relative Gene Expression Data Using Real-Time Quantitative PCR and the $2^{-\Delta\Delta CT}$ Method. methods 25: 402–408.
95. Peng X, Zhao Y, Cao J, Zhang W, Jiang H, et al. (2012) CCCH-type zinc finger family in maize: genome-wide identification, classification and expression profiling under abscisic acid and drought treatments. PLoS One 7: e40120.
96. Bryfczynski SP, Pargas RP (2009) GraphPad: a graph creation tool for CS2/CS7. ACM SIGCSE Bulletin 41: 389–389.

A Comprehensive Analysis of the Cupin Gene Family in Soybean (*Glycine max*)

Xiaobo Wang[1], Haowei Zhang[1], Yali Gao[1], Genlou Sun[1,3], Wenming Zhang[1], Lijuan Qiu[2]*

1 School of Agronomy, Anhui Agricultural University, Hefei, China, **2** National Key Facility for Gene Resources and Genetic Improvement/Key Laboratory of Crop Germplasm Utilization, Ministry of Agriculture, Institute of Crop Science, Chinese Academy of Agricultural Science, Beijing, China, **3** Biology Department, Saint Mary's University, Halifax, NS, Canada

Abstract

Cupin superfamily of proteins, including germin and germin-like proteins (GLPs) from higher plants, is known to play crucial roles in plant development and defense. To date, no systematic analysis has been conducted in soybean (*Glycine max*) incorporating genome organization, gene structure, expression compendium. In this study, 69 putative *Cupin* genes were identified from the whole-genome of soybean, which were non-randomly distributed on 17 of the 20 chromosomes. These Gmcupin proteins were phylogenetically clustered into ten distinct subgroups among which the gene structures were highly conserved. Eighteen pairs (52.2%) of duplicate paralogous genes were preferentially retained in duplicated regions of the soybean genome. The distributions of *GmCupin* genes implied that long segmental duplications contributed significantly to the expansion of the *GmCupin* gene family. According to the RNA-seq data analysis, most of the *Gmcupins* were differentially expressed in tissue-specific expression pattern and the expression of some duplicate genes were partially redundant while others showed functional diversity, suggesting the *Gmcupins* have been retained by substantial subfunctionalization during soybean evolutionary processes. Selective analysis based on single nucleotide polymorphisms (SNPs) in cultivated and wild soybeans revealed sixteen *Gmcupins* had selected site(s), with all SNPs in *Gmcupin10.3* and *Gmcupin07.2* genes were selected sites, which implied these genes may have undergone strong selection effects during soybean domestication. Taken together, our results contribute to the functional characterization of *Gmcupin* genes in soybean.

Editor: Baohong Zhang, East Carolina University, United States of America

Funding: This study was supported by the National Natural Science Foundation of China (31201226), the National Natural Science Foundation of Anhui Province (1308085QC49) and the Agricultural Science and Technology Innovation Program of Chinese Academy of Agricultural Sciences. The funders had no role in study design, data collection and analysis, decision to publish, or preparation of the manuscript.

Competing Interests: The authors have declared that no competing interests exist.

* Email: qiulijuan@caas.cn

Introduction

The cupin superfamily of proteins, mainly consisted of germin and germin-like protein (GLP) subfamilies, is extremely diverse in plants and possess various enzymatic activities such as sugar-binding metal-independent epimerases, and metal-dependent enzymes possessing dioxygenase, and decarboxylase [1,2]. Germin, initially identified as a specific marker for germination in wheat embryos [3], has been characterized as a homopentameric glycoprotein with oxalate oxidase (OxO) activity [4]. To date, it is speculate to play significant roles in plant development and defense through oxidative breakdown of oxalate, leading to generation of H_2O_2 [5,6]. Germin like proteins (GLPs), with a high sequence and structural similarity to cereal germins, differ from germin as they mostly lack oxalate oxidase activity, and possess activity of SOD and phosphodiesterase [2,7–9]. Cupin-domain has been reported to be associated with the biological properties in plants. For instance, a group of single cupin-domain related proteins, including two phosphomannose isomerases and two epimerases involved in cell wall synthesis, were identified in *Synechocystis PCC6803* genome [10]. Moreover, a duplicated, two cupin-domain GLP protein showed close similarity in structure of an oxalate decarboxylase from the fungus Collybia velutipes and is considered as a putative progenitor of the storage proteins of land plants [10].

Until now, a total of 27 *GLP* genes have been identified in *Arabidopsis*, and their expression vary in different tissues such as roots, leaves and flowers [11–13]. Lapik et al reported a cupin-domain protein *AtPirin1* could interact with a CCAAT box binding transcription factor, and served as a downstream component of GPA1 in regulating seed germination and early seedling development [14]. Recently, another two GLP proteins (PDGLP1 and PDGLP2) in *Arabidopsis*, which could interact with *Cucurbita maxima* PHLOEM PROTEIN 16 (Cm-PP16), involved in the regulation of growth of primary root through modulating phloem-mediated resource allocation between the primary and lateral root meristems [15]. The PDGLP1 signal peptide was shown to function in localization to the plasmodesmata (PD) by a novel mechanism involving the endoplasmic reticulum-Golgi secretory pathway. Further, in plum (*Prunus salicina*), two GLP-encoding genes (designated as *Ps-GLP1* and *Ps-GLP2*) were cloned, and the regulation was studied throughout fruit develop-

Table 1. Summary of Cupin family members in soybean.

Gene Symbol	Gene Locus	Primary transcript	Gene location	Amino Acids	Extrons
Gmcupin01.1	Glyma01g04450	Glyma01g04450.1	Gm01:3990477-3991986	220	1
Gmcupin02.1	Glyma02g01085	Glyma02g01085.1	Gm02:816290-817482	147	2
Gmcupin02.2	Glyma02g03100	Glyma02g03100.1	Gm02:2414696-2415880	220	1
Gmcupin02.3	Glyma02g05010	Glyma02g05010.1	Gm02:4077995-4078612	205	1
Gmcupin03.1	Glyma03g32030	Glyma03g32030.1	Gm03: 39840052 - 39842763	495	4
Gmcupin03.2	Glyma03g38630	Glyma03g38630.1	Gm03:44934345-44933730	218	2
Gmcupin04.1	Glyma04g39040	Glyma04g39040.2	Gm04:45306333-45307231	199	3
Gmcupin05.1	Glyma05g25620	Glyma05g25620.1	Gm05:31685194-31686123	215	2
Gmcupin06.1	Glyma06g15930	Glyma06g15930.1	Gm06:12516557-12517688	228	1
Gmcupin07.1	Glyma07g04310	Glyma07g04310.1	Gm07:3163391-3164534	209	1
Gmcupin07.2	Glyma07g04320	Glyma07g04320.1	Gm07:3167203-3168078	208	1
Gmcupin07.3	Glyma07g04330	Glyma07g04330.1	Gm07:3173276-3174394	208	1
Gmcupin07.4	Glyma07g04340	Glyma07g04340.1	Gm07:3179749-3180833	225	1
Gmcupin07.5	Glyma07g04400	Glyma07g04400.1	Gm07:3202414-3203532	208	1
Gmcupin08.1	Glyma08g08600	Glyma08g08600.1	Gm08:6134739-6134464	215	2
Gmcupin08.2	Glyma08g24320	Glyma08g24320.1	Gm08:18508831-18509842	211	1
Gmcupin09.1	Glyma09g03010	Glyma09g03010.1	Gm09:2110529-2109847	217	2
Gmcupin09.2	Glyma09g08030	Glyma09g08030.1	Gm09:7066672-7066667	135	1
Gmcupin10.1	Glyma10g08360	Glyma10g08360.1	Gm10:7201264-7200871	226	2
Gmcupin10.2	Glyma10g11935	Glyma10g11935.1	Gm10:12509357-12509734	125	1
Gmcupin10.3	Glyma10g28010	Glyma10g28010.1	Gm10:36807794-36807686	221	2
Gmcupin10.4	Glyma10g28020	Glyma10g28020.1	Gm10:36812065-36811696	220	2
Gmcupin10.5	Glyma10g28190	Glyma10g28190.1	Gm10:36981942-36980791	223	2
Gmcupin10.6	Glyma10g31200	Glyma10g31200.2	Gm10:39762112-39761524	198	3
Gmcupin10.7	Glyma10g31210	Glyma10g31210.1	Gm10:39768393-39768098	232	2
Gmcupin10.8	Glyma10g42611	Glyma10g42611.1	Gm10:49519941-49520574	177	3
Gmcupin12.1	Glyma12g09630	Glyma12g09630.2	Gm12:7391558-7392181	207	1
Gmcupin12.2	Glyma12g09640	Glyma12g09640.2	Gm12:7398088-7398957	212	2
Gmcupin12.3	Glyma12g09760	Glyma12g09760.2	Gm12:7531728-7532351	207	1
Gmcupin12.4	Glyma12g31110	Glyma12g31110.1	Gm12:34711894-34712517	207	1
Gmcupin13.1	Glyma13g16960	Glyma13g16960.2	Gm13:20815056-20816399	199	2
Gmcupin13.2	Glyma13g18450	Glyma13g18450.2	Gm13: 22109247 - 22113254	226	4
Gmcupin13.3	Glyma13g22050	Glyma13g22050.1	Gm13:25624544-25624397	239	2
Gmcupin13.4	Glyma13g40360	Glyma13g40360.1	Gm13:40856942-40859117	483	5
Gmcupin15.1	Glyma15g05040	Glyma15g05040.2	Gm15: 3611464 - 3613757	351	7
Gmcupin15.2	Glyma15g13960	Glyma15g13960.1	Gm15:10534152-10533514	215	2
Gmcupin15.3	Glyma15g19510	Glyma15g19510.1	Gm15:16833879-16835243	213	1
Gmcupin15.4	Glyma15g35130	Glyma15g35130.1	Gm15:39672342-39673419	231	1
Gmcupin16.1	Glyma16g00980	Glyma16g00980.1	Gm16:652712-653968	209	1
Gmcupin16.2	Glyma16g00990	Glyma16g00990.1	Gm16:656546-657171	181	2
Gmcupin16.3	Glyma16g01000	Glyma16g01000.1	Gm16:660570-661190	206	1
Gmcupin16.4	Glyma16g06500	Glyma16g06500.1	Gm16:5853241-5853122	221	2
Gmcupin16.5	Glyma16g06520	Glyma16g06520.1	Gm16:5858235-5858104	221	2
Gmcupin16.6	Glyma16g06530	Glyma16g06530.1	Gm16:5860996-5862104	220	2
Gmcupin16.7	Glyma16g06630	Glyma16g06630.1	Gm16:5947078-5946961	221	2
Gmcupin16.8	Glyma16g06640	Glyma16g06640.1	Gm16:5951136-5952363	215	2
Gmcupin16.9	Glyma16g07550	Glyma16g07550.1	Gm16:6838751-6839383	210	1
Gmcupin16.10	Glyma16g07560	Glyma16g07560.2	Gm16:6844140-6843743	188	1
Gmcupin16.11	Glyma16g07580	Glyma16g07580.1	Gm16:6860067-6860840	214	1
Gmcupin17.1	Glyma17g05760	Glyma17g05760.1	Gm17:4052453-4053577	208	1

Table 1. Cont.

Gene Symbol	Gene Locus	Primary transcript	Gene location	Amino Acids	Extrons
Gmcupin19.1	Glyma19g09370	Glyma19g09370.2	Gm19:11189789-11189486	181	3
Gmcupin19.2	Glyma19g09810	Glyma19g09810.1	Gm19:11530120-11531183	221	2
Gmcupin19.3	Glyma19g09830	Glyma19g09830.1	Gm19:11554699-11555709	221	2
Gmcupin19.4	Glyma19g09840	Glyma19g09840.1	Gm19:11601504-11602281	221	2
Gmcupin19.5	Glyma19g09860	Glyma19g09860.1	Gm19:11630024-11631147	221	2
Gmcupin19.6	Glyma19g09990	Glyma19g09990.1	Gm19:11922733-11922620	221	2
Gmcupin19.7	Glyma19g24840	Glyma19g24840.1	Gm19:30509520-30510472	212	2
Gmcupin19.8	Glyma19g24850	Glyma19g24850.1	Gm19:30514099-30513955	221	2
Gmcupin19.9	Glyma19g24870	Glyma19g24870.2	Gm19:30532519-30533617	221	2
Gmcupin19.10	Glyma19g24900	Glyma19g24900.1	Gm19:30559228-30560036	221	2
Gmcupin19.11	Glyma19g24910	Glyma19g24910.1	Gm19:30576453-30577383	219	2
Gmcupin19.12	Glyma19g27580	Glyma19g27580.1	Gm19:34882234-34884637	212	2
Gmcupin19.13	Glyma19g34780	Glyma19g34780.1	Gm19: 42366324 - 42369290	481	4
Gmcupin19.14	Glyma19g41070	Glyma19g41070.2	Gm19:47390045 - 47391131	184	3
Gmcupin19.15	Glyma19g41220	Glyma19g41220.1	Gm19:47524955-47524382	219	2
Gmcupin20.1	Glyma20g22180	Glyma20g22180.1	Gm20:32106832-32105935	224	2
Gmcupin20.2	Glyma20g25430	Glyma20g25430.1	Gm20:35111115-35111738	207	1
Gmcupin20.3	Glyma20g36300	Glyma20g36300.1	Gm20:44453211-44454264	232	2
Gmcupin20.4	Glyma20g36320	Glyma20g36320.1	Gm20:44458621-44459718	222	2

ment and during maturity of early and late cultivars. All these demonstrated that GLPs may involve in certain developmental stages in plants.

Expression of Cupin proteins could be modulated by abiotic or biotic stresses, suggesting their multifunctional roles in plant defense response. For instance, a 66-kDa cupin protein BspA (for "boiling-stable protein"), highly expressed in cultured shoots of aspen (*Populus tremula*) in the presence of water stress, was considered to contribute to membrane stability [16]. However, the mechanism of how cupin proteins involve in the plant defense is still not well defined. One germin-like gene (*CchGLP*) cloned from geminivirus-resistant pepper, induced by ethylene and salicylic acid other than jasmonic acid, encoded an enzyme with manganese superoxide dismutase (Mn-SOD)activity [17]. Also, Mn-SOD activity was identified in GLPs isolated from tobacco and Barbula unguiculata [18–20]. Considering plant Mn-SODs was distributed extracellularly as well as in mitochondria and peroxisomes and associated with defense against biotic stress in plants [18,21,22], it is probably to speculate that Cupin protein may involve in the plant defense through scavenging free radicals.

The ubiquitous distribution of GLPs implies their indispensable and fundamental roles in plants [23,24]. In soybean, rare studies have been performed on the functional characterization of Cupin proteins [25]. Completion of the soybean genome greatly facilitated the identification of gene families at the whole-genome level [26]. In the present study, a genome-wide identification of Cupin domain was performed in soybean, and detailed analysis of the sequence phylogeny, genome organization, gene structure, expression profiling and selective effects of *Gmcupin* genes during soybean domestication was performed. Our data contributes to the evolutionary and functional analysis of the *Cupin* gene family in soybean.

Materials and Methods

Sequence retrieval and phylogenetic analysis

Amino-acid sequence of the Cupin domain was used to search for potential Dof-domain homolog hits in the whole-genome sequence of *Glycine max* with BLASTP at the Phytozome database (http:/www.phytozome.net) [27]. All non-redundant hits with expected values of <1E-5 were collected. Subsequently, manual analysis was performed to confirm the presence of Cupin domain using InterProScan program (http://www.ebi.ac.uk/Tools/InterProScan/) [28].

Sequence alignments of the full-length protein sequences were performed using Clustal X software (version 1.8) [29]. Phylogenetic trees were constructed with MEGA 5.0 using Neighbor-Joining (NJ) method with 1000 replicates of bootstrap analysis [30]. The evolutionary distances were computed using the p-distance method. WebLogo was used to create the distribution of amino-acid residues at the corresponding positions in domain profiles for the conserved Cupin domain of Gmcupins [28].

Identification of conserved motifs

For the motif analysis, deduced amino-acid sequences of the Gmcupins were analysis by Multiple EM for Motif Elicitation version 4.9.1 (http://meme.nbcr.net/meme/cgi-bin/meme.cgi) [31]. Structural motif annotation was performed using the Pfam (http://pfam.sanger.ac.uk), NCBI (http://www.ncbi.nlm.nih.gov/) and SMART (http://smart.embl-heidelberg.de) database.

Genomic structure and chromosomal location of *Gmcupins*

The exon/intron organization for individual cupin gene was illustrated with Gene structure display server program (GSDS) (http://gsds.cbi.pku.edu.cn/) [32] by comparing the cDNAs with their corresponding gDNA sequences in the Phytozome database (http://www.phytozome.net/gmax). The chromosomal locations of soybean cupins were mapped to the duplicated blocks using the Chromosome Visualization Tool (CViT) genome search and synteny viewer at the Legume Information System (http://comparative-legumes.org/) [33,34].

Calculation of Ka/Ks values

Clustal X (version 1.8) was used for the pairwise alignments of the paralogous nucleotide sequences [29]. Ka (non-synonymous substitution rate) and Ks (synonymous substitution rate) were estimated using the DnaSp v5 program [35]. Divergence time (T) was calculated using as the formula: $T = Ks/2\lambda$, where the synonymous mutation rate λ was 6.1×10^{-9} for soybean [26,36,37].

Expression analysis of *GmCupin* genes

Genome-wide transcriptome data of seeds during various developmental stages were downloaded from Soybase database (http://soybase.org/). The transcript data were obtained from vegetative tissues (e.g. young leaf, root and nodule), seed of seven developmental stages (10, 14, 21, 25, 28, 35, and 42 days after flowering), and reproductive tissues (e.g. flower, one cm pod, pod shell of 10 and 14 days after flowering). All transcript data were analyzed with Cluster 3.0 [38] and the heat map was viewed in Java Treeview [39].

Evolutionary analysis of *Gmcupin* genes

Single nucleotide polymorphisms (SNPs) of the *Gmcupin* genes were downloaded from the SoyKB database (http://soykb.org/) based on the resequencing of wild and cultivated soybean genomes [40]. The ratio of each SNP in wild and cultivated soybean populations was analyzed respectively. The SNP site with reverse distribution ratio in different types of soybean population was defined as a putative selective site throughout domestication.

Results and Discussion

Identification of *Cupin* gene family in soybean

In order to identify the *Cupin* gene family in soybean genome, BLASTP was performed against the *G. max* v1.1 genome using the conserved Cupin domain. Afterwards, the obtained sequences were used as secondary queries. A total of 69 non-redundant *Cupin* genes were identified in the soybean genome (Table 1). To identify the conserved Cupin domain, all candidates were subjected to functional analysis using InterproScan program (http://www.ebi.ac.uk/Tools/InterProScan/). Soybean *Cupin* genes were numbered from *Gmcupin01.1* to *Gmcupin 20.4* according to their localization on chromosomes. Peptides consisted of 125–495 (average 224) amino acids were encoded by the identified Gmcupin genes in soybean.

Multiple alignment analysis was performed to discover the features of the homologous domain sequence and the frequency of

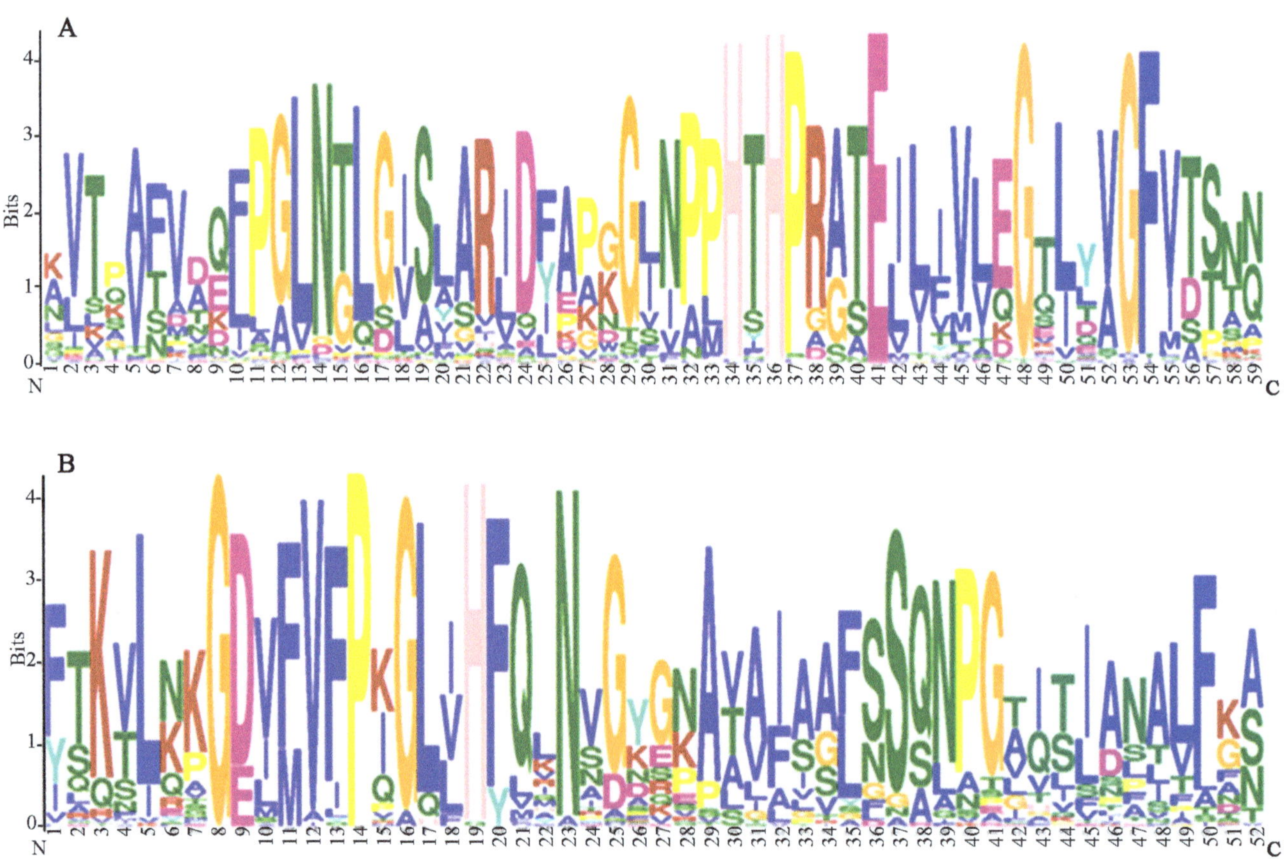

Figure 1. Conserved domains across cupin proteins in soybean. The sequence logos are based on alignments of 69 Gmcupin domains. Multiple alignment analysis of all typical Gmcupin domains (A: Gmcuppin 1; B: Gmcupin 2) were performed with Clustal W. The bit score indicates the information content for each position in the sequence.

the amino-acids at each position of the Gmcupin domains. Multiple EM for Motif Elicitation was used to identify the putative cupin motif. Two conserved domains, designated as Gmcupin 1 and Gmcupin 2, were found in these Gmcupins, and were formed by 59 amino acids and 52 amino acids, respectively. In Gmcupin 1, seven highly conserved residues were identified, including H-34, H-36, P-37, E-41, Gly-48, Gly-53 and F-54. In Gmcupin 2, four conserved residues were identified such as Gly-8, P-14, H-19 and N-23 (Figure 1). In the previous reports, the histidines and glutamic acid(s) have been reported to act as ligands for the active-site metal [18,19,41]. Additionally, studies showed that a set of conserved histidine residues employed in sugar-binding in the ancestral non-enzymatic domain evolved into the metal-coordinating histidine residues in oxalate oxidase [42] and oxalate decarboxylase [43].

Phylogenetic relationships and gene structure of soybean *Cupin* genes

The abundance of *Gmcupin* genes may derive from multiple gene duplication events, which was represented by a whole-genome duplication following multiple segmental and tandem duplications [44]. In this study, an unrooted tree was constructed to examine the phylogenetic relationships among the Cupin domains using alignments of the full-length amino-acid sequences in all Gmcupin proteins (Figure 2). The *Gmcupin* gene family was classified into ten subgroups (I-X) with 2-22 members in each subgroup. The very high bootstrap value in each subgroup suggested a common origin for the *Gmcupin* gene in each group except subgroup I. Surprisingly, 12 *Gmcupin* genes (80%) on chromosome 19 were classified into subgroup I with five genes (*Gmcupin19.2*, *Gmcupin19.3*, *Gmcupin19.4*, *Gmcupin19.5* and *Gmcupin19.6*) showed the same base composition. Phylogenetic

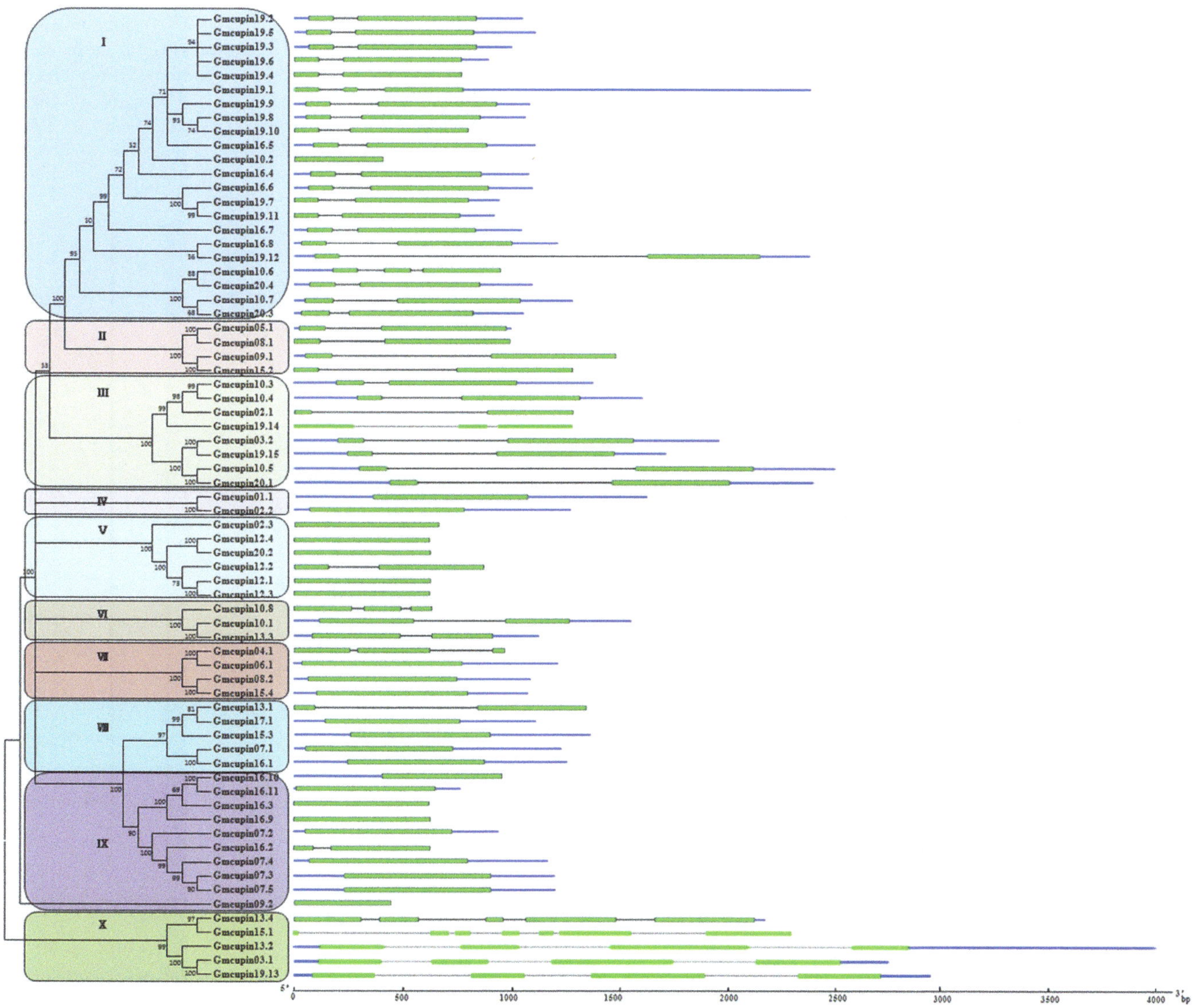

Figure 2. Phylogenetic relationships and gene structure of Gmcupin genes. The phylogenetic tree of Gmcupin proteins constructed from a complete alignment of 69 Gmcupin proteins using MEGA 5.0 by the neighbor-joining method. The bootstrap consensus tree inferred from 1000 replicates is taken to represent the evolutionary history of the taxa analyzed. Percentage bootstrap scores of>50% are indicated on the nodes. Ten major phylogenetic subgroups (designated as I to X) are indicated. Exons of *Gmcupin* genes are represented by green boxes and introns and untranslated region (UTR) by black and blue lines. The sizes of exons and introns can be estimated using the scale below.

tree topology revealed that 22 Gmcupin pairs located at the terminal nodes shared high similarities. Thus, they were assigned as paralogous pairs (homologous genes that diverged by gene duplication, Figure 2). These paralogous pairs of *Gmcupin* genes, accounted for more than 63% of the entire Gmcupin family, and showed a sequence similarity of 77.2%~100% (Table S1). This implied that these genes may evolve from a recent soybean genome duplication event [45].

As gene structural diversity is a possible explanation to the evolution of multigene families, the exon/intron organization in the coding sequences of each *Gmcupin* gene was compared. According to their predicted structures, extremely similar gene structures were observed in most of the closely related *Gmcupin* members within the same. In addition, the position and length of intron were almost completely conserved (Fig. 2A). For instance, most of the *Gmcupin* genes in subgroup I, II and III contained one intron respectively, except *Gmcupin 19.4*, *Gmcupin 10.2*, *Gmcupin 10.6* in subgroup I and *Gmcupin 19.14* in subgroup III. Meanwhile, no introns were identified in 23 of the *Gmcupin* genes (23/30) in subgroup IV, V, VI, VII, VIII and IV, respectively. In contrast, the gene structures in *Gmcupin* subfamily X appeared to be more variable and displayed the largest number of exon/intron structure variants compared with the other *Gmcupin* genes. The dissimilarity of intron phases between subfamilies and the conservation within *Gmcupin* subfamilies may reciprocally support to the results of phylogenetic analysis and genome duplication.

Chromosomal location and duplication of soybean *Cupin* genes

As revealed in Figure 3, *Gmcupin* genes were non-randomly distributed on 17 of the 20 chromosomes. Fifteen *Cupin* genes were localized on chromosome 19, while eleveen genes were localized on chromosome 16. In contrast, no more than two *Gmcupins* genes were localized on eleven chromsomes. What's more, no *Cupin* genes were distributed on chromosome 11, 14 and 18, respectively. Most *Gmcupins* presented substantial clustering on several chromosomes especially on those with high densities of the genes. To be exact, 10 *Gmcupin* genes on chromosome 16 were arranged in four clusters, with each in less than 9-kb (*Gmcupin16.1*, *Gmcupin16.2*, and *Gmcupin16.3* located within 8.5-kb; *Gmcupin16.4*, *Gmcupin16.5* and *Gmcupin16.6* located within 8.8-kb; *Gmcupin16.7* and *Gmcupin16.8* located within 5.3-kb; *Gmcupin16.9* and *Gmcupin16.10* located within 5-kb), the other *Gmcupin* gene *Gmcupin16.11* is also close to its neighbor *Gmcupin16.10* within a 1.7-kb segment. Similarly, *Gmcupin19.7* and *Gmcupin19.8* located within a 4.5-kb segment, while *Gmcupin19.10* and *Gmcupin19.11* located within a 19-kb segment on the same chromosome.

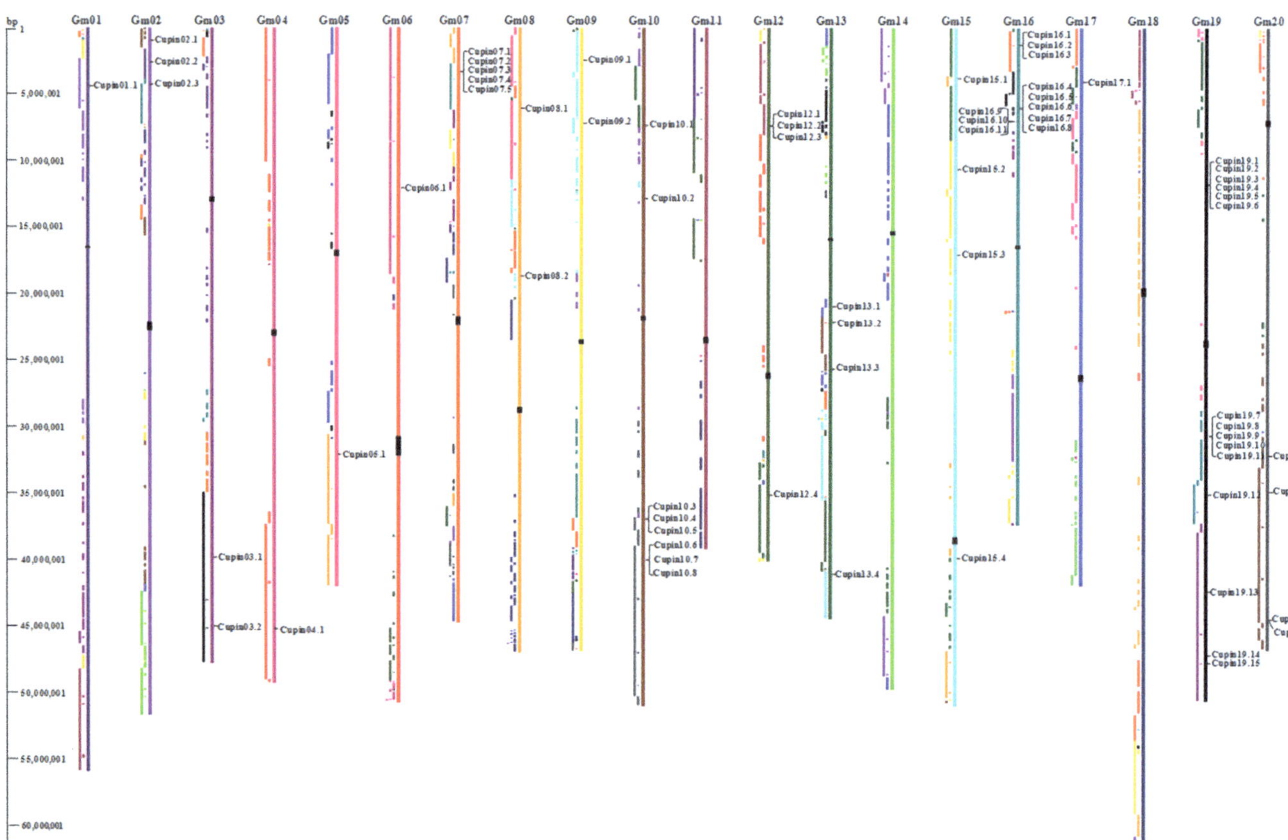

Figure 3. Chromosomal locations and predicted clusters for *Gmcupin* genes. The schematic diagram of genome-wide chromosome organization and segmental duplication arising from the genome duplication event in soybean was derived from the CViT genome search and synteny viewer at the Legume Information System (http://comparative-legumes.org). The chromosomal positions of all *Gmcupin* genes were mapped on each chromosome. Colored blocks to the left of each chromosome show duplications with chromosomes of the same color. The chromosome numbers are indicated at the top of each bar and sizes of chromosomes are represented by the vertical scale. The locations of centromeric repeats are shown as black rectangles over the chromosomes.

Table 2. Duplicated Cupin genes in soybean and the dates of the duplication blocks.

Gene1	Gene2	Duplication	Fragment size	Ka	Ks	Ka/Ks	Date(Mya)
Gmcupin5.1	Gmcupin8.1	Fragment	Large	0.0370	0.1177	0.3144	9.65
Gmcupin9.1	Gmcupin15.1	Fragment	Large	0.0328	0.1441	0.2276	11.81
Gmcupin3.1	Gmcupin19.13	Fragment	Large	0.0576	0.1479	0.3895	12.12
Gmcupin10.5	Gmcupin20.1	Fragment	Small	0.0181	0.1073	0.1687	8.80
Gmcupin1.1	Gmcupin2.2	Fragment	Large	0.0161	0.1972	0.0816	16.16
Gmcupin10.1	Gmcupin13.3	Fragment	Large	0.0457	0.1493	0.3061	12.24
Gmcupin13.1	Gmcupin17.1	Fragment	Small	0.1186	0.1838	0.6453	15.07
Gmcupin7.1	Gmcupin16.1	Fragment	Large	0.0173	0.1582	0.1094	12.97
Gmcupin16.2	Gmcupin7.4	Fragment	Large	0.0404	0.1916	0.2109	15.70
Gmcupin4.1	Gmcupin6.1	Fragment	Large	0.1079	0.2469	0.4370	20.24
Gmcupin8.2	Gmcupin15.4	Fragment	Small	0.0290	0.0965	0.3005	7.91
Gmcupin10.7	Gmcupin20.3	Fragment	Large	0.0270	0.1925	0.1403	15.78
Gmcupin10.6	Gmcupin20.4	Fragment	Large	0.0330	0.1511	0.2184	12.39
Gmcupin16.8	Gmcupin19.12	Fragment	Small	0.2215	0.3497	0.6334	28.66
Gmcupin19.13	Gmcupin3.1	Fragment	Large	0.0576	0.1479	0.3895	12.12
Gmcupin13.4	Gmcupin15.1	Fragment	Large	0.0658	0.1742	0.3777	14.28
Gmcupin19.15	Gmcupin3.2	Fragment	Large	0.0239	0.1799	0.1329	14.75
Gmcupin10.3	Gmcupin10.4	Tandem Repeat		0.0351	0.0909	0.3861	7.45

	young_ leaf	flower	one cm pod	pod shell 10DAF	pod shell 14DAF	seed 10DAF	seed 14DAF	seed 21DAF	seed 25DAF	seed 28DAF	seed 35DAF	seed 42DAF	root	nodule
Gmcupin01.1	0	1	0	0	1	0	0	0	0	0	0	0	6	0
Gmcupin02.2	1	2	3	1	0	0	0	0	0	0	0	0	2	0
Gmcupin02.3	6	5	2	4	8	0	1	3	2	1	4	0	0	0
Gmcupin03.1	0	0	0	0	0	0	0	1	3072	5909	23378	30041	0	0
Gmcupin03.2	9	11	3	3	4	2	3	4	2	1	4	1	62	9
Gmcupin04.1	0	0	0	0	0	0	0	0	0	0	0	0	0	0
Gmcupin05.1	0	0	0	0	0	0	0	0	0	0	0	0	0	0
Gmcupin06.1	0	0	0	1	0	0	0	0	0	0	0	0	3	0
Gmcupin07.1	17	1	0	1	1	0	0	0	0	1	3	1	0	0
Gmcupin07.2	0	0	0	0	0	0	0	0	0	0	0	0	12	0
Gmcupin07.3	0	0	0	0	0	0	0	0	0	0	0	0	0	0
Gmcupin07.4	0	0	0	0	0	0	0	0	0	0	0	0	1	0
Gmcupin07.5	0	0	0	0	0	0	0	0	0	0	0	0	0	0
Gmcupin08.1	0	0	0	0	0	0	0	0	0	0	0	0	0	0
Gmcupin08.2	0	2	1	0	0	0	0	0	0	0	0	0	2	0
Gmcupin09.1	0	0	0	0	0	0	0	0	0	0	0	0	0	0
Gmcupin09.2	12	0	0	2	1	0	0	1	0	0	0	0	0	0
Gmcupin10.1	4	1	1	2	2	1	1	0	1	1	3	1	13	0
Gmcupin10.3	1	2	0	1	1	0	0	0	0	0	0	0	5	2
Gmcupin10.4	1	2	1	1	11	0	0	0	0	0	0	0	10	2
Gmcupin10.5	0	0	0	0	1	0	0	0	0	0	0	0	1	1
Gmcupin10.6	0	0	0	0	0	0	0	0	0	0	0	0	0	0
Gmcupin10.7	0	0	0	0	0	0	0	0	0	0	0	0	13	0
Gmcupin12.1	0	0	0	0	0	0	0	0	0	0	0	0	0	0
Gmcupin12.2	0	0	0	0	0	0	0	0	0	0	0	0	0	0
Gmcupin12.3	0	0	0	0	0	0	0	0	0	0	0	0	0	0
Gmcupin12.4	0	0	0	0	0	0	0	0	0	0	0	0	0	0
Gmcupin13.1	4	0	1	0	0	0	0	0	0	0	0	0	0	0
Gmcupin13.2	0	0	0	0	0	0	0	0	703	1494	10794	18772	0	0
Gmcupin13.3	3	2	1	1	0	0	0	1	0	0	1	1	6	3
Gmcupin13.4	0	0	0	0	0	0	0	0	0	0	0	0	0	0
Gmcupin15.1	0	0	0	0	0	0	0	0	0	0	0	0	0	0
Gmcupin15.2	0	0	0	0	0	0	0	0	0	0	0	0	0	0
Gmcupin15.3	332	27	69	69	60	16	8	7	15	8	7	0	0	1
Gmcupin15.4	0	0	0	0	0	0	0	0	0	0	0	0	2	0
Gmcupin16.1	63	0	0	0	2	0	0	1	1	1	1	1	0	0
Gmcupin16.2	0	0	0	0	0	0	0	0	0	0	0	0	0	0
Gmcupin16.3	0	0	0	0	0	0	0	1	0	0	0	0	0	0
Gmcupin16.4	0	0	0	0	0	0	0	0	0	0	0	0	14	1
Gmcupin16.5	0	0	0	0	0	0	0	0	0	0	0	0	78	4
Gmcupin16.6	0	0	0	0	0	0	0	0	0	0	0	0	12	0
Gmcupin16.7	0	0	0	0	0	0	0	0	0	0	0	0	0	0
Gmcupin16.8	52	42	219	150	101	0	0	0	0	0	0	0	0	0
Gmcupin16.9	0	0	0	0	0	0	0	0	0	0	0	0	0	0
Gmcupin16.10	0	0	0	0	0	0	0	0	0	0	0	0	0	0
Gmcupin16.11	0	0	0	0	0	0	0	0	0	0	0	0	0	0
Gmcupin17.1	796	20	80	111	92	8	10	14	2	1	1	0	1	7
Gmcupin19.1	0	0	0	0	0	0	0	0	0	0	0	0	0	0
Gmcupin19.2	0	0	0	0	0	0	0	0	0	0	0	0	0	0
Gmcupin19.3	0	0	0	0	0	0	0	0	0	0	0	0	0	0
Gmcupin19.4	0	0	0	0	0	0	0	0	0	0	0	0	0	0
Gmcupin19.5	0	0	0	0	0	0	0	0	0	0	0	0	0	0
Gmcupin19.6	0	0	0	0	0	0	0	0	0	0	0	0	0	0
Gmcupin19.7	0	0	0	0	0	0	0	0	0	0	0	0	0	0
Gmcupin19.8	0	0	0	0	0	0	0	0	0	0	0	0	0	0
Gmcupin19.9	0	0	0	0	0	0	0	0	0	0	0	0	0	0
Gmcupin19.10	0	0	0	0	0	0	0	0	0	0	0	0	0	0
Gmcupin19.11	0	0	0	0	0	0	0	0	0	0	0	0	0	0
Gmcupin19.12	1	8	25	26	22	0	0	0	0	0	0	0	0	0
Gmcupin19.13	0	0	0	0	0	0	0	0	80	158	885	1314	0	0
Gmcupin19.14	0	0	0	0	0	0	0	0	0	0	0	0	0	0
Gmcupin19.15	1	3	4	4	5	1	1	2	4	2	4	2	5	4
Gmcupin20.1	1	0	0	1	2	0	0	0	0	0	0	0	2	0
Gmcupin20.2	0	2	0	2	1	0	2	1	2	1	3	2	0	1
Gmcupin20.3	0	0	0	0	0	0	0	0	0	0	0	0	2	0
Gmcupin20.4	0	1	0	0	0	0	0	0	0	0	0	0	61	1

magnitude of change greater than zero: 0 2 4 6 8 10 12 14 16 18 >20

Figure 4. Expression profile of *Gmcupin* genes in different tissues. The numbers in the expression profile are normalized data, which were calculated as reads/kilobase/million normalization of the raw data. All data were downloaded from the SoyBase.

Soybean genome is speculated to undergo at least two rounds of genome-wide duplication followed by multiple segmental duplication, tandem duplication, transposition events (e.g. retroposition and replicative transposition) [45]. A tandem duplication event was confirmed by the presence of two or more genes on the same chromosome, while a segmental duplication event was defined as gene duplication on different chromosomes [46]. To our knowledge, the major causes for gene-family expansion were segmental duplication, tandem duplication, and transposition events. To reveal the potential relationship between putative paralogous pairs of *Gmcupin* gens and segmental duplications, CViT genome search and synteny viewer (http://comparative-legumes.org/) were used to map the *Gmcupin* genes [33]. The distributions of *Gmcupin* genes relative to the corresponding duplicate blocks were illustrated in Figure 3. Within the identified duplicated blocks associated with a duplication event, about 18 (81.8%) of *Gmcupins* were preferentially retained duplicates that located in duplicated regions, with 13 putative paralogous pairs located in a segmental duplication of a long fragment (>1 Mb) and 4 located in a segmental duplication of a short fragment (< 1 Mb, Table 2). Meanwhile, another putative paralogous pairs (*Gmcupin10.3/Gmcupin10.4*) were formed, which was supposed to be possibly due to tandem duplication in the same orientation. Taken together, we implied that long segmental duplication was predominant for evolution of *Gmcupin* genes, which may be associated with tandem duplication.

To investigate whether Darwinian positive selection is involved in the divergence of *Gmcupin* genes after duplication and trace the dates of the duplication blocks, the substitution rate ratios (Ka/Ks) of 18 paralogous pairs are calculated using DnaSP program. Ks was used to calculate the approximate dates of duplication events. The segmental duplications of the *Gmcupin* genes in soybean was supposed to originate from 7.45 Mya (million years ago, Ks = 0.0909) to 28.66 Mya (Ks = 0.3497), with a mean value of 13.78 Mya (Ks = 0.1682, Table 2). Meanwhile, the Ks of tandem duplication of *Gmcupin10.3* and *Gmcupin10.4* was 0.0909, dating the duplication event at 7.45 Mya. Considering the fact that the soybean genome underwent two polyploidy events at 13 and 58 Mya, all the segmental duplications of the *Gmcupin* genes occurred around 13 Mya when *Glycine*-specific duplication occurred in the soybean genome [26].

Generally, a Ka/Ks of less than 1 demonstrates a functional constraint with purifying or negative selection of the genes. In this study, The Ka/Ks ratios of 8 segmental duplication pairs were less than 0.3, while the ratios of the other 9 segmental duplication pairs and one tandem duplication pair were more than 0.3, which demonstrated a possibility of significant functional divergence of some *Gmcupin* genes after the duplication events. The Ka/Ks ratios of another two paralogous gene pair (*Gmcupin16.8/19.12* and *Gmcupin13.1/17.1*) were slightly larger than 0.5 (Table 2). This suggests that they experience relatively rapid evolution following duplication. On this basis, we concluded that *Gmcupin* gene family experienced strong purifying selection pressure with limited functional divergence after segmental duplications.

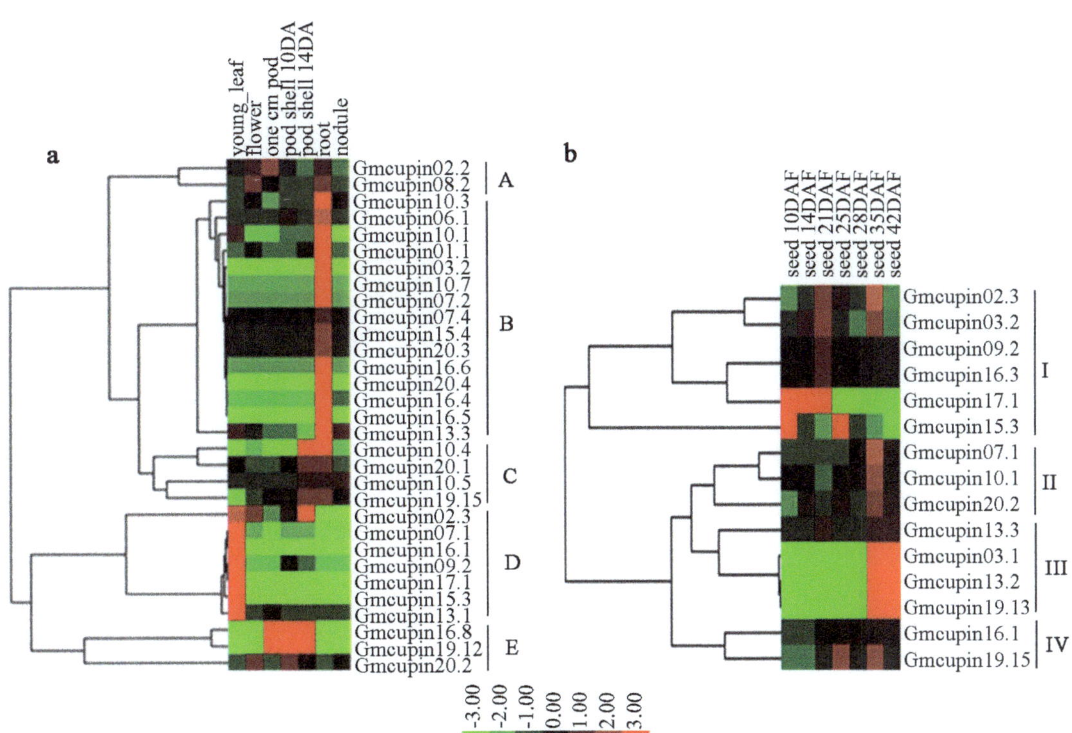

Figure 5. Expression profiles of 35 expressed *Gmcupin* genes in different tissues. a. Heatmap showing hierarchical clustering of 35 expressed *Gmcupin* genes among various tissues analyzed. b. Heatmap showing hierarchical clustering of 35 expressed *Gmcupin* genes during the development of soybean seeds.

Table 3. Selected sites of *Gmcupin* genes during soybean domestication.

Name	Position	SNPs		Name	Position	SNPs	
		Wild	Cultivar			Wild	Cultivar
Gmcupin03.1	39840871	11C/5T	5C/9T	Gmcupin10.3	36808227	9T/7A	1T/13A
	39841560	8G/7T	0G/13T		36808469	10G/5A	1G/13A
	39841777	8C/3T	3C/8T		36808488	10A/5G	1A/13G
	39842153	10T/6A	6T/8A		36808586	10A/4T	1A/13T
	39842186	9T/5C	4T/7C	Gmcupin10.4	36812510	9C/6T	1C/13T
	39842508	10A/5G	5A/9G		36812530	10G/6A	1G/13A
	39842704	10T/6A	5T/8A		36812729	8G/6C	1G/13C
Gmcupin07.1	3163748	12A/4C	4A/9C	Gmcupin10.7	39767387	14T/1C	4T/5C
	3164179	10A/5T	4A/9T		39768245	11T/1A	4T/6A
Gmcupin07.2	3167526	13T/4C	4T/10C	Gmcupin13.1	20816255	13C/4A	5C/8A
	3167819	12T/4C	4T/10C	Gmcupin13.2	22112820	9C/7G	5C/9G
	3167944	14G/3T	4G/10T	Gmcupin16.4	5853162	12C/4G	6C/8G
	3168029	13T/2C	4T/8C		5853378	11A/6G	6A/7G
	3168069	13T/2A	4T/8A		5853406	11G/5T	6G/8T
Gmcupin07.4	3179898	14T/2C	4T/10C	Gmcupin16.8	5951996	10T/4C	3T/7C
Gmcupin10.1	7200497	9C/6A	2C/12A	Gmcupin17.1	4052528	10A/5C	4A/9C
	7200888	13A/4G	2A/12G	Gmcupin19.13	42367147	10C/4T	6C/8T
	7200956	8G/8A	2G/12A	Gmcupin19.15	47524873	8A/8G	5A/9G
	7201734	12T/4C	2T/12C	Gmcupin20.1	32106107	9G/8T	1G/12T

Differential expression profile of *Gmcupin* genes

To highlight the expression profiles of *Gmcupin* genes, we then analyzed the previously publicly-available RNA-Seq data regarding seven soybean tissues, three pod development stages and seven seed developmental stages. Thirty-five *Gmcupin* genes had sequence reads in at least one tissue, and most of them showed a distinct tissue-specific expression pattern (Figure 4). For example, two genes (*Gmcupin17.1* and *Gmcupin15.3*) had a significantly higher transcript accumulation in the young leaf of soybean. *Gmcupin16.8* was mainly expressed during pod development, while *Gmcupin16.5*, *Gmcupin03.2* and *Gmcupin20.4* were specifically expressed in soybean root. Besides, three genes (*Gmcupin03.1*, *Gmcupin13.2* and *Gmcupin19.13*) of subfamily X were highly expressed at the later stage of seed development. Most *Gmcupin* genes showed a relative low expression level in soybean nodule (Figure 4). These genes were clustered into five groups (A–E) and four groups (I–IV) based on their expression patterns in soybean tissues (excet seeds) and the expression profiles during seven soybean seed development stages (Figure 5). The genes in clusters A–E were mainly expressed in flower/root, root, pod/root, young leaf and pod, respectively. Six genes in cluster I mainly expressed during the early stage of soybean seed development, while seven genes in cluster II and III mainly expressed during the later stage of soybean seed development. In addition, three genes in cluster III having a much higher and specific expression level during soybean seed development from 25 days after flower (DAF) to 42DAF. Further, genes of cluster IV were expressed in most stages of soybean seed development.

The evolutionary fates of duplicate genes may be classified into subfunctionalization (partition of original functions), neofunctionalization (acquisition of novel functions), or nonfunctionalization(loss of original functions) [47]. In this study, we investigated the functional redundancy of *Gmcupin* genes with high proporation of segmental/tandem duplications. Six paralogous pairs (*Gmcupin03.1/19.13*, *Gmcupin07.1/16.1*, *Gmcupin10.1/13.3*, *Gmcupin01.1/02.2*, *Gmcupin16.8/19.12* and *Gmcupin10.7/20.3*) derived from segmental duplications and one paralogous pair (*Gmcupin10.3/10.4*) derived from tandem duplication shared almost identical expression patterns. In contrast, the expression patterns of another seven paralogous pairs (*Gmcupin17.1/13.1*, *Gmcupin08.2/15.4*, *Gmcupin04.1/06.1*, *Gmcupin12.4/20.2*, *Gmcupin10.5/20.1*, *Gmcupin10.6/20.4* and *Gmcupin3.2/19.5*) diversified significantly. These findings indicated that expression profiles of *Gmcupins* have diverged substantially after segmental/tandem duplications. Therefore, we speculate that *Gmcupins* have been retained by substantial subfunctionalization during soybean evolutionary processes.

Artificial selection analysis for *Gmcupins* during soybean domestication

Thirty-five *Gmcupin* genes were analyzed for the selection effects during soybean domestication based on the sequence diversity analysis between seventeen wild soybean and fourteen cultivars. The reverse distribution of SNPs in different evolutionary type of soybeans was defined as strong selected sites, and then *Cupin* genes with one or more type of reverse distribution were assumed to undergo an artificial selection during soybean domestication. Sixteen *Gmcupins* have selected site(s), among which more than one selected sites were determined in 8 *Gmcypins* and one selected sites in 8 genes (Table 3). Additionally, all SNP sites were selected in *Gmcupin10.3* and *Gmcupin07.2* genes, which implied these genes may have undergone strong selection effects during soybean domestication. Interestingly, selected sites were identified in *Gmcupin03.1* (7 sites), *Gmcupin13.2* (1 site) and *Gmcupin19.13* (1 site) that were highly expressed at the later stages of soybean seed. The genetic diversity of most *Gmcupins* was declined sharply in cultivars compared with that of wild soybeans. However, *Gmcupin10.7* gene that specifically expressed in soybean root showed three types of haplotype in wild soybeans, while four types of haplotype were identified in cultivars. Further, a new type of haplotype in *Gmcupin10.7* appeared during soybean domestication under the pressure of artificial selection, which would endow it with new functions. These selected genes reflected the important roles of *Gmcupins* on soybean domestication and contribute to the cultivation of soybeans in order to meet the demands of human beings.

Supporting Information

Table S1 Pairwise identities between homologous pairs of Cupin genes from soybean.

Author Contributions

Conceived and designed the experiments: XBW LJQ. Performed the experiments: XBW HWZ YLG. Analyzed the data: XBW GLS LJQ. Contributed reagents/materials/analysis tools: XBW YLG WMZ. Wrote the paper: XBW LJQ GLS.

References

1. Dunwell JM, Purvis A, Khuri S (2004) Cupins: the most functionally diverse protein superfamily? Phytochemistry 65: 7–17.
2. Dunwell JM (1998) Cupins: a new superfamily of functionally diverse proteins that include germins and plant storage proteins. Biotechnol Genet Eng Rev 15: 1–32.
3. Thompson EW, Lane BG (1980) Relation of protein synthesis in imbibing wheat embryos to the cell-free translational capacities of bulk mRNA from dry and imbibing embryos. J Biol Chem 255: 5965–5970.
4. Jaikaran AS, Kennedy TD, Dratewka-Kos E, Lane BG (1990) Covalently bonded and adventitious glycans in germin. J Biol Chem 265: 12503–12512.
5. Lane BG, Dunwell JM, Ray JA, Schmitt MR, Cuming AC (1993) Germin, a protein marker of early plant development, is an oxalate oxidase. J Biol Chem 268: 12239–12242.
6. Patnaik D, Khurana P (2001) Germins and germin like proteins: an overview. Iindian J Exp Biol 39: 191–200.
7. Dunwell JM, Khuri S, Gane PJ (2000) Microbial relatives of the seed storage proteins of higher plants: conservation of structure and diversification of function during evolution of the cupin superfamily. Microbiol Mol Biol Rev 64: 153–179.
8. Dunwell JM, Culham A, Carter CE, Sosa-Aguirre CR, Goodenough PW (2001) Evolution of functional diversity in the cupin superfamily. Trends Biochem Sci 26: 740–746.
9. Dunwell JM, Gibbings JG, Mahmood T, Naqvi SMS (2008) Germin and germin-like proteins: Evolution, structure, and function. Crit Rev Plant Sci 27: 342–375.
10. Dunwell JM (1998) Sequence analysis of the cupin gene family in Synechocystis PCC6803. Microb Comp Genomics 3: 141–148.
11. Carter C, Graham RA, Thornburg RW (1998) Arabidopsis thaliana contains a large family of germin-like proteins: characterization of cDNA and genomic sequences encoding 12 unique family members. Plant Mol Biol 38: 929–943.
12. Membre N, Bernier F, Staiger D, Berna A (2000) Arabidopsis thaliana germin-like proteins: common and specific features point to a variety of functions. Planta 211: 345–354.
13. Membre N, Berna A, Neutelings G, David A, David H, et al. (1997) cDNA sequence, genomic organization and differential expression of three Arabidopsis genes for germin/oxalate oxidase-like proteins. Plant Mol Biol 35: 459–469.
14. Lapik YR, Kaufman LS (2003) The Arabidopsis cupin domain protein AtPirin1 interacts with the G protein alpha-subunit GPA1 and regulates seed germination and early seedling development. Plant Cell 15: 1578–1590.
15. Ham BK, Li G, Kang BH, Zeng F, Lucas WJ (2012) Overexpression of Arabidopsis plasmodesmata germin-like proteins disrupts root growth and development. Plant Cell 24: 3630–3648.
16. Pelah D, Shoseyov O, Altman A (1995) Characterization of BspA, a major boiling-stable, water-stress-responsive protein in aspen (Populus tremula). Tree physiol 15: 673–678.

17. Leon-Galvan F, Joaquin-Ramos AD, Torres-Pacheco I, de la Rosa APB, Guevara-Olvera L, et al. (2011) A Germin-Like Protein Gene (CchGLP) of Capsicum chinense Jacq. Is Induced during Incompatible Interactions and Displays Mn-Superoxide Dismutase Activity. Int J Mol Sci 12: 7301–7313.
18. Carter C, Thornburg RW (2000) Tobacco nectarin I - Purification and characterization as a germin-like, manganese superoxide dismutase implicated in the defense of floral reproductive tissues. J Biol Chem 275: 36726–36733.
19. Woo EJ, Dunwell JM, Goodenough PW, Marvier AC, Pickersgill RW (2000) Germin is a manganese containing homohexamer with oxalate oxidase and superoxide dismutase activities. Nat Struct Biol 7: 1036–1040.
20. Nakata M, Shiono T, Watanabe Y, Satoh T (2002) Salt stress-induced dissociation from cells of a germin-like protein with manganese superoxide dismutase activity and an increase in its mRNA in a moss, Barbula unguiculata. Plant Cell Physiol 43: S55–S55.
21. van Loon LC, Rep M, Pieterse CMJ (2006) Significance of inducible defense-related proteins in infected plants. Annu Rev Phytopathol 44: 135–162.
22. Manosalva PM, Davidson RM, Liu B, Zhu XY, Hulbert SH, et al. (2009) A Germin-Like Protein Gene Family Functions as a Complex Quantitative Trait Locus Conferring Broad-Spectrum Disease Resistance in Rice. Plant Physiol 149: 286–296.
23. Druka A, Kudrna D, Kannangara CG, von Wettstein D, Kleinhofs A (2002) Physical and genetic mapping of barley (Hordeum vulgare) germin-like cDNAs. Proc Natl Acad Sci U S A 99: 850–855.
24. Manosalva PM, Davidson RM, Liu B, Zhu X, Hulbert SH, et al. (2009) A germin-like protein gene family functions as a complex quantitative trait locus conferring broad-spectrum disease resistance in rice. Plant Physiol 149: 286–296.
25. Lu M, Han YP, Gao JG, Wang XJ, Li WB (2010) Identification and analysis of the germin-like gene family in soybean. Bmc Genomics 11.
26. Schmutz J, Cannon SB, Schlueter J, Ma J, Mitros T, et al. (2010) Genome sequence of the palaeopolyploid soybean. Nature 463: 178–183.
27. Goodstein DM, Shu S, Howson R, Neupane R, Hayes RD, et al. (2012) Phytozome: a comparative platform for green plant genomics. Nucleic acids research 40: D1178–1186.
28. Quevillon E, Silventoinen V, Pillai S, Harte N, Mulder N, et al. (2005) InterProScan: protein domains identifier. Nucleic acids research 33: W116–120.
29. Thompson JD, Gibson TJ, Plewniak F, Jeanmougin F, Higgins DG (1997) The CLUSTAL_X windows interface: flexible strategies for multiple sequence alignment aided by quality analysis tools. Nucleic acids research 25: 4876–4882.
30. Tamura K, Peterson D, Peterson N, Stecher G, Nei M, et al. (2011) MEGA5: molecular evolutionary genetics analysis using maximum likelihood, evolutionary distance, and maximum parsimony methods. Mol Biol Evol 28: 2731–2739.
31. Bailey TL, Williams N, Misleh C, Li WW (2006) MEME: discovering and analyzing DNA and protein sequence motifs. Nucleic acids research 34: W369–373.
32. Guo AY, Zhu QH, Chen X, Luo JC (2007) GSDS: a gene structure display server. Yi chuan 29: 1023–1026.
33. Cannon EK, Cannon SB (2011) Chromosome visualization tool: a whole genome viewer. International journal of plant genomics 2011: 373875.
34. Cannon SB, Shoemaker RC (2012) Evolutionary and comparative analyses of the soybean genome. Breeding science 61: 437–444.
35. Librado P, Rozas J (2009) DnaSP v5: a software for comprehensive analysis of DNA polymorphism data. Bioinformatics 25: 1451–1452.
36. Lavin M, Herendeen PS, Wojciechowski MF (2005) Evolutionary rates analysis of Leguminosae implicates a rapid diversification of lineages during the tertiary. Systematic biology 54: 575–594.
37. Lynch M, Conery JS (2000) The evolutionary fate and consequences of duplicate genes. Science 290: 1151–1155.
38. de Hoon MJ, Imoto S, Nolan J, Miyano S (2004) Open source clustering software. Bioinformatics 20: 1453–1454.
39. Page RD (1996) TreeView: an application to display phylogenetic trees on personal computers. Computer applications in the biosciences: CABIOS 12: 357–358.
40. Lam HM, Xu X, Liu X, Chen W, Yang G, et al. (2010) Resequencing of 31 wild and cultivated soybean genomes identifies patterns of genetic diversity and selection. Nature genetics 42: 1053–1059.
41. Yamahara T, Shiono T, Suzuki T, Tanaka K, Takio S, et al. (1999) Isolation of a germin-like protein with manganese superoxide dismutase activity from cells of a moss, Barbula unguiculata. J Biol Chem 274: 33274–33278.
42. Woo EJ, Bauly J, Chen JG, Marshall J, Macdonald H, et al. (2000) Crystallization and preliminary X-ray analysis of the auxin receptor ABP1. Acta crystallographica Section D, Biological crystallography 56: 1476–1478.
43. Anand R, Dorrestein PC, Kinsland C, Begley TP, Ealick SE (2002) Structure of oxalate decarboxylase from Bacillus subtilis at 1.75 angstrom resolution. Biochemistry-Us 41: 7659–7669.
44. Freeling M (2009) Bias in plant gene content following different sorts of duplication: tandem, whole-genome, segmental, or by transposition. Annual review of plant biology 60: 433–453.
45. Schlueter JA, Lin JY, Schlueter SD, Vasylenko-Sanders IF, Deshpande S, et al. (2007) Gene duplication and paleopolyploidy in soybean and the implications for whole genome sequencing. Bmc Genomics 8: 330.
46. Liu Y, Jiang HY, Chen WJ, Qian YX, Ma Q, et al. (2011) Genome-wide analysis of the auxin response factor (ARF) gene family in maize (Zea mays). Plant Growth Regul 63: 225–234.
47. Prince VE, Pickett FB (2002) Splitting pairs: the diverging fates of duplicated genes. Nature reviews Genetics 3: 827–837.

11

Genome-Wide Association Studies Identifies Seven Major Regions Responsible for Iron Deficiency Chlorosis in Soybean (*Glycine max*)

Sujan Mamidi[1,2], Rian K. Lee[1,2], Jay R. Goos[3], Phillip E. McClean[1,2]*

1 Genomics and Bioinformatics Program, North Dakota State University, Fargo, North Dakota, United States of America, **2** Department of Plant Sciences, North Dakota State University, Fargo, North Dakota, United States of America, **3** Department of Soil Science, North Dakota State University, Fargo, North Dakota, United States of America

Abstract

Iron deficiency chlorosis (IDC) is a yield limiting problem in soybean (*Glycine max* (L.) Merr) production regions with calcareous soils. Genome-wide association study (GWAS) was performed using a high density SNP map to discover significant markers, QTL and candidate genes associated with IDC trait variation. A stepwise regression model included eight markers after considering LD between markers, and identified seven major effect QTL on seven chromosomes. Twelve candidate genes known to be associated with iron metabolism mapped near these QTL supporting the polygenic nature of IDC. A non-synonymous substitution with the highest significance in a major QTL region suggests soybean orthologs of *FRE1* on Gm03 is a major gene responsible for trait variation. NAS3, a gene that encodes the enzyme nicotianamine synthase which synthesizes the iron chelator nicotianamine also maps to the same QTL region. Disease resistant genes also map to the major QTL, supporting the hypothesis that pathogens compete with the plant for Fe and increase iron deficiency. The markers and the allelic combinations identified here can be further used for marker assisted selection.

Editor: Swarup Kumar Parida, National Institute of Plant Genome Research (NIPGR), India

Funding: This work is supported by North Central Soybean Research Program. The funders had no role in study design, data collection and analysis, decision to publish, or preparation of the manuscript.

Competing Interests: The authors have declared that no competing interests exist.

* Email: Phillip.Mcclean@ndsu.edu

Introduction

Iron (Fe) is an essential element for multiple plant functions including photosynthesis, respiration, chlorophyll biosynthesis, and redox reactions in plants, and is a structural component in heme, the Fe-sulfur cluster, and Fe-binding sites. To avoid toxicity, plants control the uptake, utilization, and storage in response to Fe availability [1]. Despite its abundance in the soil, Fe is only slightly soluble under aerobic conditions, especially in high-pH and calcareous soils [2]. Under these conditions, iron deficiency leads to developmental defects, including chlorosis, growth retardation, and reduced crop productivity [3]. Iron deficiency chlorosis (IDC) is an important yield-limiting factor for soybeans (*G. max*) grown on calcareous soils that have a high percentage of calcium carbonate and soluble salts. This soil type is common in the north-central regions of the United States [4] where soybean is widely grown. In these soils, interactions of high pH, carbonate and high field moisture content at planting leads to early IDC symptoms that effect yield. It is estimated that the current revenue losses due to IDC in soybean are $260 million [5].

IDC results from the inability of susceptible genotypes to efficiently mobilize iron into the plant when it is growing in high pH calcareous soils. In these soils, ferrous iron is not readily released from soil particles, and subsequently iron availability is limited. Based on its response to Fe availability, soybean is considered a Strategy I plant [6] that reduces ferric chelates at the root surface and absorbs ferrous ions across the plasma membrane of the root [2]. Other processes involved in Strategy I include excretion of proton and phenolic compounds from the roots to the rhizosphere, which increases the solubility of ferric ions and support the reducing capacity of ferric Fe on the root surface [2]. Strategy I plants also increase root hair formation, thereby increasing the surface area available for iron uptake [7]. Once iron enters the root, membrane transporters move Fe into the xylem where it chelates with citrate. The chelated form of iron then moves through the xylem stream to growing leaves. Iron is then remobilized from the leaves, forms a complex with nicotianamine, and is transported via the phloem to younger leaves and seeds. Excess water leads to an elevated concentration of bicarbonates in the root apoplast that impedes the Fe^{3+}– chelate reductase activity necessary for the conversion of Fe^{3+} to Fe^{2+}. Bicarbonates also immobilize the movement of iron to young leaves once it is absorbed at the root level [8].

Given the many physiological mechanisms associated with iron metabolism, it is not surprising that IDC is a complex quantitative trait with multiple genetic factors [9]. In addition to the genes involved in Fe acquisition, translocation and compartmentalization within the organelles, there are other genes involved in Fe transcriptional and post transcriptional changes that are expressed/repressed under deficient and excess conditions or affect iron homeostasis pathways [10–17].

Genetic experiments using bi-parental mapping populations have identified multiple quantitative trait loci (QTL) associated with IDC in soybean [11–12]. O'Rourke et al. [13] identified several genes induced under iron-limiting conditions using a microarray study. Severin et al. [14] discovered multiple introgressed genomic regions responsible for IDC in near isogenic lines, and Stec et al. [15] further refined these introgression regions. Recently, Peiffer et al. [5] evaluated candidate genes on Gm03.

A genome-wide association study (GWAS) is an excellent approach to discover genetic factors in a population because of the high number of recombinant events the population represents. GWAS uses the linkage disequilibrium (LD) pattern in a large population of unrelated individuals [16] to map significant effect loci. Since IDC is a complex trait, utilizing GWAS can identify major factors controlling the IDC response in soybean. Previously, Wang et al. [17] used GWAS to identify simple sequence repeat (SSR) markers associated with IDC in two independent populations. Later, nine shared SNP were discovered and confirmed by Mamidi et al. [9] in two separate populations. The objective of this study was to identify QTL and genes involved in IDC of soybean using a GWAS population and a new high density SNP map obtained through genotype-by-sequencing (GBS). Those results were then translated into SNP markers that can be used for marker assisted selection (MAS).

Materials and Methods

Phenotype and genotyping

Two plant populations consisting of unique set of advanced soybean breeding lines developed for the north-central region of the US were grown in years 2005 (n = 132) and 2006 (n = 138). IDC screening protocols and DNA isolation procedures were described previously [9]. The trait was rated on a scale of 1 to 5, where 1 indicates no chlorosis and a normal green plant, 2 indicates slight yellowing of upper leaves and the leaf veins, 3 indicates interveinal chlorosis in the upper leaves with no stunning of growth or death of tissue (necrosis), 4 indicates interveinal chlorosis of the upper leaves along with some apparent stunning of growth or necrosis of tissue, and 5 indicates severe chlorosis plus stunned growth and necrosis in the youngest leaves and growing points [9]. Phenotypic data was adjusted based on least squares means for each population independently.

GBS libraries were prepared and analyzed at the Institute for Genomic Diversity (IGD), Cornell University, according to Elshire et al. [18], using the enzyme *Ape*KI. The GBS analysis pipeline [3.0.128], an extension of the Java program TASSEL [19], was used to call SNPs from the sequenced GBS library. Reads were aligned to the soybean reference genome v1.1 [20] using BWA 0.6.1 [21]. VCFtools (v0.1.8) [22] was used to summarize the SNP data.

The data was analyzed by combining the two populations (which are of similar size) to account for unique alleles that are at a low frequency in one or both populations. These low frequency alleles can have a significant phenotypic effect in a small set of genotypes, but that effect cannot be detected in a normal statistical framework [16].

Genome-wide association study

Imputation. fastPHASE 1.3 [23], a likelihood-based imputation software, was used with default settings to impute missing data for the combined population. Imputation analysis was performed to increase the power of the study and fine map the causal variant [24]. A minor allele frequency (MAF) of 0.05 was used as cutoff, since variants at low frequency have little power to detect association with the phenotype [16].

Linkage disequilibrium (LD). The LD between markers was estimated as the partial squared allele frequency correlation (r^2) using CORR procedure in SAS 9.3. Three principal components (PC) that explain about 25% of cumulative variation were used as cofactors. We used a partial correlation because unlinked loci can be in LD simply because of population structure [25–26]. The decay of r^2 with physical and/or genetic distance between loci is often used to determine the density of markers to use in whole genome association scans [27] whereas local LD on chromosomes is used to account for genes/QTL associated with trait variation. Overestimation of LD can lead to misinterpretations either on the extent of LD decay or on the size of the QTL [26]. LD decay graphs were plotted with physical distance (Mbp) vs. r^2 for all intra-chromosomal comparisons using nonlinear regression as described by Remington et al. [28]. The expected decay of LD was estimated as described by Pyhajarvi et al. [29]. We fit this equation into a nonlinear regression model using NLIN procedure in SAS 9.3.

Population structure, kinship, and model testing. A principal component analysis was utilized to control for population structure [30]. PCs were estimated using the PRINCOMP procedure in SAS 9.3. The PCs that explained 25% (PC_{25}) and 50% (PC_{50}) cumulative variation were selected for the analysis. In addition an identity-by-state matrix [31] was used to control for relatedness estimated as a centered relatedness matrix in Gemma 0.92 [32]. Gower's similarity coefficient was calculated using the DISTANCE procedure in SAS 9.3 to measure the relatedness by state of the individuals.

To test for marker-trait associations, the MIXED procedure in SAS for the three linear models that do not have kinship (Naïve, PC_{25} and PC_{50}) was used. Since approximate tests compromise the analysis of complex genetic architectures when one genetic factor masks another [33], we used the exact test implemented in Gemma 0.92 to estimate marker significance [32] for the other three models with kinship (Kinship, PC_{25}+kinship, PC_{50}+kinship). As the standard errors follow a uniform distribution [34], we used a rank based mean squared difference (MSD) for model selection as described in Mamidi et al [9].

Marker-trait associations. Markers are defined as significant at two levels, within 0.01 percentile, and 0.1 percentile tails of the empirical p-value distribution of 10,000 bootstraps. This method of using an empirical distribution of data as a replacement for a population with an unknown distribution has the advantage of simplicity and provides an efficient and precise estimation [35–36]. This approach is similar to choosing an arbitrary value but instead it is based on choosing a predefined percentile tail from an empirical distribution. Here, the confidence intervals for an empirical distribution were obtained from 10,000 bootstraps over ~35,000 markers distributed throughout the genome. We chose this approach over a cutoff because p-value is dependent on the distribution of phenotype, the variation explained by the marker, structure and relatedness of the population, and heritability of the trait.

The means for the alleles were calculated using MEANS procedure in SAS. The percent of variation explained by the marker is calculated using a simple regression (GLM procedure) in SAS. The additive effect of the variant allele is calculated as half the difference between IDC mean of the variant allele and IDC mean of the reference allele.

To determine the minimal number of SNPs independently associated with IDC, stepwise regression analysis was applied to all significant markers from a single-SNP analysis at a significance level of 0.1 percentile [37]. In addition, a multiple R^2 value for all these loci was also estimated [9]. A significant p-value of 0.05 was necessary for both marker and model for stepwise inclusion of the marker in REG procedure in SAS 9.3.

Annotation of SNP. To characterize SNPs as intergenic, UTR, intron, or coding site, snpEFF 3.3 [38] was used. Coding sites were further differentiated into synonymous or non-synonymous substitutions based on amino acid changes. Coding regions were also differentiated into transitions and transversions based on changes between or among purines and pyrimidines.

Significant QTL and genes. A QTL region is defined as the region around significant stepwise markers that has a partial LD (r^2) >0.6 with the adjacent marker (corresponds to partial correlation of 0.77). Adjacent blocks were combined if the distance is less than 10 kb and includes up to four markers that are not in LD with either of the block. Candidate genes for iron homeostasis were identified from the research literature, and gene symbols are used from the annotation file available at phytozome (http://www.phytozome.net/soybean.php). A candidate iron gene is considered to be significant in our population if the gene is within the QTL region.

Epistatic interactions. To detect epistatic interactions between significant markers, we used a general linear model with both significant markers and their interaction term in the model. We used PC_{25} as covariates in the regression model and an interaction term was considered significant at a p-value of≤1E-03. Genes that are in LD with these significant markers are used to identify the gene-by-gene interactions.

Significant allelic combinations. We compared the allelic combinations of the significant markers (0.1 percentile) that were included in stepwise regression. Allelic combinations with a mean IDC<2.5 were termed tolerant, and those with a mean IDC>3.5 were considered susceptible. We chose these values based on standard genotypes grown in the field instead of a statistical approach of mean differentiation, because not all allelic combinations are present in the population. With this approach, the phenotypic distribution does not need to be continuous, where statistically two allelic combinations can be considered significantly different in the absence of intermediate phenotypes.

Results

Phenotype and genotyping

For the combined population, the range of IDC is 1.46 to 3.84 with an average of 2.76. IDC scores were normally distributed with a p-value of 0.15 for Kolmogorov–Smirnov test (Figure 1). An average of 1,645,563 reads was obtained for each genotype, and ~92% of reads mapped to the reference genome. A total of 79,000 putative SNPs that are bi-allelic were identified in the population using the TASSEL-GBS analysis pipeline after removing SNPs that have greater than 50% missing data. The mean coverage depth for each identified SNP was 6.85±2.72.

Association Mapping

Imputation. Missing data (37.5%) was estimated using imputation. Of the 79,000 SNPs, 34,428 had a MAF>0.05 (Table S1). Based on the genome size of *Glycine max*, a polymorphic marker was located, on average, every 31 kbp. In addition, the maximum distance between two adjacent markers did not exceed 500 Kbp providing a good coverage of the SNPs. Of these polymorphic SNPs, 6,520 mapped to a coding region, 6,898 are located in the non-coding region of a gene model and 21,010 are in an intergenic region. The transition to transversions ratio was 1.67.

Linkage Disequilibrium Decay. A nonlinear regression model that fits the partial r^2 of pairwise intra-chromosomal comparisons to physical distance was developed. The average decay of LD (r^2) in terms of physical distance declined to $r^2 = 0.7$ at ~500 kb (Figure S1). The physical distance at which $r^2 = 0.2$ is 3.2 Mbp and at $r^2 = 0.1$ was 8.0 Mbp.

Population structure, Kinship and model selection. For the population, three and 13 PC explained 26.8% and 50.5% of the cumulative genotypic variation. Based on the first two PCs, two distinct clusters of populations were observed which indicated the presence of a subpopulation structure (Figure 2). In the population, 80% of the Gowers similarity coefficients are within the range of 0.6 to 0.7 (Figure 3) indicating the presence of relatedness among individuals. This is expected given the narrow genetic base of the breeding population. For the six models tested, the best linear model for the population included 13 PCs (PC_{50}) and kinship matrix (MSD = 6.06E-06; Figure 4).

Marker Trait associations. Five markers were significant at the 0.01 percentile level (p≤4.0E-05). One located on Gm03 at 45.03 Mbp, one on Gm05 at 8.82 Mbp and three on Gm11 at 0.53 Mbp (Table 1, Fig. 5). The variation explained by these markers ranged from 5% to 16%, and the mean allelic difference is between 0.31 and 0.38. Stepwise regression included three markers (one from each of the chromosome). Together these three markers explained 23.2% of phenotypic variation.

At a significance level of 0.1 percentile (p≤2.12E-04), 33 significant markers were distributed on 10 of the 20 chromosomes (Table 1, Figure 5). Eight of these markers included in stepwise regression are on Gm03 (45.03 Mbp), Gm05 (8.87 Mbp), Gm07 (6.39 Mbp), Gm11 (0.53 Mbp), Gm16 (27.30 Mbp), Gm17 (25.85 Mbp), Gm18 (28.14 Mbp), and Gm19 (40.19 Mbp). These markers together explained 46.3% variation.

Annotation of SNP. Of the significant markers at 0.1 percentile, five SNP are present in a coding region (three are transitions and two are transversions), 11 in introns, two in 3′UTR regions, and 15 in intergenic regions. Of the five SNPs from a coding region, two are non-synonymous substitutions. For FRE1 (Glyma03G38620), one of these amino acid changes is associated with both a polarity and a charge change of the protein. This SNP at position 44,927,455 bp of Gm03 changes an asparagine to aspartic acid. The R-square explained by this marker alone is 15.2% for the population. The variant allele ‘A’ is more tolerant than reference allele. The other non-synonymous substitution is located in the gene model (Glyma03g39151) close to the above non-synonymous substitution at 45,343,612 bp. This SNP explains about 18.85% of the phenotypic variation in the population and has a partial r^2 of 0.80 with the above marker. These two markers together explain 19.0% of phenotypic variation.

Significant QTL and genes. At the significance level of 0.01 percentile, three markers were included in the stepwise regression. One of these marker is in LD ($r^2 \geq 0.6$) with the adjacent markers, and the QTL is located on Gm03 between 44.798 Mbp and 47.140 Mbp (Table 2; Figure 5; Figure 6). The two major candidate genes in this QTL are phytochrome A (FRE1; Glyma03g38620) and nicotianamine synthase 3 (NAS3; Glyma03g39050). In addition to these genes, nitrate transporter 1:2 (NRT1:2; Glyma03g38640) and ammonium transporter 2 (AMT2; Glyma03g40700), each with a role in iron homeostasis, are also located within the QTL region. The QTL around the significant marker on Gm05 is between 7.837 and 10.627 Mbp (Table 2; Figure 5; Figure 6). The candidate genes in this QTL

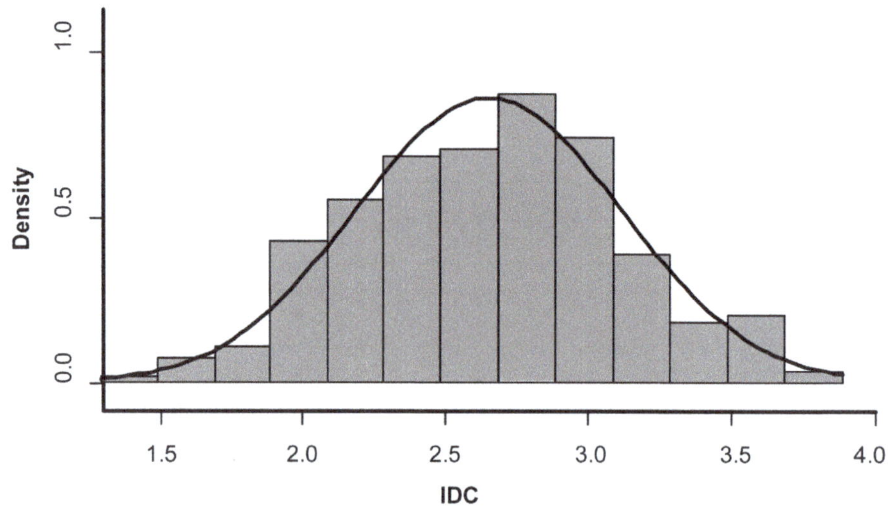

Figure 1. Phenotypic distribution of iron deficiency chlorosis (IDC). The rating of IDC (1–5) is on x-axis and the number of genotypes having the IDC score (represented as density) are presented as y-axis. The curve over the histogram represents the normal distribution of the IDC scores in the population.

include stabilizer of iron transporter (AGO10, PNH, ZLL; Glyma05g08170), RAS-related nuclear protein-1 (RAN1; Glyma05g08260) and H (+)-ATPase 11 (AHA11; Glyma05g01460). The third significant marker on Gm11 is located in the 5′ UTR region of the CVP2 like 1 gene (CVL1; Glyma11g00990). The marker is not in LD with any of the adjacent markers.

At a significance level of 0.1 percentile, eight markers are included in the stepwise regression model. These eight markers are located in eight QTL regions. The QTL on Gm03, Gm05 and Gm11 are the same as the ones identified at the 0.01 percentile level. The other QTL at this significance level include Gm07 (6.074 Mbp–7.101 Mbp), Gm16 (26.914–27.597 Mbp), Gm17 (25.629–27.007 Mbp), Gm18 (27.547–29.978 Mbp), and Gm19 (39.831–41.320 Mbp) (Table 2; Figure 5; Figure 6). On Gm07, the significant candidate genes include ferric reduction oxidase 2 (FRO2; Glyma07g07380), and signal recognition particle receptor

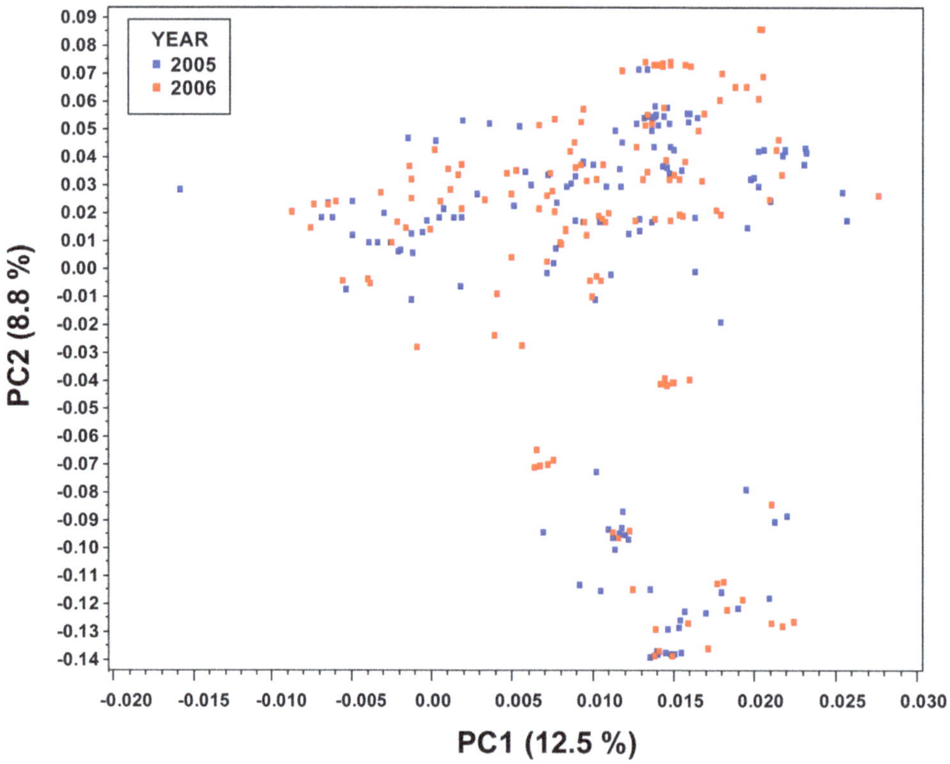

Figure 2. A graph of the first two principal components obtained from all the 34,428 polymorphic SNPs. This graph explains the similarities between the two independent populations and the overall population structure.

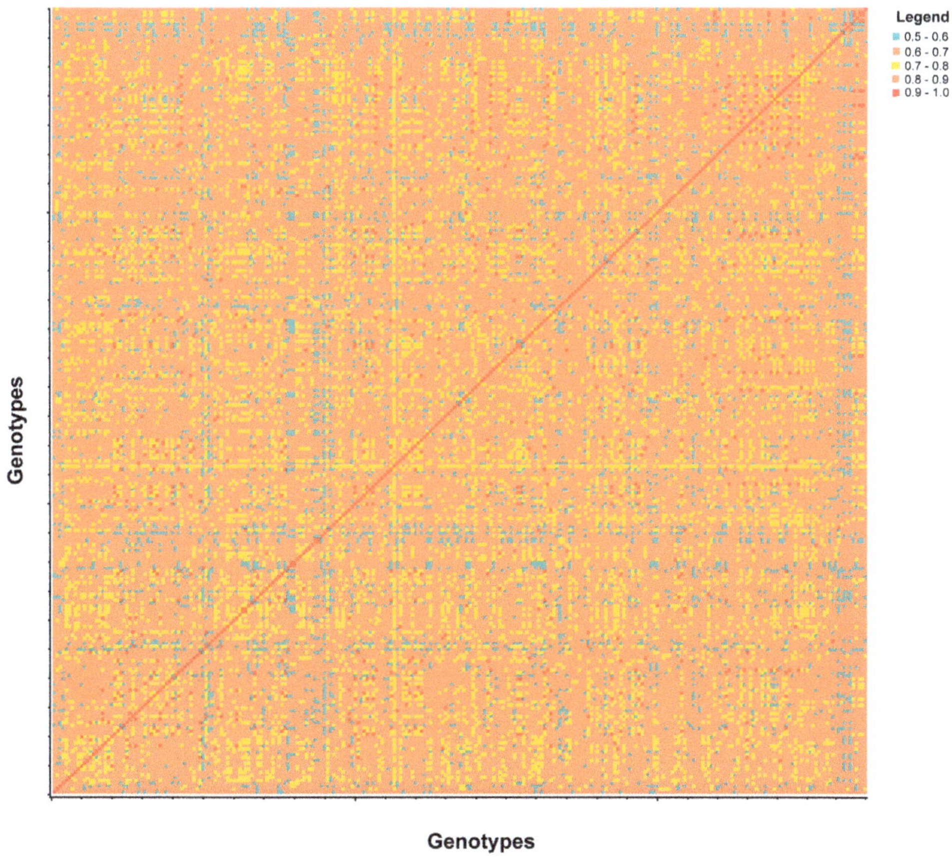

Figure 3. Heatmap of Gower's similarity distance among the individuals of the population. Based on the colors distribution, it is evident that relatedness can have a major effect on identifying associations to IDC in the population.

protein on chloroplast (FRD4; Glyma07g08240). On Gm16, the significant candidate gene ferrodoxin NADP (+) oxidoreductase 1 (FNR1, Glyma16g23710) is within the QTL region. On Gm19, heavy metal atpase-5 (HMA5; Glyma19g32190), and oligopeptide transporter-5 (OPT5; Glyma19g32401) are significant.

Other than the genes involved in iron metabolism, several genes involved in disease resistance are also identified in the QTL regions. For example, QTL on Gm03 has two RING/U-box superfamily proteins that have a role in plant defense. Similarly on Gm05, two CC-NBS-LRR, one NBS and one LRR genes are present. On Gm07 within the QTL region, three NBS genes are present. On Gm16, seven LRR proteins, and on Gm19, six NBS proteins, two LRR and one RING/U-box superfamily proteins are present.

Epistatic interactions. With 33 significant markers at 0.1 percentile, 528 two way interactions were tested with a population structure component, and five interactions were significant at p-value$<$0.001. Three of these interactions are between Gm07 (around 6.4 Mbp) and Gm18 (59.28 Mbp). The other significant interactions are between Gm16 (27.30 Mbp) and Gm18 (59.28 Mbp), Gm02 (46.939) and Gm15 (21.42 Mbp). The genes within LD regions include cytochrome P450 like protein (Glyma18g50051) on Gm18 with FRO2 (Glyma07g07380) on Gm07 and FNR1 (Glyma16g23710) on Gm16. The genes on Gm02 include AUX1 (Glyma02g42290) and FER4 (Glyma02g43040) which interact with CCC1 (Glyma15g22520) on Gm15.

Significant Marker Allelic combinations. Allelic combinations of SNPs can be used to infer the tolerant/susceptible phenotype of a genotype. For the significant markers (0.1 percentile) that were included in stepwise regression in the population, 14 allelic combinations are predicted as tolerant (IDC mean$\leq$2.5) and 6 allelic combinations indicate susceptibility (IDC mean$\geq$3.5) (Table 3).

Discussion

A higher marker density facilitates identification of high resolution QTL and gene discovery [39,40]. One method of generating a high frequency of SNPs is next generation sequencing [41,42]. Using GBS, we identified ~35,000 polymorphic markers with a minor allele frequency$>$0.05. This provided good coverage, one marker at every 31 kb. This density is useful for characterizing most genetic factors responsible for soybean IDC given that LD decayed here to $r^2 = 0.7$ at 500 kbp.

LD decay assesses the depth of markers required for GWAS. Partial LD was calculated using population structure as a cofactor, and as expected we observed a variation in LD decay when including this as a cofactor. For example, LD decay for the population is at 10 Mbp at $r^2 = 0.1$ whereas adjustments for confounding effect of population structure reduced LD decay to 7 Mbp. The use of partial LD (r^2) helps overcome the bias between adjacent markers and leads to smaller QTL regions.

The population consisted of breeding lines developed by public and private programs targeting the north central states of the US. The germplasm is expected to be narrow given the limited range of maturity group (0–1), the short history of soybean breeding in USA, and the limited number of traits selected in the breeding

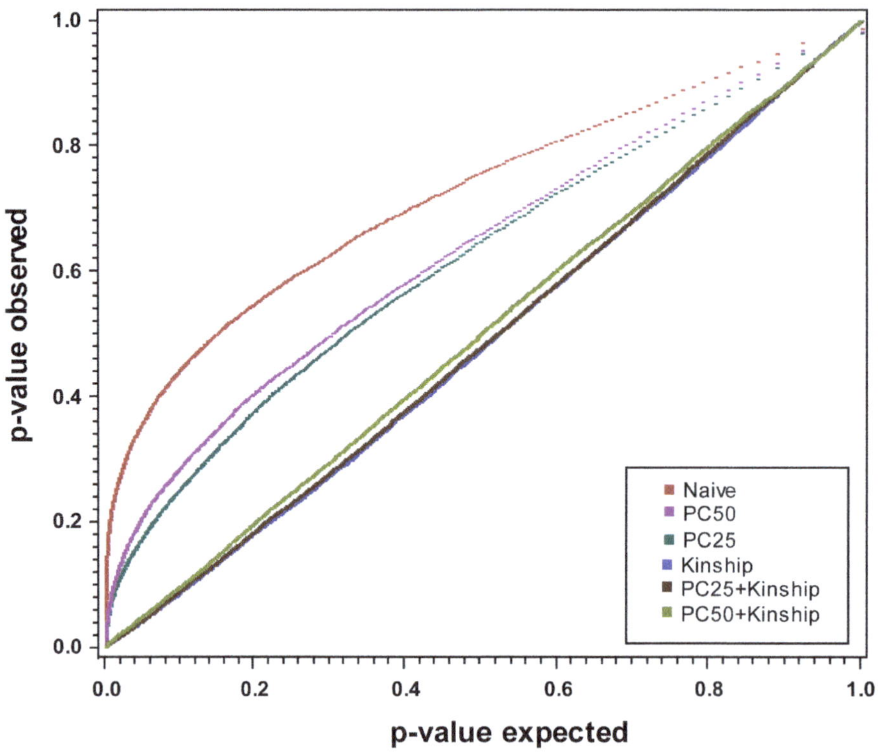

Figure 4. QQ plot for the six models tested. P-value observed is plotted on the y-axis and P-expected is plotted on the x-axis. Each color represents a model of regression used. The best model is the one that is close to the diagonal line.

programs. Here we used multiple mixed models [31,34,43–45] that control for confounding effect of population structure and population relatedness [16,46].

The percentile p-value cutoffs derived from empirical distribution were all within the $p = 0.001$ error level, and the significant markers were distributed on eleven of the twenty chromosomes. This is expected given the complex nature of iron homeostasis in plants. At a significance level of one percentile ($p \leq 2.12E\text{-}04$), many more markers were significant (345 markers) and explained the majority of the phenotypic variation in the population ($R^2 = 85\%$). However, owing to the complexity of the trait, we limited the interpretations to the 0.1 percentile to find the major QTL effects.

To determine if any significant SNP is a potential causal variant, we searched for non-synonymous substitutions. One was discovered in Far red elongated 1 (FRE1) which encodes a protein that increase reductase activity and provides a higher level of tolerance in high pH calcareous soils [47–49]. The other non-synonymous substitution was discovered in a gene with no known function related to iron homeostasis. This can be expected because the functions of many genes are still unknown [50–52].

Stepwise regression facilitates the selection of markers that have a major effect in a QTL region and simultaneously masks the effects of other minor QTL. As an example, the QTL regions on Gm02 and Gm15 that have a significant interaction were not included in the stepwise regression given these QTL have a minor effect. Additionally, the partial LD between the 33 significant markers is 0.09 (±0.247). But for the eight markers included in stepwise regression, partial LD (r^2) is 0.008 (±0.021). This subset of markers included in stepwise regression explains 23.2% and 46.3% of the 0.01 and 0.1 percentile, respectively. This is similar to the variation explained by all significant markers (23.2% and 48.6% at the 0.01 and 0.1 percentile, respectively).

Comparing with independent populations, the combined population analysis was able to identify markers from eight addition QTL regions that are not discovered in independent populations at significance level of 0.1 percentile. Only six QTL regions are common with either one of the population. Further using a stepwise regression to identify major effect QTL, the combined analysis was able to identify three unique QTL regions (On Gm05, Gm17, and Gm18) that were not identified in the independent populations. Of the other five QTL regions, two are common with 2005 population (on Gm03, Gm11) and three are common with 2006 population (Gm07, Gm16, Gm19).

For the population, we identified seven QTL on seven chromosomes. Given the complex quantitative nature of iron metabolism, multiple QTL distributed throughout the genome are expected. The region on Gm03 is in the same region identified earlier by Severin et al. [14] and Stec et al. [15] using an introgression analysis and is a significant QTL in many biparental studies [53–55]. In addition to this major QTL, other QTL regions were also identified earlier. For example, QTL region on Gm05 were identified earlier by Severin et al. [14]. The QTL on Gm16 is close to an introgression region identified earlier [14,15]. Similarly Lin et al. [53] identified QTL on Gm18 and Gm19 and Lin et al.[56] identified QTL on Gm18. Most of these regions were also identified earlier using the same population with about 1000 polymorphic markers [9]. The major difference of this study from the previous one is the ability to narrow the QTL region and the ability to study the entire genome given a polymorphic marker every 31 kb. In addition we were able to identify significant SNP markers within the candidate genes.

Table 1. Significant markers associated with IDC in the population of soybean.

Chromosome	Position	-log10 (p-value)	Reference Allele	Variant Allele	R^2§	MAF§§	Mean Difference	Additive Effect (Variant allele)	Included in stepwise regression§§§
Gm02	46,939,135	3.784**	A	T	5.20	21.48	0.26	−0.13	
Gm03	44,927,455	4.165**	A	G	15.20	38.89	0.37	−0.19	
Gm03	44,930,187	4.165**	A	T	15.20	38.89	0.37	−0.19	
Gm03	44,970,860	4.275**	A	G	14.42	39.26	0.36	−0.18	
Gm03	44,985,022	4.165**	G	A	15.20	38.89	0.37	−0.19	
Gm03	45,031,929	5.127***	T	C	15.77	38.89	0.38	−0.19	x
Gm03	45,038,912	3.684**	C	T	14.71	39.26	0.36	−0.18	
Gm03	45,107,141	3.905**	C	A	13.87	45.56	0.35	−0.17	
Gm03	45,343,317	3.689**	G	T	17.28	42.59	0.39	−0.19	
Gm03	45,343,612	4.231**	A	T	18.85	41.48	0.41	−0.20	
Gm05	8,817,714	3.912**	G	A	10.81	49.26	0.30	0.15	
Gm05	8,823,135	4.617***	G	A	11.43	49.63	0.31	0.16	
Gm05	8,877,264	3.855**	G	T	8.30	45.93	0.27	0.13	x
Gm05	17,397,118	3.704**	G	A	0.08	39.63	0.03	0.01	
Gm05	17,397,130	3.704**	T	C	0.08	39.63	0.03	0.01	
Gm05	17,397,135	3.704**	C	T	0.08	39.63	0.03	0.01	
Gm07	6,397,319	3.961**	C	T	8.80	25.56	0.32	−0.16	x
Gm07	6,405,421	3.961**	T	C	8.80	25.56	0.32	−0.16	
Gm07	6,540,541	4.128**	G	A	8.10	26.30	0.30	−0.15	
Gm07	6,575,306	3.961**	A	G	8.80	25.56	0.32	−0.16	
Gm11	530,116	4.401***	A	G	4.73	8.89	0.35	0.18	x
Gm11	530,117	4.401***	A	C	4.73	8.89	0.35	0.18	
Gm11	530,118	4.401***	G	T	4.73	8.89	0.35	0.18	
Gm12	20,406,669	3.697**	T	C	0.02	20.74	0.01	0.01	
Gm15	21,422,057	4.152**	A	G	2.10	20.37	0.17	0.08	
Gm16	27,300,116	3.779**	T	G	4.80	8.52	0.36	0.18	x
Gm17	25,859,992	4.322**	T	C	0.66	6.67	0.15	0.08	x

Table 1. Cont.

Chromosome	Position	-log10 (p-value)	Reference Allele	Variant Allele	$R^{2\S}$	MAF[§§]	Mean Difference	Additive Effect (Variant allele)	Included in stepwise regression[§§§]
Gm17	25,859,995	4.322**	T	C	0.66	6.67	0.15	0.08	
Gm18	15,905,372	3.781**	T	C	3.25	28.52	0.18	0.09	
Gm18	28,141,888	3.786**	G	A	3.78	22.59	0.22	0.11	x
Gm18	59,284,729	4.006**	C	T	0.70	45.93	0.08	0.04	
Gm19	34,914,546	3.769**	A	G	0.81	5.19	0.19	0.09	
Gm19	40,193,564	3.955**	C	A	15.62	44.07	0.37	−0.18	x

***Significant at 0.01 percentile (4.0E-05).
**Significant at 0.1 percentile (2.1E-04).
[§]R^2– Calculated using a simple linear regression and is represented as percentage.
[§§]MAF = Minor Allele Frequency.
[§§§]The markers that have an 'x' are included in stepwise regression. All other significant markers are in LD with one of these eight markers. This is the subset of markers around which QTLs were identified, and allelic combinations that have applications for MAS are characterized.

Iron is a structural component of many important processes in plants, and multiple physiological steps are used by strategy-I plants, such as soybean, to manage plant growth and development in iron-deficient conditions [56]. Protons release results in the acidification of the rhizosphere which increases the solubility of iron. Ferric iron is reduced at the root surface, and ferrous iron uptake increases [57–59]. Here we discovered multiple genes located in major QTL associated with these major activities.

AHA11 (Gm05) is an abundant protein found in the seedlings and leaves [60] and limits the transport of iron to leaves resulting in chlorotic plants. Ferric-chelate reductase oxidase (FRO2; Gm07) is expressed in the epidermal cells of Fe-deficient roots and is one of the main Fe (III) chelate reductase proteins. It reduces iron at the cell surface before transport into roots [61]. FRO2 also interacts with IRT1 under Fe sufficiency conditions to increase iron uptake [62,63]. FRE1 (Gm03) increases reductase activity and tolerance to high pH in calcareous soils [47,48]. Based on transgenic lines in *Oryza*, Ishimaru et al. [47,48] suggested that this gene is regulated by the IRT1 since transgenic lines have a higher level of Fe uptake and enhanced tolerance to low Fe availability in both hydroponic culture and calcareous soils.

Ferric chelate reductase defective (FRD4; Gm07) is a gene that acts post-translationally and reduces Fe (III) chelate reductase activity in response to iron deficiency [56]. In addition, *frd4* mutants are chlorotic, grow more slowly, have smaller chloroplasts, and possess fewer thylakoid membranes and grana stacks. Map-based cloning revealed that this gene encodes a component that is responsible for the insertion of proteins into the thylakoid membranes of the chloroplast [56]. Ferredoxin NADP reductase 1 (FNR1; Gm16) is an enzyme that catalyzes reversible electron transfer in photosynthetic electron transport. In iron stress conditions, its activity is decreased resulting in reduced chlorophyll biosynthesis [64].

Nicotianamine (NA) is an iron chelator involved in iron (Fe) acquisition, transport, and homeostasis [65,66], and NA is synthesized by nicotianamine synthase that is encoded by *NAS3* (Gm03) gene. Overexpression of NAS genes in rice increased Fe and Zn uptake by at least twofold [67,68]. Nicotianamine synthase genes are induced by nitrate levels [69,70]. Ammonium and nitrate transporters (Gm03) increase the uptake of nitrate from rhizosphere and increase the concentration of nitrate in plants and lead to increased growth and extension of roots [71,72]. In addition, nitrates enhance H+ extrusion and acidify the rhizosphere leading to increased bioavailability of Fe [71–75].

Several transporters such as RAN1 (Gm05), OPT5 (Gm19), and HMA (Gm19) are in the significant QTL regions. RAN1 is involved in copper trafficking and iron transport [76]. OPT5, an oligopeptide transporter is highly expressed in multiple organs of adult plants [77], and HMA sequesters Cd, Zn, and Ni in vacuoles under Fe deficiency to maintain metal balance and to detoxify heavy metals [78–80].

Argonaute (AGO) like proteins are down-regulated under conditions of low Fe and phosphorus [81]. Vaucheret [82] suggested that these genes have functions in both development and stress response by regulating gene expression at various levels and has additional functions including chromatin remodeling, DNA methylation, translational repression, and RNA cleavage. One such *AGO* gene, *AGO10* (Gm05) is in a major QTL.

Disease resistant genes (R genes) are distributed throughout the *G.max* genome and are a key component of gene interactions between plants and pathogens [83]. We identified multiple R genes in IDC QTL regions. Iron is an important element for all living organisms and pathogens and hosts actively compete for available metals. Plants need metals for defensive generation of

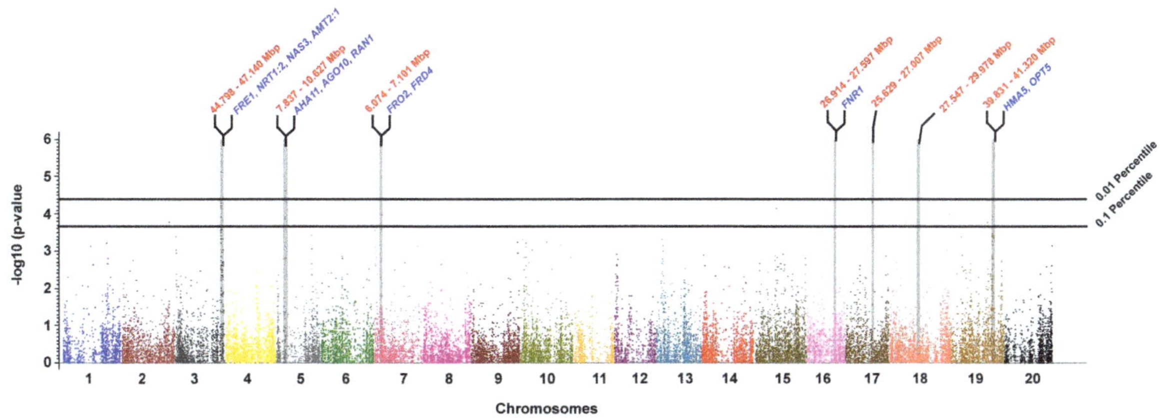

Figure 5. Manhattan plots with significant QTL that are associated with IDC. Chromosomes (1–20) ordered on x-axis and each chromosome is represented by a different color. The $-\log_{10}$(p-value) is presented on the y-axis. The cutoff horizontal lines indicate 0.01 and 0.1 percentile tails of the empirical distribution obtained using 10,000 bootstraps. Vertical grey blocks indicate QTL regions that have major effect on IDC trait identified by stepwise regression. The boundaries of the QTL region (based on the position of markers) and significant candidate genes are represented on the top of QTL.

reactive oxygen species (ROS) and other plant defenses that limit pathogen growth whereas pathogens use low-metal conditions as a signal to recognize and respond to the host environment [84]. Pathogens have developed sophisticated strategies to acquire metal during plant growth that include production of multiple siderophores. Chen et al. [85] studied the interaction IDC and soybean cyst nematode (SCN) resistance and discovered SCN susceptible varieties had higher IDC rating. Since there exists a significant interaction between diseases and IDC, selecting for disease resistance may improve IDC tolerance.

For many traits, estimating epistatic interactions in plants [86,87] has been restricted to bi-parental populations [50,88–90]. For the population, FRO2, a proton with reductase activity interacts with Cytochrome P450 like proteins. Cytochrome P450

Table 2. Significant QTL regions associated with IDC that have Candidate iron genes.

Gene Model*	Chrom	Start	End	Gene symbol*	Annotation*	Function
Gm03: 44.798–47.140						
Glyma03g38620	Gm03	44,925,730	44,930,360	FRE1	Phytochrome A	Increased reductase activity and better tolerance in high pH soils [47–48].
Glyma03g38640	Gm03	44,943,759	44,948,193	NRT1:2	Nitrate Transporter 1:2	Increase concentration of nitrate in plants and lead to increased growth and root extension [71–72].
Glyma03g39050	Gm03	45,279,743	45,281,178	NAS3	Nicotianamine Synthase 3	Synthesis of iron chelator, nicotiamine [67–68].
Glyma03g40700	Gm03	46,372,699	46,374,428	AMT2;1	Ammonium Transporter 2	Increase concentration of nitrate in plants and lead to increased growth and root extension [71–72].
Gm05: 7.837–10.627						
Glyma05g01460	Gm05	960,779	967,307	AHA11	H(+)-ATPase 11	Limits transport of iron in plants [60].
Glyma05g08170	Gm05	8,130,422	8,137,850	AGO10	Stabilizer of iron transporter SufD/Polynucleotidyl transferase	Stress response [82].
Glyma05g08260	Gm05	8,219,986	8,222,377	RAN1	RAS-related nuclear protein-1	Involved in iron transport [76].
Gm07: 6.074–7.101						
Glyma07g07380	Gm07	6,059,979	6,064,461	FRO2	Ferric reduction oxidase 2	Reduce iron at cell surface before transport [61]
Glyma07g08240	Gm07	6,837,072	6,844,618	FRD4	Signal recognition particle receptor protein, chloroplast (FTSY)	Reduces iron in response to iron deficiency [56].
Gm16: 26.914–27.597						
Glyma16g23710	Gm16	27,560,086	27,563,676	FNR1	Ferrodoxin NADP (+) oxidoreductase 1	Decreased chlorophyll biosynthesis [64].
Gm19: 39.831–41.320						
Glyma19g32190	Gm19	39,950,268	39,955,547	HMA5	Heavy Metal Atpase 5	Maintain metal balance under Fe deficiency [78–80].
Glyma19g32401	Gm19	40,141,923	40,144,157	OPT5	Oligopeptide transporter 5	Oligopeptide transporter [77].

*Gene Model, gene symbol and annotation obtained from phytozome (www.phytozome.net/soybean, V8.0).

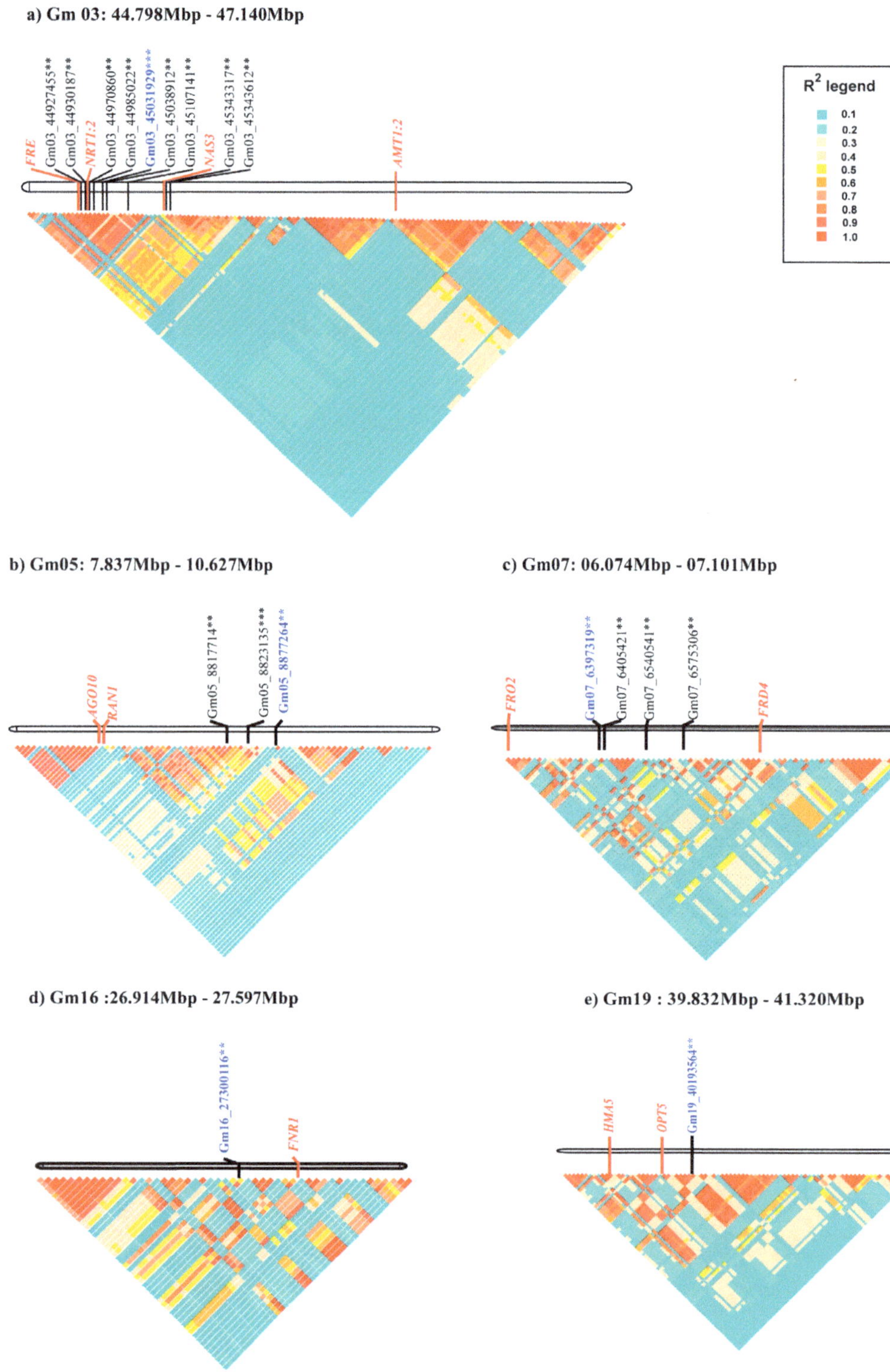

Figure 6. Heatmap of major QTL regions. QTL that has known candidate gene are only represented. Only significant markers at the 0.1 percentile bootstrapped empirical distribution were included in the figure. Significance of the markers is indicated as *** if fall in the 0.01 percentile tail and ** if fall in the 0.1 percentile tail of the bootstrapped empirical distribution. The stepwise included markers are marked in blue. r^2 values are represented in different colors. Marker pairs with r^2 values>0.6 are considered to be in linkage disequilibrium. a) is the QTL on Gm03 (44.798 Mbp–47.140 Mbp), b) is the QTL on Gm05 (7.837 Mbp–10.627 Mbp), c) is the QTL on Gm07 (6.074 Mbp–7.101 Mbp), d) is the QTL on Gm16 (26.914 Mbp–27.597 Mbp) and e) is QTL region on Gm19 (39.832 Mbp–41.320 Mbp).

proteins accumulate in response to iron deficiency [91]. If iron uptake is limited by the function of FRO2, it would lead to iron deficiency and limit AUX1 and initiate cytochrome P450 activities. Similar to Cytochrome P450 proteins, activity of FNR is reduced under iron deficient conditions [64]. AUX1 is a gene that is involved in auxin transport, a plant hormone that

Table 3. Allelic combinations and IDC means for the allele combinations that have a mean<2.5 (Tolerant) and>3.5 (Susceptible).

Allelic combination[§]	# of genotypes	IDC			
		Minimum	Maximum	Mean	Standard Deviation
CTTATTGA	3	1.63	2.23	1.88	0.31
TGTATTGA	7	1.46	2.58	2.17	0.37
CGTAGTGA	2	2.18	2.34	2.26	0.11
CTTATTGC	6	1.96	2.86	2.29	0.33
TTTATTGA	11	1.95	3.25	2.32	0.40
CGCATTGA	9	2.08	2.83	2.38	0.26
CTCGTTAA	1	2.4	2.4	2.40	
CTTATTAA	2	2.16	2.66	2.41	0.35
CTCATTGA	26	1.8	3.01	2.41	0.33
CTTGTTGC	2	2.16	2.69	2.43	0.37
CTCATTGC	9	2.11	3.03	2.44	0.33
CGTATTGA	5	2.31	2.57	2.47	0.13
CTTATCGC	2	2.42	2.52	2.47	0.07
CGTATTGC	1	2.5	2.5	2.50	
TTCGTTAC	1	3.56	3.56	3.56	
TTCAGTAC	1	3.58	3.58	3.58	
TGCAGCGC	1	3.68	3.68	3.68	
TGCGTTGC	1	3.7	3.7	3.70	
TGCGTTAA	1	3.72	3.72	3.72	
TGCGTTAC	1	3.79	3.79	3.79	

[§]The order of the markers for the allelic combination are Gm03_45031929, Gm05_8877264, Gm07_6397319, Gm11_530116, Gm16_27300116, Gm17_25859992, Gm18_28141888, and Gm19_40193564.

accumulates in Fe-deficient Arabidopsis roots and acts upstream of nitric oxide [92]. FER4 is a gene responsible for cellular iron homeostasis and subcellular iron trafficking [93]. CCC is a transporter molecule that transports iron [94]. If the activity of the transporter is reduced, auxin accumulates more and the activity of FER4 is reduced.

Evaluating allelic combinations of markers can define appropriate groups of markers for MAS. It is well understood, that there is no single universal marker that is significant in all populations. This is because of the altered effect of the genes and the QTL. It is the combination of these markers that have a significant difference on the phenotype. Of the many allelic combinations possible for multiple SNP (2^n, where n is the number of biallelic markers), only a few of them represent the extremes of phenotype. We provide allelic combination details for the extreme phenotypes for the significant markers that are included in the stepwise regression.

Conclusion

We used a high density GBS data set with a large plant GWAS population to map genes/QTL involved in IDC of soybean. Stepwise regression was used to select a subset of markers that have a major effect on IDC. We identified several significant QTL and candidate iron genes confirming the polygenic nature of the trait. Among these, a non-synonymous substitution identifies FRE1 as a candidate gene. Stepwise regression was effective for selecting the subset of markers that have major effect. This subset of markers can further be used for validation and testing in other populations. A high throughput study on expression differences of genes and the variants responsible for these differences would further help breeders identify varieties tolerant to IDC.

Supporting Information

Figure S1 Genome-wide linkage disequilibrium (LD) decay plot for the population. Linkage disequilibrium, measured as partial R^2, between pairs of polymorphic marker loci (intra-chromosomal comparisons) is plotted against the physical distance (Mbp).

Table S1 Detailed information of SNPs used in the study. SNP postions, alleles based on soybean reference genome V1.1 (phytozome.org) and minor allele frequency for this population (in %).

Author Contributions

Conceived and designed the experiments: PEM. Performed the experiments: SM RKL JRG. Analyzed the data: SM. Contributed reagents/materials/analysis tools: SM RKL JRG. Contributed to the writing of the manuscript: SM PEM.

References

1. Pil JS, Jungmin P, Mi-Jeong P, Youn-Sung K, Sang-Gyu K, et al. (2012) A Golgi-localized MATE transporter mediates iron homoeostasis under osmotic stress in Arabidopsis. Biochemical Journal 442: 551–561.
2. Kobayashi T, Nishizawa NK (2012) Iron uptake, translocation, and regulation in higher plants. Annual Review of Plant Biology 63: 131–152.
3. Briat J-F, Curie C, Gaymard F (2007) Iron utilization and metabolism in plants. Current Opinion in Plant Biology 10: 276–282.
4. Franzen DW, Richardson JL (2000) Soil factors affecting iron chlorosis of soybean in the Red River Valley of North Dakota and Minnesota. Journal of Plant Nutrition 23: 67–78.
5. Peiffer GA, King KE, Severin AJ, May GD, Cianzio SR, et al. (2012) Identification of candidate genes underlying an iron efficiency quantitative trait locus in soybean. Plant Physiology 158: 1745–1754.
6. Marschner H, Romheld V, Kissel M (1986) Different strategies in higher plants in mobilization and uptake of iron. Journal of Plant Nutrition 9: 695–713.
7. Schmidt W (1999) Mechanisms and regulation of reduction-based iron uptake in plants. New Phytologist 141: 1–26.
8. Barker AV, Pilbeam DJ (2010) Handbook of plant nutrition. CRC press.
9. Mamidi S, Chikara S, Goos RJ, Hyten DL, Annam D, et al. (2011) Genome-wide association analysis identifies candidate genes associated with iron deficiency chlorosis in soybean. The Plant Genome 4: 154–164.
10. Curie C, Briat J-F (2003) Iron transport and signaling in plants. Annual Review of Plant Biology 54: 183–206.
11. Diers BW, Cianzio SR, Shoemaker RC (1992) Possible identification of quantitative trait loci affecting iron efficiency in soybean. Journal of Plant Nutrition 15: 2127–2136.
12. King KE, Peiffer GA, Reddy M, Lauter N, Lin SF, et al. (2013) Mapping of Iron and Zinc Quantitative Trait Loci in Soybean for Association to Iron Deficiency Chlorosis Resistance. Journal of Plant Nutrition 36: 2132–2153.
13. O'Rourke J, Charlson D, Gonzalez D, Vodkin L, Graham M, et al. (2007) Microarray analysis of iron deficiency chlorosis in near-isogenic soybean lines. BMC Genomics 8: 476.
14. Severin AJ, Peiffer GA, Xu WW, Hyten DL, Bucciarelli B, et al. (2010) An integrative approach to genomic introgression mapping. Plant Physiology 154: 3–12.
15. Stec AO, Bhaskar PB, Bolon Y-T, Nolan R, Shoemaker RC, et al. (2013) Genomic heterogeneity and structural variation in soybean near isogenic lines. Frontiers in Plant Science 4: 104.
16. Myles S, Peiffer J, Brown PJ, Ersoz ES, Zhang Z, et al. (2009) Association mapping: critical considerations shift from genotyping to experimental design. The Plant Cell Online 21: 2194–2202.
17. Wang J, McClean PE, Lee R, Goos RJ, Helms T (2008) Association mapping of iron deficiency chlorosis loci in soybean (*Glycine max* L. Merr.) advanced breeding lines. Theoretical and Applied Genetics 116: 777–787.
18. Elshire RJ, Glaubitz JC, Sun Q, Poland JA, Kawamoto K, et al. (2011) A robust, simple genotyping-by-sequencing (GBS) approach for high diversity species. PloS One 6: e19379.
19. Bradbury PJ, Zhang Z, Kroon DE, Casstevens TM, Ramdoss Y, et al. (2007) TASSEL: software for association mapping of complex traits in diverse samples. Bioinformatics 23: 2633–2635.
20. Schmutz J, Cannon SB, Schlueter J, Ma J, Mitros T, et al. (2010) Genome sequence of the palaeopolyploid soybean. Nature 463: 178–183.
21. Li H, Durbin R (2009) Fast and accurate short read alignment with Burrows Wheeler transform. Bioinformatics 25: 1754–1760.
22. Danecek P, Auton A, Abecasis G, Albers CA, Banks E, et al. (2011) The variant call format and VCFtools. Bioinformatics 27: 2156–2158.
23. Scheet P, Stephens M (2006) A fast and flexible statistical model for large-scale population genotype data: applications to inferring missing genotypes and haplotypic phase. The American Journal of Human Genetics 78: 629–644.
24. Marchini J, Howie B (2010) Genotype imputation for genome-wide association studies. Nature Reviews Genetics 11: 499–511.
25. Mangin B, Siberchicot A, Nicolas S, Doligez A, This P, et al. (2011) Novel measures of linkage disequilibrium that correct the bias due to population structure and relatedness. Heredity 108: 285–291.
26. Long Q, Rabanal FA, Meng D, Huber CD, Farlow A, et al. (2013) Massive genomic variation and strong selection in *Arabidopsis thaliana* lines from Sweden. Nature genetics 45: 884–890.
27. Stram DO (2004) Tag SNP selection for association studies. Genetic Epidemiology 27: 365–374.
28. Remington DL, Thornsberry JM, Matsuoka Y, Wilson LM, Whitt SR, et al. (2001) Structure of linkage disequilibrium and phenotypic associations in the maize genome. Proceedings of the National Academy of Sciences 98: 11479–11484.
29. Pyhäjärvi T, García-Gil MR, Knürr T, Mikkonen M, Wachowiak W, et al. (2007) Demographic history has influenced nucleotide diversity in European *Pinus sylvestris* populations. Genetics 177: 1713–1724.
30. Price AL, Patterson NJ, Plenge RM, Weinblatt ME, Shadick NA, et al. (2006) Principal components analysis corrects for stratification in genome-wide association studies. Nature Genetics 38: 904–909.
31. Zhao K, Aranzana MJ, Kim S, Lister C, Shindo C, et al. (2007) An Arabidopsis example of association mapping in structured samples. PLoS Genetics 3: e4.
32. Zhou X, Stephens M (2012) Genome-wide efficient mixed-model analysis for association studies. Nature Genetics 44: 821–824.
33. Rakitsch B, Lippert C, Stegle O, Borgwardt K (2013) A Lasso multi-marker mixed model for association mapping with population structure correction. Bioinformatics 29: 206–214.
34. Yu J, Pressoir G, Briggs WH, Bi IV, Yamasaki M, et al. (2005) A unified mixed-model method for association mapping that accounts for multiple levels of relatedness. Nature Genetics 38: 203–208.
35. Hall P, Miller H (2010) Bootstrap confidence intervals and hypothesis tests for extrema of parameters. Biometrika 97: 881–892.
36. Li J, Tai BC, Nott DJ (2009) Confidence interval for the bootstrap P-value and sample size calculation of the bootstrap test. Journal of Nonparametric Statistics 21: 649–661.
37. Meyer KB, O'Reilly M, Michailidou K, Carlebur S, Edwards SL, et al. (2013) Fine-scale mapping of the FGFR2 breast cancer risk locus: putative functional variants differentially bind FOXA1 and E2F1. The American Journal of Human Genetics 93: 1046–1060.
38. Cingolani P, Platts A, Coon M, Nguyen T, Wang L, et al. (2012) A program for annotating and predicting the effects of single nucleotide polymorphisms, SnpEff: SNPs in the genome of *Drosophila melanogaster* strain w1118; iso-2; iso-3. Fly 6: 80–92.
39. Davey JW, Hohenlohe PA, Etter PD, Boone JQ, Catchen JM, et al. (2011) Genome-wide genetic marker discovery and genotyping using next-generation sequencing. Nature Reviews Genetics 12: 499–510.
40. Deschamps S, Campbell MA (2010) Utilization of next-generation sequencing platforms in plant genomics and genetic variant discovery. Molecular Breeding 25: 553–570.
41. Ganal MW, Durstewitz G, Polley A, Berard A, Buckler ES, et al. (2011) A large maize (*Zea mays* L.) SNP genotyping array: development and germplasm genotyping, and genetic mapping to compare with the B73 reference genome. PloS One 6: e28334.
42. Mandel JR, Nambeesan S, Bowers JE, Marek LF, Ebert D, et al. (2013) Association Mapping and the Genomic Consequences of Selection in Sunflower. PLoS Genetics 9: e1003378.
43. Gurung S, Mamidi S, Bonman JM, Jackson EW, Del Rio LE, et al. (2011) Identification of novel genomic regions associated with resistance to *Pyrenophora tritici-repentis* races 1 and 5 in spring wheat landraces using association analysis. Theoretical and Applied Genetics 123: 1029–1041.
44. Iqbal MJ, Mamidi S, Ahsan R, Kianian SF, Coyne CJ, et al. (2012) Population structure and linkage disequilibrium in *Lupinus albus* L. germplasm and its implication for association mapping. Theoretical and Applied Genetics 125: 517–530.
45. Patel JS, Mamidi S, Bonman JM, Adhikari TB (2013) Genome-wide Association Analysis in Spring Wheat Accessions Identifies QTL Associated with Resistance to a 2 Novel Isolate of *Pyrenophora tritici-repentis* 3. Crop Science 53: 842–852.
46. Astle W, Balding DJ (2009) Population structure and cryptic relatedness in genetic association studies. Statistical Science 24: 451–471.
47. Ishimaru Y, Kim S, Tsukamoto T, Oki H, Kobayashi T, et al. (2007) Mutational reconstructed ferric chelate reductase confers enhanced tolerance in rice to iron deficiency in calcareous soil. Proceedings of the National Academy of Sciences 104: 7373–7378.
48. Oki H, Kim S, Nakanishi H, Takahashi M, Yamaguchi H, et al. (2004) Directed evolution of yeast ferric reductase to produce plants with tolerance to iron deficiency in alkaline soils. Soil Science and Plant Nutrition 50: 1159–1165.
49. Samuelsen AI, Martin RC, Mok DWS, Mok MC (1998) Expression of the yeast FRE genes in transgenic tobacco. Plant Physiology 118: 51–58.
50. Grillo MA, Li C, Hammond M, Wang L, Schemske DW (2013) Genetic architecture of flowering time differentiation between locally adapted populations of *Arabidopsis thaliana*. New Phytologist 197: 1321–1331.
51. Pineau C, Loubet S, Lefoulon C, Chalies C, Fizames C, et al. (2012) Natural Variation at the FRD3 MATE Transporter Locus Reveals Cross-Talk between Fe Homeostasis and Zn Tolerance in *Arabidopsis thaliana*. PLoS Genetics 8: e1003120.
52. Zhang N, Gur A, Gibon Y, Sulpice R, Flint-Garcia S, et al. (2010) Genetic analysis of central carbon metabolism unveils an amino acid substitution that alters maize NAD-dependent isocitrate dehydrogenase activity. PloS One 5: e9991.
53. Lin S, Cianzio S, Shoemaker R (1997) Mapping genetic loci for iron deficiency chlorosis in soybean. Molecular Breeding 3: 219–229.
54. Lin SF, Grant D, Cianzio S, Shoemaker R (2000) Molecular characterization of iron deficiency chlorosis in soybean. Journal of Plant Nutrition 23: 1929–1939.
55. Lin S-F, Baumer JS, Ivers D, de Cianzo SR, Shoemaker RC (1998) Field and nutrient solution tests measure similar mechanisms controlling iron deficiency chlorosis in soybean. Crop Science 38: 254–259.
56. Durrett TP, Connolly EL, Rogers EE (2006) Arabidopsis cpFtsY mutants exhibit pleiotropic defects including an inability to increase iron deficiencyinducible root Fe (III) chelate reductase activity. The Plant Journal 47: 467–479.
57. Eide D, Broderius M, Fett J, Guerinot ML (1996) A novel iron-regulated metal transporter from plants identified by functional expression in yeast. Proceedings of the National Academy of Sciences 93: 5624–5628.

58. Robinson NJ, Procter CM, Connolly EL, Guerinot ML (1999) A ferric-chelate reductase for iron uptake from soils. Nature 397: 694–697.
59. Vert G, Grotz N, Dédaldéchamp F, Gaymard F, Guerinot ML, et al. (2002) IRT1, an Arabidopsis transporter essential for iron uptake from the soil and for plant growth. The Plant Cell Online 14: 1223–1233.
60. Alsterfjord M, Sehnke PC, Arkell A, Larsson H, Svennelid F, et al. (2004) Plasma membrane H+-ATPase and 14-3-3 isoforms of Arabidopsis leaves: evidence for isoform specificity in the 14-3-3/H+-ATPase interaction. Plant and cell physiology 45: 1202–1210.
61. Kim SA, Guerinot ML (2007) Mining iron: iron uptake and transport in plants. FEBS letters 581: 2273–2280.
62. Connolly EL, Campbell NH, Grotz N, Prichard CL, Guerinot ML (2003) Overexpression of the FRO2 ferric chelate reductase confers tolerance to growth on low iron and uncovers posttranscriptional control. Plant Physiology 133: 1102–1110.
63. Connolly EL, Fett JP, Guerinot ML (2002) Expression of the IRT1 metal transporter is controlled by metals at the levels of transcript and protein accumulation. The Plant Cell Online 14: 1347–1357.
64. Miller GW, Pushnik JC, Welkie GW (1984) Iron chlorosis, a world wide problem, the relation of chlorophyll biosynthesis to iron. Journal of Plant Nutrition 7: 1–22.
65. Pich A, Manteuffel R, Hillmer S, Scholz G, Schmidt W (2001) Fe homeostasis in plant cells: does nicotianamine play multiple roles in the regulation of cytoplasmic Fe concentration? Planta 213: 967–976.
66. von Wiren N, Klair S, Bansal S, Briat J-F, Khodr H, et al. (1999) Nicotianamine chelates both FeIII and FeII. Implications for metal transport in plants. Plant Physiology 119: 1107–1114.
67. Lee S, Jeon US, Lee SJ, Kim Y-K, Persson DP, et al. (2009) Iron fortification of rice seeds through activation of the nicotianamine synthase gene. Proceedings of the National Academy of Sciences 106: 22014–22019.
68. Masuda H, Usuda K, Kobayashi T, Ishimaru Y, Kakei Y, et al. (2009) Overexpression of the barley nicotianamine synthase gene HvNAS1 increases iron and zinc concentrations in rice grains. Rice 2: 155–166.
69. Wang R, Okamoto M, Xing X, Crawford NM (2003) Microarray analysis of the nitrate response in Arabidopsis roots and shoots reveals over 1,000 rapidly responding genes and new linkages to glucose, trehalose-6-phosphate, iron, and sulfate metabolism. Plant Physiology 132: 556–567.
70. Wang Y-H, Garvin DF, Kochian LV (2001) Nitrate-induced genes in tomato roots. Array analysis reveals novel genes that may play a role in nitrogen nutrition. Plant Physiology 127: 345–359.
71. Bai H, Euring D, Volmer K, Janz D, Polle A (2013) The Nitrate Transporter (NRT) Gene Family in Poplar. PloS One 8: e72126.
72. Gu R, Duan F, An X, Zhang F, von Wiren N, et al. (2013) Characterization of AMT-Mediated High-Affinity Ammonium Uptake in Roots of Maize (*Zea mays* L.). Plant and Cell Physiology 54: 1515–1524.
73. Bloom AJ, Meyerhoff PA, Taylor AR, Rost TL (2002) Root development and absorption of ammonium and nitrate from the rhizosphere. Journal of Plant Growth Regulation 21: 416–431.
74. Taylor AR, Bloom AJ (1998) Ammonium, nitrate, and proton fluxes along the maize root. Plant, Cell & Environment 21: 1255–1263.
75. Zou C, Shen J, Zhang F, Guo S, Rengel Z, et al. (2001) Impact of nitrogen form on iron uptake and distribution in maize seedlings in solution culture. Plant and Soil 235: 143–149.
76. Woeste KE, Kieber JJ (2000) A strong loss-of-function mutation in RAN1 results in constitutive activation of the ethylene response pathway as well as a rosette-lethal phenotype. The Plant Cell Online 12: 443–455.
77. Stacey MG, Osawa H, Patel A, Gassmann W, Stacey G (2006) Expression analyses of Arabidopsis oligopeptide transporters during seed germination, vegetative growth and reproduction. Planta 223: 291–305.
78. Arrivault S, Senger T, Kramer U (2006) The Arabidopsis metal tolerance protein AtMTP3 maintains metal homeostasis by mediating Zn exclusion from the shoot under Fe deficiency and Zn oversupply. The Plant Journal 46: 861–879.
79. Morel M, Crouzet J, Gravot A, Auroy P, Leonhardt N, et al. (2009) AtHMA3, a P1B-ATPase allowing Cd/Zn/Co/Pb vacuolar storage in Arabidopsis. Plant Physiology 149: 894–904.
80. Schaaf G, Honsbein A, Meda AR, Kirchner S, Wipf D, et al. (2006) AtIREG2 encodes a tonoplast transport protein involved in iron-dependent nickel detoxification in *Arabidopsis thaliana* roots. Journal of Biological Chemistry 281: 25532–25540.
81. Zheng L, Huang F, Narsai R, Wu J, Giraud E, et al. (2009) Physiological and transcriptome analysis of iron and phosphorus interaction in rice seedlings. Plant Physiology 151: 262–274.
82. Vaucheret H (2008) Plant argonautes. Trends in Plant Science 13: 350–358.
83. Kang Y, Kim K, Shim S, Yoon M, Sun S, et al. (2012) Genome-wide mapping of NBS-LRR genes and their association with disease resistance in soybean. BMC Plant Biology 12: 139.
84. Fones H, Preston GM (2012) Reactive oxygen and oxidative stress tolerance in plant pathogenic Pseudomonas. FEMS microbiology letters 327: 1–8.
85. Chen S, Kurle JE, Stetina SR, Miller DR, Klossner LD, et al. (2007) Interactions between iron-deficiency chlorosis and soybean cyst nematode in Minnesota soybean fields. Plant and Soil 299: 131–139.
86. Mackay TFC (2014) Epistasis and quantitative traits: using model organisms to study gene-gene interactions. Nature Reviews Genetics 15: 22–33.
87. Shang J, Zhang J, Sun Y, Liu D, Ye D, et al. (2011) Performance analysis of novel methods for detecting epistasis. BMC Bioinformatics 12: 475.
88. Caicedo AL, Stinchcombe JR, Olsen KM, Schmitt J, Purugganan MD (2004) Epistatic interaction between Arabidopsis FRI and FLC flowering time genes generates a latitudinal cline in a life history trait. Proceedings of the National Academy of Sciences 101: 15670–15675.
89. Jiang Z, Han Y, Teng W, Zhang Z, Sun D, et al. (2010) Identification of QTL underlying the filling rate of protein at different developmental stages of soybean seed. Euphytica 175: 227–236.
90. Lin H, Liang Z-W, Sasaki T, Yano M (2003) Fine mapping and characterization of quantitative trait loci Hd4 and Hd5 controlling heading date in rice. Breeding Science 53: 51–59.
91. Colangelo EP, Guerinot ML (2004) The essential basic helix-loop-helix protein FIT1 is required for the iron deficiency response. The Plant Cell Online 16: 3400–3412.
92. Chen WW, Yang JL, Qin C, Jin CW, Mo JH, et al. (2010) Nitric oxide acts downstream of auxin to trigger root ferric-chelate reductase activity in response to iron deficiency in Arabidopsis. Plant Physiology 154: 810–819.
93. Tarantino D, Santo N, Morandini P, Casagrande F, Braun H-P, et al. (2010) AtFer4 ferritin is a determinant of iron homeostasis in *Arabidopsis thaliana* heterotrophic cells. Journal of Plant Physiology 167: 1598–1605.
94. Conte SS, Walker EL (2011) Transporters contributing to iron trafficking in plants. Molecular Plant 4: 464–476.

12

Bayesian Inference of Baseline Fertility and Treatment Effects via a Crop Yield-Fertility Model

Hungyen Chen[1], Junko Yamagishi[2], Hirohisa Kishino[1]*

1 Graduate School of Agricultural and Life Sciences, The University of Tokyo, Tokyo, Japan, **2** Institute for Sustainable Agro-ecosystem Services, The University of Tokyo, Tokyo, Japan

Abstract

To effectively manage soil fertility, knowledge is needed of how a crop uses nutrients from fertilizer applied to the soil. Soil quality is a combination of biological, chemical and physical properties and is hard to assess directly because of collective and multiple functional effects. In this paper, we focus on the application of these concepts to agriculture. We define the baseline fertility of soil as the level of fertility that a crop can acquire for growth from the soil. With this strict definition, we propose a new crop yield-fertility model that enables quantification of the process of improving baseline fertility and the effects of treatments solely from the time series of crop yields. The model was modified from Michaelis-Menten kinetics and measured the additional effects of the treatments given the baseline fertility. Using more than 30 years of experimental data, we used the Bayesian framework to estimate the improvements in baseline fertility and the effects of fertilizer and farmyard manure (FYM) on maize (*Zea mays*), barley (*Hordeum vulgare*), and soybean (*Glycine max*) yields. Fertilizer contributed the most to the barley yield and FYM contributed the most to the soybean yield among the three crops. The baseline fertility of the subsurface soil was very low for maize and barley prior to fertilization. In contrast, the baseline fertility in this soil approximated half-saturated fertility for the soybean crop. The long-term soil fertility was increased by adding FYM, but the effect of FYM addition was reduced by the addition of fertilizer. Our results provide evidence that long-term soil fertility under continuous farming was maintained, or increased, by the application of natural nutrients compared with the application of synthetic fertilizer.

Editor: Wenju Liang, Chinese Academy of Sciences, China

Funding: The Grant-in-Aid for Scientific Research (grant number 59440009, PI: Wataru Sunohara) by the Ministry of Education, Science and Culture (MEXT, http://www.mext.go.jp) supported the initial phase (1984–1987) of the long-term fertilizer experiment. The University Farm continued the experiment until 2010 by the regular budget supplied by the University of Tokyo. This work was supported by Grants-in-Aid for Scientific Research (grant number 25280006) by the Japan Society for the Promotion of Science (JSPS, http://www.jsps.go.jp/) to H. K. The funders had no role in study design, data collection and analysis, decision to publish, or preparation of the manuscript.

Competing Interests: The authors have declared that no competing interests exist.

* Email: kishino@lbm.ab.a.u-tokyo.ac.jp

Introduction

Agricultural crop production is directly related to food supply, so agricultural soil productivity must be maintained. Balanced fertilization provides all the essential nutrients for crops to remain healthy and grow productively [1–3]. In a world facing increasing population pressure [4], our highest priority must be to increase crop productivity to ensure food security [2,5–7]. In this context, there has been increasing concern about the long-term productivity of soils on a global scale [8–11].

The relationship between crop yield and soil fertility under different fertilization regimes has been studied for decades. Aref and Wander [12] investigated the long-term trends of corn yield and soil organic matter in different crop sequences and soil fertility treatments. Merick and Németh [13] used results from 60-year field experiments to provide information on the relationship between fertilization and yields of rye and potato. Hallin et al. [14] investigated the relationship between microbial communities and total crop yield and nitrogen content in the crop in a 50-year-old fertilization experiment. Fan et al. [15] studied the trends in grain yields and soil organic carbon (SOC) in a 26-year dryland fertilization trial. By carrying out a 27-year experiment with various fertilization treatments in a rotation cropping system with wheat and maize in a red soil, Zhang et al. [16] investigated trends in SOC, soil nitrogen, and grain yield.

Many researchers have suggested that soil fertility under continuous farming is maintained, or increased, by the application of farmyard manure (FYM) compared with the application of synthetic fertilizer [2,11,17–19]. To ensure gains in crop productivity, it is necessary to maintain and even improve the soil fertility of a field under continuous farming practice [20–23]. In some cases, soils do not contain sufficient amounts of the essential nutrients required for rapid crop growth and high productivity [2,7,24]. As a result, supplemental nutrients, applied as fertilizers, manure or compost, are needed [25–27]. Many studies have analyzed the effects of manure [28–30], chemical fertilizer [31,32], and both [33,34] on soil fertility and crop yield.

Previous models have predicted the crop-yield response to fertilizer application [19] and climate change [35] and have greatly contributed to the software used in agricultural systems research [36,37]. Myers [38] used a static model to estimate the nitrogen fertilizer requirements of cereal crops. Deng et al. [39] proposed a theoretical framework that predicts the optimum planting density and maximal yield for an annual crop plant. Cong et al. [40] indicated that the CENTURY model [41] can simulate fertilization effects on SOC dynamics under different climate and soil conditions.

Soil quality is a complex combination of biological, chemical, and physical properties. In this paper, we focus on the application of these concepts to agriculture and define the baseline fertility as the level of soil fertility that crops utilize for their growth. Therefore, our soil quality is an interaction of the above properties and the crop activities. With this strict definition of baseline fertility, we propose a new crop yield-fertility model that enables quantification of the process of improving baseline fertility and the effects of treatments solely from the time series of crop yields. The model was modified from Michaelis-Menten kinetics and measured the additional effects of the treatments given the baseline fertility. Using more than 30 years of experimental data of maize (*Zea mays*), barley (*Hordeum vulgare*), and soybean (*Glycine max*) yield, we estimated in the Bayesian framework the improvements in baseline fertility and the effects of fertilizer and farmyard manure (FYM). We compared the efficiency of separate applications of fertilizer or FYM to the three crops (i.e., maize, barley and soybean) using estimated model parameters. The temporal variations in the baseline fertility of each crop were estimated and compared for six treatments and two soil types.

Materials and Methods

Field experiments

The long-term fertilizer experiment was conducted between 1980 and 2010 at the University Farm at the Institute for Sustainable Agro-ecosystem Services at the University of Tokyo, Nishitokyo, Tokyo, Japan (35°43′ N latitude and 139°32′ E longitude and an altitude of 53 m above mean sea level). The field site was located in the Kantō Plain, where the soil is covered with pyroclastic material from volcanoes that surrounded the western Kantō region 126,000 years ago. The soil parent material is Tachikawa loam. The surface soil is a black-colored fertile andosol containing a high percentage of humus and the subsurface soil (25 cm below the surface) is red-colored barren clay. Andosols are soils found in volcanic areas formed in volcanic tephra. Andosols have a different composition from chernozems and are not commonly found outside the Pacific "ring of fire". At the beginning of the experiment, the percent nitrogen was 2.23 g/kg in the surface soil and 0.97 g/kg in the subsurface soil. A 2-year crop rotation of maize-barley-soybean-barley was maintained throughout the experimental period and the crop yields were measured. The maize crop was sown at the beginning of July and harvested at the end of September every second year from 1980 onwards. The barley crop was sown in the first half of November and harvested either at the end of May or at the beginning of June of the next year each year from 1980 onwards. The soybean crop was sown at the end of June and harvested in the second half of October every second year from 1981 onwards. The seeds of the three crops were ridge sown using a seeding machine. For maize and soybean, one seed was sown in each hole, and the widths between the ridges and stocks were 71 cm and 23 cm, and 71 cm and 17.5 cm, respectively. For barley, row seeding was applied using a width of 17.8 cm between the ridges.

Six treatment plots containing the NPK fertilizer applications and two levels of farmyard manure (FYM) combined with compost were established in both of the fields with the surface soil (64 m^2 per plot, 8×8 m) and the subsurface soil (56 m^2 per plot, 7×8 m) and were replicated four times. The percent nitrogen, phosphorus, and potassium in the fertilizer treatments were N:P:K = 12:7.9:13.3 for maize and barley and 3:4.4:8.3 for soybean. Fertilizer was added to the soil at a rate of 1 t ha^{-1} for all crops. The FYM comprised wheat straw and cow dung. The percent nitrogen in the FYM ranged from 0.2–0.4. FYM was added to the plots at two levels, i.e., 20 and 60 t ha^{-1}. The FYM treatments were added to the soil twice a year from 1980 to 1990 before the crops were sown and once a year (before the barley was sown) from 1991 onwards. After the barley harvest in 2008, fertilizer and FYM applications were discontinued for all treatments. Phosphate fertilizer was added simultaneously with the NPK fertilizer and FYM at a rate of 2 t ha^{-1} after the soybean harvest from 1980 until 2001. All aboveground components were removed from the field during harvesting. The dead foliage, below-ground components, and remnant stem sections (i.e., 15 cm for maize and 10 cm for barley and soybean) remained in the field. The harvested material from the six treatments for both the surface and subsurface soils were weighed after drying at 80°C. The dry weight (g m^{-2}) of the aboveground maize components, the barley spike, and the soybean seed were measured.

Testing significance of yield differences in different soils

The crop yields of the surface and subsurface soils without the addition of fertilizer or farmyard manure (FYM) reflect the yield differences that occurred purely as a result of the soil. The crop yields observed for the subsurface soil were 39±25%, 49±20%, and 19±31% (±SD) lower than the yields observed for the surface soil for maize (Fig. S1A), barley (Fig. S1B), and soybean (Fig. S1C), respectively. There was a difference in yield between the two types of soils but it was only significant for barley ($t = -3.69$, $P<0.001$; for maize, $t = -1.58$, $P = 0.13$ and for soybean, $t = -0.94$, $P = 0.35$). Although the crops were planted in the same fields with the same soil, they obtained and used nutrients from the soil in a manner unique to each crop.

The maize and soybean yields did not differ significantly ($P>0.2$) between the surface and subsurface soils when fertilizer or FYM or a combination of fertilizer and FYM were applied (Fig. S1A and C for maize and soybean, respectively). The barley yields differed significantly between the surface and subsurface soils when both fertilizer and FYM ($t = 2.27$, $P<0.05$) or FYM alone ($t = 2.18$, $P<0.05$) were added, but did not differ significantly when only fertilizer was added ($t = -0.63$, $P = 0.52$) (Fig. S1B). The difference in yields between the surface and subsurface soils was reduced when either fertilizer or FYM was added to the soil. The application of fertilizer and FYM improved the yields for both the surface and subsurface soils and reduced the yield differences between the soil types. This finding indicates that productivity can be increased to an average level by fertilization even in a field with barren soil. While this may have been prior knowledge, the manner in which fertilizer and FYM contribute nutrients to the soil and how the crops make use of the nutrients obtained from these sources remains unclear.

Testing significance of yearly variance

The maize and soybean yields increased significantly ($P<0.05$) for both soil types over the 30-year period. The barley yield increased significantly in the subsurface soil ($P<0.05$) but not in the surface soil ($P = 0.15$). The soils were fertilized by crop residues (i.e., roots, fallen leaves, and stem sections (15 cm above ground for maize and 10 cm for barley and soybean)) that remained in and on the soil after harvesting. Organic matter accumulated in the soil over time, providing a long-term, slow-release source of nitrogen, phosphorus, sulfur, and other important nutrients for crop growth, which resulted in increased soil productivity.

Testing significance of the treatment effects

We used a nested analysis of variance to compare the yields of the three crops in the surface and subsurface soils in response to the different treatments. The surface soil produced significantly greater barley crop yields compared with the subsurface soil (Fig. S1B, $F = 14.68$, $P < 0.001$). Maize and soybean yields (Fig. S1A, $F = 2.91$, $P = 0.09$ and Fig. S1C, $F = 2.19$, $P = 0.14$, respectively) did not differ significantly between the two soil types. The interaction effects of fertilizer and FYM were significant for maize ($F = 3.32$, $P < 0.05$) and barley ($F = 10.71$, $P < 0.001$), but not for soybean ($F = 0.69$, $P = 0.60$). The soybean yield did not differ significantly among the treatments ($P > 0.1$ for both fertilizer and FYM).

The derivation of the crop yield-fertility model

Our model is based on the general pattern of the observed crop yield. The temporal variation in maize yield in response to four treatments applied to the surface soil over 30 years is shown in Fig. 1A. There were clear differences in the yield between the four treatments prior to 1982 (the first two data points). The response of the crop yield descended in the following order: the addition of NPK fertilizer and FYM (F:1, M:1), the addition of NPK fertilizer only (F:1, M:0), the addition of FYM only (F:0, M:1), and no nutrient application (F:0, M:0). The yield differences between the treatments receiving NPK fertilizer and FYM (F:1, M:1) and NPK fertilizer only (F:1, M:0) could not be detected from 1984 onwards. The effect of FYM could no longer be detected once the fertilizer application was continued. From 1994, similar crop yields were observed in the treatments with fertilizer, FYM or a combined application of fertilizer and FYM. In the last year of the study, there were negligible differences between treatments because fertilizer and FYM applications were discontinued in 2008. The soybean data are presented in Fig. 1C. The pattern of the soybean yield differed from maize. There were almost no differences in crop yield among the three treatments that received fertilizer, FYM or a combined application of fertilizer and FYM in the first and later years. The differences in crop yield between the treatments with and without fertilizer were negligible from 1995 onwards. The inherent soil fertility may be sufficient to grow soybean without the addition of fertilizer. The pattern of barley yield (Fig. 1B) was similar to maize but was not as well defined.

Our model can predict crop yields given the total fertility level of the soil. The model can be used to describe the variations in crop yield and the effect of different fertilizer treatments over different years for each crop using the estimated soil fertility. The crop yield-fertility curves for maize, barley, and soybean are presented in Fig. 2A, B, and C, respectively. Soil fertility was increased by the application of fertilizer over a 20-year period. Furthermore, the effect of fertilizer decreased over time for all three crops from 1980–2000.

The crop yield-fertility model

We developed a simple mathematical model to estimate the soil fertility level and to quantify the contributions of fertilizer and FYM to the improvements in crop yields. The model was modified from Michaelis-Menten kinetics [42]:

$$Y_{i,t} \cong \frac{V}{1+\dfrac{K}{BF_{i,t}+a\times \mathrm{F}+b\times \mathrm{M}}}. \tag{1}$$

We did not apply the model to describe the kinetics. Rather, we used a form of the model to interpret the effects of baseline fertility, fertilizer, and FYM on the crop yield and the extent of saturation of additional inputs. The model related crop yields ($\mathrm{Y}_{\mathrm{i,t}}$) to the total fertility ($BF_{i,t}+a\times \mathrm{F}+b\times \mathrm{M}$) of the soil for treatment i at time t (Fig. 2). The value of the baseline fertility (BF) for the subsurface soil without the addition of fertilizer or FYM in the first year was normalized by setting this value to 1. V is the maximum yield in response to the maximum fertilizer input; K is the fertility level before fertilization that is required to produce half of the maximum yield, V. A large K indicates that more nutrients need to be added to the soil for the crop to grow, so K can infer the sterility of the soil prior to fertilization. $BF_{i,t}$ is the baseline fertility of the crop for treatment i at time t and is assumed to vary gradually over time; F is the level of fertilizer application (0 and 1), M is the level of FYM application (0, 1/3, and 1), and a and b represent the contributions of F and M relative to the baseline fertility, respectively. Hereafter, we refer to V, K, a, and b as the maximum yield, half-saturated fertility, fertilizer contribution, and FYM, respectively. A Bayesian framework was adopted for parameter estimation assuming a gradual change in the baseline fertility.

The likelihood and priors

The likelihood of the yield for treatment i at time t followed a normal distribution with the mean $\frac{V}{1+\frac{K}{BF_{i,t}+a\times \mathrm{F}+b\times \mathrm{M}}}$ and the variance δ:

$$\mathrm{Yield}_{\mathrm{i,t}} \sim N\left(\frac{V}{1+\dfrac{K}{BF_{i,t}+a\times \mathrm{F}+b\times \mathrm{M}}},\delta\right) \tag{2}$$

This value was normalized by setting the BF of the subsurface soil without the addition of fertilizer or FYM in the first year to 1. The priors of the BF for the other treatments in the first year followed a gamma distribution with a shape parameter of 1 and a scale parameter of 1. The smoothness priors of the BF from the second year followed a normal distribution with the mean equal to the value of the BF in the preceding year and the variance τ:

$$BF_{i,t} \sim N(BF_{i,t-1},\tau), \quad t>1. \tag{3}$$

The inverse of δ followed a gamma distribution with a shape parameter of 0.1 and a scale parameter of 10. The inverse of τ followed a gamma distribution with a shape parameter of 0.1 and a scale parameter of 10. The prior of V followed a normal distribution with a mean of 0 and a standard deviation of 1000. The prior of K followed a gamma distribution with a shape parameter of 0.1 and a scale parameter of 10. The priors of a and b followed a gamma distribution with a shape parameter of 1 and a scale parameter of 1. The priors of the estimates were designed to be as non-informative as possible within a realistic range of the

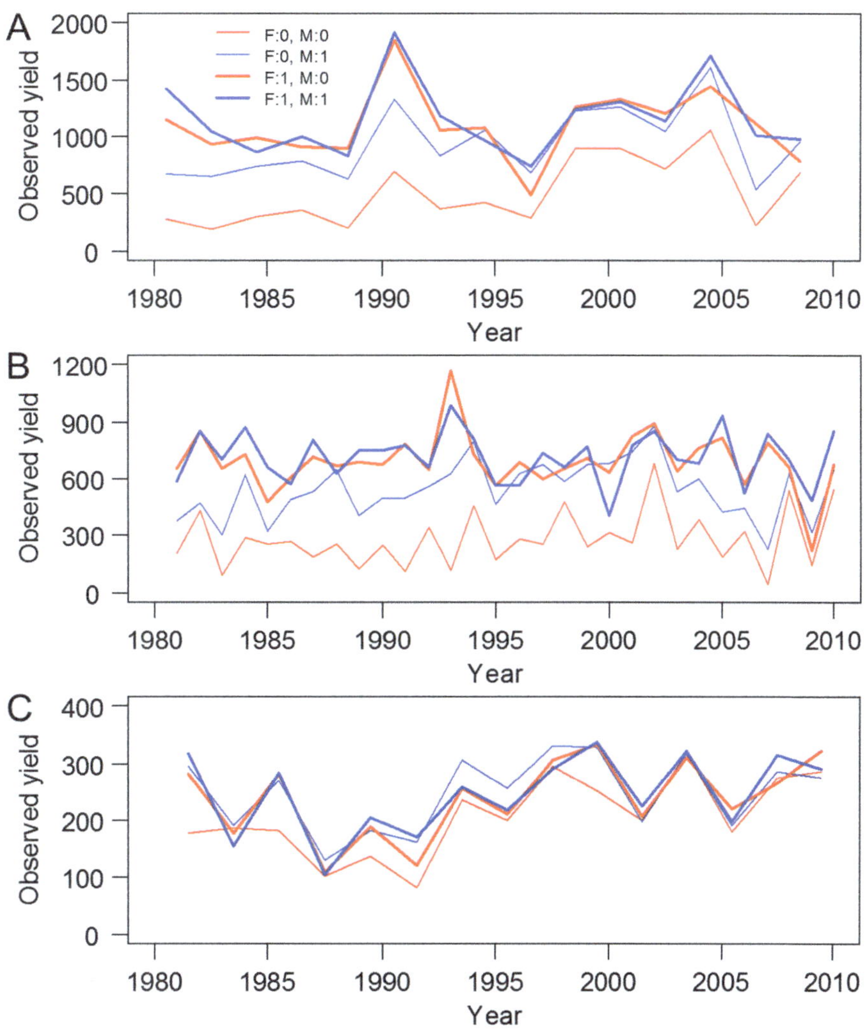

Figure 1. Temporal variations in observed crop yield (g m^{-2}) for four treatments in the surface soil. (A) Maize; (B) Barley; (C) Soybean; F, level of fertilizer; M, level of farmyard manure. See Fig. S1 for all treatments in surface and subsurface soils.

parameter values. All calculations and data analyses were performed using R v2.13.2 [43]. The raw data is available as (Data S1).

Results

Increasing trend of soil fertility

The baseline fertility of the surface and subsurface soils in all treatments increased over the 30-year period for all three crops (Figs. 3, S5–6). The baseline fertility of the subsurface soil was very low for maize and barley but was close to half-saturated for soybean when fertilization was initiated.

The treatment differences in the baseline fertility (i.e., maximum minus minimum) were increased by continuous cropping over the experimental period. The difference in the baseline fertility between the treatments was 0.89 in 1980 and 3.24 in 2008 for maize; for barley, it was 1.20 in 1980 and 2.54 in 2010, and for soybean, it was 0.96 in 1980 and 3.19 in 2009.

We used a 2×2 factorial analysis of variance to compare the baseline fertility of the three crops in response to the different treatments. The interaction effects of fertilizer and FYM were significant for maize ($F = 15.09$, $P<0.001$) and barley ($F = 50.67$, $P<0.001$), but not for soybean ($F = 1.77$, $P = 0.19$). The baseline fertility of soybean differed significantly for fertilizer ($F = 10.26$, $P<0.01$) but not for FYM ($F = 4.00$, $P = 0.05$).

The effects of fertilizer and FYM

The Bayesian estimates of the model for the three crops are shown in Table 1. The traces of the MCMC (Markov Chain Monte Carlo) samples show the well-mixing and convergence to the posterior distributions (Fig. S7). We compared the mean values of the parameters for the three crops. The maize crop had the highest half-saturated fertility level and maximum yield. The soybean crop had the lowest half-saturated fertility level and maximum yield. For maize, the size of the fertilizer contribution was almost equal to the size of the half-saturated fertility, and FYM contribution was 81% smaller than the half-saturated fertility. For barley, the fertilizer contribution was 40% higher than the half-saturated fertility and the FYM contribution was 65% lower than the half-saturated fertility. This indicates that the amount of fertilizer applied in the experiment was 40% more than the fertility level at which the barley yield was half of the maximum yield, but the amount of FYM applied was 65% less than required. For soybean, the values of the fertilizer and FYM

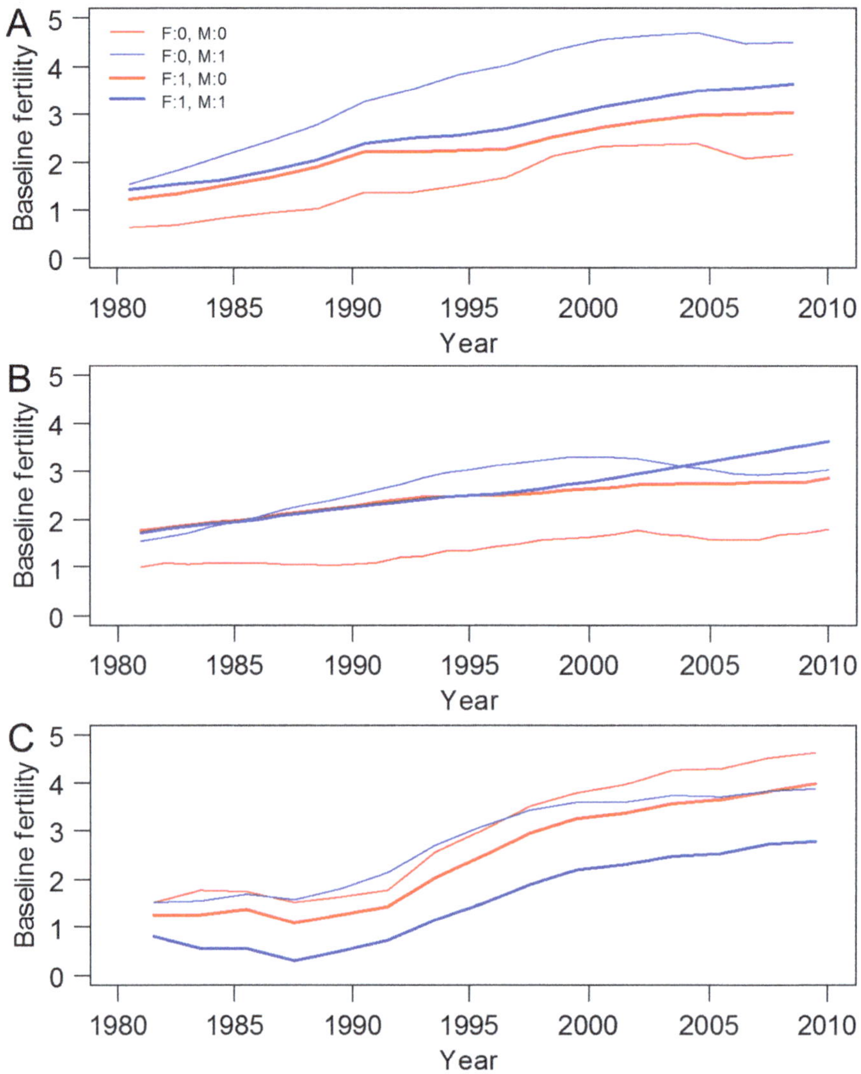

Figure 2. Crop yield-fertility curves for (A) Maize, (B) Barley, and (C) Soybean. The value of 1 for fertility is the fertility without the addition of fertilizer or FYM in the initial year for the subsurface soil. F, level of fertilizer. See Figs. S2A, S3A, and S4A for the band that corresponds to the standard deviation of the curves.

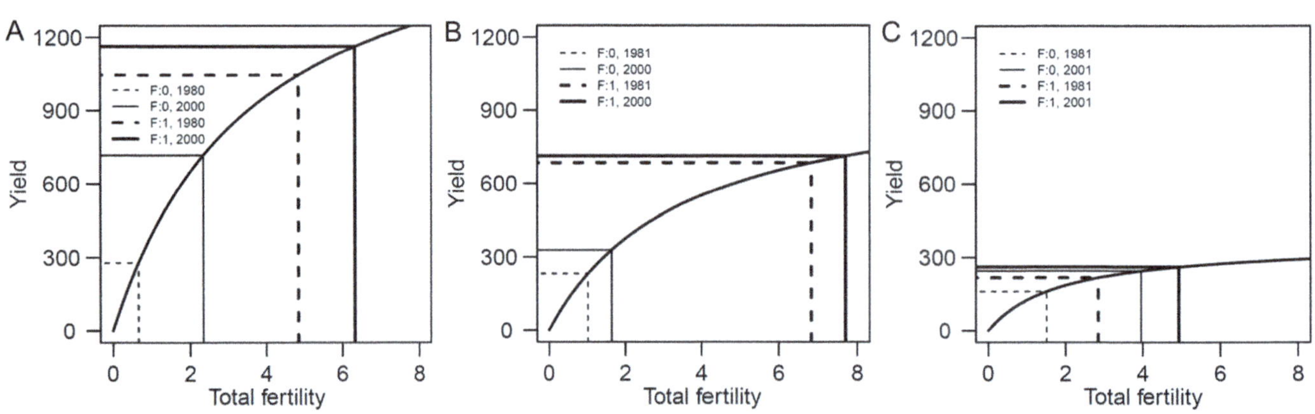

Figure 3. Temporal variations in the baseline fertility for four treatments in the surface soil. (A) Maize; (B) Barley; (C) Soybean; F, level of fertilizer; M, level of farmyard manure. See Fig. S5 for all treatments in surface and subsurface soils.

Table 1. The posterior mean and standard deviation (SD) of the Bayesian estimates.

Crop	V (g m^{-2})		K		A		B	
	Mean	SD	Mean	SD	Mean	SD	Mean	SD
Maize	1833.01	316.56	3.65	1.68	3.61	1.27	0.68	0.53
Barley	1042.87	134.97	3.56	1.18	5.07	1.44	1.25	0.63
Soybean	360.17	65.54	1.87	1.01	1.58	0.58	1.62	0.94

V, maximum yield; K, half-saturated fertility; a, fertilizer contribution; b, FYM contribution.

contributions were similar, i.e., 16 and 13%, respectively, and were lower than the half-saturated fertility.

In contrast with maize and barley, the value of the FYM contribution was higher than the fertilizer contribution for the soybean crop. The application rate of fertilizer and FYM approximated the fertility level at which the soybean yield was half of the maximum yield. The half-saturated fertility level for maize was 3.65, and the half-saturated fertility levels for barley and soybean were 2 and 49% lower, respectively. The fertilizer contribution for barley was 5.07, and the fertilizer contributions for maize and soybean were 29 and 69% lower, respectively. The FYM contribution for soybean was 1.62, and the FYM contributions for maize and barley were 58 and 23% lower, respectively. The model estimates revealed how the different crops used the nutrients from fertilizer and FYM applications to the soil. Maize needed the highest level of fertility to reach half of the maximum yield (i.e., the largest half-saturated fertility) compared with barley and soybean. Fertilizer contributed the maximum amount to the barley yield, and the FYM contributed the maximum amount to the soybean yield.

The total fertility levels among the treatments for both surface and subsurface soils during the 30-year period ranged from 0.54 to 7.22 for maize (Fig. S2A), 0.41 to 9.78 for barley (Fig. S3A), and 0.90 to 5.56 for soybean (Fig. S4A). Barley may be the most efficient crop in terms of fertilizer use among the three crops. The maximum total fertility obtained for soybean was 2.97× the half-saturated fertility, which is larger than the values obtained for maize (1.98) and barley (2.75).

Using the total fertility range, we estimated the predicted range in crop yield using the crop yield-fertility model for each crop. The predicted crop yields ranged from 236.24 to 1217.51 g m^{-2} for maize, 107.70 to 764.56 g m^{-2} for barley, and 117.02 to 269.52 g m^{-2} for soybean. Because of fluctuations in crop yield that could not be described by the model, the predicted yields covered 49, 57, and 42% of the observed yield ranges for maize (35 and 2023 g m^{-2}), barley (14 and 1168 g m^{-2}), and soybean (16 and 381 g m^{-2}), respectively.

The decreasing effect of fertilization over time

In the sterile subsurface soil in the initial year, the application of fertilizer resulted in 153, 189, and 62% increases (the FYM resulted in 42, 65, and 61% increases) in crop yield for maize, barley, and soybean, respectively. In the fertile surface soil in the initial year, the application of fertilizer resulted in 275, 196, and 35% increases (FYM resulted in 149, 98, and 40% increases) in crop yield for maize, barley, and soybean, respectively. After 20 years of farming, the application of fertilizer resulted in 89, 219, and 11% increases (FYM resulted in 74, 150, and 13% increases) in the subsurface soil, and 62, 113, and 7% increases (FYM resulted in 50, 75, and 8% increases) in the surface soil for maize, barley, and soybean, respectively, because of the increased baseline fertility. The greatest yield increases were observed in the initial year in the fertile surface soil with the application of both fertilizer and FYM (compared with the treatment without fertilizer or FYM), and then the yield increment decreased each year (Fig. 4). In the sterile subsurface soil, it took 5–10 years for the yield increase to reach the maximum value.

Predicted crop yields reflected the trend of the observed yields

The estimated fertility levels were used to predict the crop yield using the yield-fertility curves for each crop (Figs. S2A, S3A, and S4A). The trends of the predicted yield were similar to the trends of the observed yield for all three crops. The correlation

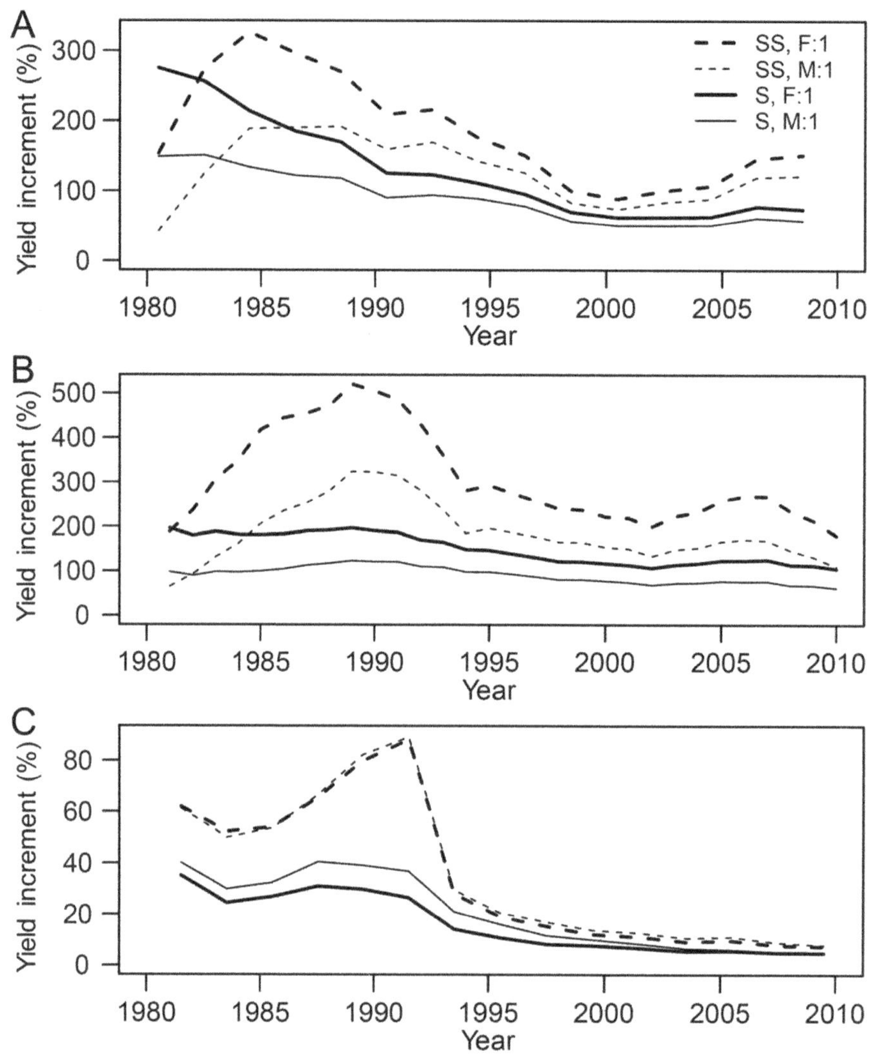

Figure 4. Temporal variations in the yield increment resulting from the application of fertilizer and farmyard manure. (A) Maize; (B) Barley; (C) Soybean; SS, subsurface soil; S, surface soil; F, level of fertilizer; M, level of farmyard manure.

coefficients were 0.758 (Fig. S2B, $P<0.001$) for maize, 0.801 (Fig. S3B, $P<0.001$) for barley, and 0.622 (Fig. S4B, $P<0.001$) for soybean.

Yield fluctuations were observed among the treatments but these fluctuations were not reflected by the predicted crop yields. For maize, the peak yields occurred in 1990 and 2004, and a reduced yield occurred in 1996. For barley, the peak yield for the fertilizer-treated plots occurred in 1993, but the peak yield occurred 1 year later in the plots without fertilizer treatment. The FYM treatments were discontinued in 1991 for the summer harvested crops (maize and soybean). Clear reductions in maize yield in 1991 and soybean yield in 1992 are shown in Fig. 1A and C, respectively, after which the yields for both crops increased until 2008. The estimated fertility levels and predicted yields represented the temporal variations that coincided with treatment changes during the experimental period (Fig. 1). Termination of the phosphate fertilizer application in 2001 was well described by the temporal variations in the baseline fertility of maize and barley. The baseline fertility levels declined, especially in the soil treated with FYM only (Fig. 4). Similar patterns were not observed for the fertilizer-treated soils. A decline in fertility was not observed for soybean. As previously mentioned, soybean yields were not significantly different among treatments.

The effect of climatic factors

Crop yield can be affected by climatic factors [35] such as precipitation [44,45], temperature [46,47] and recent warming [48]. The impact of climate change may damage crop yields on a global scale [48,49] and lead to decreases in crop production [50]. However, weather variation may play a small role in crop yield on a regional scale [45] and the effect may depend on fertilization and soil type [45,50]. Monthly precipitation and average monthly temperature recorded at the experimental site are shown in Fig. S8A and B. We performed a correlation analysis between the residual of the crop yield (i.e., predicted yield minus observed yield) and the climatic variables for the three crops (Table S1 in File S1). The simple residual analysis indicated that temperature had some effect on maize and soybean yields. We then conducted additional analyses that included the effect of temperature in the crop yield-fertility model (see File S1 and Table S2 in File S1). Based on the results of the minor contribution of temperature, we concluded that the annual fluctuations in crop yields were mainly caused by unknown factors other than the climatic factors.

Discussion

The baseline fertility

The baseline fertility of the subsurface soil was very low for maize and barley. This suggests that the fertility in the subsurface soil prior to the application of fertilizer or FYM was not sufficient for maize or barley.

The increasing baseline fertility trend was also observed in soil without additional fertilizer or FYM. The increased fertility may have resulted from the addition of crop residues to the soil and nitrogen fixation by the soybean crop. These fertility increases may indicate soil maturation (i.e., the accumulation of nutrients) in the experimental fields.

The baseline fertility of maize indicated that when FYM was applied alone, soil fertility increased more than when both FYM and fertilizer were applied, for both soil types (Fig. 3A). The application of FYM increased the organic matter content of the soil, which resulted in increased soil fertility. Similar trends were observed for barley and soybean (Fig. 3B and C), supporting the results observed for maize. The results for soybean suggested that the application of fertilizer could decrease the soil fertility when soybean was planted in a soil with a saturated level of fertilization (Fig. 3C).

The effects of fertilizer and FYM

Our result indicates that the amount of fertilizer applied in the experiment reached the fertility level at which the maize yield was half of the maximum yield, but the amount of FYM required to reach this value was much lower. Barley obtained the highest fertility from the soil and fertilization despite the fact that maize produced the greatest crop yield and had the largest half-saturated fertility level. By using nutrients from the soil and fertilizer, soybean was the crop that most often approached the maximum yield.

Our findings indicate that the sterile subsurface soil needed more time to accumulate nutrients than the surface soil. The percentage increments in crop yield were larger in the subsurface soil than in the surface soil, which indicates that fertilizer application effectively improved the crop yield in the barren field. The fertilizer-induced increment in maize and barley crop yield increased after 2001. This is because the phosphate fertilizer application was discontinued in 2001 and the baseline soil fertility subsequently declined. The soybean crop did not show a similar reduction in yield after 2001.

Long-term field crop fertilization experiment and crop models

Long-term experiments examining fertilizer treatments and crop yield have been widely conducted to assess the effects of applying fertilizer [51], manure [52], or a combination of fertilizer and manure [14,53,54]. In addition, experiments have been conducted to determine the impacts of nutrient addition on the fertility status of different soils [55–57]. However, it is important to note that our long-term record of crop yields with various types of treatments has made this the first study to simultaneously examine the temporal variation in baseline fertility and the contribution of nutrient applications.

Conclusions

By allowing the gradual temporal variation in baseline fertility, it was possible to estimate the process of soil fertilization and the short-term effects of different fertilizer treatments. Our results support the proposal that naturally derived nutrients should be used to maintain soil fertility and synthetic fertilizer should be used to maintain productivity. Our results provide a clear description of the relationship between crop yield and soil, which can be easily understood.

Supporting Information

Figure S1 Temporal variations in observed crop yield (g m^{-2}) for six treatments in a field with a fertile surface soil and a field with a barren subsurface soil from 1980 to 2010. (A) Maize; (B) Barley; (C) Soybean; SS, subsurface soil; S, surface soil; F, level of fertilizer; M, level of farmyard manure.

Figure S2 Temporal variations in the observed and predicted yields (g m^{-2}) of maize for six treatments in a field with a fertile surface soil and a field with a barren subsurface soil presented every second year from 1980 to 2008. (A) A crop yield-fertility map. The crop yield-fertility model translates total fertility into predicted yield. The gray band represents the band that corresponds to the standard deviation (±SD) of the crop yield-fertility curve. (B) The relationship between the observed and predicted yields. SS, subsurface soil; S, surface soil; F, level of fertilizer; M, level of farmyard manure; BF, baseline fertility.

Figure S3 As for Figure S2 but for barley every year from 1980 to 2010.

Figure S4 As for Figure S2 but for soybean every 2 years from 1981 to 2009.

Figure S5 Temporal variations in the baseline fertility estimated using the crop yield-fertility model for six treatments in a field with a fertile surface soil and a field with a barren subsurface soil from 1980 to 2010. (A) Maize; (B) Barley; (C) Soybean; SS, subsurface soil; S, surface soil; F, level of fertilizer; M, level of farmyard manure.

Figure S6 Temporal variations in the posterior means and the bands that correspond to the standard deviation (±SD) of the Bayesian estimates of baseline fertility for six treatments in a field with a fertile surface soil and a field with a barren subsurface soil. (A) Maize; (B) Barley; (C) Soybean; SS, subsurface soil; S, surface soil; F, level of fertilizer; M, level of farmyard manure.

Figure S7 Traces of the MCMC samples of V (g m^{-2}), K, a, and b. (A) Maize; (B) Barley; (C) Soybean. The chain length was set to 1,000,000 steps logging every 100th step.

Figure S8 Temporal variation of the climatic variables recorded at the experimental site from January 1980 to December 2011. (A) Monthly precipitation. (B) Average monthly temperature.

File S1 Contribution of the climatic factor. The analyses by including a temperature effect in the crop yield-fertility model.

Data S1 The raw data of the crop yields of maize, barley, and soybean from the long-term fertilizer experiment from 1980 to 2010. As for the detail of the experiment, see text.

Author Contributions

Conceived and designed the experiments: HC HK. Performed the experiments: JY. Analyzed the data: HC. Wrote the paper: HC HK. Confirmed the manuscript: HC JY HK.

References

1. Mäder P, Fliessbach A, Dubois D, Gunst L, Fried P (2002) Soil fertility and biodiversity in organic farming. Science 296: 1964–1967.
2. Tilman D, Cassman KG, Matson PA, Naylor R, Polasky S (2002) Agricultural sustainability and intensive production practices. Nature 418: 671–677.
3. Baligar VC, Fageria NK, He ZL (2001) Nutrient use efficiency in plants. Commun Soil Sci Plant 32: 921–950.
4. Holdren JP, Ehrlich PR (1974) Human population and the global environment: population growth, rising per capita material consumption, and disruptive technologies have made civilization a global ecological force. Am Sci 62: 282–292.
5. Glover JD, Reganold RP, Cox CM (2012) Plant perennials to save Africa's soils. Nature 489: 359–361.
6. Alexandratos N (1999) World food and agriculture: outlook for the medium and longer term. PNAS 96: 5908–5914.
7. Cassman KG (1999) Ecological intensification of cereal production systems: yield potential, soil quality, and precision agriculture. PNAS 96: 5952–5959.
8. Isbella F, Reichb PB, Tilmana D, Hobbiea SE, Polasky S, et al. (2013) Nutrient enrichment, biodiversity loss, and consequent declines in ecosystem productivity. PNAS 110: 11911–11916.
9. Cordell D, Drangert J-O, White S (2009) The story of phosphorus: global food security and food for thought. Glob Environ Change 19: 292–305.
10. Rasmussen PE, Goulding KWT, Brown JR, Grace PR, Janzen H, et al. (1998) Long-term agroecosystem experiments: assessing agricultural sustainability and global change. Science 282: 893–896.
11. Matson PA, Parton WJ, Power AG, Swift MJ (1997) Agricultural intensification and ecosystem properties. Science 277: 504–509.
12. Aref S, Wander MM (1997) Long-term trends of corn yield and soil organic matter in different crop sequences and soil fertility treatments on the morrow plots. Adv Agron 62: 153–197.
13. Mercik S, Németh K (1985) Effects of 60-year N, P, K and Ca fertilization on EUF-nutrient fractions in the soil and on yields of rye and potato crops. Plant Soil 83: 151–159.
14. Hallin S, Jones CM, Schloter M, Philippot L (2009) Relationship between N-cycling communities and ecosystem functioning in a 50-year-old fertilization experiment. ISME J 3: 597–605.
15. Fan T, Xu M, Song S, Zhou G, Ding L (2008) Trends in grain yields and soil organic C in a long-term fertilization experiment in the China Loess Plateau. J Plant Nutr Soil Sci 171: 448–457.
16. Zhang H, Xu M, Zhang F (2009) Long-term effects of manure application on grain yield under different cropping systems and ecological conditions in China. J Agr Sci 147: 31–42.
17. Power AG (2010) Ecosystem services and agriculture: tradeoffs and synergies. Philos Soc Trans R Soc Lond B 365: 2959–2971.
18. Altieri MA, Nicholls CI (2003) Soil fertility management and insect pests: harmonizing soil and plant health in agroecosystems. Soil Till Res 72: 203–211.
19. Clark MS, Horwath WR, Shennan C, Scow KM (1998) Changes in soil chemical properties resulting from organic and low-input farming practices. Agron J 90: 662–671.
20. Fan R, Zhang X, Liang A, Shi X, Chen X, et al. (2012) Tillage and rotation effects on crop yield and profitability on a Black soil in northeast China. Can J Soil Sci 92: 463–470.
21. Yanni YG, Dazzo FB (2010) Enhancement of rice production using endophytic strains of *Rhizobium leguminosarum* bv. trifolii in extensive field inoculation trials within the Egypt Nile delta. Plant Soil 336: 129–142.
22. Neupane RP, Thapa GB (2001) Impact of agroforestry intervention on soil fertility and farm income under the subsistence farming system of the middle hills, Nepal. Agric Ecosyst Environ 84: 157–167.
23. Peoples MB, Herridge DF, Ladha JK (1995) Biological nitrogen fixation: an efficient source of nitrogen for sustainable agricultural production? Plant Soil 174: 3–28.
24. Chapin HS III (1980) The mineral nutrition of wild plants. Ann Rev Ecol Syst 11: 233–60.
25. Altieri MA (2002) Agroecology: the science of natural resource management for poor farmers in marginal environments. Agric Ecosyst Environ 93: 1–24.
26. Watson CA, Atkinson D, Gosling P, Jackson LR, Rayns FW (2002) Managing soil fertility in organic farming systems. Soil Use Manage 18: 239–247.
27. Watson CA, Bengtsson H, Ebbesvik M, Lùes A-K, Myrbeck A (2002) A review of farm-scale nutrient budgets for organic farms as a tool for management of soil fertility. Soil Use Manage 18: 264–273.
28. Khan AUH, Iqbal M, Islam KR (2007) Dairy manure and tillage effects on soil fertility and corn yields. Bioresource Technol 98: 1972–1979.
29. Zhang H, Xu M, Zhang F (2009) Long-term effects of manure application on grain yield under different cropping systems and ecological conditions in China. J Agr Sci 147: 31–42.
30. Yadvinder-Singh Bijay-Singh, Ladha JK, Khind CS, Gupta RK, et al. (2004) Long-term effects of organic inputs on yield and soil fertility in the rice-wheat rotation. Soil Sci Soc Am J 68: 845–853.
31. Li J, Xu M, Qin D, Li D, Yasukazu H, et al. (2005) Effects of chemical fertilizers application combined with manure on ammonia volatilization and rice yield in red paddy soil. Plant Nutrit Fertilizer Sci 11: 51–56.
32. Malhi SS, Lemke R (2007) Tillage, crop residue and N fertilizer effects on crop yield, nutrient uptake, soil quality and nitrous oxide gas emissions in a second 4-yr rotation cycle. Soil Till Res 96: 269–283.
33. Min DH, Islam KR, Vough LR, Weil RR (2003) Dairy manure effects on soil quality properties and carbon sequestration in alfalfa–orchardgrass systems. Commun Soil Sci Plant 34: 781–799.
34. Körschensa M, Albert E, Armbruster M, Barkusky D, Baumecker M, et al. (2013) Effect of mineral and organic fertilization on crop yield, nitrogen uptake, carbon and nitrogen balances, as well as soil organic carbon content and dynamics: results from 20 European long-term field experiments of the twenty-first century. Arch Agron Soil Sci 59: 1017–1040.
35. Lobell DB, Burke MB (2010) On the use of statistical models to predict crop yield responses to climate change. Agric For Meteorol 150: 1443–1452.
36. McCown RL, Hammer GL, Hargreaves JNG, Holzworth DP, Freebairn DM (1996) APSIM: a novel software system for model development, model testing and simulation in agricultural systems research. Agric Syst 50: 255–271.
37. International Benchmark Sites Network for Agrotechnology Transfer (1993) The IBSNAT decade. Honolulu, Hawaii: Department of Agronomy and Soil Science, College of Tropical Agriculture and Human Resources, University of Hawaii.
38. Myers RJK (1984) A simple model for estimating the nitrogen fertilizer requirement of a cereal crop. Fertil Res 5: 95–108.
39. Deng J, Ran J, Wang Z, Fan Z, Wang G (2012) Models and tests of optimal density and maximal yield for crop plants. PNAS 109: 15823–15828.
40. Cong R, Wang X, Xu M, Ogle SM, Parton WJ (2014) Evaluation of the CENTURY model using long-term fertilization trials under corn-wheat cropping systems in the typical croplands of China. PLOS ONE 9: e95142.
41. Denef K, Six J, Merckx R, Paustian K (2004) Carbon sequestration in microaggregates of no-tillage soils with different clay mineralogy. Soil Sci Soc Am J 68: 1935–1944.
42. Michaelis L, Menten ML (1913) Die Kinetik der Invertinwirkung. Biochem Z 49: 333–369.
43. R Development Core Team (2011) R: a language and environment for statistical computing. Vienna, Austria: R Foundation for Statistical Computing.
44. Spiecker H (1995) Growth dynamics in a changing environment – long-term observations. Plant Soil 168–169: 555–561.
45. Drury CF, Tan CS (1994) Long-term (35 years) effects of fertilization, rotation and weather on corn yields. Can J Soil Sci 75: 355–362.
46. Lobell DB, Hammer GL, McLean G, Messina C, Roberts MJ (2013) The critical role of extreme heat for maize production in the United States. Nat Clim Chang 3: 497–501.
47. Welch JR, Vincent JR, Auffhammerc M, Moyae PF, Dobermann A (2010) Rice yields in tropical/subtropical Asia exhibit large but opposing sensitivities to minimum and maximum temperatures. PNAS 107: 14562–14567.
48. Lobell DB, Field CB (2007) Global scale climate–crop yield relationships and the impacts of recent warming. Environ Res Lett 2: 014002.
49. Schlenker W, Roberts MJ (2009) Nonlinear temperature effects indicate severe damages to U.S. crop yields under climate change. PNAS 106: 15594–15598.
50. Rosenzweig C, Parry ML (1994) Potential impact of climate change on world food supply. Nature 367: 133–138.
51. Shen J, Li R, Zhang F, Fan J, Tang C (2004) Crop yields, soil fertility and phosphorus fractions in response to long-term fertilization under the rice monoculture system on a calcareous soil. Field Crops Res 86: 225–238.
52. Canali S, Trinchera A, Intrigliolo F, Pompili L, Nisini L, et al. (2004) Effect of long term addition of composts and poultry manure on soil quality of citrus orchards in Southern Italy. Biol Fertil Soils 40: 206–210.
53. Poulton PR (1996) The Rothamsted long-term experiments: are they still relevant? Can J Soil Sci 76: 559–571.

54. Jenkinson DS (1991) The Rothamsted long-term experiments: are they still of use? Agron J 83: 2–10.
55. Huang S, Peng X, Huang Q, Zhang W (2010) Soil aggregation and organic carbon fractions affected by long-term fertilization in a red soil of subtropical China. Geoderma 154: 364–369.
56. Zhu P, Ren J, Wang L, Zhang X, Yang X (2007) Long-term fertilization impacts on corn yields and soil organic matter on a clay-loam soil in Northeast China. J Plant Nutr Soil Sci 170: 219–223.
57. Vanlauwe B, Diels J, Sanginga N, Merckx R (2005) Long-term integrated soil fertility management in South-western Nigeria: crop performance and impact on the soil fertility status. Plant Soil 273: 337–354.

13

A Metabolic Profiling Strategy for the Dissection of Plant Defense against Fungal Pathogens

Konstantinos A. Aliferis[1], **Denis Faubert**[2], **Suha Jabaji**[1]*

1 Department of Plant Science, Macdonald Campus of McGill University, Sainte-Anne-de-Bellevue, Quebec, Canada, **2** Institut de Recherches Cliniques de Montréal, Montréal, Quebec, Canada

Abstract

Here we present a metabolic profiling strategy employing direct infusion Orbitrap mass spectrometry (MS) and gas chromatography-mass spectrometry (GC/MS) for the monitoring of soybean's (*Glycine max* L.) global metabolism regulation in response to *Rhizoctonia solani* infection in a time-course. Key elements in the approach are the construction of a comprehensive metabolite library for soybean, which accelerates the steps of metabolite identification and biological interpretation of results, and bioinformatics tools for the visualization and analysis of its metabolome. The study of metabolic networks revealed that infection results in the mobilization of carbohydrates, disturbance of the amino acid pool, and activation of isoflavonoid, α-linolenate, and phenylpropanoid biosynthetic pathways of the plant. Components of these pathways include phytoalexins, coumarins, flavonoids, signaling molecules, and hormones, many of which exhibit antioxidant properties and bioactivity helping the plant to counterattack the pathogen's invasion. Unraveling the biochemical mechanism operating during soybean-*Rhizoctonia* interaction, in addition to its significance towards the understanding of the plant's metabolism regulation under biotic stress, provides valuable insights with potential for applications in biotechnology, crop breeding, and agrochemical and food industries.

Editor: Kyoung-Heon Kim, Korea University, Republic of Korea

Funding: The work was financially supported by a Discovery grant from the Natural Sciences and Engineering Research Council of Canada (NSERC, http://www.nserc-crsng.gc.ca/index_eng.asp). The funders had no role in study design, data collection and analysis, decision to publish, or preparation of the manuscript.

Competing Interests: The authors have declared that no competing interests exist.

* Email: suha.jabaji@mcgill.ca

Introduction

Metabolomics is a robust bioanalytical tool for the comprehensive analysis and monitoring of plant metabolome [1–3]. However, its application for monitoring the regulation of the global plant metabolism in response to biotic stresses is still in its infancy, receiving increasing attention [4–6]. This, could provide valuable information for applications in plant biotechnology, biomarker-assisted selection, and agrochemical, food, and pharmaceutical industries [7], and in turn could boost agricultural production. The recent advances in bioanalytical protocols, analyzers, metabolite databases, and bioinformatics software enable the recording of a vast number of chemical features in the analyzed plant samples, whose identification and biological interpretation is challenging. Moreover, there is an increasing demand for standardization of data reporting for large-scale metabolomics [8], which will help researchers to cross-reference results from different studies with profound benefits. Within this context, we have undertaken the task of developing a high-throughput metabolomics/bioinformatics protocol for the robust dissection of plant-fungal pathogen interaction using the pathosystem; soybean [*Glycine max* (L.) Merrill, Leguminosae] and its soil-borne fungal pathogen-*Rhizoctonia solani* Kühn (anastomosis group 4, AG4). For the analysis of soybean's metabolome direct infusion Orbitrap mass spectrometry (DIMS) and gas chromatography-MS (GC/MS) analyzers were employed, which exhibit complimentary capabilities for metabolite detection and identification.

Soybean is a crop grown on almost 6% of arable land [9] and among the most important plant sources of human food, animal feed protein, and cooking oil [10], phytoestrogens [11], and biodiesel [12]. It is the first legume species with a complete sequence [13], and therefore, a key reference for the development of high-throughput plant metabolomics protocols. Various biotic constraints such as, bacteria, fungi, nematodes, and insects threaten its production by directly reducing seed yield and/or quality [14]. Among them is the soil-borne fungal pathogen *R. solani*, which causes Rhizoctonia rot of young seedlings and is characterized by post-emergence symptoms of brown necrotic lesions observed on roots, hypocotyls and stems. In North America, severe yield losses of up to 48% result from stand reduction in newly planted fields and premature death of diseased plants that produce undersize seeds [15]. Few management options exist since no resistant or tolerant varieties are available.

During plant-pathogen interaction a sequence of chemical, molecular, and mechanical events occur, implicated in a distinctive "dialogue" leading to the development of the disease [16]. Apart from recent "omics" studies on the responses of soybean to stresses [17,18], no study exists on the regulation of the global soybean metabolome in response to pathogen infection. In the present study, we explore whole-cellular processes at the metabolome level under a pathogen attack and present a comprehensive overview of soybean's primary and secondary metabolism regulation. This is accomplished by combining multidimensional metabolomics data in an effort to detect specific pathways indispensable for the regulation of metabolism. Key components in the approach are the construction of a standardized soybean library and the use of bioinformatics software for the visualization and analyses of soybean metabolome, which can significantly accelerate the steps of metabolite identification and biological interpretation of results.

Results and Discussion

Monitoring Soybean's Metabolome and its Temporal Perturbation in Response to *Rhizoctonia* Infection

The complexity of plants' metabolome makes their deconvolution challenging, requiring often the utilization of more than one analyzer. For DIMS-based metabolomics, ion suppression can impact the validity of analysis, however, information on its dynamics is yet fragmented. Here, analysis of samples with similar metabolite profiles resulted in consistent ion suppression as revealed by the tight clustering among biological replications performing multivariate analyses (Fig. S1). The latter confirms the potential of DIMS for high-throughput metabolomics applications in line with recent studies [19,20]. The developed protocol (Fig. S2) enabled the in-depth deconvolution of DIMS data, as confirmed by the large number of obtained frames using the software SIEVE (Table S1). On the other hand, GC/MS analyses facilitated the construction of a matrix composed of 135 features, reproducibly detected across treatments. In total, 377 putatively or absolutely identified metabolites were statistically significant different between controls and infected soybean sprouts (Data Set S1). MS spectra of identified metabolites of biological origin from GC/MS analysis, and MS/MS spectra from DIMS Orbitrap analysis provided in the Data Set S2 and Data Sets S3 and S4, respectively. Sets of original GC/MS and DIMS Orbitrap data can be found at the public repository of Metabolights (http://www.ebi.ac.uk/metabolights/) (Accession # MTBLS118 and MTBLS117, respectively). The complexity of undergoing biochemical events during soybean-*Rhizoctonia* interaction (Fig. S3) is indicated by the diversity of chemical groups and biosynthetic pathways involved (Figs. 1 and 2, Fig. S4). Up-regulated metabolites also detected in fungal profiles, which could have leverage on data interpretation, were omitted from analyses.

Principal component analysis (PCA) was performed initially for the whole dataset revealing no outliers (data not shown). In a second step, PLS-DA revealed a strong discrimination between metabolite profiles of control and *Rhizoctonia*-inoculated seedlings (Fig. S1), reflecting their response as early as 24 h post-inoculation. Furthermore, the tight clustering among biological replications confirms the robustness and reproducibility of the experimental protocol.

Metabolomics Data Reporting and Biological Interpretation for the Global Overview and Biological Interpretation of Fluctuations of Soybean Metabolome

Here, a protocol was developed for the robust data reporting for metabolic profiling experiments, which provides an overview of metabolome regulation (Fig. S2). Central to this approach is the construction of a soybean-specific in-house built library, in which entries for metabolites and pathways follow the coding of freely available repositories. The results reveal its applicability for robust interpretation of large metabolomics data sets (Figs. 1, 2, Table 1). Also, the use of standardized chemical classification allowed the robust synopsis of the soybean's response to pathogen attack based on the chemical groups of metabolites (Fig. 1).

The vast majority of identified signatory metabolites of the infection belong to carboxylic and amino acids, carbohydrates, and flavonoids (Fig. 1*A*). Clear patterns of fluctuation in the number of increased and decreased metabolites are observed at both time points (Fig. 1*B*). Based on this pattern, the chemical classes of the identified signatory metabolites are categorized into three groups: i) group 1, for which the number of metabolites that increased following inoculation was substantially higher at 24 h, and higher (or equal) at 48 h post-inoculation, compared to those that were decreased, ii) group 2, for which their number of metabolites that increased following inoculation was substantially lower at 24 h and substantially higher (or equal) at 48 h post-inoculation, compared to those that were decreased, and ii) group 3, for which their number of metabolites that increased following inoculation was higher at 24 h and lower at 48 h post-inoculation, compared to those that were decreased.

Since a given metabolite could be component of different pathways and influence multiple biological processes, the monitoring of plant's metabolic networks becomes challenging. To overcome such challenge and simplify the process, the standardized overview of pathways was achieved by adapting the KEGG classification for soybean pathways (Fig. 2 and Table 1). A general disturbance of the plant's metabolism was observed in response to infection, with the largest number of identified signatory metabolites involved in amino acid and carbohydrate metabolism, and biosynthesis of secondary metabolites. In the vast majority of instances, metabolic pathways of infected seedlings are much more active compared to those in controls (Table 1). Such observation indicates that the plant responds at early stages of infection and invokes a rapid metabolic response involving primary and secondary biosynthetic pathways.

Visualization of the Soybean Metabolome, Metabolic Sub-Networks, and their Perturbation using Bioinformatics Software

The use of Cytoscape and the cellular overview tool of SoyCyc made possible the visualization of the global metabolome of soybean and *in-silico* dissection of its sub-networks. Using Cytoscape's plug-in BisoGenet [21], selected metabolites such as phytoalexins and flavonoids, and biosynthetic precursors, whose relative concentrations significantly increased at 48 h post-inoculation and possible interconnecting paths between them are highlighted (Fig. 3*A*). Also, metabolites (Data Set 4), whose relative concentrations significantly increased at 48 h post-inoculation, are displayed within the metabolome of soybean using the SoyCyc tools (Fig. 3*B*). The visualization of pathogen-mediated perturbations of soybean's metabolome using bioinformatics software facilitates not only the global overview and monitoring of the undergoing biochemical processes, but also, applying a reverse genetics approach, the detection of genes that encode the biosynthesis of detected signatory metabolites (Figs. 4 and 5).

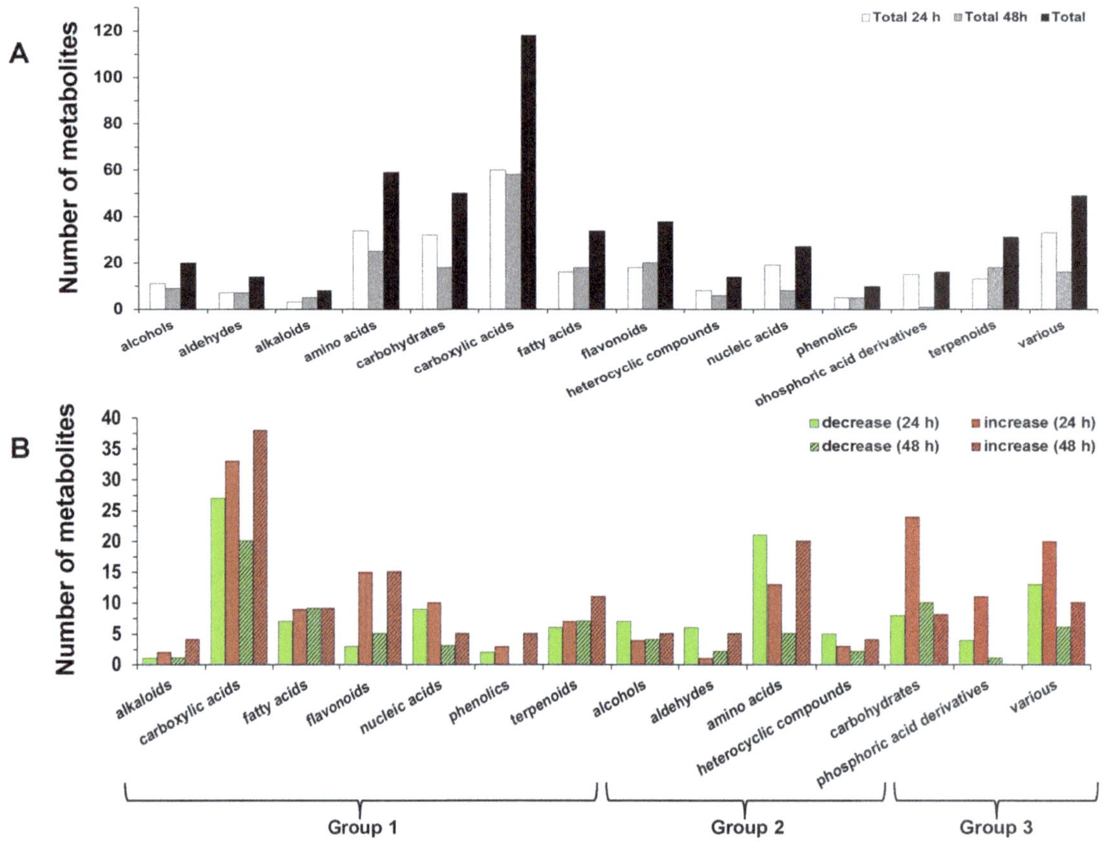

Figure 1. Classification of soybean metabolites into chemical groups in response to *Rhizoctonia* infection. The total number of signatory metabolites at 24 h and 48 h post-inoculation (*A*), and the total number of signatory metabolites classified into increased and decreased in infected seedlings compared to controls at 24 h and 48 h post-inoculation (*B*), are displayed. For the chemical classification of metabolites information was retrieved from the database PubChem.

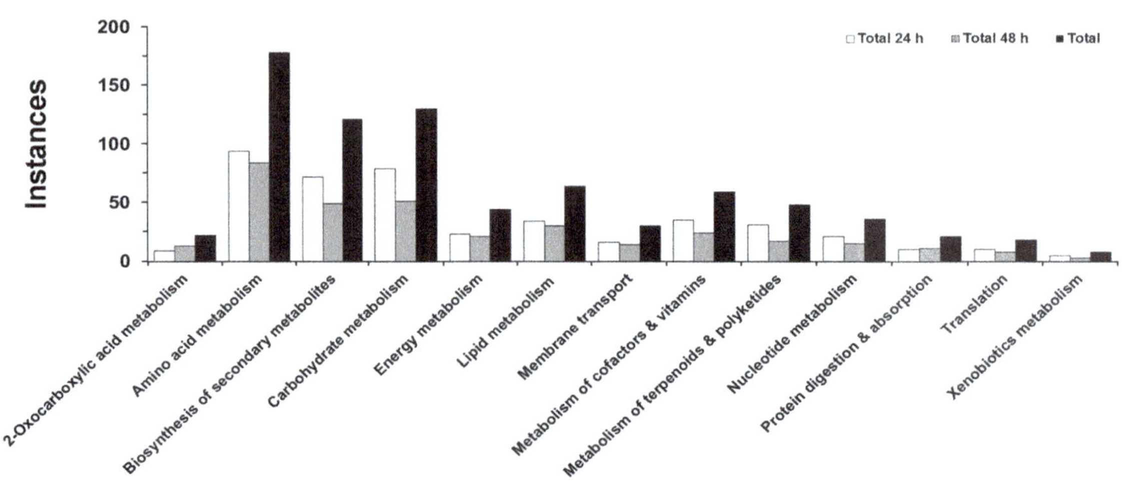

Figure 2. Classification of metabolites signatory of the soybean's response to *Rhizoctonia solani* invasion based on their participation in plant metabolic pathways/functions, measured as instances, since a metabolite can be involved in more than one pathway. For standardization purposes the KEGG coding system for soybean pathways is adapted. The total number of instances at 24 h and 48 h post-inoculation (*A*), and the total number of instances classified into increased and decreased in infected seedlings at 24 h and 48 h post-inoculation (*B*), are displayed.

Table 1. Classification of signatory metabolites of soybean response to *Rhizoctonia solani* invasion at 24 h and 48 h post-inoculation based on their participation in biosynthetic pathways, measured as instances, since a metabolite can be involved in more than one pathway.

Biosynthetic pathway	Decrease (24 h)	Increase (24 h)	Decrease (48 h)	Increase (48 h)	Function
map01210 2-Oxocarboxylic acid metabolism	6	3	1	12	2-Oxocarboxylic acid metabolism
map00280 Valine leucine & isoleucine degradation	1	4	0	4	Amino acid metabolism
map00290 Valine leucine & isoleucine biosynthesis	3	4	1	6	Amino acid metabolism
map00300 Lysine biosynthesis	2	1	2	5	Amino acid metabolism
map00310 Lysine degradation	0	1	0	4	Amino acid metabolism
map00330 Arginine & proline metabolism	1	4	6	3	Amino acid metabolism
map00360 Phenylalanine metabolism	6	0	1	4	Amino acid metabolism
map00380 Tryptophan metabolism	2	1	1	4	Amino acid metabolism
map00400 Phenylalanine tyrosine & tryptophan	6	3	3	5	Amino acid metabolism
map00410 b-Alanine metabolism	3	3	4	1	Amino acid metabolism
map00940 Phenylpropanoid biosynthesis	8	8	0	5	Biosynthesis of secondary metabolites
map00941 Flavonoid biosynthesis	1	5	0	1	Biosynthesis of secondary metabolites
map00943 Isoflavonoid biosynthesis	2	8	4	11	Biosynthesis of secondary metabolites
map00945 Stilbenoid diarylheptanoid & gingerol	1	3	0	1	Biosynthesis of secondary metabolites
map00960 Tropane piperidine & pyridine alkaloid biosynthesis	3	0	1	5	Biosynthesis of secondary metabolites
map00966 Glucosinolate biosynthesis	5	4	0	10	Biosynthesis of secondary metabolites
map00010 Glycolysis-Gluconeogenesis	1	5	1	1	Carbohydrate metabolism
map00030 Pentose phosphate pathway	1	5	0	3	Carbohydrate metabolism
map00040 Pentose & glucuronate interconversions	1	4	1	1	Carbohydrate metabolism
map00052 Galactose metabolism	3	4	0	3	Carbohydrate metabolism
map00053 Ascorbate & aldarate metabolism	2	5	3	3	Carbohydrate metabolism
map00500 Starch & sucrose metabolism	2	4	1	4	Carbohydrate metabolism
map00520 Amino sugar & nucleotide sugar metabolism	3	4	0	3	Carbohydrate metabolism
map04973 Carbohydrate digestion & absorption	1	3	0	3	Carbohydrate metabolism
map00073 Cutin suberine & wax biosynthesis	4	0	5	1	Lipid metabolism
map00592 α-Linolenic acid metabolism	0	4	0	7	Lipid metabolism
map00561 Glycerolipid metabolism	0	3	1	1	Lipid metabolism
map00564 Glycerophospholipid metabolism	1	5	1	0	Lipid metabolism
map00561 Glycerolipid metabolism	0	3	1	1	Lipid metabolism
map02010 ABC transporters	6	10	4	10	Membrane transport
map00750 Vitamin B6 metabolism	3	6	2	1	Metabolism of cofactors & vitamins
map00770 Pantothenate & CoA biosynthesis	1	5	2	3	Metabolism of cofactors & vitamins
map00780 Biotin metabolism	1	0	0	3	Metabolism of cofactors & vitamins
map04977 Vitamin digestion & absorption	1	4	2	1	Metabolism of cofactors & vitamins
map00908 Zeatin biosynthesis	2	7	1	0	Metabolism of terpenoids & polyketides
map00230 Purine metabolism	6	6	4	2	Nucleotide metabolism
map00240 Pyrimidine metabolism	5	5	5	4	Nucleotide metabolism
map04974 Protein digestion & absorption	6	4	2	9	Protein digestion & absorption
map00970 Aminoacyl-tRNA biosynthesis	7	3	2	6	Translation

For standardization purposes the KEGG classification system for soybean pathways was adapted. Pathways with the highest number of identified **signatory metabolites** are displayed.

Key Elements of Etiolated Soybean Seedling's Metabolome and Defense Mechanism Against *Rhizoctonia*

Analysis focused not only on temporal metabolite changes but also on global metabolic network regulation in response to fungal infection (Fig. 6). The most up- and down-regulated soybean pathways are summarized in Table 1.

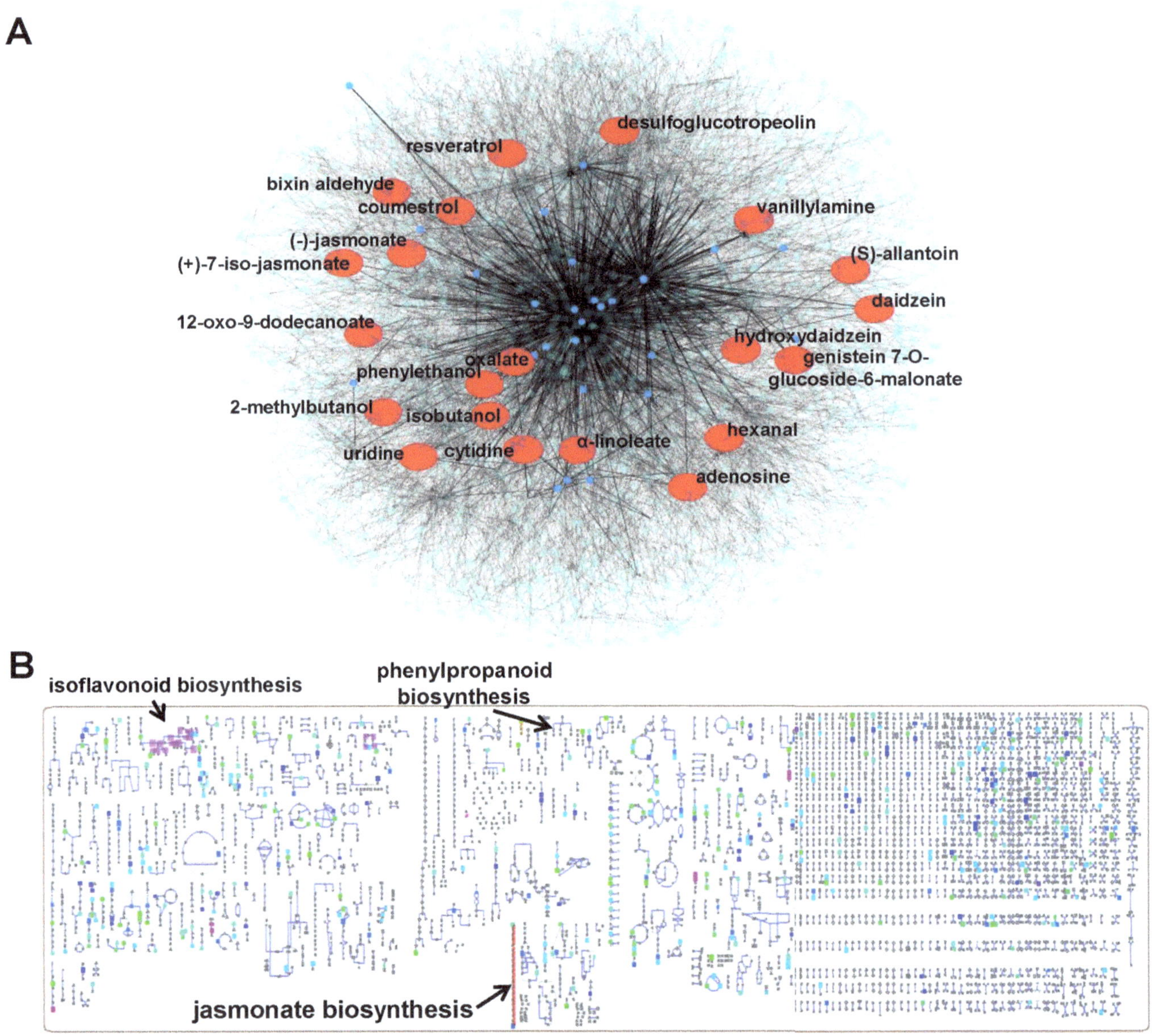

Figure 3. Global soybean metabolome and its perturbation in response to *Rhizoctonia solani* at 48h post-inoculation (*A*) and sub-network generated by connecting possible paths between the displayed signatory metabolites (*B*) using the software Cytoscape and the cellular overview tool of SoyCyc (*C*). Representative metabolites are indicated in red color (*A*) and (*B*), whereas in the panel (*C*), using the Data Set S2, metabolites with KEGG identifiers are highlighted.

Rhizoctonia Infection Substantially Alters the Primary Metabolism of Soybean Seedlings

The pathogen attack substantially altered soybean's primary metabolism, as seen by the fluctuation in carbohydrate, amino and carboxylic acids contents (Fig. 4) and instances for up- and down-regulation of amino acid and carbohydrate metabolism at 24 h and 48 h post-inoculation (Table 1, Data Set S1).

Rhizoctonia triggered soybean's carbohydrate metabolism causing accumulation of hexose sugars, mainly glucose and fructose, which is in line with previous reports on plant-pathogen pathosystems [22]. This is an indication of altered carbohydrate metabolism following stimulation, with a portion of available hexoses possibly diverted towards the biosynthesis of secondary metabolites [23], and another portion used as an energy source to support the induction of stress-related genes [24].

The protein amino acid pool of seedlings was also substantially affected in two distinct phases, one at 24 h and the second at 48 h post-inoculation. In the first phase, the decrease possibly indicates utilization of amino acids for the biosynthesis of PR proteins and metabolites. A similar trend was reported in the potato-*Rhizoctonia* pathosystem [4]. The situation was reversed in the second phase.

Among non-protein amino acids, the rapid increased levels of *β*-alanine at early stages of infection, also seen in soybean parasitized with aphids [25] and potato sprouts infected with *Rhizoctonia* [4], provide evidence that *β*-alanine may be required in the increased biosynthesis of the coenzyme A (CoA), which plays a central role in various metabolite biosyntheses [26] and contributes to soybean resistance to biotic stress. Early and rapid accumulation of the non-protein amino

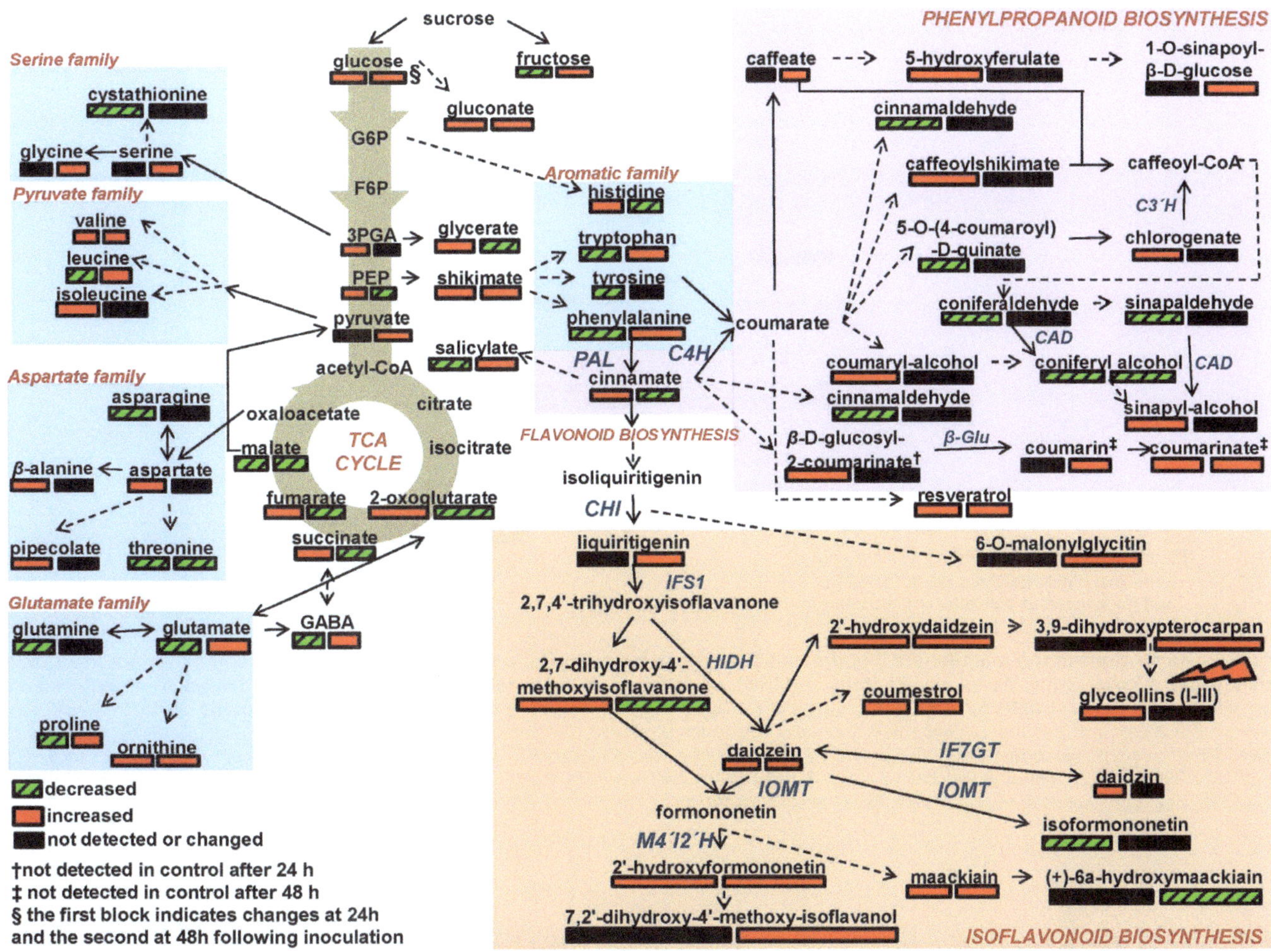

Figure 4. Color-coded fluctuation of the soybean's seedling metabolic network caused by *Rhizoctonia solani* infection at 24 h and 48 h post-inoculation, including portions of the amino acid biosynthesis, and the isoflavonoid and phenylpropanoid biosynthetic pathways, which were detected as important components of its defense mechanism. Genes in the reference pathway of KEGG for soybean that are hyperlinked to gene entries are displayed. Metabolite fluctuations are coded using a color code based on *P*-values ($P<0.05$) performing the Student's t-test and means of scaled and centered PLS regression coefficients (CoeffCS) from five biological replications and a quality control sample per treatment. Dashed lines symbolize multi-step or not fully elucidated reactions and solid lines one-step reactions. [3PGA; 3-phosphoglycerate, isoflavone synthase 1; C3′H; coumaroylquinate (coumaroylshikimate) 3′-monooxygenase, EC: 1.14.13.36, NCBI-GeneID: 100811080, C4H; coumarate 4-hydroxylase, EC: 1.14.13.11, NCBI-GeneID: 100499623, CAD; cinnamyl-alcohol dehydrogenase, EC: 1.1.1.195, NCBI-GeneID: 100777558, CHI; chalconeisomerase (5 isozymes), F6P; fructose-6-phosphate, G6P; glucose-6-phosphate, β-Glu; glucosidase β-glucosidase, EC: 3.2.1.21, NCBI-GeneID: 100779642, HIDH; 2-hydroxyisoflavanone dehydratase, EC: 4.2.1.105, NCBI-GeneID: 547489, IF7GT; isoflavone 7′-O-glucosyltransferase, EC: 2.4.1.170, NCBI-GeneID: 100101902, IFS1; isoflavone synthase 1, EC: 5.4.99.-, NCBI-GeneID: 100037450, IOMT; isoflavone 7′-O-Methyltransferase (14 isozymes), PAL; phenylalanine ammonia-lyase, EC: 4.3.1.24, NCBI-Gene ID: 100787902, PEP; phosphoenolpyruvate].

acid pipecolate, known to act as an osmoprotectant [27] in infected seedlings and a regulator of inducible plant immunity [28] in pathogen-induced systems, is indicative of soybean's response to osmotic stress caused by cell membrane ruptures as early as 24h post-inoculation and priming of the plant to early defense against *Rhizoctonia*.

Rhizoctonia Infection Substantially Alters the Secondary Metabolism of Soybean Seedlings

The secondary metabolism of soybean was also considerably affected by the pathogen's attack as it is indicated by its effect on α-linolenate metabolism, isoflavonoid, glucosinolate, and phenylpropanoid biosynthesis, ABC transporters, and protein digestion and absorption (Fig. 2 and Table 1).

Regulation of the Isoflavonoid Biosynthetic Pathway. A large number of flavonoids, which are components of the isoflavonoid biosynthetic pathway, were clearly responsive to *Rhizoctonia* infection (Fig. 4, Table 1, Data Set S1) designating it as a primary soybean's defense mechanism against *Rhizoctonia*. Among them, the levels of daidzein, 2′-hydroxydaidzein, 2′-hydroxyformononetin, and maackiain, significantly increased in a concerted fashion following infection.

Daidzein is a central component of the isoflavonoid biosynthetic pathway and it is considered precursor of the soybean's phytoalexins glyceollins and coumestrol and other bioactive flavonoids [29,30]. Its increased levels led to increase of the phytoalexins coumestrol, and the pterocarpans glyceollins (I-III) (Fig. 4, Data Set S1). The lack of chromatographic separation performing DIMS does not allow separation between glyceollins I-III, therefore from here and throughout the

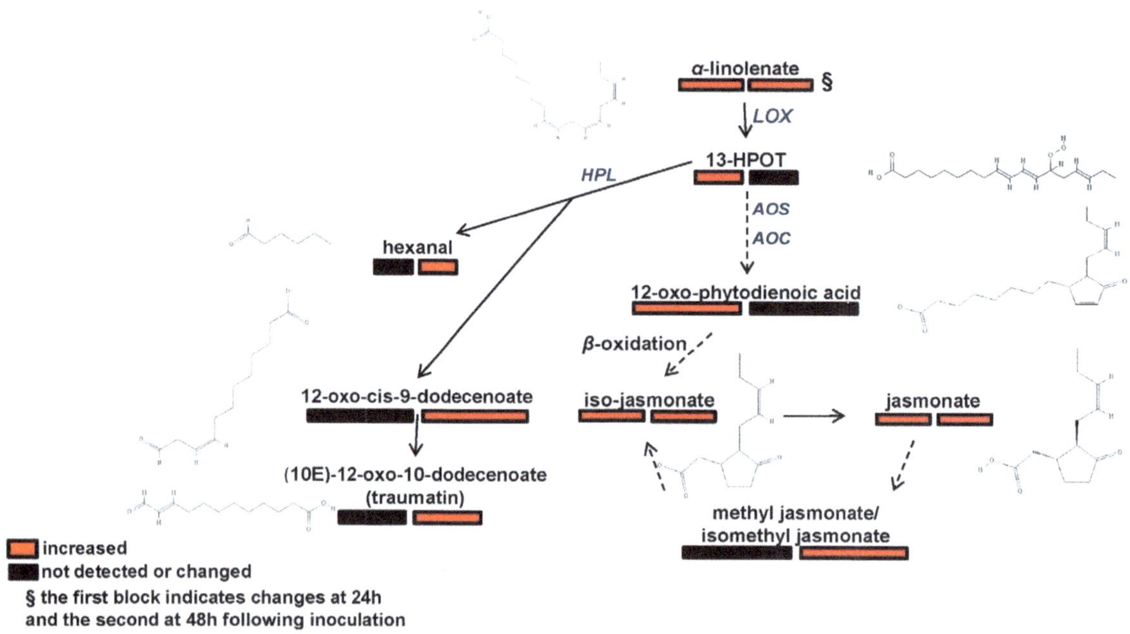

Figure 5. Fluctuations in the α-linolenate metabolism of soybean in response to *Rhizoctonia solani* at 24 h and 48 h post-inoculation coded using a color code based on *P*-values ($P<0.05$) performing the Student's t-test or means of scaled and centered PLS regression coefficients (CoeffCS) from five biological replications and a quality control sample per treatment. Genes in the reference pathway of KEGG for soybean that are hyperlinked to gene entries are displayed. Dashed lines symbolize multi-step or not fully elucidated reactions and solid lines one-step reactions [13-HPOT; (9Z,11E,15Z)-(13S)-hydroperoxyoctadeca-9,11,15-trienoate, AOC; allene oxide cyclase, EC: 5.3.99.6, NCBI-GeneID:100800036, AOS; hydroperoxidedehydratase, EC: 4.2.1.92, NCBI-GeneID: 100037481, HPL; hydroperoxidelyase, EC: 4.1.2.-, NCBI-GeneID: 100784395, LOX; lipoxygenase, EC: 1.13.11.12, NCBI-GeneID: 100785480].

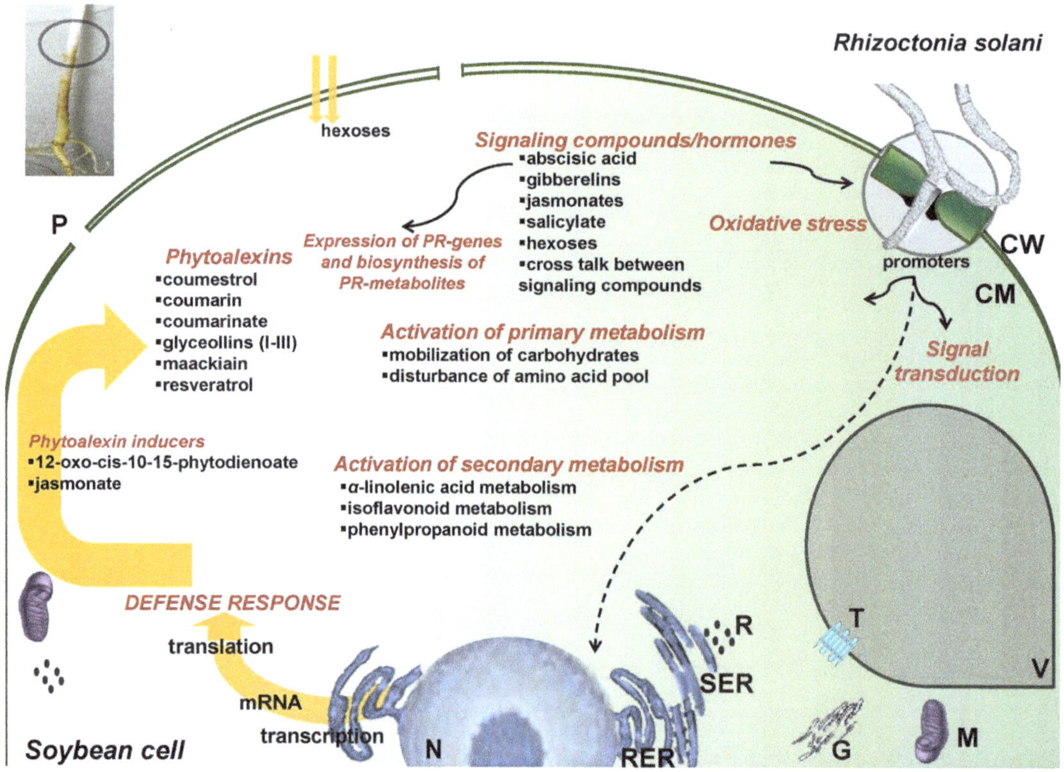

Figure 6. Proposed graphical model for the role of *Rhizoctonia solani* in activation of soybean defense mechanisms. [CM; cell membrane, CW; cell wall, G; Golgi body, M; mitochondrion, N; nucleus, P; plasmalemma, R; ribosomes, RER; rough endoplasmic reticulum, SER; soft endoplasmic reticulum, T; transporters, V; vacuole].

manuscript; the term glyceollins refers to their total relative content. Compared to control, infected seedlings had significantly higher levels of phytoalexins at both time points, with the exception of glyceollins, for which significant difference was observed only at 24 h post-inoculation. In contrast, the less common glyceollin IV was detected at significantly lower levels.

The accumulation of glyceollins and coumestrol in soybean tissues is signal to early defense response to pathogens [31]. Glyceollins are fungitoxic known to restrict pathogen colonization [32,33]. Also, since they have analogous structures to isoflavonoids, it is likely that their antioxidant activities are similar, thus playing a protective role against reactive oxygen species (ROS) during pathogen attack and conferring partial resistance. Interestingly, glyceollins' levels in *Rhizoctonia*-infected seedlings were not significantly different 48h following infection compared to controls. This, combined with the inability to detect 7-hydroxyglyceollin, a product of glyceollin degradation by pathogens [33], indicates their reduced biosythesis. As with other soybean pathogens [34], challenging soybean with *Rhizoctonia* substantially induces coumestrol biosynthesis whose protective role is attributed to ROS scavenging. Apart from their profound role in soybean physiology, both glyceollins and coumestrol exhibit anti-estrogen and antioxidant activities [35]. Maackiain is another isoflavonoid phytoalexin playing a key role in disease resistance of leguminous crops [36] providing that the attacking pathogen is not able to overcome this defense reaction by degrading it [37]. Here no products of maackiain metabolism were detected.

Regulation of the Phenylpropanoid Biosynthetic Pathway. The general disturbance of soybean metabolism is also evident by the fluctuation of metabolites of the phenylpropanoid pathway (Fig. 4, Table 1, and Data Set S1), which plays an important role in plant's physiology, including defense responses, and several of its steps are CyP450-depended [38,36]. Up-regulation of its branch that leads to the biosynthesis of coumarinate from coumarin was among the most important discoveries. Interestingly, the identified aldehydes of the pathway displayed the same fluctuation pattern, increasing at 24 h post-inoculation, whereas none was detected 48 h later.

Accumulation of several of these metabolites as early as 24 h post-inoculation indicates their involvement in early defense response of soybean. Intriguingly, metabolites of the phenylpropanoid pathway branch that leads to coumarinate were not detected in controls at 48 h after treatments, possibly indicating *de novo* biosynthesis. Coumarins contribute essentially to the persistence of plants being involved in processes such as defense against phytopathogens, response to abiotic stresses, regulation of oxidative stress, and probably hormonal regulation [39]. Caffeate and chlorogenate are phenolics of the pathway with important roles in plant defense mechanism [38], which play a role in the early response to *Rhizoctonia*, plausibly through ROS scavenging [40].

Regulation of the α-Linolenic acid Metabolism. The precursor of jasmonate, α-linolenate, was detected as another key component of soybean's defense (Fig. 5, Table 1, Data Set S1). The activation of the pathway requires increased levels of α-linolenate, which is likely released by complex membrane lipids [41]. Although no discrimination can be achieved between iso-jasmonate and jasmonate (JA), and methyl-jasmonate (MeJA) and isomethyl-jasmonate by DIMS, it is evident that their biosynthesis is an important defense response of soybean. A distinct pattern is observed between the two displayed branches of the pathway. The one that leads to the biosynthesis of JA is activated at 24 h post-inoculation and only JA and MeJA derivatives are detected as signatory metabolites at 48 h. The second that leads to the biosynthesis of traumatin, exhibits a late response to *Rhizoctonia*, since signatory metabolites were detected only at 48 h.

The recognition of elicitor molecules by soybean cell receptors triggers a signal-transduction cascade leading to, among others, the production of signaling compounds such as, JA and its derivatives, 12-oxo-cis-10-15-phytodienoate (OPDA), and traumatin through α-linolenate metabolism. Accumulation of jasmonates triggers the biosynthesis of phytoalexins, mediates systemic wound responses, and affects the expression of PR genes synergistically or by antagonizing the action of other plant hormones. JA biosynthesis is regulated by a positive feedback and catalyzed by fatty acid β-oxidation enzymes and 4-Cl-like CoA ligases [42]. The biosynthetic intermediate OPDA exhibits stronger phytoalexin-inducing activity compared to JA and MeJA [43]. In this study, elevated levels of OPDA at 24 h post-inoculation indicate its role as an early signal of phytoalexin, JA, and MeJA biosynthesis. Moreover, the accumulation of traumatin in soybeans is possibly associated with lesion healing process of infected seedlings [44].

Effect on Various. Metabolites with Important Role in Plant Physiology and Defense. Soybean defense against *Rhizoctonia* is under the regulation of signaling/hormone molecules such as, salicylate (SA), abscisic acid (ABA), JAs, and GAs (Fig. 4, Data Set S1), which are involved in a cross talk [45]. SA is involved in functions related to the systemic acquired resistance (SAR), regulation of defense genes, and programmed cell death. Interestingly, its levels in infected seedlings were lower at 24 h but higher at 48 h post-inoculation, compared to controls. Low levels of SA have been linked to cell death of soybean cell suspensions [45]. The complexity of soybean defense regulation by signaling metabolites/hormones is further confirmed by the increased levels of GAs, which regulate plant growth through the regulation of DELLA family proteins [46,47], that in turn, affect the balance between SA and JA [48]. Mounting evidence suggests that ABA plays an indecisive role in plant defense responses to pathogens, acting as both a positive and negative regulator of resistance [41,45]. Here, the increased ABA levels in combination with decreased levels of SA at 48 h, possibly indicates the suppression of SA biosynthesis. Taken together, our results are in line with findings on the involvement of ABA in pathogenesis in the soybean-*Phytophthora sojae* pathosystem [49].

The detection of nucleic acids and components of nucleotide metabolism as signatory metabolites is indicative of the effect of the pathogen on nucleic acid synthesis, energy equilibrium, and biosynthesis of primary and secondary metabolites [50]. As a result of extensive metabolic compartmentation [51], plants require efficient mechanism for resource allocation, which involves passage through various membranes facilitated by ABC transporters [52,53]. Plants' ABC transporters are not only involved in the transport of plant metabolites and xenobiotics but also contribute to plant–pathogen interactions and the modulation of ion channels [52,53]. The high number of instances for ABC transporters observed in infected seedlings confirms the intensified operation of the mechanism during soybean-Rhizoctonia interaction.

Evidence of reactive oxygen species (ROS) involvement. The biosynthesis of ROS scavengers was also highlighted as an essential component of pathogenesis, confirming ROS involvement during a plant-pathogen interaction, which is not yet fully elucidated [54]. The recognition of the pathogen by plants triggers ROS production, which also plays an important role in virulence and development of pathogens [54]. In addition to the abovementioned metabolites exhibiting antioxidant activities, other metabolites with well-established scavenging properties synthesized during interactions in pathosystems such as, resveratrol [55], ascorbate [56], ornithine [57], and members of the B6 complex [58] were involved in the soybean-*Rhizoctonia* interaction. Furthermore, this is mirrored by increases in glutamate, proline and γ-aminobutyric acid (GABA) that can, directly or indirectly, exhibit ROS scavenging activity [59,60].

Conclusion

Although the biochemical basis of pathogenesis is extensively studied in plant-pathogen pathosystems, the recent developments in metabolomics now facilitate the comprehensive monitoring of the plant's metabolome and metabolism regulation in response to stimuli, and their study as a whole rather as fragmented pathways. Additionally, although the general knowledge exists for a given plant-pathogen pathosystem, factors related to their genotypic composition could alter the final outcome. Based on its potential, the developed approach could provide new insights and could fill gaps in the knowledge related to the metabolic responses of plants during pathogen invasion.

Additionally, metabolomics data reporting and biological interpretation are facing challenges arising from the inconsistency of chemical names and lack of standardized nomenclature for metabolic pathways across public repositories. The metabolomics strategy reported here enables the robust reporting and biological interpretation of data from untargeted metabolomics experiment by providing standardized overview of the soybean's metabolism regulation during fungal infection, and could be adapted in similar studies. In addition to its significance for plant pathology, results provide information that could be exploited in biomarker-assisted crop breeding, agrochemical, food, and pharmaceutical industries.

Materials and Methods

Chemicals and Reagents

Chemicals and reagents of the highest grade commercially available were used. Methoxylamine hydrochloride, *N*-methyl-*N*-(trimethyl-silyl)trifluoroacetamide (MSTFA), and pyridine, for GC/MS sample preparation, and analytical standards, were purchased from Sigma-Aldrich Canada Ltd. (Oakville, ON, Canada). Ethyl acetate, methanol, formic acid, ammonium acetate (Optima grade), and water (HPLC grade), were purchased from Fisher Scientific Company (Ottawa, ON, Canada). For the calibration of analyzers, the ProteoMass ESI Calibration Kit MSCAL5 and MSCAL6 (Sigma-Aldrich), and the DRO/GRO Range Calibration Standard (Restek Corporation, Bellefonte, PA, USA), which is a mixture of 12 alkanes in methylene chloride, were used.

Biological Material and Inoculation of Etiolated Soybean Seedlings

Seeds of the soybean cultivar AC Proteina (Semican, Plessisville, QC, Canada), which is susceptible to *R. solani*, *were* sterilized by immersion in ethanol (70%, v/v) for 1 min, followed by rinsing three times in sterile distilled water. For pre-germination, seeds were placed on sterilized moistened filter paper (Whatman No1, 11 cm diameter) in Petri plates, and kept at 24°C in the dark. Uniformly pre-germinated seeds were planted in sterile mix of turf:perlite (1:1, v/v) in plastic trays in the dark in a growth chamber at 22±2°C and relative humidity of 85%. Sterile water was added every second day. Seedlings were etiolated in order to provide longer lengths of hypocotyls for inoculation. They were grown for one week to approximately 15 cm height before they were gently removed and inoculated.

For the production of the inoculum, starter cultures of the highly pathogenic *R. solani* AG4 isolate A76, provided by M. Cubeta (N. Carolina University, USA), were grown on potato dextrose agar (PDA) for three days. Uniform and healthy one-week-old seedlings were transferred to sterile Pyrex glass trays (38×27×6 cm) on filter paper (Whatman No1), arranged horizontally, and their roots were covered with sterile moistened cotton. For the analysis of hyphae, the fungus was grown on cellophane membrane.

Inoculation was performed by sandwiching hypocotyls between two PDA strips (2 cm×8 cm) of 3-day-old *Rhizoctonia* cultures or clean strips (control) (Fig. S3) in sterile environment under photosynthetically inactive black light (365 nm) on designated marked sites of the hypocotyls, 3 cm away from the roots. Trays were wrapped with plastic membrane, and returned to growth chambers for incubation at 24°C, in the dark. The development of necrotic lesions was stereoscopically observed and plants were harvested 24 and 48 h post-inoculation, and prepared for metabolite extraction. These time points were specifically chosen in order to capture the onset of infection and the development of infection cushions. Each treatment (infected or control) was consisted of five biological replications, each obtained by pooling the hypocotyls of four seedlings.

Sampling and Metabolite Extraction

Samples (100 mg of fresh weight, fw) were harvested from the edges of necrotic lesions at 24 h and 48 h post-inoculation and corresponding portions of control seedlings (Fig. S3). Additionally, mycelia (50 mg, fw) from 3-day-old *R. solani* cultures were harvested. Samples were pulverized to a fine powder in a mortar with a pestle under liquid nitrogen and kept at −80°C in glass autosampler vials (2 ml) until further processing. Metabolite extraction was performed as previously described [4]. Briefly, 1 ml of a mixture of methanol:ethyl acetate (50:50, v/v) was added to the pulverized samples followed by sonication in an ultrasonic bath Branson 5510 (Branson, Connecticut, USA) for 25 min. Samples were then transferred to an orbital shaker (Infors AG, Bottmingen, Switzerland) for extraction under continuous agitation (150 rpm) for 2 h at 24°C. Extracts were filtered through 0.2 μm filters (Millex-FG, Millipore, MA, USA) to remove debris and the volume of samples was adjusted to 1.0 ml. Consecutively, extracts were divided into two equal portions (0.5 ml) in glass autosampler vials for GC/MS and direct infusion Orbitrap MS analysis (DIMS). The latter were sub-divided into equal portions of 0.25 ml for analysis in positive (ESI^+) and negative (ESI^-) electrospray modes, respectively. Finally, extracts were dried using a Labconco CentriVap refrigerated vacuum concentrator (Labconco, Kansas City, MO, USA) equipped with a cold trap.

Chemical Analyses and Data Pre-processing

Direct Infusion Orbitrap MS (DIMS) Analysis. For analyses in ESI^+, 160 μL of a mixture of methanol:formic acid (0.1% v/v) (50–50, v/v) were added to the dried samples, whereas for analyses in ESI^-, 160 μL of methanol:ammonium acetate (2.5 mM) were added. Finally, extracts were transferred into microinserters, which were consecutively placed in glass autosampler vials. An LTQ-Orbitrap MS Classic (Thermo Scientific, San Jose, CA, USA) mass spectrometer was employed and experimental events were controlled by the software Xcalibur v.2.2 (Thermo Scientific). The analyzer was equipped with a quadrupole linear ion trap, an Orbitrap electrostatic Fourier transform mass spectrometer (FTMS) with a heated electrospray ionization probe (HESI-II, Thermo Scientific), and an Accela pump (Thermo Scientific). The source voltage was set to 3.2 kV (ESI^+) and 4.0 kV (ESI^-) and the capillary voltage to 5.0 V (ESI^+) and −35 V (ESI^-). The capillary temperature was set to 275°C. Sheath gas flow was set to 10 (ESI^+), and 20 (ESI^-) whereas no auxiliary and sweep gases were used. Samples (10 μL) were injected manually performing direct infusion at a flow rate of 10 μl min^{-1} using a 100 μl syringe (Hamilton, Reno, NV, USA). DIMS analysis was performed at a mass resolution of 60,000 at m/z 400 and spectra were acquired over the range of 50–1,800 Da for 3.5 min. For selected samples MS/MS analyses were performed with the normalized collision energy set at 35 eV, the activation q set to 0.25 and the activation time to 30 ms. Target ions already selected for MS/MS were dynamically excluded for 15 s. AGC target value was set to $5E^5$ for the Orbitrap and to $1E^4$ for the MS/MS analysis in the linear ion trap.

Acquired data (*.raw format) were processed using the software SIEVE v1.3 (Thermo Scientific) installed in a 12-core 64-bit PC workstation (Windows 7 Ultimate SP1, Intel Core i7 CPU, 24 GB RAM). Although the software is designed for the analysis of LC/MS data, with appropriate changes in the settings, we were able to analyse data obtained by DIMS. The acquired mass spectra between 0.6 and 1.2 min were processed (Fig. S5). Obtained frames [rectangular regions of *m/z* vs retention time (RT)] were filtered based on their coefficient of variation distribution (CV) in the reconstructed ion chromatograms (RIC). Frame intensities were normalized to the total ion chromatogram (TIC) and frames with CV>0.5 were excluded from further analysis. The obtained data matrices were exported to Simca P+, after the removal of peaks corresponding to isotopes, for multivariate analyses.

Gas Chromatography-Mass Spectrometry (GC/EI/MS) Analysis. Derivatization of samples for GC/MS was performed as previously described [4]. Briefly, methoxymation was performed by adding to the dried samples 80 μL of methoxylamine hydrochloride (20 mg/ml in pyridine) followed by incubation at 30°C for 120 min. In a second step, silylation was performed by adding 80 μL of MSTFA (37°C for 90 min). Derivatized samples were then added to microinserters (150 μL, Fisher Scientific Company), which were placed in autosampler vials (2 ml). For analyses, an Agilent 7890A GC platform (Agilent Technologies Inc. Santa Clara, CA, USA) coupled with a 5975C series mass selective detector (MSD) and a 7693A series autosampler was employed. Electron impact (EI) at 70 eV was used. Full scan mass spectra were acquired at the mass range of 50 to 800 Da at 1 scan s^{-1} rate with a 10.0-min solvent delay. The temperature for the ion source was set to 150°C, for the transfer line to 230°C, and for the injector to 230°C. Samples (1 μL) were injected using a split ratio of 10:1 into a HP-5MS ultra inert (UI) capillary column (30 m×250 μm I.D, 0.25 μm film thickness; Agilent Technologies Inc.). Helium was used as the carrier gas at a constant flow rate of 1 ml min^{-1}. The temperature of the oven was initially 70°C stable for 5 min, followed by a 5°C min^{-1} increase to 310°C and finally stable for 1 min.

For data pre-processing (peak deconvolution, metabolite identification, and peak integration) the Agilent MSD Chemstation E.02.00.493 was used. Automatic integration was performed after optimization of the ChemStation Integrator parameters based on the total ion chromatogram (TIC). Processed chromatograms were exported to Microsoft excel for the creation of data matrices. Following retention time (RT) alignment, chromatograms were normalized to the TIC and the constructed matrix was exported to Simca P+ v.12.0.1 for the discovery of signatory metabolites and trends. To ensure the quality and validity of analyses, standard operating procedures (SOP) and quality control (QC) measures were applied throughout analyses.

Metabolite Identification and Construction of Standardized Metabolite Library for Soybean. In order to develop a high-throughput metabolomics protocol, identification was performed following a biologically-driven approach. Central to this approach is the species-specific in-house built library for soybean composed of 1339 metabolites, which includes information on metabolite monoisotopic mass, molecular formula, biosynthetic pathways, and chemical classification. For the construction of the library, information was retrieved from the publicly available databases of SoyCyc (http://www.soybase.org:8082/SOY/class-instances?object=Compounds), the Kyoto Encyclopedia of Genes and Genomes (KEGG, http://www.genome.jp/kegg/), PubChem (http://pubchem.ncbi.nlm.nih.gov/), the European Bioinformatics Institute (EMBL-EBI, http://www.ebi.ac.uk/), KNApSAcK (http://kanaya.naist.jp/KNApSAcK/), and the literature. For standardization purposes, the KEGG coding system for pathways was adapted, which enables the comprehensive grouping of pathways and is compatible with bioinformatics tools such as the pathway tools of BioCyc (http://bioinformatics.ai.sri.com/ptools/). For the chemical classification of metabolites, information was retrieved from the database PubChem (http://pubchem.ncbi.nlm.nih.gov/).

For Orbitrap MS analyses, using the software SIEVE, searches for the constructed frames were performed against the constructed library with a mass error (Δppm) <2 ppm. Putative identification of metabolites was based on mass accuracy and, where available, MS/MS fragmentation and isotope patterns, following heuristic rules [61]. MS/MS spectra deconvolution was performed using the software MetWorks 1.3 SP3 and Mass Frontier 7.0 SR1 (Thermo Scientific).

For GC/MS, the guidelines of the metabolomics standards initiative (MSI) [8] were followed. Mass spectra searches were performed against the library of the National Institute of Standards and Technology (Gaithersburg, MD, USA) NIST 08 and for selected signatory metabolites, absolute identification was based on fragmentation patterns and retention times (RT) analyzing authentic chemical standards on the same GC/MS system with the same analytical method [8]. In cases in which a very good fit (e.g,>90%) could be achieved, tentative identification was performed. Metabolites of non-biological origin were excluded from further analysis.

Statistical Analyses. For Orbitrap MS data, one-way ANOVA was executed using the integrated statistical function of SIEVE v.1.3 performing the Student's t-test ($P<0.05$). GC/MS data matrices were subjected to multivariate analyses using the software SIMCA-P+ v.12.0 (Umetrics, MKS Instruments Inc. Andover, MA, USA) as previously described (4) and one-way ANOVA using the software JMP 11.0 (SAS Institute Inc, NC, USA) performing the Student's t-test ($P<0.05$). The discovery of signatory metabolites performing multivariate analysis was based on partial least square-discriminant analysis (PLS-DA) regression coefficients ($P<0.05$). Standard errors were calculated using Jack-knifing, which is based on the variability in the model parameters encountered in the different cross-validation cycles with 95% confidence interval [62]. The performance of the models was assessed by the cumulative fraction of the total variation of the *X*'s that could be predicted by the extracted components

[$Q^2_{(cum)}$] and the fraction of the sum of squares of all X's (R^2X) and Y's (R^2Y) explained by the current component.

Analysis of metabolic networks The global soybean metabolome was visualized using the software Cytoscape v.2.8.2 (http://www.cytoscape.org/) [63] and the SoyCyc database as previously described [4]. By integrating information from analyses, enzymes catalyzing the biosynthesis of metabolites that are signatory of the soybean's response to *Rhizoctonia* infection and corresponding encoding genes were inferred and highlighted. Additionally, the soybean's metabolome fluctuation in response to pathogen attack was visualized using the cellular overview tool of SoyCyc (http://www.soybase.org:8082/overviewsWeb/celOv.shtml) and information from the in-house built library. The Data Set S5 was used.

Quality Control of Metabolomics Analyses. For each treatment, a quality control (QC) sample was obtained by pooling 25 μL from each of the five biological replicates (total 125 μL), and analyzed. In addition to biological replications and QC samples, technical one replication for randomly selected samples was performed per treatment. Following each sample's analysis, a blank (solvent system only) was injected to ensure that no carry over is observed between samples. Additionally, blank samples were used for the detection of possible contamination during the different experimental steps, such as material and reagent impurities and column stationary phase bleeding (for GC/MS). For this purpose, 1.5 mL of the extraction solvent mixture was processed alongside the experimental samples and was subjected to identical handling. Results were taken into account for the detection of metabolites not related to the biological material that was analyzed, which were subsequently excluded from analyses. Additionally, the metabolic composition of fungal samples was taken into consideration, in order to remove fungal-derived metabolites that could influence the biological interpretation of results. In order to optimize the performance of analyzers, calibration was performed at the beginning of analyses following the recommended by the manufacturers' procedures and using calibration solutions.

For GC/MS, tuning of the MS detector was performed automatically using the AutoTune function of the Agilent MSD Chemstation. The DRO/GRO Range Calibration Standard was injected every six samples in order to monitor the performance of the system and additionally, samples were spiked by adding 20 μL of ribitol solution (0.2 mg/mL) to methanol-water (50–50, v/v), which served as the internal standard.

For Orbitrap MS, the ProteoMass ESI Calibration Kit MSCAL5 (Sigma-Aldrich), which covers the range between 138 and 1822 Da, was used for ESI^+ analyses. For ESI^-, calibration was performed using the ProteoMass ESI Calibration Kit MSCAL6 (Sigma-Aldrich), which covers the range between 265 to 1880 Da. Internal standards were not used but instead, mass errors were estimated based on accurate masses of common identified metabolites such as amino (ESI^+) and lipid (ESI^-) acids.

Supporting Information

Figure S1 Partial least squares-discriminant analyses (PLS-DA) PC1/PC2 score plots of direct infusion Orbitrap MS and GC/MS metabolite profiles of control (■) and *Rhizoctonia solani*-infected (•) soybean seedlings, at 24 h and 48 h post-inoculation. The ellipse represents the Hotelling T^2 with 95% confidence interval. Five (5) biological replications were used per treatment and one quality control sample (QC) [$Q^2_{(cum)}$; cumulative fraction of the total variation of the X's that can be predicted by the extracted components, R^2X and R^2Y; the fraction of the sum of squares of all X's and Y's explained by the current component, respectively, *PCs*; principal components].

Figure S2 Pipeline for the dissection of plant-pathogen pathosystems performing high-throughput metabolomics using as model the pathosystem soybean-*Rhizoctonia solani*.

Figure S3 Experimental set up for the infection of soybean seedlings with *Rhizoctonia solani* AG4. For the inoculation, three-day old cultures of *Rhizoctonia* grown on PDA were used (*A*). The basal portion of seedlings was sandwiched between two strips of PDA (control) (*B*) or *Rhizoctonia* culture (*C*) and the roots were sandwiched between two layers of sterile moistened cotton. Seedlings were kept in sterilized Pyrex trays on filter paper and sealed with cellophane membrane. Samples were taken at 24 h (*D*) and 48 h (*E*) post-inoculation from the edges of the necrotic lesions and corresponding segments of control seedlings.

Figure S4 Chemical structures of representative soybean's signatory metabolites in response to *Rhizoctonia solani* at 24 h and/or 48 h post-inoculation.

Figure S5 Representative total ion chromatograms (TIC) of control and *Rhizoctonia solani*-infected soybean seedlings at 24 h and 48 h post-inoculation performing direct infusion Orbitrap MS analysis. Mass spectra that correspond to the area between the dashed lines were used in metabolomics analyses.

Table S1 Number of frames of direct infusion Orbitrap MS using the software SIEVE v.1.3.

Data Set S1 Identified metabolites signatory of the soybean defense against *Rhizoctonia solani* based on LTQ Orbitrap and/or GC/MS analyses. With red color metabolites whose relative concentration is increased in infected seedlings and with green color, those whose relative concentration is decreased compared to controls.

Data Set S2 GC/MS spectra library. GC/MS mass spectra library (*.MSL) of identified metabolites of biological origin. The library can be visualized using the freely available software AMDIS (http://chemdata.nist.gov/mass-spc/amdis/downloads/). Information on the retention times and analytical conditions are provided. The library should be downloaded and then viewed by selecting "Build one library" under the "Library" tab of the software.

Data Set S3 DIMS Orbitrap MS/MS spectra. Recorded MS/MS spectra of representative chromatogram from Soybean seedlings infected by *Rhizoctonia solani* at 48 h post-infection in the positive electrospray mode between 0.6 and 1.2 min.

Data Set S4 DIMS Orbitrap MS/MS spectra. Recorded MS/MS spectra of representative chromatogram from Soybean seedlings infected by *Rhizoctonia solani* at 48 h post-infection in the negative electrospray mode between 0.6 and 1.2 min.

Data Set S5 Data set for the visualization of identified signatory metabolites within the global metabolome of soybean using the cellular overview tool of SoyCyc. Kegg identifiers for metabolites (first column), and relative composition of metabolites in control (second column) and infected by *Rhizoctonia solani* soybean seedlings (third column), at 48 h post-inoculation are displayed.

Author Contributions

Conceived and designed the experiments: KAA SJ. Performed the experiments: KAA SJ. Analyzed the data: KAA SJ DF. Contributed reagents/materials/analysis tools: KAA SJ DF. Wrote the paper: KAA SJ DF.

References

1. De Vos RC, Moco S, Lommen A, Keurentjes JJ, Bino RJ, et al. (2007) Untargeted large-scale plant metabolomics using liquid chromatography coupled to mass spectrometry. Nat Prot 2: 778–791.
2. Nakabayashi R, Saito K (2013) Metabolomics for unknown plant metabolites. Anal Bioanal Chem: 1–7.
3. Schauer N, Fernie AR (2006) Plant metabolomics: towards biological function and mechanism. Trends Plant Sci 11: 508–516.
4. Aliferis KA, Jabaji S (2012) FT-ICR/MS and GC-EI/MS metabolomics networking unravels global potato sprout's responses to *Rhizoctonia solani* infection. PLoS One 7: e42576.
5. Allwood JW, Ellis DI, Goodacre R (2008) Metabolomic technologies and their application to the study of plants and plant–host interactions. Physiol Plant 132: 117–135.
6. Balmer D, Flors V, Glauser G, Mauch-Mani B (2013) Metabolomics of cereals under biotic stress: current knowledge and techniques. Frontiers Plant Sci 4: 82.
7. Aliferis KA, Chrysayi-Tokousbalides M (2011) Metabolomics in pesticide research and development: review and future perspectives. Metabolomics 7: 35–53.
8. Fiehn O, Robertson D, Griffin J, van der Werf M, Nikolau B, et al. (2007) The metabolomics standards initiative (MSI). Metabolomics 3: 175–178.
9. Hartman GL, West ED, Herman TK (2011) Crops that feed the World 2. Soybean-worldwide production, use, and constraints caused by pathogens and pests. Food Security 3: 5–17.
10. Ribeiro APB, Grimaldi R, Gioielli LA, Gonçalves LAG (2009) Zero trans fats from soybean oil and fully hydrogenated soybean oil: physico-chemical properties and food applications. Food Res Int 42: 401–410.
11. Shahidi F (2009) Nutraceuticals and functional foods: Whole versus processed foods. Trends Food Sci Technol 20: 376–387.
12. Hill J, Nelson E, Tilman D, Polasky S, Tiffany D (2006) Environmental, economic, and energetic costs and benefits of biodiesel and ethanol biofuels. Proc Natl Acad Sci USA 103: 11206–11210.
13. Schmutz J, Cannon SB, Schlueter J, Ma J, Mitros T, et al. (2010) Genome sequence of the palaeopolyploid soybean. Nature 463: 178–183.
14. Wrather JA, Anderson TR, Arsyad DM, Tan Y, Ploper LD, et al. (2001) Soybean disease loss estimates for the top ten soybean-producing counries in 1998. Can J Plant Pathol 23: 115–121.
15. OMAFRA (2009) Ontario Ministry of Agriculture, Food and Rural Affairs. Agronomy guide for field crops. Publication 811, Chapter 2-Soybeans, Toronto. http://www.omafra.gov.on.ca/english/crops/pub811/p811toc.pdf. Accessed July 2014.
16. Dodds PN, Rathjen JP (2010) Plant immunity: towards an integrated view of plant–pathogen interactions. Nat Rev Genet 11: 539–548.
17. Cheng J, Yuan C, Graham TL (2011) Potential defense-related prenylated isoflavones in lactofen-induced soybean. Phytochemistry 72: 875–881.
18. Komatsu S, Yamamoto A, Nakamura T, Nouri M-Z, Nanjo Y, et al. (2011) Comprehensive analysis of mitochondria in roots and hypocotyls of soybean under flooding stress using proteomics and metabolomics techniques. J Prot Res 10: 3993–4004.
19. Draper J, Lloyd AJ, Goodacre R, Beckmann M (2013) Flow infusion electrospray ionisation mass spectrometry for high throughput, non-targeted metabolite fingerprinting: a review. Metabolomics 9: 4–29.
20. Kirwan JA, Broadhurst DI, Davidson RL, Viant MR (2013) Characterising and correcting batch variation in an automated direct infusion mass spectrometry (DIMS) metabolomics workflow. Anal Bioanal Chem 405: 5147–57.
21. Martin A, Ochagavia M, Rabasa L, Miranda J, Fernandez-de-Cossio J, et al. (2010) BisoGenet: a new tool for gene network building, visualization and analysis. BMC Bioinformatics 11: 91.
22. Scharte J, Schön H, Weis E (2005) Photosynthesis and carbohydrate metabolism in tobacco leaves during an incompatible interaction with *Phytophthora nicotianae*. Plant Cell Environ 28: 1421–1435.
23. Hendrawati O, Yao Q, Kim HK, Linthorst HJM, Erkelens C, et al. (2006) Metabolic differentiation of *Arabidopsis* treated with methyl jasmonate using nuclear magnetic resonance spectroscopy. Plant Sci 170: 1118–1124.
24. Smeekens S (2000) Sugar-induced signal transduction in plants. Annual Review of Plant Physiol Plant Mol Biol 51: 49–81.
25. Sato D, Akashi H, Sugimoto M, Tomita M, Soga T (2013) Metabolomic profiling of the response of susceptible and resistant soybean strains to foxglove aphid, *Aulacorthum solani* Kaltenbach. J Chromatogr B 925: 95–103.
26. Raman SB, Rathinasabapathi B (2004) Pantothenate synthesis in plants. Plant Sci 167: 961–968.
27. Moulin M, Deleu C, Larher F (2000) L-Lysine catabolism is osmo-regulated at the level of lysine-ketoglutarate reductase and saccharopine dehydrogenase in rapeseed leaf discs. Plant Physiol Biochem 38: 577–585.
28. Návarová H, Bernsdorff F, Döring A-C, Zeier J (2012) Pipecolic acid, an endogenous mediator of defense amplification and priming, is a critical regulator of inducible plant immunity. The Plant Cell 24: 5123–5141.
29. Dixon RA, Pasinetti GM (2010) Flavonoids and isoflavonoids: from plant biology to agriculture and neuroscience. Plant Physiol 154, 453–457.
30. Lee CH, Yang L, Xu JZ, Yeung SYV, Huang Y, et al. (2005) Relative antioxidant activity of soybean isoflavones and their glycosides. Food Chem 90: 735–741.
31. Ahuja I, Kissen R, Bones AM (2012) Phytoalexins in defense against pathogens. Trends Plant Sci 17: 73–90.
32. Kim HJ, Suh H-J, Kim JH, Park S, Joo YC, et al. (2010) Antioxidant activity of glyceollins derived from soybean elicited with *Aspergillus sojae*. J Agric Food Chem 58: 11633–11638.
33. Lygin AV, Hill CB, Zernova OV, Crull L, Widholm JM, et al. (2010) Response of soybean pathogens to glyceollin. Phytopathology 100: 897–903.
34. Lozovaya VV, Lygin AV, Zernova OV, Li S, Hartman GL, et al. (2004) Isoflavonoid accumulation in soybean hairy roots upon treatment with *Fusarium solani*. Plant Physiol Biochem 42: 671–679.
35. Jeon HY, Seo DB, Shin HJ, Lee SJ (2012) Effect of *Aspergillus oryzae*-challenged germination on soybean isoflavone content and antioxidant activity. J Agric Food Chem 60: 2807–2814.
36. Naoumkina MA, Zhao Q, Gallego-Giraldo L, Dai X, Zhao PX, et al. (2010) Genome-wide analysis of phenylpropanoid defence pathways. Mol Plant Pathol 11, 829–846.
37. Enkerli J, Bhatt G, Covert SF (1998) Maackiain detoxification contributes to the virulence of *Nectria haematococca* MP VI on chickpea. Mol Plant-Microbe In 11: 317–326.
38. Dixon RA, Achnine L, Kota P, Liu CJ, Reddy M, et al. (2002) The phenylpropanoid pathway and plant defence-a genomics perspective. Mol Plant Pathol 3: 371–390.
39. Bourgaud F, Hehn A, Larbat R, Doerper S, Gontier E, et al. (2006) Biosynthesis of coumarins in plants: a major pathway still to be unravelled for cytochrome P450 enzymes. Phytochemistry Rev 5: 293–308.
40. Sakihama Y, Cohen MF, Grace SC, Yamasaki H (2002) Plant phenolic antioxidant and prooxidant activities: phenolics-induced oxidative damage mediated by metals in plants. Toxicology 177: 67–80.
41. León J, Rojo E, Sánchez-Serrano JJ (2001) Wound signalling in plants. J Exper Bot 52: 1–9.
42. Wasternack C (2007) Jasmonates: an update on biosynthesis, signal transduction and action in plant stress response, growth and development. Ann Bot 100: 681–697.
43. Fliegmann J, Schuler G, Boland W, Ebel J, Mithofer A (2003) The role of octadecanoids and functional mimics in soybean defense responses. Biological Chem 384: 437–446.
44. Gardner HW (1998) 9-Hydroxy-traumatin, a new metabolite of the lipoxygenase pathway. Lipids 33: 745–749.
45. Vlot AC, Dempsey DMA, Klessig DF (2009) Salicylic acid, a multifaceted hormone to combat disease. Annu Rev Phytopathol 47: 177–206.
46. Bari R, Jones JDG (2009) Role of plant hormones in plant defence responses. Plant Mol Biol 69: 473–488.
47. Yang D-L, Yao J, Mei C-S, Tong X-H, Zeng L-J, et al. (2012) Plant hormone jasmonate prioritizes defense over growth by interfering with gibberellin signaling cascade. Proc Natl Acad Sci USA109: E1192–E1200.
48. Navarro L, Bari R, Achard P, Lisón P, Nemri A, et al. (2008) DELLAs control plant Immune responses by modulating the balance of jasmonic acid and salicylic acid signaling. Curr Biol 18: 650–655.
49. McDonald KL, Cahill DM (1999) Influence of abscisic acid and the abscisic acid biosynthesis inhibitor, norflurazon, on interactions between *Phytophthora sojae* and soybean (*Glycine max*). Eur J Plant Pathol 105: 651–658.
50. Stasolla C, Katahira R, Thorpe TA, Ashihara H (2003) Purine and pyrimidine nucleotide metabolism in higher plants. J Plant Physiol 160: 1271–1295.
51. Lunn JE (2007) Compartmentation in plant metabolism. J Exp Bot 58, 35–47.
52. Martinoia E, Maeshima M, Neuhaus HE (2007) Vacuolar transporters and their essential role in plant metabolism. J Exp Bot 58: 83–102.
53. Verrier PJ, Bird D, Burla B, Dassa E, Forestier C, et al. (2008) Plant ABC proteins-a unified nomenclature and updated inventory. Trends Plant Sci 13: 151–159.

54. Heller J, Tudzynski P (2011) Reactive oxygen species in phytopathogenic fungi: signaling, development, and disease. Annu Rev Phytopathol 49: 369–390.
55. Jang M, Cai L, Udeani GO, Slowing KV, Thomas CF, et al. (1997) Cancer chemopreventive activity of resveratrol, a natural product derived from grapes. Science 275: 218–220.
56. de Pinto MC, Lavermicocca P, Evidente A, Corsaro MM, Lazzaroni S, et al. (2003) Exopolysaccharides produced by plant pathogenic bacteria affect ascorbate metabolism in *Nicotiana tabacum*. Plant Cell Physiol 44: 803–810.
57. Walters D (2003) Resistance to plant pathogens: possible roles for free polyamines and polyamine catabolism. New Phytol 159: 109–115.
58. Tambasco-Studart M, Titiz O, Raschle T, Forster G, Amrhein N, et al (2005) Vitamin B6 biosynthesis in higher plants. Proc Natl Acad Sci USA 102: 13687–13692.
59. Takahashi H, Matsumura H, Kawai-Yamada M, Uchimiya H (2008) The cell death factor, cell wall elicitor of rice blast fungus (*Magnaporthe grisea*) causes metabolic alterations including GABA shunt in rice cultured cells. Plant Signal Behav 3: 945–953.
60. Verslues PE, Sharma S (2010) Proline metabolism and its implications for plant-environment interaction. The Arabidopsis Book 8: e0140. doi:10.1199/tab.0140.
61. Kind T, Fiehn O (2007) Seven golden rules for heuristic filtering of molecular formulas obtained by accurate mass spectrometry. BMC Bioinformatics 8: 105.
62. Efron B, Gong G (1983) A leisurely look at the bootstrap, the jackknife, and cross-validation. American Statistician 37: 36–48.
63. Smoot M, Ono K, Ruscheinski J, Wang P.-L, Ideker T (2011) Cytoscape 2.8: new features for data integration and network visualization Bioinformatics 27: 431.

Overexpression of a Soybean Ariadne-Like Ubiquitin Ligase Gene *GmARI1* Enhances Aluminum Tolerance in Arabidopsis

Xiaolian Zhang, Ning Wang, Pei Chen, Mengmeng Gao, Juge Liu, Yufeng Wang, Tuanjie Zhao, Yan Li*, Junyi Gai*

National Key Laboratory of Crop Genetics and Germplasm Enhancement, National Center for Soybean Improvement, Key Laboratory for Biology and Genetic Improvement of Soybean (General, Ministry of Agriculture), Nanjing Agricultural University, Nanjing, Jiangsu, China

Abstract

Ariadne (ARI) subfamily of RBR (Ring Between Ring fingers) proteins have been found as a group of putative E3 ubiquitin ligases containing RING (Really Interesting New Gene) finger domains in fruitfly, mouse, human and Arabidopsis. Recent studies showed several RING-type E3 ubiquitin ligases play important roles in plant response to abiotic stresses, but the function of ARI in plants is largely unknown. In this study, an ariadne-like E3 ubiquitin ligase gene was isolated from soybean, *Glycine max* (L.) Merr., and designated as *GmARI1*. It encodes a predicted protein of 586 amino acids with a RBR supra-domain. Subcellular localization studies using Arabidopsis protoplast cells indicated GmARI protein was located in nucleus. The expression of *GmARI1* in soybean roots was induced as early as 2–4 h after simulated stress treatments such as aluminum, which coincided with the fact of aluminum toxicity firstly and mainly acting on plant roots. In vitro ubiquitination assay showed GmARI1 protein has E3 ligase activity. Overexpression of *GmARI1* significantly enhanced the aluminum tolerance of transgenic Arabidopsis. These findings suggest that *GmARI1* encodes a RBR type E3 ligase, which may play important roles in plant tolerance to aluminum stress.

Editor: Ji-Hong Liu, Key Laboratory of Horticultural Plant Biology (MOE), China

Funding: This work was supported by the National Hightech R & D Program of China (2013AA102602), the National Key Basic Research Program of China (2011CB1093), the Program for Changjiang Scholars and Innovative Research Team in University (PCSIRT13073), the Program for New Century Excellent Talents in University (NCET-12-0891), the MOE 111 Project (B08025), the Program for High-level Innovative and Entrepreneurial Talents in Jiangsu Province, and the Jiangsu Higher Education PAPD Program. The funders had no role in study design, data collection and analysis, decision or preparation of the manuscript.

Competing Interests: The authors have declared that no competing interests exist.

* Email: yanli1@njau.edu.cn (YL); sri@njau.edu.cn (JG)

Introduction

Ubiquitination is an enzymatic, protein post-translational modification by which proteins are selectively targeted for a variety of cellular processes including DNA transcription and repair, cell cycle and division, response to stresses and many others [1]. This process is carried out by three types of enzyme, including an ubiquitin-activating enzyme (E1), an ubiquitin-conjugating enzyme (E2), and an ubiquitin protein ligase (E3) [2]. Encoded by a large gene family of widely divergent isoforms [3], E3 ligases play important roles in governing the ubiquitin signaling pathway by transferring ubiquitin from E2 conjugation to specific protein substrates. E3 ligases are generally divided into two families, with either a HECT or RING-finger domain(s) [1], [4]. The RING-type E3 ubiquitin ligases are generally identified by the presence of conserved cysteine- and histidine- rich RING finger motifs that coordinate zinc atoms [5]. Recently several RING-type E3 ubiquitin ligases were found to play important roles in plant responses to abiotic and biotic stresses. The pepper E3 ubiquitin ligase RING1 gene, *CaRING1*, is required for cell death and the salicylic acid (SA)-dependent defense response [6]. *AtAIRP1* and *AtAIRP2* play roles in abscisic acid (ABA)-mediated drought stress responses in Arabidopsis [7]. In soybean, a RING-finger protein encoded by *GmRFP1* was identified and shown to be involved in ABA signaling and stress responses through the ubiquitin-proteasome pathway [8].

RBR (Ring Between Ring fingers) proteins are characterized by the presence of their RING1 – IBR – RING2 supra-domain, which is composed of two RING finger domains plus an IBR (In Between Rings) domain [9]. Many RBR proteins are known to have E3 ubiquitin ligase activity [10]. ARIADNE (ARI) proteins, a subclass of RBRs, have been identified in fruitfly [11], mouse [12], [13], human [14–16], and Arabidopsis [17], [18]. ARI proteins are characterized by the presence of an N-terminal acid-rich cluster, followed by a C3HC4 RING-finger motif, a central IBR or B-box, a second C3HC4 RING-finger structure, and Leu-rich domain at the C terminus. ARI proteins share their RBR domain with PARKIN, a protein involved in autosomal recessive familial Parkinson's disease [9]. PARKIN functions as E2-dependent ubiquitin-protein ligase [17]. Recent studies suggest that the ARI/PARKIN proteins define a new class of RING-finger E3 ligases [19]. There are only few studies on ARI proteins in plants.

Figure 1. The protein structure of GmARI1. Acid: acid-rich cluster; **Leu**: Leu-rich cluster; **RING1**: a C3HC4 RING-finger; **IBR**: In Between RING fingers (IBR); **RING2**: a second C3HC4 RING finger; **Coile~**: Coiled coil.

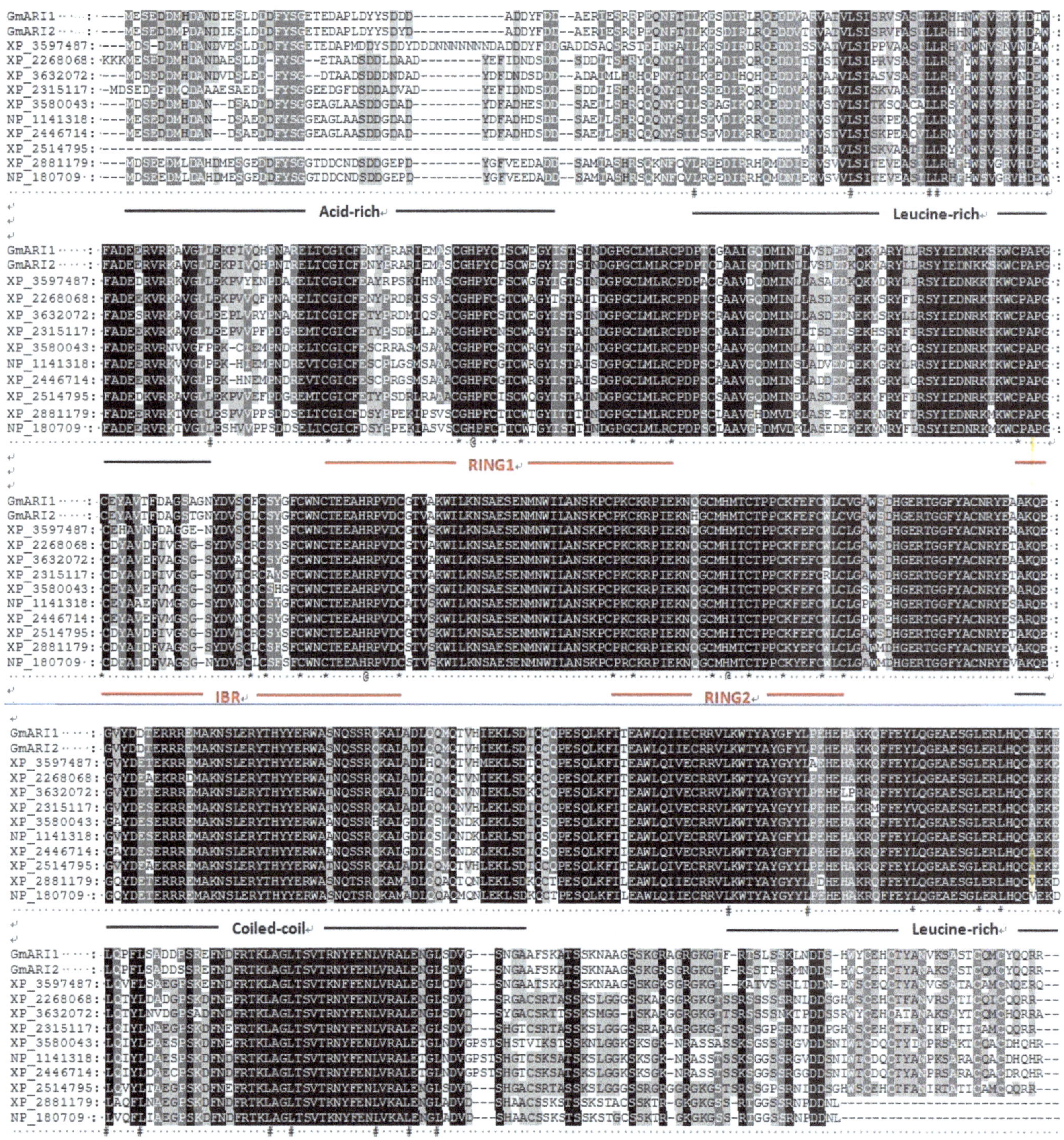

Figure 2. Multiple sequence alignment of GmARI1 with other RBR domain containing proteins. XP_3597487: *Medicago truncatula*; XP_2268068 and XP_3632072: *Vitis vinifera*; XP_2315117: *Populus trichocarp*; XP_3580043: *Brachypodium distachyon*; NP_1141318: *Zea mays*; XP_2446714: *Sorghum bicolor*; XP_2514795: *Ricinus communis*; XP_2881179: *Arabidopsis lyrata*; NP_180709: *Arabidopsis thaliana*; *: Cys; @: His; #: Leu and Ile.

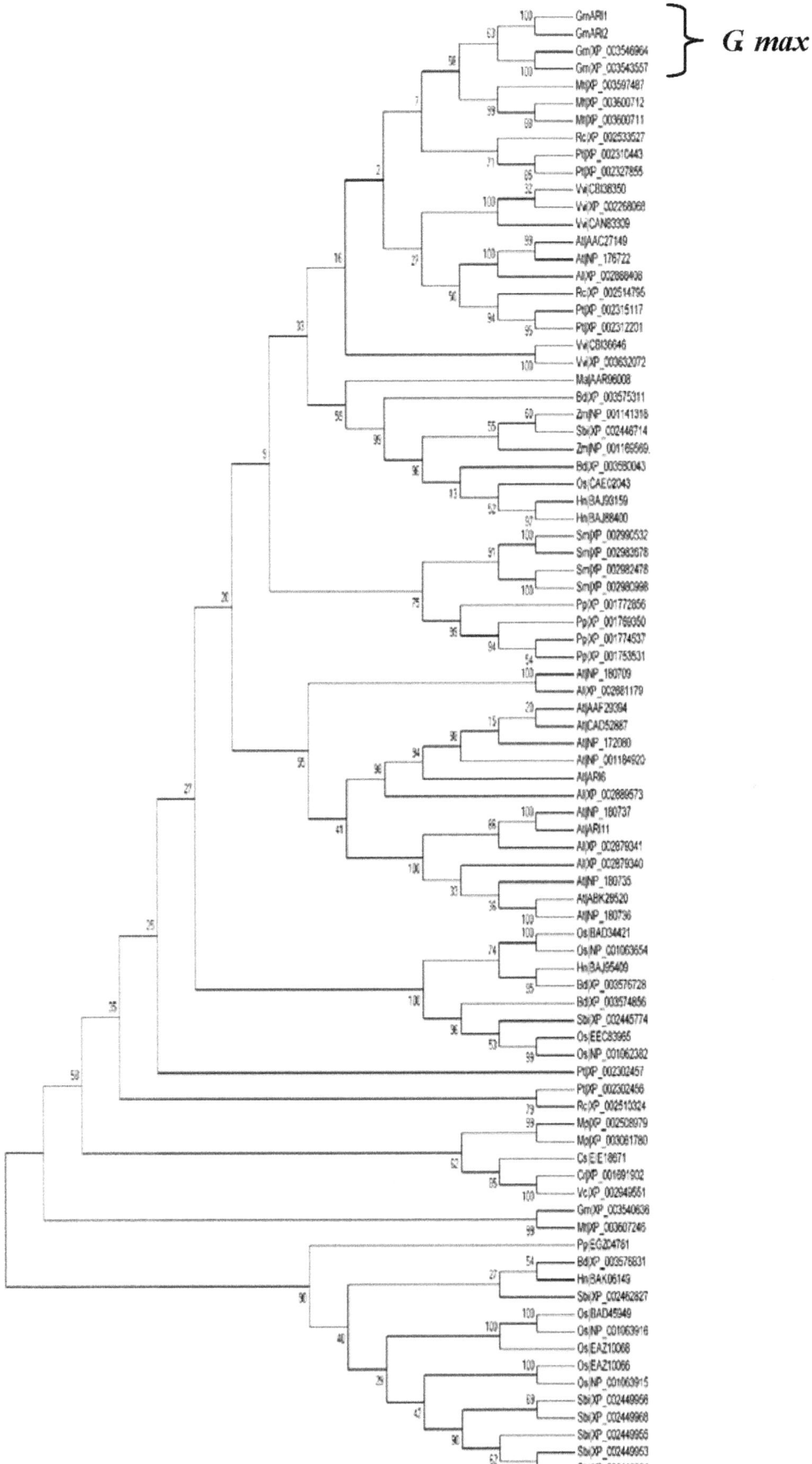
G. max

Figure 3. Phylogenetic relationships among soybean ARI proteins and other RBR domain containing proteins. The numbers on the tree indicate percent bootstrap values. The species abbreviations are listed as follows; *Al: Arabidopsis lyrata; At: Arabidopsis thaliana; Bd: Brachypodium distachyon; Cr: Chlamydomonas reinhardtii; Cs: Coccomyxa subellipsoidea; Hn: Hordeum vulgare; Ma: Musa acuminata; Mp: Micromonas pusilla; Mt: Medicago truncatula; Nt: Nicotana plumbaginifolia; Os: Oryza sativa subsp. Japonica; Pp: Physcomitrella patens; Pt: Populus trichocarpa; Rc: Ricinus communis; Sbi: Sorghum bicolor; Sm: Selaginella moellendorffii; Vc: Volvox carteri, Vvi: Vitis vinifera.*

Recently *AtARI12* in Arabidopsis was identified to be involved in UV-B signaling pathway [20].

Soybean (*Glycine max* [L.] Merr.) is widely grown as a major source of vegetable oil and protein. Soybean quality and yield are affected by various abiotic and biotic stresses. Soybean is also an important crop grown in South China, where acid soils comprise approximately 21% of the total land area [21], [22]. Aluminum (Al) toxicity is a major limiting factor of plant growth and crop production on acidic soils. There is large variation in Al tolerance among soybean varieties, and most of the Al tolerant varieties are from China [23]. Another study showed many Al tolerant varieties are from South China [24]. However, the genes underlying the Al tolerance in soybean remain largely unknown, except recently a soybean malate transporter gene *GmALMT1* which was shown to mediate root malate efflux which underlies soybean Al tolerance [25].

Increasing evidence indicates that RING-type E3 ubiquitin ligases play important roles in plant response to abiotic stresses. However, to date, there are no reports on the functions of soybean ARI proteins. Therefore, in this study, an ariadne-like E3 ubiquitin ligase gene *GmARI1* was cloned from soybean, and its gene expression patterns in different soybean tissues were studied. The transcriptional changes of *GmARI1* in response to various stress such as aluminum (Al) and plant hormone treatments were investigated using real-time quantitative PCR (qRT-PCR). We further characterized the *GmARI1* gene function by its subcellular location, in vitro ubiquitination assay, and performance of the transgenic Arabidopsis overexpressing *GmARI1* under Al stress. The possible mechanisms and signal pathways involved in soybean response to Al are also discussed.

Materials and Methods

Plant materials

Seeds of the soybean (*Glycine max* [L.] Merr.) cultivar Nannong 1138-2, provided by the National Center for Soybean Improvement (Nanjing, China), were germinated in sand under 25°C, 60% relative humidity (RH) and a photoperiod of 16 h/8 h (light/dark) cycle (light intensity was about 110 μmol photons. $m^{-2}s^{-1}$). Nannong 1138-2 is a released cultivar adapted to South China, which has good agronomic traits and moderate Al tolerance. The soybean plants at VE stage (emergence) were transferred to the 'standard' nutrient solution [26], and grown for another ten days before various stress and hormone treatments. The nutrient solution was renewed every five days.

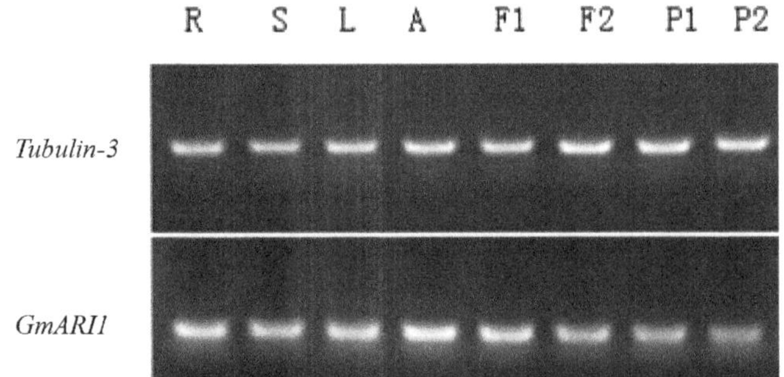

Figure 4. Tissue expression patterns of *GmARI1* in soybean. R: Roots; S: Stems; L: Leaves; A: Apex (SAMs); F1, F2: Flowers; P1, P2: Pods.

Isolation of the *GmARI1* gene from soybean

The full-length opening reading frame of the *GmARI1* gene was obtained by RT-PCR using soybean RNA. Total RNA was isolated using Trizol Reagent (Invitrogen, USA) according to the user's manual. 0.2 μg of the purified total RNA was used to synthesize first-strand cDNA by the MMLV-reverse transcriptase (TaKaRa). The primers: 5′-TCCCAATTCTTCTTCTGCCC-TAG-3′ and 5′-GCAACCTTTCTTCCAAG CCTTAC -3′ were designed to amplify the *GmARI1* gene located on Chromosome 11. The PCR products were cloned into the pGEM-T vector (Promega) and sequenced (Invitrogen). The sequencing results showed that two *ARI* genes, *GmARI1* and *GmARI2*, were isolated using above primers, due to their high similarity of 97%.

Sequence analysis

Protein domains were analyzed by the SMART (Simple Modular Architecture Research Tool) (http://smart.embl-heidelberg.de/) and Pfam (Protein families database of alignments and HMMs). The molecular mass, isoelectric point and secondary structure were predicted using ProtParam and SOPMA on the ExPASY(http://www.expasy.org/tools/). The BLASTP program at GenBank (http://www.ncbi.nlm.nih.gov/blast) was used to search the homologous sequences of GmARI1/GmARI2 from Non-Redundant (NR) database. Alignment was performed with ClustalW2 (http://www.ebi.ac.uk/Tools/msa/clustalw2/) and MUSCLE (http://www.ebi.ac.uk/Tools/msa/muscle/) using the default settings. The phylogenetic tree was constructed by the neighbor-joining algorithm (NJ) using MEGA version 5 with 1000 bootstraps.

Semi-quantitative RT-PCR

To study the tissue expression pattern of *GmARI1*, soybean roots, stems, leaves, and shoot apical meristem (SAM) were collected from 15-day-old plants, flowers and pods were collected from plants at 20 days after flowering (DAF). All tissues were frozen immediately in liquid nitrogen and stored at −80°C. The semi-quantitative RT-PCR assay was performed with 0.1 μg RNA as template for cDNA synthesis. Primers 5′-CTCCATTCTC-CATTCTCCTCCTTTGC-3′ and 5′-GTCGTCGTCGCTG-TAGTAGT CC -3′ were used for *GmARI1*. Primers:5′-ATCTCATTCCCTTCCCTCGTCTG-3′ and 5′-CTGCCT-CTGTGAACTCCATCTCG -3′ were used for *Tubulin-3* (GeneBank accession No. U12286) as the internal control. The PCR products were examined by electrophoresis in 2.0% agarose gel.

Transient expression of the GmARI1-GFP fusion protein

GmARI1-GFP was cloned into pMDC83 vector, with the expression driven by the cauliflower mosaic virus 35S promoter. The ORF of *GmARI1* was amplified by PCR using primers:5′-GGGGACAAGTTTGTACAAAAAA GCAGGCTTCCCAAT-TCTTCTTCTGCCCTAG-3′ and 5′-GGGGACCACTTTG-

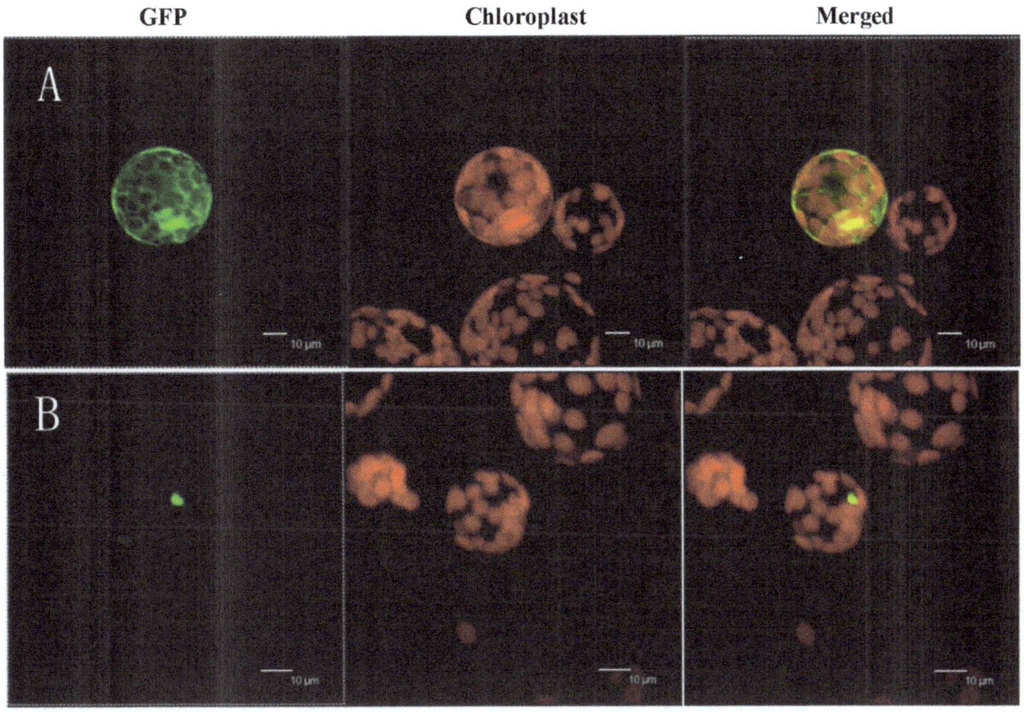

Figure 5. Sub-cellular localizations of GmARI1 protein. A: 35S-GFP; **B**: 35S-GmARI1-GFP.

TACA AGAAAGCTGGGTCTCGACGTTGTTGATAGCA-CATCTG -3′, without the stop codon. The *35S-GmARI1-GFP* in-frame fusion construct and control vector of *35S-GFP* were introduced into the Arabidopsis protoplasts cells by PEG-mediated protocol [27], [28]. After culturing in dark at 23°C, the localization of GFP was observed with a confocal microscope the next day (Leica TCS SP2).

Stress treatments

The plants were put in the 'standard' nutrient solution [26] with one of the following chemicals for various treatments: Al stress (10 µM $Al(NO_3)_3$, pH 4.3), drought (osmotic) stress (400 mM mannitol), salt stress (200 mM NaCl), absicisic acid (ABA,100 µM), indoleacetic acid (IAA, 100 µM), jasmonic acid (JA, 100 µM), and salicylic acid (SA, 150 µM), with 'standard' nutrient solution (Mg^{2+} was withdrew for Al treatment) as control. The leaves and roots were harvested at 0, 0.5, 1, 2, 4, 6, 8, 12, 24, and 48 h after each treatment. Each sample was the mixture of three seedlings and each treatment was repeated three times. All samples were immediately frozen in liquid nitrogen and stored at −80°C for later use.

Real-time quantitative PCR

RNAs of different treatments were extracted using Trizol reagent (Invitrogen) and purified with RNase-free DNase I. The cDNA was synthesized from 0.2 µg RNA in a 10 µl reaction volume using PrimeScript R 1st Strand cDNA Synthesis kit (TaKaRa). Primers for *Tubulin-3* (GeneBank accession No. U12286) were 5′- TCATTCCCTTCCCTCGTCTGC-3′ and 5′-CCTCCTTGGTGCTCATCTTGC-3′. Primers for *GmARI1* were 5′-CGCTGGTTCCTGAATTTCCCTTG-3′ and 5′- GTCGTCGTCGCTGTAGTAGTCC-3′. Quantitative real-time PCR was performed with SYBR Green method on ABI 7500 Fast Real-Time PCP system. The following procedure was used for qPCR: 95°C for 5 min; 40 cycles of 95°C for 3 sec; 60°C for 30 sec; 72°C for 30 sec. Data was analyzed using the $2^{-\Delta\Delta CT}$ method as described by Livak and Schmittgen [29].

In vitro E3 ubiquitin ligase activity assay of GmARI1 protein

The full-length cDNA of GmARI1 with SalI/XhoI restriction enzyme sites was amplified by PCR, using primers ARI1-F: 5′-ACGCGTCGACATGGAGTCAGAGGATATGCAC-3′ and ARI1-R: 5′-CCGCTCGAGTCGACGTTGTTGATAGCA-CATCTG -3′. The fragment was cloned into the vector pET28a via the SalI/XhoI restriction sites, with 6×His tag fused to GmARI1 at the N-terminal. The expression construct (pET28a-His_6-GmARI1) was transformed into *E.coli* BL21 (DE3) cells to produce a recombinant His_6-GmARI1 fusion protein with an expected mass of about 66 kDa. The transformed cells harboring pET28a-His_6-GmARI1 were grown at 37°C with vigorous shaking until an OD600 of 0.4–0.6 is reached and induced with 0.1 mM isopropylthio-b-galactoside (IPTG) for 12 h at 16°C. The overexpressed His_6-GmARI1 was purified using Ni–NTA resin according to the supplier's instructions (GE life sciences). The protein concentration was determined as described by Bradford [30] using BSA as a standard.

For the autoubiquitination assay, each reaction (30 µl final volume) contained 10 µg of recombinant ubiquitin (Ub, Sigma), 0.1 µg rabbit E1 (Boston Biochemicals), 0.2 µg human E2 (UbcH5b, Boston Biochemicals), 2 mM ATP, 50 mM Tris–HCl (pH 7.5), 5 mM MgCl2, and 2 mM DTT contained 500 ng purified His6-GmARI1 [31]. After incubation at 30°C for 2–3 h, the reactions were stopped with sodium dodecyl sulfate-polyacrylamide gel electrophoresis (SDS-PAGE) loading buffer at 95°C for 5 min. The reaction samples were electrophoretically separated on 12% SDS-PAGE gels and transferred on two PVDF membranes separately. The two membranes were blocked and thereafter

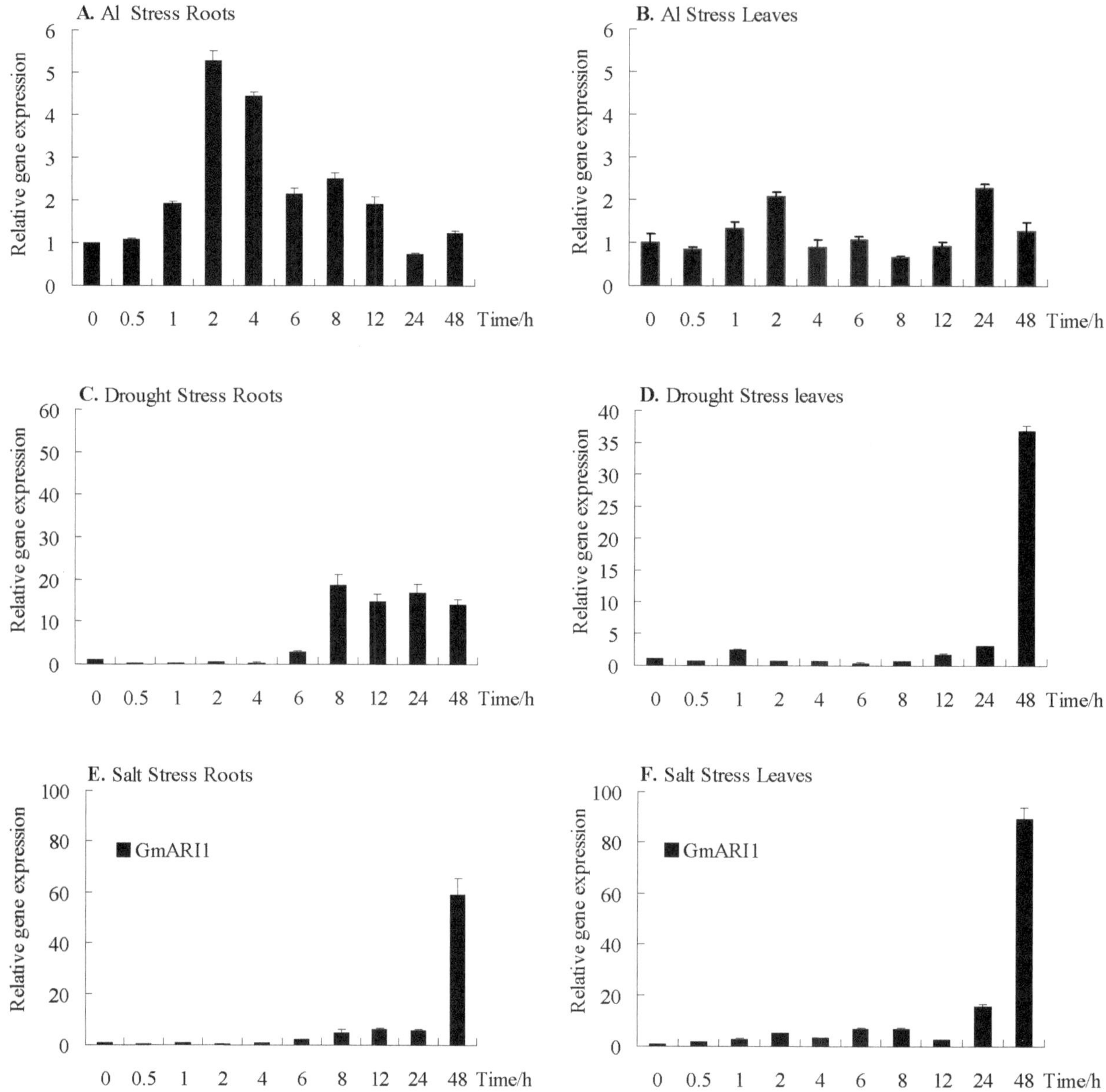

Figure 6. Relative gene expression levels of *GmARI1* in soybean under simulated stress treatments. A. relative gene expression in roots under Al stress (10 μM $Al(NO_3)_3$, pH4.3) **B**. relative gene expression in leaves under Al stress (10 μM $Al(NO_3)_3$, pH4.3) **C**. relative gene expression in roots under drought stress (400 mM mannitol) **D**. relative gene expression in leaves under drought stress (400 mM mannitol) **E**. relative gene expression in roots under salt stress (200 mM NaCl) **F**. relative gene expression in leaves under salt stress (200 mM NaCl). Error bars represent the standard error of three replicates.

blotted with an anti-ubiquitin monoclonal antibody (Santa Cruz Biotechnology) and an anti-His_6 monoclonal antibody (Sigma, USA) for 6 h at a 1:3000 dilution, respectively. After extensive washing, each of the bound primary antibody was detected with a horseradish peroxidase-conjugated goat anti-rabit IgG secondary antibody using the 3,3′-diaminobenzidine (DAB) development kit according to the manufacturer's protocol (Bio Basic Inc, Canada).

Generation of *GmARI1* transgenic Arabidopsis

The *GmARI1* gene was amplified by RT-PCR as described above and cloned into the plant expression vector pMDC83 under the control of CaMV 35S promoter by Gateway technology (Invitrogen), and the recombined plasmid was transferred into *A. tumefaciens* strain EHA105. Arabidopsis plants (Col-0 ecotype) were transformed using the floral dip method [32]. Twenty transgenic lines of *GmARI1* were obtained. Eight T3 lines of the transgenic *GmARI1* Arabidopsis were examined by RT-PCR to select positive transgenic lines for further analyses.

Al-tolerance of the *GmARI1* transgenic Arabidopsis

Homozygous T_3 seeds of the transgenic lines and wild type plants were used for Al-tolerance analysis. Seeds were surfaced-sterilized as described before [33] and germinated on 1/2 MS medium for 7 days. The seedlings were transferred to 1/2 MS without Mg^{2+} but with 0 or 15 μM $Al(NO_3)_3$ and 8% Agar, pH 4.3, and then put on the medium with the plates placed

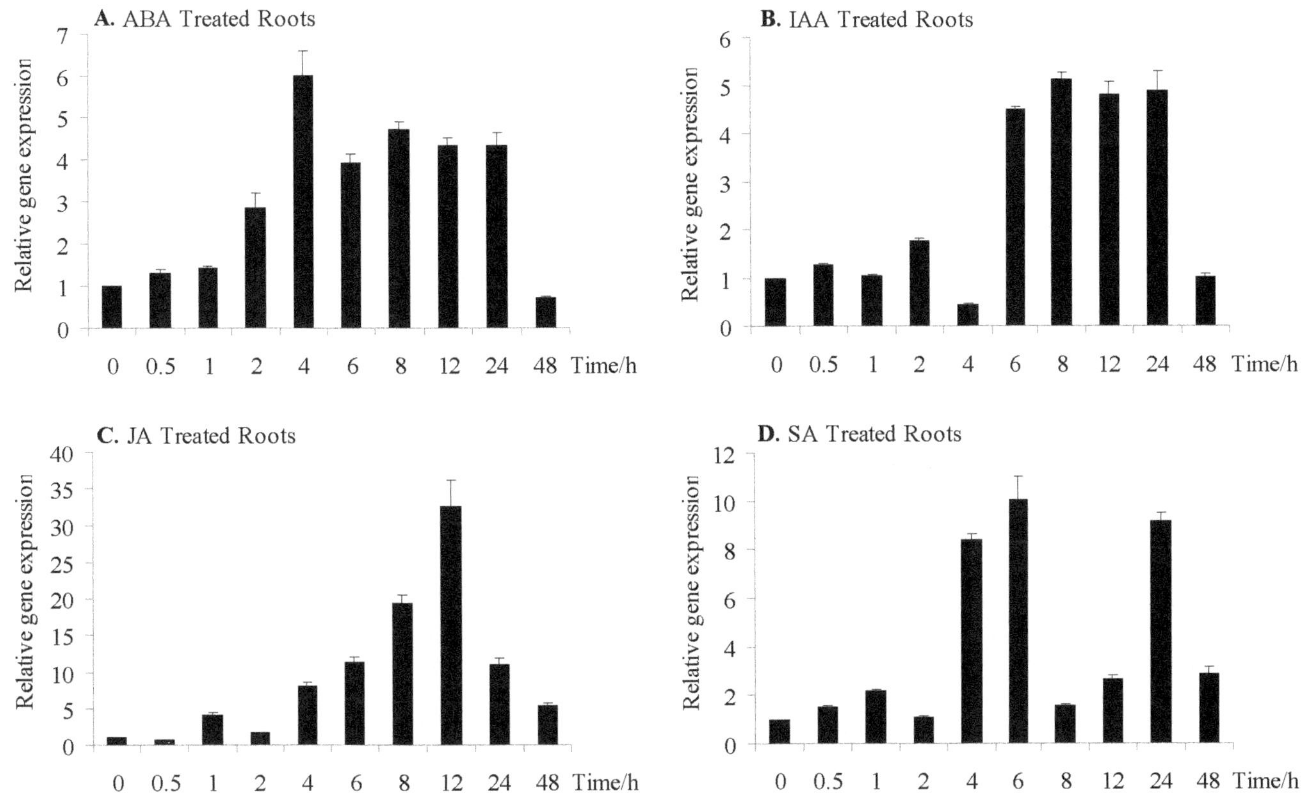

Figure 7. Relative gene expression levels of *GmARI1* in the roots of soybean after exogenous application of plant hormones. A. 100 μM ABA treatment **B.** 100 μM IAA treatment **C.** 100 μM JA treatment **D.** 150 μM SA treatment. Error bars represent the standard error of three replicates.

vertically. After 15 days, the lengths of the roots were measured, and the Relative Root Growth (RRG) in each line was calculated as: RRG (%) = $(RL_{Alt} - RL_{Al0})/(RL_{ct} - RL_{c0})$, in which RL_{Al0} represent the root length before Al treatment, RL_{Alt} represent the root length after 15 days of Al treatment (15 μM $Al(NO_3)_3$, pH 4.3), RL_{c0} represent the root length before growth on the control medium, and RL_{ct} represent the root length after growth on the control medium (0 μM $Al(NO_3)_3$, pH 4.3). The statistical analysis of the experimental data was conducted by t-test with SPSS [34].

Results

Isolation and sequence analysis of *GmARI1* gene

The cDNA of *GmARI1* gene is 2043 bp in length, containing an open reading frame (ORF) of 1761 bp, a 112 bp 5′-untranslated region (UTR), and a 170 bp 3′-UTR. Its homologous gene, *GmARI2*, shares 97% of similarity in the nucleotide sequences of the ORF region. The *GmARI1* and *GmARI2* nucleotide sequence and the predicted amino acid sequence have been deposited in GenBank (accession number JX392390 and JX392391). The genomic sequence of *GmARI1* and *GmARI2* variants in cultivar Williams 82 have 15 exons and 14 introns (http://www.phytozome.net/soybean.php), which located on chromosome 11 and 12, respectively.

The deduced protein of *GmARI1* comprises 586 amino acids with the predicted molecular mass of 66.99 kDa and isoelectric point of 5.37. GmARI1 protein has a RBR domain, which contains an IBR (C6HC) domain flanked by RING1 and RING2 (C3HC4) (Fig. 1). The secondary structure of GmARI1 protein is predicted to be composed of 45.56% alpha helix, 11.09% extended strand, 41.64% random coil, and 1.71% beta turn.

In order to determine the relationship of GmARI1 and other RBR proteins, Blastp on Uniprot was used to search the homologues proteins of GmARI1. The top 87 amino acid sequences with RBR conserved domain were selected from different plant species, including *Arabidopsis thaliana*, *Medicago truncatula*, *Ricinus communis*, and *Zea mays L.* Multiple sequence alignment of these 87 amino acid sequences was performed. In addition to the ARI proteins from soybean which showed high similarity with *GmARI1*, the protein from *M. truncatula* showed 80% similarity with *GmARI1* (Fig. 2) The phylogenetic tree was drawn using MEGA 5.0 program based on Neighbor-Joining (NJ) with 1000 bootstrap replications (Fig. 3), which also showed ARI proteins from *G. max* was closely related to the protein from *M. truncatula*.

Tissue expression pattern of *GmARI1* and the subcellular localization of its protein

GmARI1 gene was expressed in roots, stems, leaves, SAMs, flowers, and pods (Fig. 4). We determined the subcellular localization of the GmARI1 protein by transient expression of *35S-GmARI1-GFP* in Arabidopsis protoplasts. GmARI1 was located in the nucleus, while the GFP control was mainly located in the cytoplasm (Fig. 5).

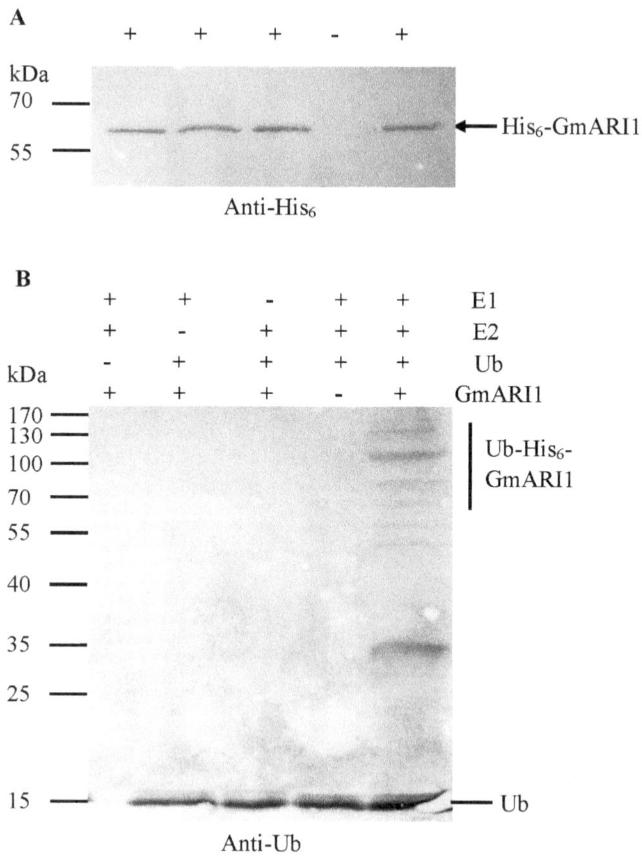

Figure 8. Western blot analysis of GmARI1 protein. A. Western blot using anti-His6 monoclonal antibody in the presence (+) or absence (−) of the purified His6 tag-GmARI1 proteins. **B.** In vitro E3 ubiquitin ligase activity assay of GmARI1 protein. His6 tag-GmARI1 fusion proteins were assayed for E3 activity in the presence of rabbit E1, human E2 (UbcH5b), and ubiquitin (Ub). The left numbers denote the molecular masses of marker proteins in kilodaltons.

Expression pattern of *GmARI1* under stress and plant hormone treatments

Real-time quantitative PCR was carried out to examine the expression pattern of *GmARI1* gene in soybean under different treatments including various abiotic stresses and plant hormones (Fig. 6 and 7). Under 10 μM Al (pH4.3) stress, the transcripts of *GmARI1* in soybean roots increased and peaked during 2 h to 4 h, but the transcripts in leaves did not change significantly (Fig. 6A, B). When treated with 400 mM mannitol (osmotic stress), the *GmARI1* gene expression was induced from 6 h to 48 h in roots, while showed delayed induction in leaves (Fig. 6C, D). Under 200 mM NaCl stress, the *GmARI1* gene expression showed a high induction after 24 h in both roots and leaves (Fig. 6E, F).

The transcriptional changes of *GmARI1* in response to ABA, IAA, JA and SA were also investigated in soybean roots. The transcripts of *GmARI1* were up-regulated during 2-24 h after ABA treatment (Fig. 7A). *GmARI1* was induced after 6 h IAA treatment and its transcript abundance was maintained at a higher level until 24 h (Fig.7B). The expression of *GmARI1* showed continual increase from 4 h and peaked around 12 h after JA treatment (Fig. 7C). Increased transcripts of *GmARI1* were detected at 4 h, 6 h and 24 h after SA treatment (Fig. 7D).

E3 ubiquitin ligase activity of GmARI1 protein

To test if GmARI1 has the E3 ubiquitin ligase activity, a full-length GmARI1 protein with maltose binding protein (6 × His tag) was expressed in *E. coli* and subsequently affinity-purified (His$_6$-GmARI1) from the soluble fraction (Fig. S1). The purified recombinant His$_6$-GmARI1 protein was about 66 kDa as expected, and the western blotting with anti-His$_6$ monoclonal antibody also showed the purified target recombinant protein had the right molecular weight of 66 kDa (Fig. 8A). In vitro self-ubiquitination assays were performed in the presence of rabbit E1, human E2 (UbcH5b), and Ub (Fig. 8B). Polyubiquitination was detected only in the presence of E1, E2, Ub and His$_6$-GmARI1. A negative result was observed if either E1, E2, Ub or His$_6$-GmARI1 was omitted in the reaction. These results indicate that GmARI1 has E3 ubiquitin ligase activity.

Performance of transgenic plants over-expressing *GmARI1* under Al stress

The gene expression of *GmARI1* was induced early by Al stress, therefore we further studied its function in transgenic plants. An expression plasmid vector of *pMDC83-GmARI1* was constructed and introduced into Arabidopsis plants using floral dip method. Transgenic T_3 Arabidopsis over-expressing *GmARI1* were generated and the positive transgenic lines were identified by RT-PCR (Fig. S2). Seeds of three T_3 homozygous transgenic lines (*GmARI1-1, GmARI1-2*, and *GmARI1-3*) and wild type Col-0 were germinated on 1/2 MS medium. After 10 days, the seedlings were transferred to 1/2 MS (pH4.3) containing 15 μM Al or 0 μM Al as a control. Fifteen days later, the root growth of the wild type Col-0 was severely inhibited by 15 μM Al as compared with the control medium (0 μM), while the transgenic lines were little affected by Al (Fig. 9 A). The relative root growth (RRG) of the transgenic lines was significantly ($p<0.01$) longer than the wild type under Al treatment (Fig. 9B). The relative abundance of *GmARI1* in the transgenic lines of *GmARI1-1* and *GmARI1-2* is higher than *GmARI1-3*, which coincided with RRG result (Fig. S3 and Fig. 9B).

Discussion

RING-type E3 ubiquitin ligases play important roles in plant responses to abiotic stresses [7], [8], [35]. RBR subclass of RING-containing E3 ligases were recently shown as an important group of proteins since the discovery of *parkin*, a mutation causing the familial autosomal-recessive juvenile parkinsonism (AR-JP) [36]. However, there is little research on the plant RBR family. The RBR family was classified into 14 subfamilies, including Ariadne (ARI), ARA54, Dorfin, parkin, PlantI, PlantII, and XAP3 [9], but the function of ARI class is largely unknown. In this study, *GmARI1* gene was isolated and characterized from soybean. Based on the analysis of predicted protein domains, *GmARI1* gene belongs to the Ariadne subfamily, characterized by the presence of an N-terminal acid-rich cluster, followed by a RBR domain and a Coiled coil region at the C-terminus [17]. The origin of Ariadne can be traced back earlier in time, as it was found not only in animal, fungal, and plant, but also in some protist [37]. The blastp result indicated among the RBR domain containing protein from other species, *Medicago truncatula* had the highest similarity (80%) with *GmARI1* and other ARI proteins from soybean.

GmARI1 gene was expressed ubiquitously in roots, stems, SAMs, leaves, flowers, and pods of soybean plants. The GmARI1 protein was located in the nucleus of the cell. Real-time quantitative PCR showed that the expression level of *GmARI1* in soybean root under Al (10 μM, pH 4.3) treatment reached the

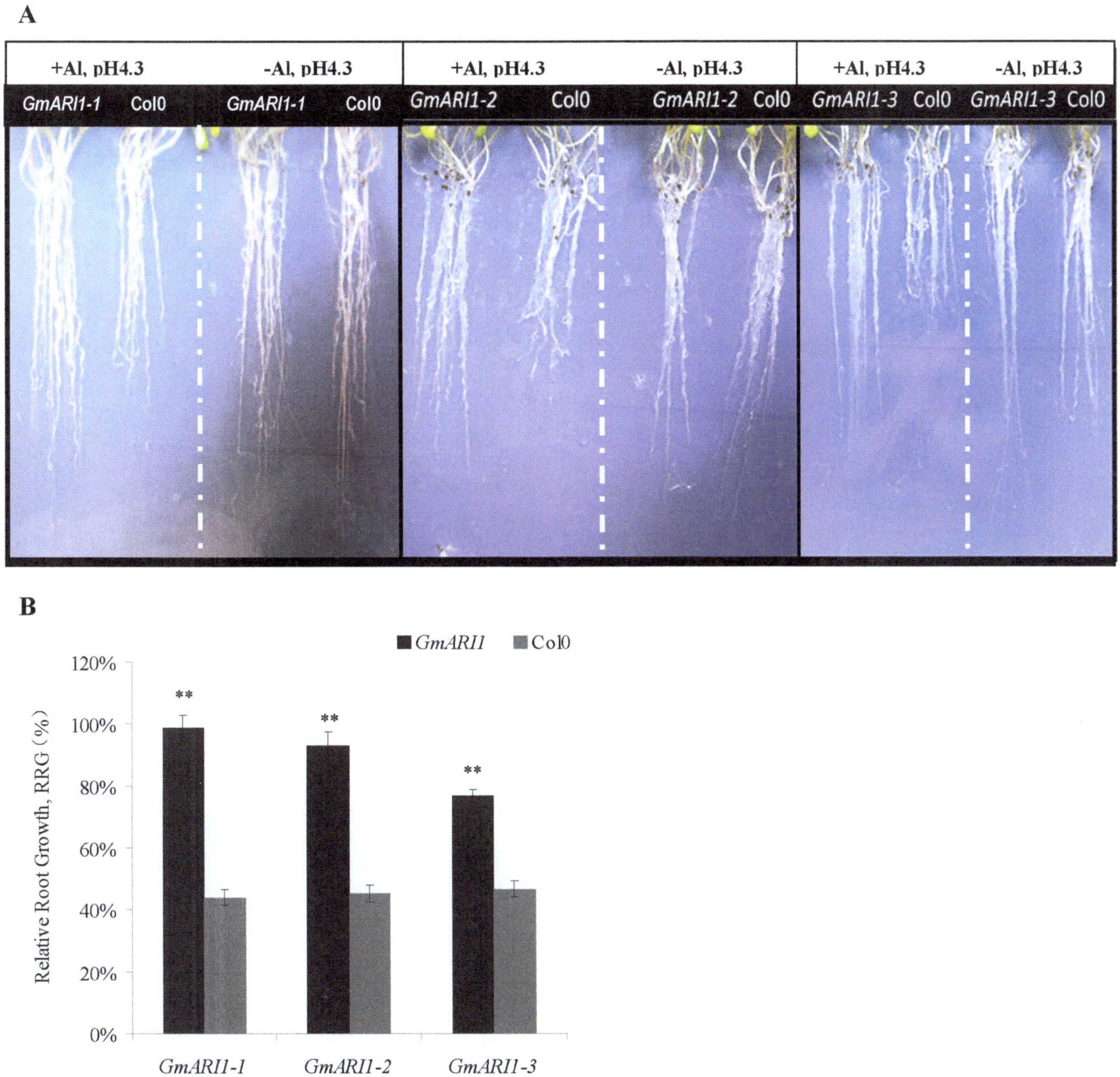

Figure 9. Performance of the transgenic Arabidopsis under 15 μM Al treatment. A. Root phenotypes of the *35S: GmARI1* overexpression lines *GmARI1-3* and the wild type Col-0 grown on 15 μM Al in 1/2 MS (+Al, pH4.3), and 0 μM Al in 1/2 MS (-Al, pH4.3). **B**. Relative root growth (RRG, %) of the transgenic Arabidopsis lines (*GmARI1-1, GmARI1-2, GmARI1-3*) and the wild type (Col-0). RRG was calculated by the root growth length under Al treatment (15 μM Al in 1/2 MS, pH4.3) divided by the root growth length under control (0 μM Al in 1/2 MS, pH4.3). Error bars represent the standard error (SE), ** indicate the significance level of 0.01 by t-tests.

peak during 2 to 4 hours, but no significant change was detected in the leaves. This coincides with the fact that Al toxicity was first and mainly acting on plant roots [38], limiting water and nutrition absorption [39], which further inhibiting the development of whole plant and reducing yield [40], [41]. Therefore in the early stage (less than 4 hours) of Al stress, induced gene expression of *GmARI1* in soybean roots may play important roles to trigger downstream signaling pathways to protect root cells from Al toxicity.

It has been reported that Al induces oxidative stress and DNA damage in plant cells [42–44]. Reactive oxygen species (ROS) constantly attack DNA, leading to oxidative DNA damage [45]. The cell cycle checkpoint regulators could detect and respond to such damage, leading to inhibition of root growth [46]. Protein ubiquitination is involved in DNA transcription and repair, cell cycle and division [1], and is emerging as a critical regulatory mechanism of DNA damage response [47]. Several RING domain-containing E3 ubiquitin ligases play an essential role in response to DNA damage [48]· In vitro ubiquitination assay showed GmARI1 has E3 ligase activity (Fig. 8), therefore we hypothesize that GmARI might be involved in the oxidative DNA damage repair to confer Al tolerance. We investigated the coexpression pattern of *GmARI* (http://bioinformatics.cau.edu.cn/SFGD/), and found Glyma02g15070 was on the top list of coexpression genes with *GmARI*. The homolog gene of Glyma02g15070 in arabidopsis is AT1G49670, which was shown to be involved in oxidative stress tolerance (http://www.arabidopsis.org/). Suppression of oxidative stress might help plants reduce the damage or root growth inhibition [49], [50]. Therefore, the Al tolerance observed in *GmARI* overexpression lines might be due to the improved tolerance to oxidative stress, or/and other signaling cascades.

The activities of cell wall-bound peroxidases in the annual legume, *Cassia tora*, significantly increased with Al concentrations, and were regulated by JA [51]. Another study showed that the SA-signaling and SA-dependent expression of a respiratory burst oxidase homolog gene is involved in Al responsive oxidative burst in Arabidopsis [52]. Here in this study, the expression of *GmARI1* was induced by Al stress, as well as JA and SA treatments in soybean roots. These suggested *GmARI1* might mediate soybean response to Al through oxidative species signals, which may overlap with plant hormone signaling pathways. The T_3 transgenic Arabidopsis plants over-expressing the *GmARI1* gene showed significant improvement in Al tolerance comparing with wild type plants, which further support the important role of *GmARI1* gene in plant response to Al stress.

Supporting Information

Figure S1 Expression and purification of the recombinant GmARI1 proteins. The recombinant His6-GmARI1 proteins were expressed in *E.coli* BL21 (DE3) and analyzed by SDS–PAGE. Lane 1, total proteins from *E. coli* cells before IPTG induction; lane 2, total proteins containing pET28a-GmARI1 from *E. coli* cells after induction by IPTG; lane 3, purified recombinant His6-GmARI1 protein.

Figure S2 RT-PCR confirmation of the transgenic Arabidopsis T3 lines GmARI 1 to 8. (−): Arabidopsis wild ecotype Col-0; (+): plasmid pMDC83-GmARI1.

Figure S3 Expression of the *GmARI1* gene in 2-week-old Arabidopsis plants quantified by qRT-PCR using actin (*AtACT2*) as the reference gene. The Arabidopsis plants were germinated and grown on 1/2 MS medium (pH5.8) for two weeks and then transferred to 1/2 MS medium with 25 μM $AlCl_3$ (pH4.3). Two hours later, tissues were sampled from the wild type Arabidopsis Col0 and homozygous transgenic lines separately (each sample was the mixture of four plants). Error bars are the standard errors from three replications.

Acknowledgments

We thank the staff at JiangPu Experimental Station for their assistance with the soybean seed production, and PLoS One editors and reviewers for their valuable comments and suggestions.

Author Contributions

Conceived and designed the experiments: JG YL XZ NW PC YW. Performed the experiments: XZ NW MG JL. Analyzed the data: XZ NW MG PC YL JG. Contributed reagents/materials/analysis tools: JG TZ YL. Wrote the paper: XZ NW YL JG.

References

1. Hershko A, Ciechanover A (1998) The ubiquitin system. Annu Rev Biochem 67: 425–479.
2. Pickart CM (2001) Mechanisms underlying ubiquitination. Annu Rev Biochem 70: 503–533.
3. Stone SL, Hauksdottir H, Troy A, Herschleb J, Kraft E, et al. (2005) Functional analysis of the RING-type ubiquitin ligase family Arabidopsis. Plant Physiol 137: 13–30.
4. Moon J, Parry G, Estelle M (2004) The ubiquitin-proteasome pathway and plant development. Plant Cell 16: 3181–3195.
5. Barlow PN, Luisi B, Milner A, Elliott M, Elliott R (1994) Structure of the C3H4 domain by 1H-nuclear magnetic resonance spectroscopy: a new structural class of zinc-finger. J Mol Biol 237: 201–211.
6. Lee DH, Choi HW, Hwang BK (2011) The Pepper E3 ubiquitin ligase RING gene *CaRING1* is required for cell death and the Salicylic Acid-dependent defense response. Plant Physiol 156: 2011–2025.
7. Cho SK, Ryu MY, Seo DH, Kang BG, Kim WT (2011) The Arabidopsis RING E3 ubiquitin ligase AtAIRP2 plays combinatory roles with AtAIRP1 in Abscisic Acid-mediated drought stress responses. Plant Physiol 157: 2240–2257.
8. Du QL, Cui WZ, Zhang CH, Yu DY (2010) *GmRFP1* encodes a previously unknown RING-type E3 ubiquitin ligase in soybean (*Glycine max*). Mol Biol Rep 37: 685–693.
9. Marin I, Ferrus A (2002) Comparative genomics of the RBR family, including the Parkinson's disease-related gene *parkin* and the genes to the Ariadne subfamily. Mol Biol Evol 19: 2039–2050.
10. Eisenhaber B, Chumak N, Eisenhaber F, Hauser MT (2007) The ring between ring fingers (RBR) protein family. Genome Biol 8: 209.
11. Aguilera M, Oliveros M, Matinez-Padron M, Barbas JA, Ferrus A (2000) Ariadne-1: A vital drosophila gene is required in development and defines a new conserved family of RING-finger proteins. Genetics 155: 1231–1244.
12. Itier JM, Ibanez P, Mena MA, Abbas N, Cohen-Salmon C, et al. (2003) Parkin gene inactivation alters behaviour and dopamine neuro-transmission in the mouse. Hum Mol Genet 12: 2277–2291.
13. Bromann PA, Weiner JA, Apel ED, Lewis RM, Sanes JR (2004) A putative ariadne-like E3 ubiquitin ligase (PAUL) that interacts with the muscle-specific kinase(MuSK). Gene Expr Patterns 4: 77–84.
14. Moynihan TP, Ardley HC, Nuber U, Rose SA, Jones PF, et al. (1999) The ubiquitin-conjugating enzymes UbcH7 and UbcH8 interact with RING finger/IBR motif-containing domains of HHARI and H7-AP1. J Biol Chem 274(43): 30963–30968.
15. Ardley HC, Tan NG, Rose SA, Markham AF, Robinson PA (2001) Features of the parkin/ariadne-like ubiquitin ligase, HHARI, that regulate its interaction with the ubiquitin-conjugating enzyme, Ubch7. J Biol Chem 276(22): 19640–19647.
16. Marteijn JA, van Emst L, Erpelinck-Verschueren CA, Nikoloski G, Menke A, et al. (2005) The E3 ubiquitin-protein ligase *Triad1* inhibits clonogenic growth of primary myeloid progenitor cells. Blood 106: 4114–4123.
17. Mladek C, Guger K, Hauser MT (2003) Identification and characterization of the *ARIADNE* gene family in Arabidopsis. A group of puptative E3 ligase. Plant Physiol 131: 27–40.
18. Marin I (2010) Diversification and specialization of plant RBR ubiquitin ligase. PLoS ONE 5(7): e11579.
19. Wenzel DM, Klevit RE (2012) Following Ariadne's thread: a new perspective on RBR ubiquitin ligases. BMC Biol 10: 24.
20. Lang-Mladek C, Xie L, Nigam N, Chumak N, Binkert M, et al. (2012) UV-B signaling pathways and fluence rate dependent transcriptional regulation of ARIADNE12. Physiol Plant 145: 527–539.
21. Huang BQ, Bai JH, Xue XQ (2001) Advances in studies on aluminum toxicity and tolerance in plants. Chinese Bulletin of Botany 18(4): 385–395.
22. Liu Q, Zheng SJ, Lin XY (2004) Plant physiological and molecular biological mechanism in response to aluminium toxicity. Chinese Journal of Applied Ecology 15: 1641–1649.
23. Sapra VT, Mebrahtu T, Mugwira LM (1982) Soybean germplasm and cultivar aluminum tolerance in nutrient solution and bladen clay loam soil. Agron J 74: 687–690.
24. Nian H, Huang H, Yan XL, Lu YG (1998) Studies on adaptability of soybean to acidic and aluminum toxicity I Study on the screening and identification for acidic and aluminum-resistance soybeans. Soybean Science 16(3): 191–196.
25. Liang CY, Pineros M, Tian J, Yao ZF, Sun LL, et al. (2013) Low pH, aluminum and phosphorus coordinately regulate malate exudation through GmALMT1 to improve soybean adaptation to acid soils. Plant Physiol 161: 1347–61.
26. Tocquin P, Corbesier L, Havelange A, Pieltain A, Kurtem E, et al. (2003) A novel high efficiency, low maintenance, hydroponic system for synchronous growth and flowering of Arabidopsis thaliana. BMC Plant Biol 3: 2.
27. Yoo SD, Cho YH, Sheen J (2007) Arabidopsis mesophyll protoplasts: a versatile cell system for transient gene expression analysis. Nat Protoc 2: 1565–1572.
28. Wu FH, Shen SC, Lee LY, Lee SH, Chan MT, et al. (2009) Tape-Arabidopsis Sandwich - a simpler Arabidopsis protoplast isolation method. Plant Methods 5: 16.

29. Livak KJ, Schmittgen TD (2001) Analysis of Relative Gene Expression Data Using Real- Time Quantitative PCR and the $2^{-\Delta\Delta CT}$ method. Methods 25: 402–408.
30. Bradford MM (1976) A rapid and sensitive method for the quantification of microgram quantities of protein utilizing the principal of protein-dye binding. Anal. Biochem 72: 248–254.
31. Zhang Y, Yang C, Li Y, Zheng N, Chen H, et al. (2007) SDIR1 is a RING finger E3 ligase that positively regulates stress-responsive abscisic acid signaling in Arabidopsis. Plant Cell 19(6): 1912–1929.
32. Clough SJ, Bent AF (1998) Floral dip: a simplified method for Agrobacterium-mediated transformation of Arabidopsis thaliana. Plant J 16: 735–743.
33. Ardie SW, Xie L, Takahashi R, Liu S, Tahano T (2009) Cloning of a high-affinity K+ transporter gene PutHKT2; 1 from Puccinellia tenuiflora and its functional comparison with OsHKT2; 1 from rice in yeast and Arabidopsis. J Exp Bot 60: 3491–3502.
34. Norusis MJ (1990) SPSS advanced statistics user's guide: SPSS Chicago.
35. Cheng MC, Hsieh EJ, Chen JH, Chen HY, Lin TP (2012) Arabidopsis RGLG2, Functioning as a RING E3 Ligase, Interacts with AtERF53 and Negatively Regulates the Plant Drought Stress Response. Plant Physiol 158: 363–375.
36. Beasley SA, Hristova VA, Shaw GS (2007) Structure of the Parkin in-between-ring domain provides insights for E3-ligase dysfunction in autosomal recessive Parkinson's disease. Proc Natl Acad Sci USA 104: 3095–3100.
37. Marin I, Lucas JI, Gradilla AC, Ferrus A (2004) Parkin and relatives: the RBR family of ubiquitin ligases. Physiol Genomics 17: 253–263.
38. Taylor GJ (1988) The physiology of aluminum tolerance in higher plants. Commun. Soil Sci. Plant Anal 19: 1179–1194.
39. Barcelo J, Poschenrieder C (2002) Fast root growth responses, root exudates, and internal detoxification as clues to the mechanisms of aluminium toxicity and resistance: A review. Environ Exp Bot 48: 75–92.
40. Foy CD (1996) Tolerance of barley cultivars to an acid, aluminum-toxic subsoil related to mineral element concentrations in their shoots. J Plant Nutr 19: 1361–1380.
41. Ma JF, Furukawa J (2003) Recent progress in the research of external Al detoxification in higher plants: a minireview. J Inorg Biochem 97: 46–51.
42. Achary VM, Jena S, Panda KK, Panda BB (2008) Aluminium induced oxidative stress and DNA damage in root cells of Allium cepa L. Ecotoxicol. Environ. Saf 70: 300–310.
43. Achary VM, Panda BB (2010) Aluminium-induced DNA-damage and adaptive response to genotoxic stress in plant cells are mediated through reactive oxygen intermediates. Mutagenesis 25: 201–209.
44. Yamamoto Y, Kobayashi Y, Devi SR, Rikiishi S, Matsumot H (2002) Aluminum toxicity is associated with mitochondrial dysfunction and the production of reactive oxygen species in plant cells. Plant Physiol 128: 63–72.
45. Markkanen E, Hübscher U, van Loon B (2012) Regulation of oxidative DNA damage repair: the adenine: 8-oxo-guanine problem. Cell Cycle 11: 1070–1075.
46. Nezames CD, Sjogren CA, Barajas JF, Larsen PB (2012) The Arabidopsis cell cycle checkpoint regulators TANMEI/ALT2 and ATR mediate the active process of aluminum-dependent root growth inhibition. Plant Cell 24(2): 608–21.
47. Pinder JB, Attwood KM, Dellaire G (2013) Reading, writing, and repair: the role of ubiquitin and the ubiquitin-like proteins in DNA damage signaling and repair. Front Genet 4: 45.
48. Bartocci C, Denchi EL (2013) Put a RING on it: regulation and inhibition of RNF8 and RNF168 RING finger E3 ligases at DNA damage sites. Front Genet 4: 128.
49. Panda SK, Sahoo L, Katsuhara M, Matsumoto H (2013) Overexpression of alternative oxidase gene confers aluminum tolerance by altering the respiratory capacity and the response to oxidative stress in Tobacco Cells. Mol Biotechnol 54: 551–63.
50. Yin L, Mano J, Wang S, Tsuji W, Tanaka K (2010) The involvement of lipid peroxide-derived aldehydes in aluminum toxicity of tobacco roots. Plant Physiol 152: 1406–17.
51. Xue YJ, Tao L, Yang ZM (2008) Aluminum-induced cell wall peroxidase activity and lignin synthesis are differentially regulated by jasmonate and nitric oxide. J Agric Food Chem 56 (20): 9676–9684.
52. Kunihiro S, Hiramatsu T, Kawano T (2011) Involvement of salicylic acid signal transduction in aluminum-responsive oxidative burst in Arabidopsis thaliana cell suspension culture. Plant Signal Behav 6: 611–616.

Methylation Affects Transposition and Splicing of a Large CACTA Transposon from a MYB Transcription Factor Regulating Anthocyanin Synthase Genes in Soybean Seed Coats

Gracia Zabala, Lila O. Vodkin*

Department of Crop Sciences, University of Illinois, Urbana, Illinois, United States of America

Abstract

We determined the molecular basis of three soybean lines that vary in seed coat color at the *R* locus which is thought to encode a MYB transcription factor. RM55-r^m is homozygous for a mutable allele (r^m) that specifies black and brown striped seeds; RM30-*R** is a stable black revertant isoline derived from the mutable line; and RM38-*r* has brown seed coats due to a recessive *r* allele shown to translate a truncated MYB protein. Using long range PCR, 454 sequencing of amplicons, and whole genome re-sequencing, we determined that the variegated RM55-r^m line had a 13 kb CACTA subfamily transposon insertion (designated *TgmR**) at a position 110 bp from the beginning of Intron2 of the *R* locus, Glyma09g36983. Although the MYB encoded by *R* was expressed at only very low levels in older seed coats of the black revertant RM30-*R** line, it upregulated expression of anthocyanidin synthase genes (*ANS2, ANS3*) to promote the synthesis of anthocyanins. Surprisingly, the RM30-*R** revertant also carried the 13 kb *TgmR** insertion in Intron2. Using RNA-Seq, we showed that intron splicing was accurate, albeit at lower levels, despite the presence of the 13 kb *TgmR** element. As determined by whole genome methylation sequencing, we demonstrate that the *TgmR** sequence was relatively more methylated in RM30-*R** than in the mutable RM55-r^m progenitor line. The stabilized and more methylated RM30-*R** revertant line apparently lacks effective binding of a transposae to its subterminal repeats, thus allowing intron splicing to proceed resulting in sufficient MYB protein to stimulate anthocyanin production and thus black seed coats. In this regard, the *TgmR** element in soybean resembles McClintock's *Spm*-suppressible and change-of-state alleles of maize. This comparison explains the opposite effects of the *TgmR** element on intron splicing of the MYB gene in which it resides depending on the methylation state of the element.

Editor: Zongbin Cui, Institute of Hydrobiology, Chinese Academy of Sciences, China

Funding: The research was funded by grants from the United Soybean Board, the Illinois Soybean Association and the USDA. The funders had no role in the design of the study, the data collection and analysis, the decision to publish nor the manuscript preparation.

Competing Interests: United Soybean Board and the Illinois Soybean Association provided funding towards this study. There are no patents, products in development or marketed products to declare. This does not alter the authors' adherence to all the PLOS ONE policies on sharing data and materials.

* Email: l-vodkin@illinois.edu

Introduction

Anthocyanins are end products of three branches of the flavonoid pathway with important functions in plant defense against pathogens and protection from UV light. Because of their antioxidant properties, seeds and vegetables with anthocyanin pigments have added nutritional and health value. Understanding the regulation and expression of all regulatory genes in the anthocyanin metabolic pathway in each major crop plant species is of significance as a model system for gene expression and for improving agronomic and nutritional properties.

We have identified the molecular basis of some of the classical loci leading to anthocyanin production in soybean (*Glycine max*) seed and plant parts. The majority of the cultivated soybean varieties have yellow seeds with hila of various colors but ancestral soybeans (*Glycine soja*) have pigmented seed coats. The pigmentation in seed coat, hilum and pubescence (trichome hairs) is determined by three independent loci, *I* (*Inhibitor*), *R*, and *T* (*Tawny*). The *I* locus controls distribution of anthocyanin and proanthocyanidin pigments and comprises the multigenic, inverted repeat region of chalcone synthase (CHS) genes (*CHS1*, *CHS3* and *CHS4*) [1]. In its dominant form *I*, silences the expression of all nine CHS gene family via short interfering RNAs in a tissue-specific manner in the seed coat [2]. Since CHS is the first committed enzyme in the anthocyanin pathway, the *I* genotypes result in yellow seed coats. Colored soybeans have recessive *I* alleles and the various colors of the seed coats are influenced by the *R* and *T* loci. Black (*i,R,T*), imperfect-black (*i,R,t*), brown (*i,r,T*) and buff (*i,r,t*) seed coats. The *T* locus encodes a flavonoid

3′hydroxylase (*F3′H*) gene, the expression of which drives the synthesis of the anthocyanin cyanidin pathway branch [3].

Recently it has been proposed that *R* maps to a locus encoding a MYB transcription factor which may positively regulate the expression of a *UDP-glucose: flavonoid 3-O-glucosyltransferase* (*UF3GT*) gene that functions in the last step of anthocyanin synthesis [4]. These authors identified the recessive, brown *r* allele to affect the expression of a seed coat-specific R2R3 MYB transcription factor gene, Glyma09g36983 from the sequenced soybean genome as a strong candidate for the *R* locus. In addition, their qRT-PCR analysis supported a correlation between the mRNA levels of the putative *R* gene and a UDP-glucose: flavonoid 3-O-glucosyltransferase (*UF3GT*) gene (Glyma08g07130) that catalyzes the glycosylation of cyanidins to anthocyanins. The recessive *r* (brown) allele at Glyma09g36983 contained a single base deletion that would result in a truncated protein and did not upregulate this *UF3GT* gene. Based on a combined analysis of transcriptome and metabolome data, another recent study identified 20 anthocyanin, flavonoid and phenylpropanoid isogenes that were differentially expressed between black (*iRT*) and brown (*irT*) seeded soybean isolines [5]. This same study suggested that *R* locus candidate might be a transcription factor at the distal end of chromosome Gm09 at or near Glyma09g36840, 120 kb from the Glyma09g36983 MYB factor mapped by Gillman et al [4].

The r^m allele of the *R* locus conditions a variegated or mutable distribution of black spots or concentric rings of black pigment superimposed on an otherwise brown seed coat (Figure 1). This phenotype is sometimes referred to as a striped phenotype. The r^m allele present in a plant introduction line was backcrossed for five generations into a brown seeded Clark line with homozygous *i, r, T* genotype to create isoline L72-2040 with *i* r^m, *T* genotype and released in 1972 by soybean breeder R.L. Bernard of the USDA Agricultural Research Service. We discovered that the r^m variegated seed of L72-2040 demonstrated somatic and germinal instability to yield fully black seed [6]. One of these lines, here named as RM30, is shown in Figure 1 and the gene symbol *R** is used to differentiate this stable revertant allele, derived from the r^m line, from the standard *R* allele that specifics black seed coats. The mutable line is here referred to as RM55-r^m and the brown isoline into which the r^m allele was backcrossed is RM38-*r* (Figure 1).

Using primers based on the Glyma09g36983 as the *R* locus [4], we here report finding a large 13 kb CACTA transposable element insertion in the second intron of the MYB factor in the RM55-r^m mutable soybean line along with a 929 bp PCR product representing excision products that restore the MYB transcription factor sequence. The transposon represents a second CACTA subfamily in soybean based on its transposase and subterminal repeat sequences that most resemble the first CACTA defective element described in soybean, *Tgm1* [7], and is more distant to *Tgmt* [8] and *Tgm9* [9]. Surprisingly the 13 kb CACTA transposon (designated *TgmR**) is retained in the *R** allele of a stable black-seeded revertant isoline. Using quantitative RNA-Seq data from a developmental series of seed coats from the black-seeded RM30-*R** revertant and the RM38-*r* brown-seeded isoline, we also demonstrate that the MYB factor encoded by the *R* gene appears to induce the expression of anthocyanidin synthase (*ANS*) genes at late stages of seed development in the black RM30-*R** line. Thus, our data further support Glyma09g3698 as the *R* locus in soybean and show that it regulates *ANS* genes and potentially other genes in the anthocyanin pathway despite a very low level of *R* gene transcripts. The *R* encoded MYB protein possesses four of the amino acids present in the subgroup five of MYB proteins that are involved in anthocyanin regulation by transcriptional activation [10].

Figure 1. Phenotypes of Seeds Coats at Late Developmental Stages for Three Soybean Isolines. Differential anthocyanin expression in the seed coats of three soybean isolines with variants of the *R* locus alleles. RM55-r^m with variegated black-brown seed coat; RM30-*R** a stable revertant black-seeded isoline derived from RM55-r^m; and RM38-*r* a brown seeded isoline used as the recurrent parent. See Table 1 for more information. Left, green seed of approximately 300–400 mg fresh weight; middle, dessicating seed of approximately 300–400 mg fresh weight; right, mature dry seed.

Further we show that intron splicing was accurate despite the presence of the 13kb *TgmR** element in Intron2 of the MYB gene. As determined by whole genome methylation sequencing, we show that the *TgmR** sequence was relatively more methylated in the RM30-*R** line than it is in the mutable RM55-r^m progenitor line. Following mechanisms elucidated for the *Spm*-suppressible alleles and change of state alleles in maize [11,12,13], our data suggest that a transposase expressed from an element in the hypomethylated mutable RM55-r^m line binds to the subterminal repeats and prevents intron splicing of the *R* gene (resulting in the brown seed coat background color) while it promotes excision events in some cells resulting in restoration of the MYB gene structure without the *TgmR** element and expression of the functional MYB protein leading to black sectors and stripes. The stabilized RM30-*R** revertant line, which had a higher level of methylation, either lacks the active transpose or it cannot bind to its subterminal repeats, thus allowing intron processing to proceed normally, although at a reduced level, resulting in enough of the MYB protein to stimulate anthocyanin production and thus black seed coats. In this regard, the *TgmR** element that causes the variegated black and brown seed coats in soybean resembles McClintock's genetically and molecularly characterized *Spm*-suppressible and change of state alleles [reviewed in 12,13].

Table 1. Genotypes and phenotypes of soybean cultivars and mutant isolines used in this study.

Lab name	Genotype	Seed Coat Phenotype	Cultivar/origin PI number	PI number
RM55	*i,r^m,T*	Black-brown striped	Clark isoline L72-2040	PI547559
RM30	*i,R*,T*	Black	Revertant in L72-2040	NA
RM38	*i,r,T*	Brown	Clark isoline L67-3484	PI547494
PI91073	*i,r^m,T*	Black-brown striped	Source of *r^m* allele	PI91073
Wm43	*i^i,R,T*	Black hilum	Williams	PI548631
UC55	*i,R,T*	Black	Mutation in Williams	PI518670
Sooty	*i,R,T*	Black	Sooty	PI548415
Peking	*i,R,T*	Black	Peking	PI417243
Jack	*I,r,T*	Yellow	Jack	PI540556
UC501	*I,r,T*	Yellow	Harosoy	PI548575
T157	*i,R,t*	Imperfect black	Mutation in Richland	PI548182

All cultivars are homozygous for the alleles indicated. The varieties are searchable by the PI number in the USDA Germplasm Resources Information Network (GRIN). NA, not applicable.

L numbers are isoline number designations. The r^m allele from PI91073 that specifies a variegated phenotype had been repetitively backcrossed into the Clark brown seeded isoline L67-3484 (here designated as RM38 with the *r* allele) to create the variegated Clark isoline L72-2040 (here designated as RM55 with the r^m) by R.L. Bernard of the USDA. The L72-2040 line demonstrated somatic and germinal instability to yield fully black seed. One of these lines is here designated as RM30 and the gene symbol R* is used to designated this revertants of the r^m line from the standard *R* allele that specifics black seed coats.

Results

The Mutable r^m and Revertant R^* Alleles Both Contain a Large Insertion in a MYB Transcription Factor Encoded by Glyma09g36983

Figure 1 illustrates the phenotypes of the lines homozygous for the variegated r^m allele (here designated as line RM55-r^m) that was repetitively backcrossed into the genetic background of a brown seeded Clark isoline (RM38-*r*) and of a stable revertant R^* allele (RM30-R^*). Based on the report by Gillman et al. (2011) [4] that Glyma09g36983 (Gm09: 42562649–42564660 (+ strand) of the Williams 82 genome) is a potential candidate for the *R* locus, we designed primers to amplify this region from these three lines. Successful amplification was achieved with two long PCR primers (R6990FP1 and R6990RPB) and a Long Amplification (LA) PCR method [8] (see Methods and File S1). Figure 2A shows the two larger ~14 kb DNA fragments amplified from RM30-R^* and RM55-r^m and the smaller 929 bp fragments in RM38-*r* and RM55-r^m. The 929 bp fragments are the expected size of the DNA sequence between the two oligo DNA primers (FP:R6990FP1and RP:R6990RPB). Thus, the larger band revealed the existence of a DNA insertion in the RM30-R^* and RM55-r^m isolines somewhere in the Glyma09g36983 gene located between the 5′end and the first part of Intron2. The amplification was repeated with independent plant material to confirm the large insertion in the RM30-R^* revertant line since the presence of a large insertion in the phenotype that mimics the wild type phenotype was unusual. The unadapted plant introduction source line for the r^m allele (PI91073) was also checked (data not shown) and produced similar sized amplicons of 14 kb and 929 bp as found in the Clark line RM55-r^m that is shown in Figure 2.

To examine whether such an insertion or any other macromolecular changes were present in other soybean lines carrying the alleles for black (*R*) or brown (*r*) pigments (Table 1), we amplified a large portion of the Glyma09g36983 in PCR reactions using the 37-nt forward primer R6990FP1 and a 31-nt reverse primer R6990RP1 located 491-nt downstream of the 37-nt R6990RPB primer (File S1). The expected PCR amplification product in the absence of insertions or deletions should be 1.45 kb in size. Figure 2B shows that all PCR amplified DNA fragments were of that size. Thus, the large insert seems to be specific for the r^m allele or the RM30-R^* line derived from it.

A 13 kb Insertion in the R^* and r^m Alleles is a Novel Transposon of the *Tgm1* Subfamily of CACTA Elements

The ~14 kb and 929 bp DNA fragments amplified from the mutable line RM55-r^m, the stable revertant isoline RM30-R^*, and the RM38-*r* self-brown soybean line (Figure 2A), were extracted from the gel pieces and cleaned using the Zymoclean Large Fragment DNA Recovery Kit protocol. The two larger ~14 kb DNA fragments were sequenced using the 454 method at the W. M. Keck Center for Comparative and Functional Genomics at the University of Illinois at Urbana-Champaign (see Methods section and Table S1). The 929 bp PCR fragments were sequenced using the Sanger method at the same sequencing facility. Contigs assembled from sequence reads of whole genome resequencing (Table S1) also allowed confirmation of the structure of these amplicons from RM30, RM55 and RM38.

The contig assembly resulting from sequencing the ~14 kb fragment in the RM30-R^* isoline with black seeds revealed the insertion of a 13 kb CACTA transposon near the beginning of Glyma09g36983 Intron2 (File S1) as illustrated in Figure 3A. This newly identified CACTA transposon that we named *TgmR** is 13,024 bp in length (File S2) and has the characteristic CACTA ends and a three-base-target site duplication (ATG). In addition, it features reiterated direct and inverted sequence motif in the subterminal regions able to form secondary DNA/RNA stem-loop structures (Files S3).

The consensus sequence of the motif in the repeats is shown at top of Figure 3B where it is compared to the consensus sequence motifs of other CACTA transponsons studied in *G. max* and other plant species. In a previous study of an autonomous 20.5-kb CACTA element (*Tgmt**) in *G. max* [8], we noticed a clear distinction in the consensus sequence motif between three characterized CACTA transposons (*Tgmt**, *Tgmw4m* and *Tgm-Express1*) and *Tgm1*. The *Tgm1* sequence motif diverged from all three others in six nucleotide positions shown in red type in

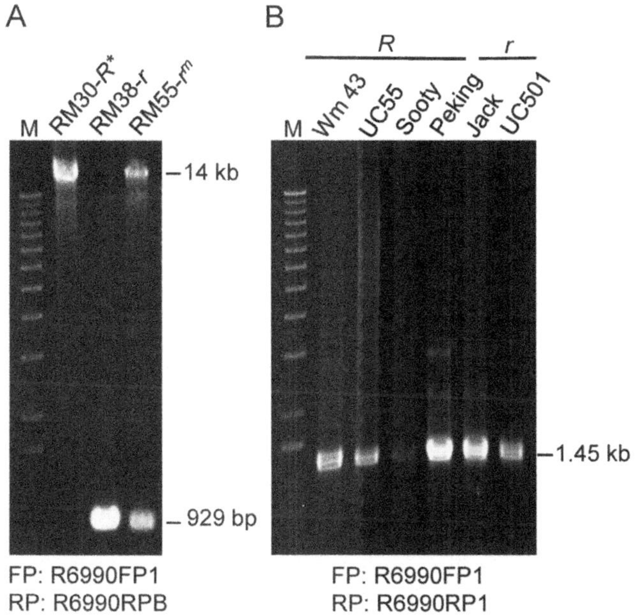

Figure 2. Amplification of a Large Insertion in the Glyma09g36983 Gene of RM55-*r*^{m} and the Revertant Black RM30-*R Isolines.** (A) Agarose gel displaying a ~14 kb PCR fragment in the RM55-*r*^{m} mutable and RM30-*R** revertant alleles of the *R* locus. A small 929 bp fragment was the only amplification product from the *r* allele of the RM38-*r* brown-seeded line and it was the expected size of the DNA fragment comprised between the two oligo DNA primers (FP, forward primer: R6990FP1 and RP, reverse primer: R6990RPB) designed to amplify the 5′end portion of the Glyma09g36983 gene encoding a putative R2R3 MYB transcription factor. The 929 bp fragment was also an amplification product from the DNA of the mutable RM55-*r*^{m}. These results predict an insertion of 13 kb in one allele of the mutable RM55-*r*^{m} line that is maintained in the revertant RM30-*R** black seed isoline. (B) Agarose gel displaying a 1.45 kb PCR fragment of the Glyma09g36983 gene amplified from four other soybean varieties with the standard *R* allele as well as from two soybean lines with an *r* allele (see Table 1 for full genotypes). The reverse primer in these instances was R6990RP1, situated 491 bp downstream from R6990RPB used in (A). The size marker "M" is a 1 kb ladder.

Figure 3B. This suggested the existence of two distinct CACTA transposon families in soybean. The *TgmR** sequence motif is almost identical to the one in *Tgm1* [7,14] with five of the six divergent nucleotides (nt) being identical in both, *TgmR** and *Tgm1*, implying that they are more closely related and belong to the same subfamily of *G. max* CACTA elements (Figure 3B). Further evidence is the more than 1,000-nt of sequence identity between *TgmR** and *Tgm1* at the 3′end (File S2) and more than 100 continuous-nt at the 5′end of the transposon. The remainder of the *Tgm1* DNA sequence, a defective element of 3,556-nt total size, deviates considerably from the *TgmR** sequence.

To determine with more accuracy the total coding capacity of this 13 kb newly identified CACTA transposon *TgmR**, we used the Softberry (http://linux1.softberry.com/berry.phtml) gene prediction program to find two putative ORFs. A large ORF1 with 10 or 11 exons (from nt 920 to 7,481) and a small ORF2 with three exons (from nt 9,347 to 12,255). The larger ORF1 predicted using *Glycine max*, *Arabidopsis* and *Vitis vinifera* genomes as models, coincides with the predicted transposase of the tnp2 family. We previously identified another potentially autonomous CACTA transposon, *Tgmt**, inserted in an *F3′H* (flavonoid 3′ hydroxylase) allele to be 20.5 kb in size and encoding a transposase with tnp2 and TNP1/EN/SPM domains [8] (Acc. No. EU190440). This same transposon was found inserted at the *W4* locus encoding dihydroflavonol-4-reductase 2 (DFR2) and was called *Tgm9* [9]. However, DNA sequence alignment between *Tgmt** and *TgmR** showed little similarity between the two elements except for the transposase sequence region where a stretch of 3,040 bp (*Tgmt**: 3373….6413; *TgmR**: 3299….6348) was 71% similar at the nucleotide level. This divergence is additional evidence in support of two distinct CACTA transposon families in soybean.

To determine the redundancy of the *TgmR** sequence in the Williams 82 genome we attempted a BLAST search in Phytozome [15] using its "unmasked" feature to allow repetitive DNA alignments. The sequence, or part of it, is so repeated that no result was returned when the entire *TgmR** sequence was used. However, when the BLAST search was done with portions of the *TgmR** sequence, then, many repeats all through the Williams 82 chromosomes showed sequence similarity. The most extensive regions of sequence similarity to *TgmR** in the Williams 82 genome were found in chromosomes Gm05 and Gm15, where Glyma05g23600 (6,412 nt) and Glyma15g21480 (5,909 nt) reside, respectively. Glyma05g23600 is annotated as related to transposase family tnp2 and 6,406 nt of its sequence are comprised in *TgmR** (Figure 3A). Glyma15g21480 has the same annotation but its sequence lacks 2,886-nt of the Glyma05g23600 3′-end sequence. The *TgmR** sequence similarity to Glyma15g21480 extends 668-nt upstream of the Glyma05g23600 5′-end sequence similarity (Figure 3A). In addition, the *TgmR** right border has similarity to a 4,191 nt sequence fragment in Gm08 (94002..98330) with no predicted gene annotation (Figure 3A).

Another previously characterized *G. max* deletion derivative transposon, *Tgm5* that is 1,002-nt in size [16] (Acc.No. X13528.1), also contains an entire tnp2 transposase domain that aligns with 89% similarity to a 1,002-nt region of *TgmR** (5,952–6,954-nt) and Glyma05g23600 (Gm05:29,250,569..29,251,571) but with 93% similarity to a 1,002-nt region of Glyma15g21480 (Gm15:19,6 88,859..19689861). Both these genes in Gm05 and Gm15 have the tnp2 transposase highly similar (89 or 93%) to the *Tgm5* tnp2 transposase domain. It was estimated [16] that *Tgm5* is 39% similar to the ORF1 of the *Zea mays En-1/Spm* transposable element which is the first autonomous CACTA element in maize and corresponds to one of McClintock's original transposable element systems. In addition to those two loci in Gm05 and Gm15, the 1,002-nt sequence portion (*Tgm5*) with the tnp2 transposase domain is repeated with some variability 81 times in Gm05 and 93 times in Gm15. In a similar fashion, that same sequence is repeated in all other 18 chromosomes of the *G. max* Williams 82 genome.

The results of sequencing the 929 bp amplification fragments derived from either the RM55-*r*^{m} mutable line or the RM38-*r* line uncovered a significant difference between the two sequences, mainly a "C"-nt deletion in Exon2 of the RM38-*r* line with the brown seed color phenotype. This base deletion causes a frame shift resulting in a prematurely terminated translation product of only 85 amino acids, 152 amino acids shorter than the mature, intact protein with 237 amino acids. In contrast, the 929 bp fragment amplified from the RM55-*r*^{m} mutable black and brown striped seeded line has the "C"-nt that is deleted in the RM38-*r* line revealing a divergence between the genetic sources of those two soybean lines. The RM38-*r* line (self-brown L67-3484) line that was used to introgress the mutable (*r*^{m}) allele has the same "C"-nt deletion in Exon2 as described by Gillman et al (2011) in mapping the *R* locus alleles.

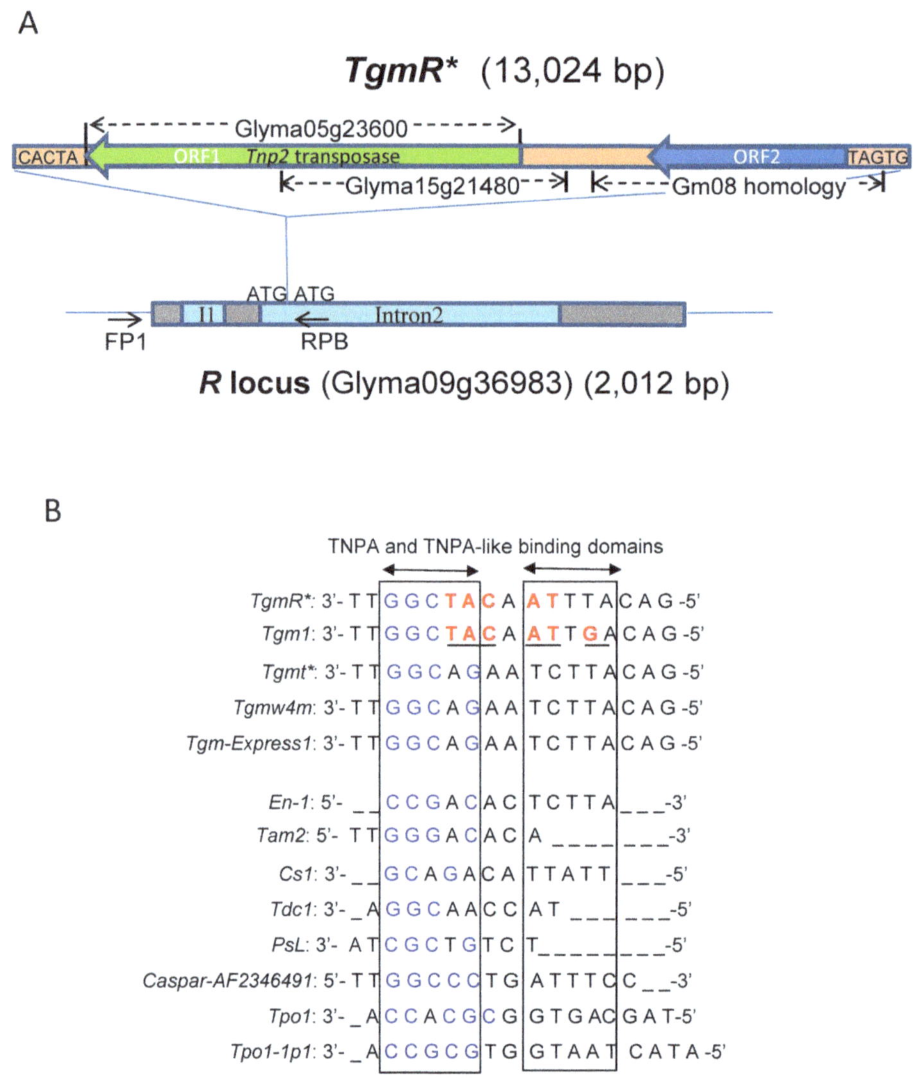

Figure 3. The 13 kb CACTA Insertion, Designated *TgmR, and its Subterminal Repeats Within the r^m and *R** Alleles.** (A) Schematics of the CACTA transposon (*TgmR**) insertion in Intron2 of the *R* locus encoding a R2R3 MYB transcription factor. The *R* locus gene, Glyma09g36983, depicted in gray (exons) and blue (introns) has a 13 kb CACTA transposon insertion 110 bp from the beginning of Intron2 in the *R* locus of both the variegated RM55-r^m and the black-seeded revertant RM30-*R** soybean isolines. The ATG duplication is shown at the site of insertion. The locations of the two primers used to amplify the 14 kb fragment containing the transposon insertion are depicted by the two small arrows, FP1 at the 5′ UTR and RPB in Intron2 near the insertion site. The 13 kb *TgmR** transposon is a CACTA element based on the terminal ends sequences and the *Tnp2* transposase it encodes. Both the location of the transposase and the end sequences are marked. Sequences of high homology to this transposase region were found in chromosomes Gm05 and Gm15 and the Glyma models predicted for the Williams 82 genome are indicated. Likewise, the 3′ end of *TgmR** has high homology to a sequence in chromosome Gm08 and to *Tgm1*, another smaller CACTA element of the soybean. Green and blue arrows represent the putative ORF1 (Tnp2 transposase) and ORF2 respectively. (B) Alignment of multiple CACTA transposon subterminal repeated motifs. The consensus sequences of the *TgmR** subterminal direct repeats were aligned to other CACTA consensus sequences from other *Glycine max* transposons (*Tgm1*, *Tgmt**, *Tgmw4m* and *Tgm-Express1*) in addition to other subterminal repeats from CACTA transposons in other plants. Portions of the sequence domains for TNPA-like protein binding are boxed. One of these motifs are GC rich (left) and the second AT rich (right). Based on the similarities of these terminal repeat motifs, it is clear that *TgmR** is more closely related to *Tgm1* than to the other three characterized *G. max* transposons and all of the other CACTA transposon motifs in the lower panel.

Differential Expression of the MYB Transcription Factor in Immature Seed Coats of the Black Revertant (RM30-*R**) and Brown (RM38-*r*) Isolines

We compared the expression Glyma09g36983 MYB transcription factor in the seed coats of the black revertant RM30-*R** soybean line to that in the defective brown RM38-*r* soybean line at various stages of seed development. Total RNA was extracted from seed coats at five stages of seed development (Figure 4A) following a modified method that prevents RNA adhesion to the proanthocyanidins (tannins) that accumulate in the vacuoles of pigmented seed coats [17,18]. The five stages of seed development were chosen based on the fresh weight of entire seeds: 100–200, 200–300, 300–400 and 400–500 mg with green cotyledons, and a later stage in which seed desiccation had initiated judging by the yellowing of cotyledons and a lower fresh weight of 300–400 mg.

The RNA-Seq results reflecting the level of transcript reads aligning to the Glyma09g36983 (Glyma09g36990 in Phytozome *G. max*_109 v.1a-1.0) sequence are given in RPKMs (reads per

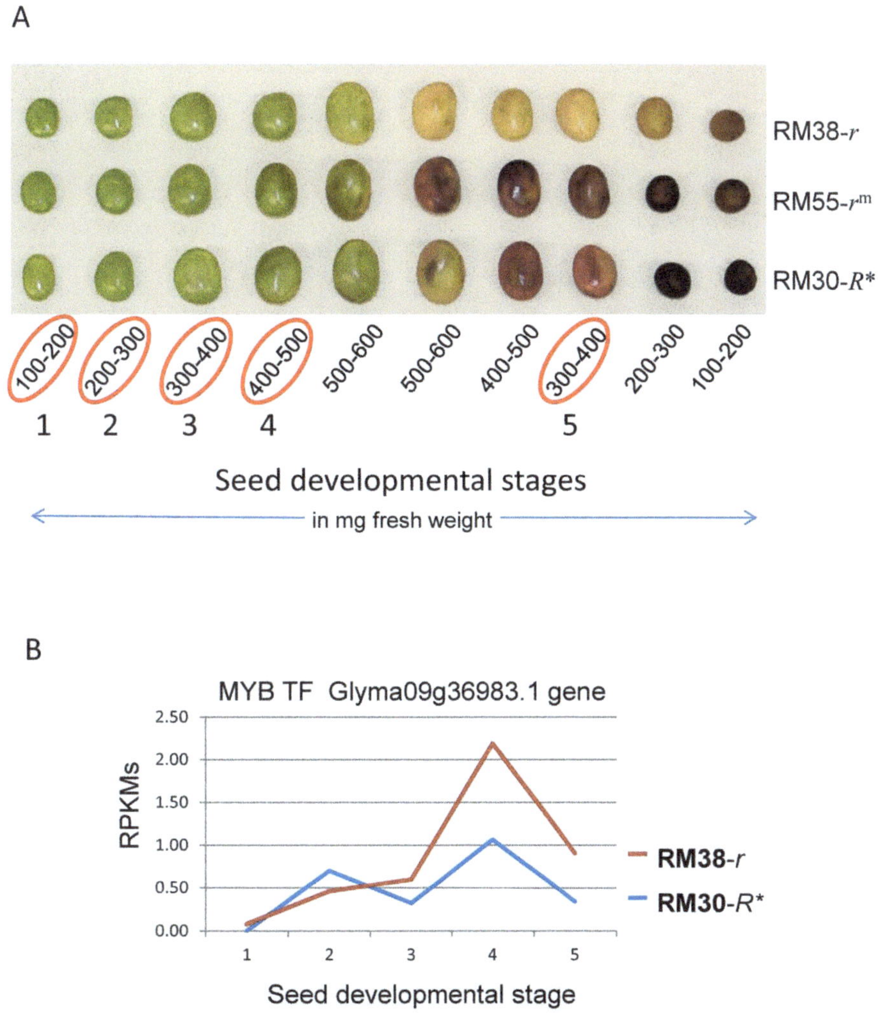

Figure 4. Differential Expression of the MYB Transcription Factor Encoded by the *R* Locus in Developing Seed Coats of a Stable Black-Seeded Soybean Revertant Line RM30-*R and a Brown-Seeded Line RM38-*r*.** (A) Soybean seed developmental stages for the RM38-*r* (brown), RM55-r^m (variegated) and RM30-*R** stable black revertant. Encircled in red are the five developmental stages (in mg seed fresh weight) chosen for the RNA-Seq analysis. See Table S1 for the sequence read counts of each sample which ranged from from 30 to 77 million. (B) Expression of Glyma09g36983 in RPKMs plotted against the five stages of seed development in mg seed fresh weight as shown in (A) above. 1: 100–200 mg, 2: 200–300 mg, 3: 300–400 mg, 4: 400–500 mg and 5: 300–400 mg as the seeds enter desiccation. The solid red line represents transcripts derived from seed coats of the RM38-*r* brown-seeded line without the *TgmR** but with a "C"-nt deletion in Exon2. The blue line represent the expression in seed coats of the RM30-*R** black revertant line with the *TgmR** insertion in Intron2.

kilobase of gene model per million mapped reads) as shown in in Table 2 (top row) and graphically in Figure 4B. This MYB transcription factor was not expressed at the early stages examined (100–200 mg) and at very low levels later in development when the anthocyanins and proanthocyanins start pigmenting the seed coat. The quantitative levels were 1.07 RPKM in the RM30-*R** black revertant that contains the *TgmR** insertion and only 2.19 in the RM38-*r* brown isoline. This RPKM ratio of 0.49 (p-value: 0.069) indicates no significant differences of expression of the Glyma09g36983 gene in the RM30-*R** (black) and RM38-*r* (brown) seed coats at the peak of its measured expression in the 400–500 mg seed weight.

A biological replicate of the 400–500 mg weight range also showed very low levels in each genotype, although the differential expression was greater with 0.38 RPKM in the RM30-*R** black revertant line versus 3.7 in the RM30-*r* brown isoline (Table 3). Thus, the *r* allele of the brown seed coats appears to be expressed at a higher level than the *R** allele that is interrupted by the *TgmR** element, although phenotypically, the RM30-*R** black revertant line more closely resembles black seed coats with the standard *R* allele. Table 3 also shows the results of RNA-Seq for two soybean lines containing the standard *R* allele with black seed coats, T157 and UC55. These lines have RPKMs ranging from 2.55 to 5.9 in two of the surveyed weight ranges near the peak of *R* gene expression. Thus, this MYB factor has relatively low expression even in the fully functional standard *R* alleles with black seed coats. The lower level of *R* gene expression in RM30-*R** could be explained by the *TgmR** insertion in Intron2 reducing functional transcript levels from the allele. However, the lower level is still sufficient to produce black seed coats.

We examined the expression of three MYB transcription factors, Glyma09g36966, Glyma09g37010, and Glyma18g49670 that have transcript sequence similarity of 90%, 87%, and 84% respectively to Glyma09g36983. The level of expression of these three genes as shown in Tables 2 and 3 is even lower or nil, and the RPKM values associated with Glyma09g36966 may reflect the

Table 2. Differential RNA expression of anthocyanin pathway genes in developing seed coats of lines RM30-*R** (black) and RM38-*r* (brown).

	RM30-*R** RPKMs in wt. range					RM38-*r* RPKMs in wt. range					RM30/RM38	p-value	Annotation	Locus
Gene Model	**100**	**200**	**300**	**400**	**300**	**100**	**200**	**300**	**400**	**300**	**400**			
Glyma09g36983.1	0.00	0.70	0.32	**1.07**	0.34	0.08	0.46	0.60	2.19	0.91	0.49	0.069	MYB	*R*
Glyma09g36966.1	0.00	0.16	0.06	0.10	0.00	0.03	0.10	0.00	0.40	0.15	0.25	0.384	MYB	
Glyma09g37010.1	0.00	0.00	0.00	0.00	0.00	0.00	0.00	0.00	0.00	0.11	0.00	1.000	MYB	
Glyma18g49670.2	0.00	0.00	0.00	0.00	0.00	0.00	0.00	0.00	0.00	0.11	0.00	1.000	MYB	
Glyma01g43880.1	509.11	212.43	210.65	122.96	19.46	526.37	120.74	41.10	44.44	20.76	2.77	7.59 E-11	CHS7	*I*
Glyma11g01350.1	382.08	135.35	138.73	75.69	13.92	395.90	76.34	25.61	28.47	14.87	2.66	6.78 E-10	CHS8	*I*
Glyma20g38560.1	5.43	13.65	11.07	17.57	11.87	7.58	17.15	9.33	12.24	7.73	1.43	0.927	CHI1A	
Glyma20g38580.1	14.65	46.66	23.13	36.87	10.31	21.70	39.93	11.63	12.26	9.30	3.01	1.75 E-10	CHI2	
Glyma06g21920.1	71.05	11.73	14.52	4.42	1.14	61.60	4.07	1.59	3.33	0.79	1.33	0.302	F3'H	*T*
Glyma13g04210.1	0.12	0.08	0.08	4.66	4.87	0.18	0.68	0.01	0.04	0.34	4.66	1.15 E-33	F3'5'H	*W1*
Glyma02g05450.1	326.86	162.99	119.24	124.64	20.30	340.30	90.65	26.45	38.72	14.52	3.21	6.24 E-14	F3H	*Wp*
Glyma14g07940.1	76.21	66.14	60.19	45.78	2.64	76.98	35.18	15.30	14.63	1.52	3.13	5.16 E-12	DFR1	*W3*
Glyma17g37060.1	4.71	1.19	0.33	3.92	2.87	3.56	1.65	0.06	0.35	0.15	3.92	4.98 E-14	DFR2	*W3*
Glyma01g42350.1	76.09	36.74	12.86	**35.41**	11.00	76.07	24.19	2.17	**3.48**	1.47	**10.17**	**2.12 E-38**	ANS2	
Glyma11g03010.1	88.80	42.81	17.37	**39.75**	9.61	91.77	26.93	2.07	**3.78**	1.03	**10.51**	**2.71 E-40**	ANS3	
Glyma08g07130.1	2.01	5.55	7.68	12.79	12.51	3.40	12.11	7.78	9.87	3.66	1.29	0.281	UGT78K2	
Glyma07g30180.1	1.43	0.67	1.39	0.97	1.04	1.93	1.45	1.04	0.94	0.32	1.03	0.915	UGT78K1	
Glyma08g06630.1	0.98	0.09	0.40	0.49	0.17	1.27	0.16	0.04	0.55	0.23	0.89	0.712	ANR1	
Glyma08g06640.1	0.06	0.00	0.02	0.04	0.00	0.06	0.00	0.00	0.03	0.00	1.33	0.972	ANR2	
Glyma05g36210.1	58.27	36.87	47.81	17.85	4.02	43.20	15.16	8.66	4.77	2.27	3.74	6.33 E-12	AOMT	

RPKMs are shown for seed coats from seed of the five seed fresh weight ranges denoted 100–200, 200–300, 300–400, 400–500 mg and 300–400 mg desiccating seed coats as shown in Figure 4A. Ratios and p-values are shown for the 400–500 mg weight range, and date for the ANS genes at this stage are in Bold.
Glyma09g36983 in Phytozome *G. max*_189 a1_1.1 was previously named Glyma09g36990 in Phytozome *G. max*_109 a1_1.0.
Glyma09g36966 in Phytozome *G. max*_189 a1_1.1 was previously named Glyma09g36970 in Phytozome *G. max*_109 a1_1.0.

Table 3. Differential RNA expression of Glyma09g36983 and related MYB transcription factor genes in seed color isolines.

Cultivar	RM30-*R**				RM38-*r*				T157-*iR*		UC55-*iR*	
Phenotype	black				brown				imp. black		black	
Weight range	300¹	300²	400¹	400²	300¹	300²	400¹	400²	300	400	300	400
Glyma09g36983.1	0.32	0.12	1.07	0.38	0.60	1.91	2.19	3.70	3.26	2.55	2.73	5.90
Glyma09g36966.1	0.06	0.04	0.1	0.04	0	0.47	0.4	0.93	0.48	0.17	0.21	0.71
Glyma09g37010.1	0	0	0	0	0	0.02	0	0	0	0	0	0
Glyma18g49670.1	0	0	0	0	0	0	0	0	0	0	0	0

RPKMs are shown for seed coats from seed of two seed fresh weight ranges denoted 300–400 and 400–500 mg as shown in Figure 4A.
Superscripts 1 and 2 for the 400–500 mg seed fresh weight represent two independent RNAseq extractions and sequencing determinations.
Glyma09g36966, Glyma09g37010, and Glyma18g49670 have transcript sequence similarities of 90%, 87%, and 84%, respectively, to Glyma09g36983.
Glyma09g36983 in Phytozome *G. max*_189 a1_1.1 was previously named Glyma09g36990 in Phytozome *G. max*_109 a1_1.0.
Glyma09g36966 in Phytozome *G. max*_189 a1_1.1 was previously named Glyma09g36970 in Phytozome *G. max*_109 a1_1.0.

number of sequence reads aligning to the 530-nt stretch with 90% transcript sequence similarity to Glyma09g36983. Three mismatches was the cutoff used to determine the number of RNA sequence reads aligning to a given gene transcript using Bowtie. A more stringent alignment search with zero mismatches supported the above statement and strongly suggested that none of the three related MYB transcription factors are expressed in the soybean seed coats.

Differential RNA Expression of Key Genes in the Anthocyanin and Proanthocyanidin Metabolic Pathways between the Black (RM30-*R**) and Brown (RM38-*r*) Seed Coats

Once we had evidence that the *TgmR** insertion slightly suppressed the expression of the *R* locus gene, Glyma09g36983, at late stages of seed development, we questioned at which point in the anthocyanin biosynthetic pathway the encoded MYB transcription factor exerted its regulatory control. Towards that end, we compared the level of mRNA expression of well characterized genes in the pathway in the RNAseq data sets obtained from the black and brown seed coats of the RM30-*R** revertant and RM38-*r* defective line, at all five stages of seed development under study (Table 2).

Contrary to the difference observed for the expression of the MYB gene in the two lines in Figure 4B, where higher levels were measured in the brown RM38-*r* seed coats, all other pathway genes analyzed had higher RPKM expression values in the black RM30-*R** line at the late stages of seed development (Figures 5 and 6, Table 2). These increases occurred at the developmental time when the *R* locus gene, Glyma09g36983, is activated. One exception is the *ANR* genes for which the low level of expression at the late stages of seed development was maintained in the seed coats of the black RM30-*R** line as well as in the brown RM38-*r* line (Table 2). This is consistent with the function of this gene's product in directing the synthesis of proanthocyanidins now in competition for its substrate that is being diverted towards anthocyanin synthesis in the black seed coats at the late stages of seed development as shown in Figure 7.

Several of the genes involved in the anthocyanin pathway (*CHS*, *F3'H*, *F3H*, *DFR*, *ANS*, *ANR1* and *AOMT*) are expressed at higher level early in seed development (100–200 mg seed fresh weight) and the amount of transcripts decline in seed coats of both lines as seed growth advances (Figures 5 and 6). However, this decline is arrested or reversed for all these genes in the seed coats of the black-seeded RM30-*R** line at the third and fourth stages of seed development. For those genes that are not expressed or expressed at very low level earlier in development (*CHI*, *F3'5'H* and *UFGT*), the expression increased to a slightly higher level in the seed coats of the black RM30-*R** seeds during the late stages of seed development. Despite the two *UFGT* genes (*UGT78K2* and *UGT78K1*) transcript sequences being 92% similar, the sequence reads from the seed coat RNA samples aligned better to the *UGT78K2* transcript and differentially between the two lines (Table 2 and Figure 6). This suggests that the *UGT78K1* gene may not be expressed in the seed coat but rather in the cotyledons as it has been reported previously [4].

Although all the mentioned genes appear to be differentially expressed in the seed coats of the black and brown seeds, the ones that show the highest RPKM ratios and lowest p-values are the *ANS* genes with an RM30/RM38 of 10.17 (p-value 2.12 E-38) and 10.51 (p-value 2.71 E-40) respectively, at the 400–500 mg seed weight (Table 2). The transcript sequences of the two *ANS* genes are 94% similar and it is not surprising that the numbers of

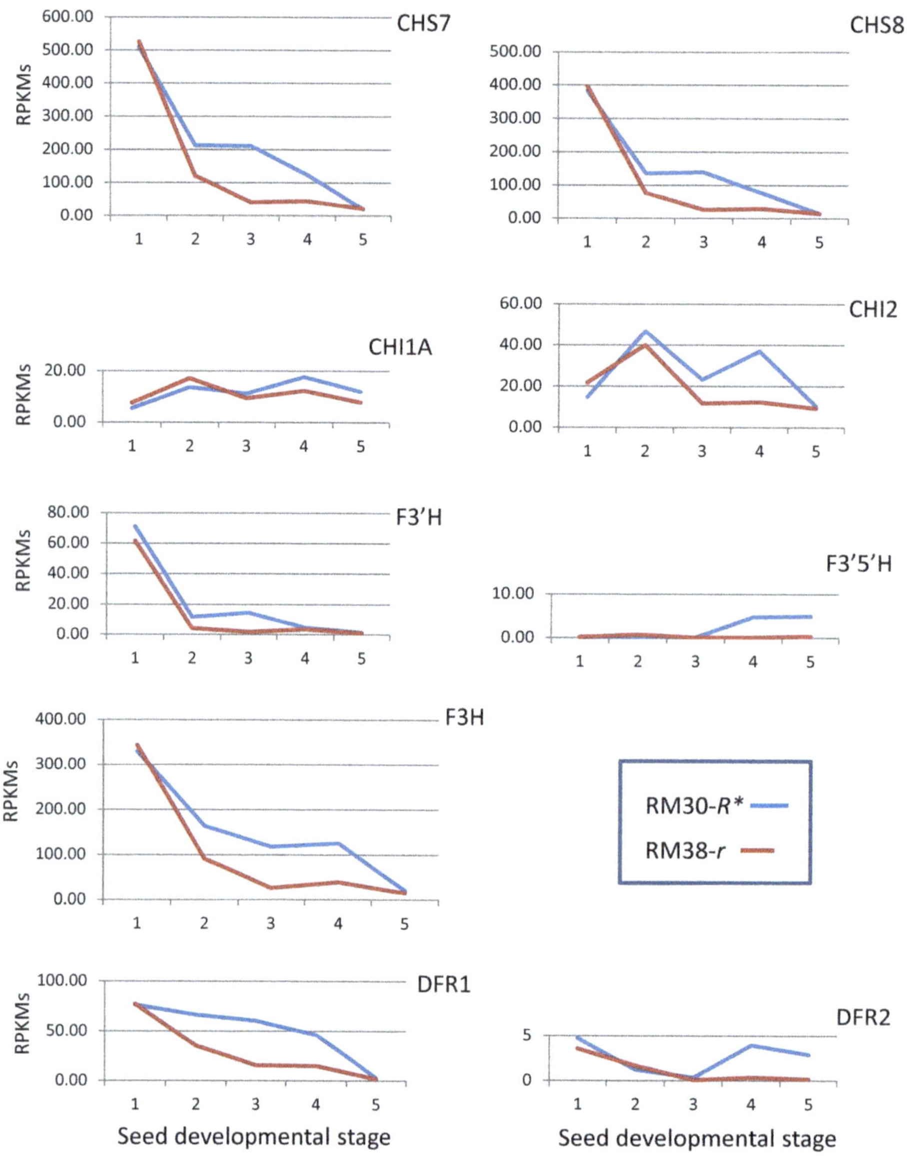

Figure 5. Differential Expression of Genes that Function in the Early Steps of the Anthocyanin and Proanthocyanidin Biosynthetic Pathways Between the Black-Seeded RM30-*R and Brown-Seeded RM38-*r* Lines.** Transcript levels are in RPKMs plotted against the same five stages of seed coat development as shown in Figure 4 for RM30-*R** where the *R** allele is interrupted by the 13 kb *TgmR** insertion (profile in blue) and the RM38-*r* line where the *r* allele is not interrupted by *TgmR** but has a "C"-nt deletion in Exon2 (red profile). Graphs have different scales. The Glyma models for the indicated genes are as follows: *CHS7* (Glyma01g43880.1); *CHS8* (Glyma11g01350.1); *CHI1A* (Glyma20g38560.1); *CHI2* (Glyma20g38580.1); *F3'H* (Glyma06g21920.1); *F3'5'H* (Glyma13g04210.1); *F3H* (Glyma02g05450.1); *DFR1* (Glyma14g07940.1); *DFR2* (Glyma17g37060.1. See Figure 7 for the pathway abbreviations.

RNA sequence reads for each gene (*ANS2* and *ANS3*) are very similar and show parallel patterns of expression in the five samples measured from the two lines (RM30-*R** and RM38-*r*). Figure 6 shows the differences in the number of sequence reads in RPKMs for the *ANS3* gene in the RM30-*R** black and the RM38-*r* brown seed coats.

Molecular Domains in the *R* Locus-Encoded MYB Transcription Factor are Predictors of its Regulatory Role in Anthocyanin Synthesis

We searched for molecular domains and motifs in the *R*-encoded MYB protein sequence which may be predictors of its mechanistic role in the activation/repression of anthocyanin synthesis. The *R* locus MYB protein sequence contains at the N-terminus the R2R3 domains characteristic of the large R2R3-MYB gene family (125 genes) in *Arabidopsis* [10]. Figure 8 shows the conserved amino acids (highlighted in black and gray) in the R2 and R3 domains of the *R* gene MYB protein sequence.

In *Arabidopsis* the 125 MYB proteins of the R2R3 family have been classified into 22 subgroups based on the sequence motifs present towards the carboxy (C)-terminus, downstream of the R2R3 domains that specify their function. One of the subgroups (#5), which includes MYBAN2, MYB75, MYB113 and MYB90, have the KPRPR[S/T]F motif and are regulators of anthocyanin biosynthesis. The *R* locus MYB contains, besides the R2R3 domains, a variant of the subgroup 5 functional motif (**KP**V**PR**AT) (Figure 8A highlighted green and red type). Simi-

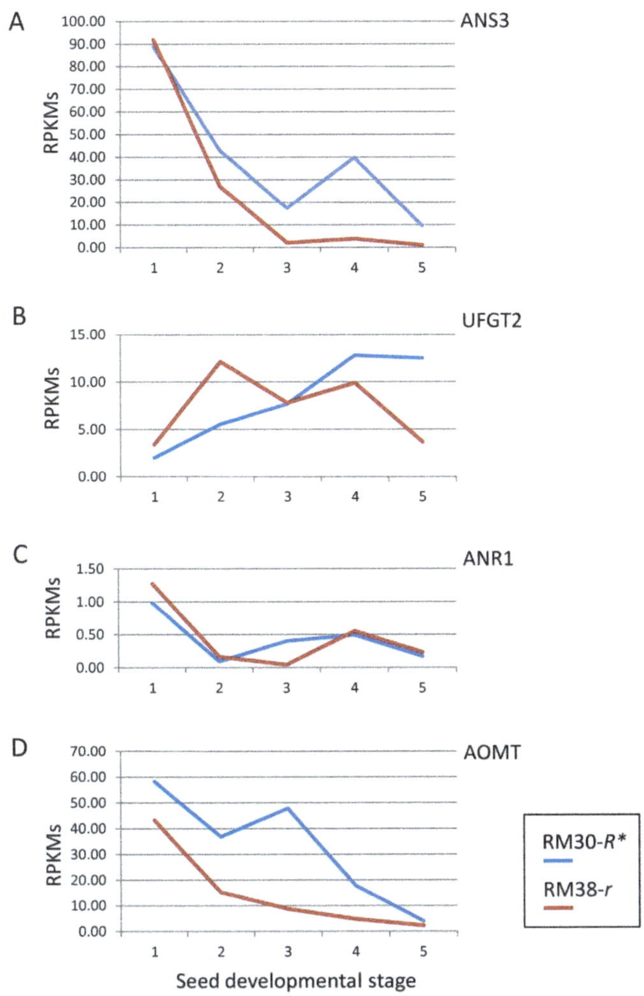

Figure 6. Differential Expression of Four Genes That Function in the Last Steps of Anthocyanin and Proanthocyanidin Synthesis Between the Black Seeded RM30-*R and Brown Seeded RM38-*R** Lines.** Transcript levels are in RPKMs plotted against the same five stages of seed coat development as shown in Figure 4 for RM30-*R** where the *R** allele is interrupted by the 13 kb *TgmR** insertion (profile in blue) and the RM38-*r* line where the *r* allele is not interrupted by *TgmR** but has a "C"-nt deletion in Exon2 (red profile). Graphs have different scales. The Glyma models for the indicated genes are as following. (A) ANS3 (Glyma11g03010.1); (B) UFGT2 (Glyma08g07130.1); (C) ANR1 (Glyma08g06630.1) and (D) AOMT (Glyma05g36210.1. See Figure 7 for the pathway abbreviations.

larly, other plant species MYB TF with anthocyanin-activating function have variations of that subgroup 5 motif ([R/K]Px[P/A/R]xx[F/Y]) [19]. *Arabidopsis* R2R3-MYB subgroup 4 (AtMYB3, AtMYB4, AtMYB7, AtMYB32) is defined by a C-terminal amino acid motif (LNL[E/D]L) predicted to be involved in transcriptional repression [20,21,22,10]. With the exception of one amino acid this motif is conserved at the C-terminus of the *R* locus R2R3-MYB amino acid sequence (**LNL**[E/**D**]**L**) (Figure 8A, fuchsia highlight).

Another molecular feature present in both those MYB subgroups (5 and 6) in *Arabidopsis* is the conserved amino acid signature ([DE]Lx_2[RK]x_3Lx_6Lx_3R) shown to be the structural basis for the interaction between many higher plants MYB and R/B-like bHLH proteins [23]. This signature is also present in the *R*-locus encoded MYB sequence (Figure 8A, aqua highlight). Except for one, all the conserved amino acids in the signature and with identical spacing are present in the R3 sequence repeat of the *R*-encoded MYB protein (burgundy type). The interaction between MYB and bHLH proteins has been studied extensively for MYB transcription factors regulating the phenylpropanoid biosynthetic pathways and seems to be conserved throughout the plant kingdom [24,21,25,26,27].

Based on the level of fidelity of all the well characterized MYB sequence domains, signatures and motifs, the MYB transcription factor encoded in the *R* locus could belong to the R2R3-MYB subgroup 5 (activation motif) and 4 (repression motif) that associates with an R/B-like bHLH protein to promote or repress the late portion of the anthocyanin biosynthetic pathway (Figure 7).

The defective *r* allele with the "C"-nt deletion in Exon2 truncates the protein prematurely preserving only the R2 domain and 16 amino acids of the R3 domain (Figure 8B). All other conserved motifs including the activation and repression motifs are lost in the resulting truncated peptide if the transcripts are processed and spliced as those transcribed from the wild type *R* gene. Interestingly, our expression data (Table 3) show that the levels of transcripts from both the standard black *R* and brown *r* alleles are about the same levels, thus the truncation of the protein in the *r* allele does not appear to reduce the levels of the low levels of cytoplasmic mRNAs as is the case for other examples of prematurely truncated proteins, presumably from destabilization of the polysomes and rapid degradation of the mRNAs [28]. Thus, the loss of function of the MYB factor in the brown *r* allele is from the C deletion prematurely truncating the protein and not from changes in transcript levels at the locus.

The 13 kb *TgmR** Element Residing in Intron2 of the RM30-*R** Stable Black Revertant Allele Is Removed by Intron Processing to Produce a Fully Functional MYB Transcription Factor from a Reduced Level of Transcripts

We inspected the alignment of the RNA-Seq reads from the Rm30-*R** and RM38-*r* lines at all five developmental stages (Table 2) against the 714-nt (3-exon gene model) transcript sequence with Bowtie 1. The results presented in Figure 9 show clearly the numbers of sequence reads and their alignments with a total coverage of the Glyma09g36983 gene transcript sequence in both lines. The expression level of the *R* gene is low in black and brown seed coats with the only significant difference manifested in seed coats of the 400–500 mg seeds, with the brown seed coats accumulating slightly more than double the level in the black seed coats (Figure 9) as previously indicated in Table 2 with the RM30-*R**/RM38-*r* ratio of 0.49. As stated earlier, these alignment results also show that the Glyma09g36983 gene is transcribed all the way to the 3′end in the RM30-*R** black seed coats (Figure 9) despite the large transposon insertion at 110 nt from the beginning of Intron2. These data indicate that despite a very low level of transcripts, they are cleanly processed at the intron exon boundaries.

On the other hand, the alignment of RNA-Seq reads from the RM38-*r* brown seed coats to the 714-nt sequence representing the standard Glyma09g36983 *R* allele transcript, shows a valley observed at 222 bp, the location of the deleted "C"-nt (Figure 9 red arrow). Bowtie 1 does not align any insertion or deletions, thus explaining the lack of alignments at the location of an intron or even a single nucleotide insertion or deletion. These data are additional evidence that RNA sequence reads from the RM38-*r* allele were missing a C at position 222. This conclusion was also directly confirmed by visual inspection showing that sequences

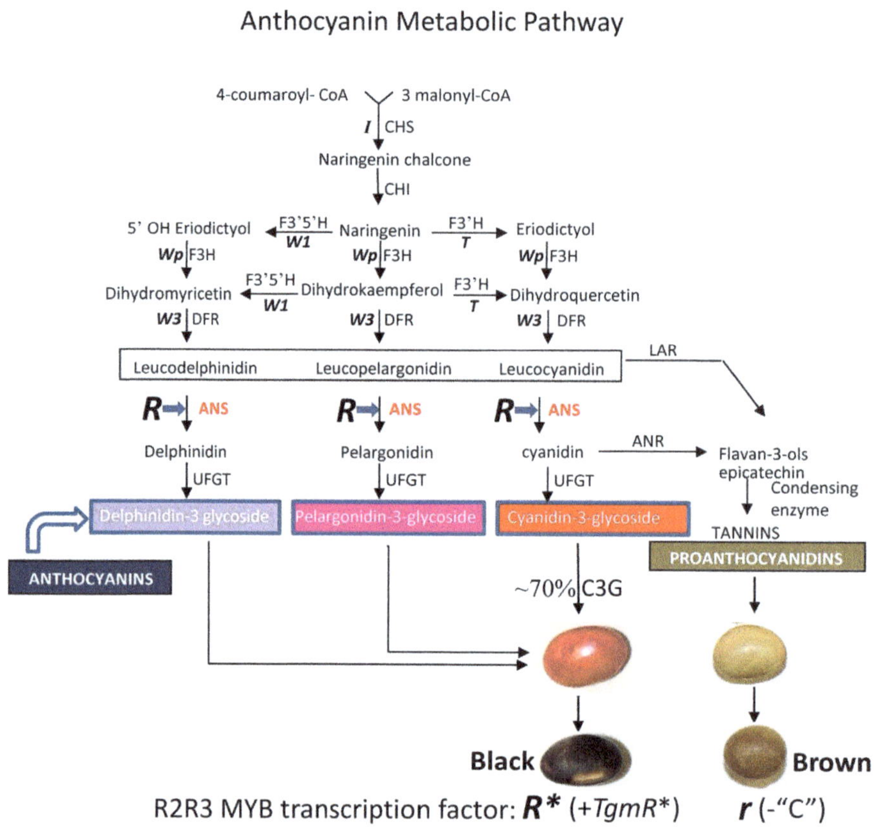

Figure 7. Diagram of the Soybean Seed Coat Anthocyanin Metabolic Pathways and Enzymatic Steps Regulated by the MYB Transcription Factor Encoded by the *R Allele in the Black Seeded Line RM30-*R**.** In display are the three branches leading to the synthesis of the purple delphinidins, pink pelargonidins, red cyanidins as well as the brown proanthocyanidins. The major anthocyanin in black soybean seed coats is cyanidin-3-glycoside (C3G) at ~70%. The brown seed coats contain mostly proanthocyanidins. Enzymes and seed coat color loci are abbreviated in uppercase letters: *I*, chalcone synthase (CHS); *W1*, flavonoid 3′5′-hydroxylase (F3′5′H); *T*, flavonoid 3′-hydroxylase (F3′H); *Wp*, flavanone 3-hydroxylase (F3H); *W3*, dihydroflavonol-4-reductase (DFR); Annthocyanidin synthase (ANS) also called leucoanthocyanidin dioxygenase (LDOX); UDP-flavonoid glucosyltransferase (UFGT); Leucoanthocyanidin reductase (LAR); Anthocyanidin reductase (ANR). The MYB transcription factor encoded by the *R** allele Glyma09g36983 gene enhances expression/accumulation of ANS transcripts as shown in Figure 6.

aligned properly to either side of the C at nucleotide 222 but none contained the C at position 222. In addition, we performed the alignment of the RM38-*r* RNA-Seq reads to the mutant transcript sequence (with the C deletion) and the valley observed at position 222 in the graph of alignments was no longer present as shown in Figure S1 indicating proper alignment over the full length transcript.

The RNAseq data obtained were also analyzed to determine the level of expression of the putative *TgmR** element in order to determine whether there are chimeric read through sequences.

Using the Bowtie 1 alignments, we determined the distribution of sequence reads along the length of the 15,450-nt genomic sequence obtained from the black seed RM30-*R** line that comprises the *R* locus gene Glyma09g36983 interrupted by the 13,024-nt *TgmR** insertion in Intron2. A representation of the distribution of sequence reads at the five stages of seed development in the black and brown seed lines is shown in Figure S2. It is clear that sequences aligned all through the *TgmR** sequence portion in RNA samples from both lines at all five stages of seed coat development examined including a low level from ORF2. Whether these sequences originate from *TgmR** or from the other genome regions containing similar elements (Figure 3) cannot be determined from the alignments. Manual inspection of the transcript sequences did not find any chimeric sequences at the intron junctions or the boundaries of the *TgmR** element, which again indicates clean processing of the introns from the gene with complete and precise removal of the 13 kb *TgmR** element located within Intron2 through the normal intron processing.

The CACTA Element is Stabilized by Increased Methylation in the RM30-*R** Black Seeded Revertant Compared to the Variegated RM55-r^m Mutable Line

The mutable allele at the *R* locus present in the RM55-r^m soybean line which confers the variegated black and brown seed coat phenotype (Figure 1) has the unusual behavior of switching between active and inactive phases somatically and germinally. Heritable changes of this r^m allele were observed in progeny plants that produced plants with all black or all brown seed [6]. Notably, the mutability of the allele resurfaced in progeny plants from the all black (*R*/R**) or all brown (*r*/r**) revertant seeds. These results indicated that the r^m revertant alleles were not stable and switched between the three forms (r^m, *R** and *r**) with high frequency.

The genomic sequences of the alleles from the variegated (r^m) and black (*R**) seed lines reported here showed that both contained the *TgmR** transposon insertion in Intron2. Could the instability of these alleles be dictated by their methylation levels in the two isolines? To answer this question we sequenced the

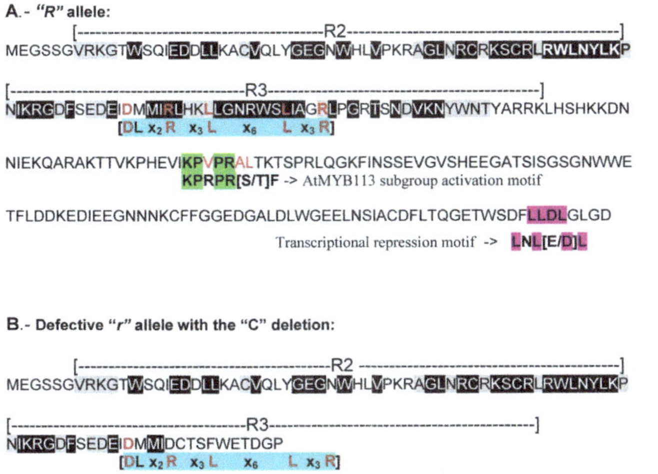

Figure 8. Features of the R2R3-MYB Transcription Factor Protein Sequence Encoded by the *R* Locus, Glyma09g36983. (A) The full length standard *R* allele. The extent of the imperfect R2 and R3 repeats is marked with the two dotted lines. Highlighted in black and gray are the conserved amino acids found in other higher plant MYB TFs of the R2R3 class (Lin-Wang et al, 2010). The amino acids highlighted black are best conserved. Aqua highlight is the [D/E]Lx_2[R/K]x_3Lx_6Lx_3R conserved amino acid signature functionally relevant in the MYB TF interaction with R/B-like bHLH proteins [23]. The conserved amino acids of this signature in the *R* locus encoded MYB are indicated in burgundy type. Green highlights the amino acids that are part of the motif found in AtMYB113 of subgroup 5 involved in anthocyanin regulation (activation) [10]. The divergent amino acids are indicated in red type. Highlighted fuchsia are four of the five amino acids constituting the transcriptional repression motif found in AtMYBs of subgroup 4 (AtMYB3, AtMYB4, AtMYB7, AtMYB32) [10]. (B) The truncated polypeptide resulting from the Exon-2 "C"-nt deletion in the defective *r* allele. It preserves only the R2 domain and 16 amino acids of the R3 domain.

genomes of three soybean lines, RM55-r^m (variegated seed), RM30-*R** (black seed) and UC44-*R* (black seed with no *TgmR** insertion), using the bisulfite method (BS-seq) first described by Cokus et al, 2008 [29] to yield 226, 209, and 235 million total reads respectively (Table S1). The resulting bisulfite sequences were processed through the three-letter aligner program, Bismark [30] to determine the type and degree of "C"-methylation of the transposon and the upstream (35,513 bp) and the downstream (25,417 bp) regions at the *R* (Glyma09g36983) loci. The results were compiled in Table 4 and revealed a higher degree of methylation for the transposon (*TgmR**) sequences, most notably, in the RM30-*R** black-seeded revertant line with an almost double (5.8%) level of methylated C in the CHH context compared to 3% in the RM55-r^m (variegated seed) and 2.9% in the UC44-*R* (black seed with no *TgmR** insertion). The percentage of CpG methylation in sequences with similarity to *TgmR** from both RM30-*R** (86.2%) and UC44-*R* (84.7%) were substantially higher than in RM55-r^m (77.6%). CHG methylation in sequences with similarity to *TgmR** was also higher for RM30-*R** (74.3%) compared to RM55-r^m (63.1%). Although UC44-*R* does not contain *TgmR** in the *R* locus, there are highly repeated regions with sequence similarity to the transposon elsewhere in the genome as previously discussed.

The distribution of differentially methylated *TgmR** regions in the black-seeded RM30-*R** and the mutable RM55-r^m soybean isolines is shown in Figure 10A as visualized by the sequence alignment tool of the Integrative genomics viewer (IGV) browser http://www.broadinstitute.org/software/igv/home [31]. Significantly lower levels of methylated sequences appear to align to the terminal repeats of the *TgmR** element and the 3′-end of the predicted ORF2 transcript in the striped black/brown-seeded mutable RM55-r^m line. The black-seeded line UC44-*R* line lacking the *TgmR** insertion in *R* presents nearly identical pattern and level of methylation in the three areas of the *TgmR** as in the RM30-*R** line.

The distribution of methylated sequences upstream and downstream of the *TgmR** element inserted in the Glyma09g36983 gene for the black-seeded RM30-*R** and the mutable RM55-r^m isolines as well as the black-seeded UC44-*R* Williams cultivar are shown in Figures 10B and 10C. A few methylation differences appear between the three lines but the most distinct are those for UC44-*R* which may represent varietal differences and not affect significantly the expression of the genes encoded upstream (A, B, C and D-5′end) and downstream (D-3′end, E and F) of the *R* locus Glyma09g36983 gene. Therefore, the methylation differences for the r^m and *R** alleles in the RM55-r^m and RM30-*R** isolines seemed to be confined primarily to the *TgmR** sequences at its extended terminal repeats and to the 3′end of the ORF2 region.

Discussion

Structural Variations in the RM55-r^m and Revertant RM30-*R** Alleles Confirm the *R* locus as Glyma09g36983 that Encodes a MYB Transciption Factor

We have previously identified and characterized the expression of several key loci controlling the synthesis of anthocyanins in the soybean seed coats. The *I* locus was found to be a cluster of *CHS* genes that produce small RNAs that downregulate CHS transcripts [2], the *T* locus encodes a flavonoid 3′hydroxylase (*F3′H*) gene [3], and the *Wp* is a flavanone 3-hydroxylase (*F3H*) gene [32] (Figure 7). However, the *R* locus that with *T* determines the various colors of the seed coat in genotypes with a recessive *I* allele (*i*) is not completely understood. Recently, it was proposed that *R* maps to a locus, Glyma09g36983, encoding a MYB transcription factor that may positively regulate the expression of a *UF3GT* gene, which catalyzes the glycosylation of cyanidins to anthocyanins in the last step of anthocyanin synthesis [4]. However, another report suggested a different transcription factor location 120 kb away on the same chromosome [5].

The results we present here support the location of the *R* locus to be the MYB encoded by Glyma09g36983 as determined recently by Gillman et al., 2011 [4] based on mapping polymorphisms associated with standard *R* and *r* alleles, one of which was a "C" deletion that truncates the MYB protein in the standard brown *r* allele. Using primers designed to Glyma09g36983, our analysis of two isolines including a mutable RM55-r^m line with black and brown striped seed phenotype and RM30-*R**, a stable black-seeded revertant isoline, revealed the existence of a 13 kb CACTA transposon insertion (*TgmR**) at the beginning of the Glyma09g36983 gene Intron2 in both of these lines (Figures 2 and 3). The transposon was absent in the brown-seeded RM38-*r* line. Thus, our data directly implicate the structural variations in Glyma09g36983 as producing the phenotypic differences of the *R* locus.

The Tgm*R** Sequence Confirms the Existence of Two CACTA Subfamilies in Soybean

Based on differences in the subterminal repeats, we previously predicted the existence of two CACTA subfamilies in soybean [8]. The novel *TgmR** element is 13 kb in length and has all the

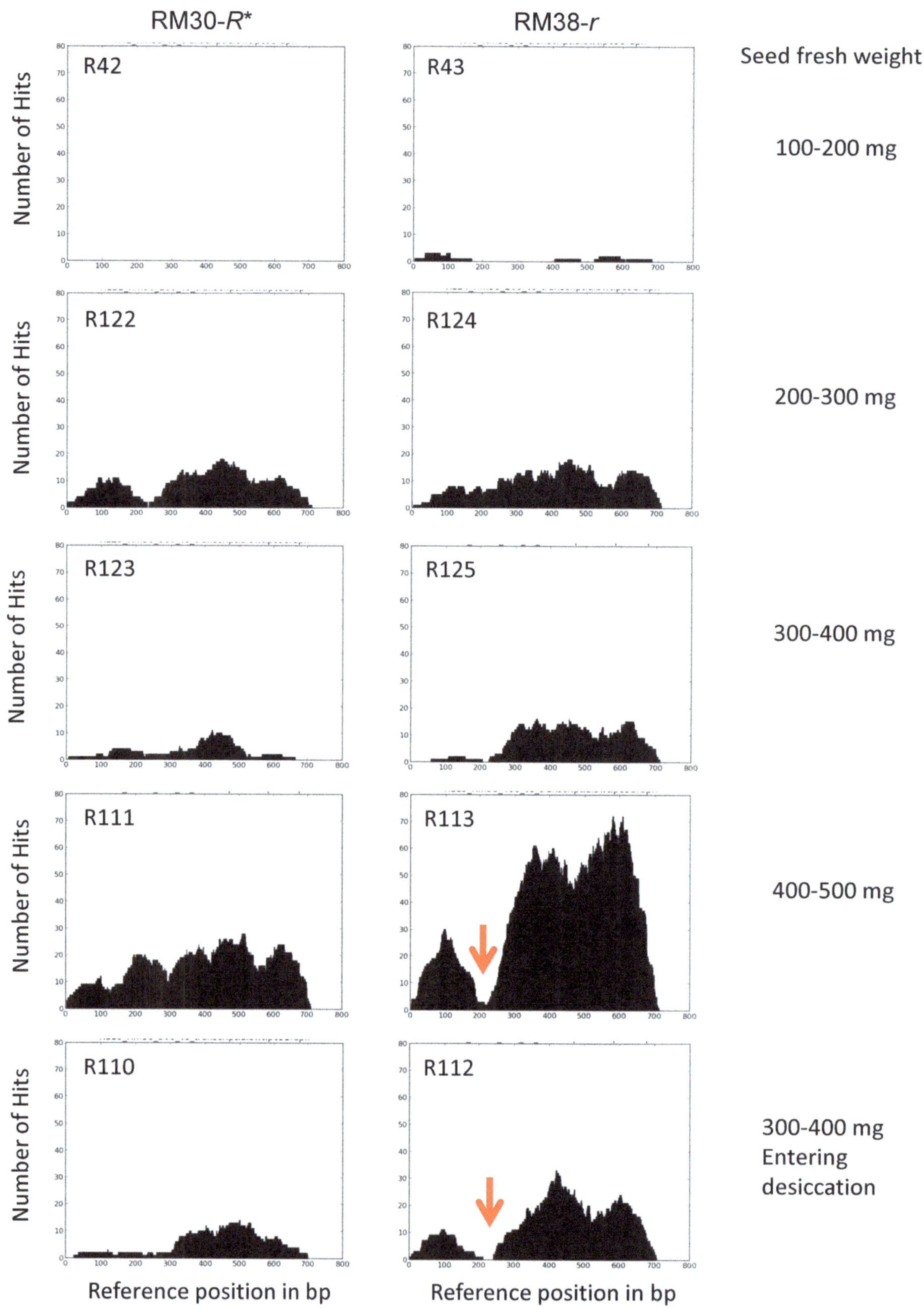

Figure 9. Distribution of RNA-Seq Reads from Seed Coats of the RM30-*R and RM38-*r* When Aligned to the Transcript Sequence of Glyma09g36983.** The pair of graphs in each row represent the distribution of non-normalized RNAseq reads that align to the 714 bp transcript sequence from the Glyma09g36983 gene at each of the five stages of seed coat development indicated in the margin to the right (described in Figure 4). The plots in the left column represent the distributions of sequence reads derived from the seed coats of the RM30-*R** that is interrupted by

the *TgmR** transposon and the plots in the right column represent the RNAseq reads derived from the RM38-*r* line with the uninterrupted *R* gene but containing the "C"-nt deletion at 222 bp of the transcript. The numbers of RNAseq reads matching a given position of the *R* gene transcript are plotted against the position along the reference Glyma09g36983 transcript sequence containing the three exons and no introns. No expression was detected in the early 100–200 mg seed developmental stage in the *RM30-R** line. The red arrow points to the location of the *r* allele "C"-nt deletion at 222 bp of the transcript.

characteristic features of a CACTA transposon based on the three base duplication (ATG) at the site of insertion, the CACTA ends and the direct and reverse repeats at the subterminal ends that can fold into complex secondary structures (Figure 3A, Files S2 and S3). *TgmR** confirms the proposed existence of two subfamilies of CACTA transposons in soybean which was based on the two distinct types of repeated motif at the subterminal ends (Figure 3B). The motif in *TgmR** is more closely related to the one in *Tgm1*, a defective element interrupting a soybean seed lectin gene [7,14], than that of three other previously characterized elements (*Tgmt**, *Tgmw4m*, *Tgm-Express1*) [32,8]. In addition, the *TgmR** 13 kb element encodes two genes, ORF1 (Tnp2-like transposase) and ORF2 of unknown function. The transposase region of *TgmR** is only 71% similar to the transposase in *Tgmt** (*Tgm9*). Thus, we propose that *TgmR** is an active element of the *Tgm1* type family of elements. Based on RNA-Seq data, both ORFs of *TgmR** or related element are expressed at low levels in the seed coats of both black RM30-*R** and RM38-*r* lines (Figure S2). Because DNA sequences with high similarity to the ORF1 gene are highly repeated throughout the entire genome, the ORF1 RNAs detected in both lines could be derived from sequences other than the one inserted in the *R** allele.

The Revertant RM30-*R** Allele Has Black Seed Coats as its Intron2 with the *TgmR** Insertion is Correctly Spliced to Produce Functional MYB Transcripts

The RNA-Seq analysis at multiple stages of seed development was crafted to determine whether the *TgmR** insertion interfered with the expression of the *R* (Glyma09g36983) gene. We contrasted the normalized RPKM reads from the black RM30-*R** revertant line to those of the brown RM38-*r* line that aligned to the Glyma09g36983. The results were congruent with *TgmR** having a negative effect on the expression of the *R** allele, since a higher number of RNAseq reads were found to align to Glyma09g36983 from the RM38-*r* line than from the RM30-*R** line at late stages of seed development (Table 2, Figures 4 and 9). The level of expression was very low in both lines, but this is common for other transcription factors. The difference in *R* expression between the two lines is more marked at the 400–500 mg seed developmental stage with a RM30-*R**/RM38-*r* ratio of 0.49 and a p-value of 0.069 (Table 2).

These results were not surprising since the 13 kb *TgmR** insertion in the *R** allele of the RM30-*R** line could reduce the level of its transcription. However, based on alignments of the RNA-Seq reads from RM30-*R** to the *R* Glyma09g36983 transcript sequence shown in Figure 9, the *TgmR** element appears to be cleanly removed with the correct intron processing as there was full coverage of the entire transcript with the aligned reads.

In contrast, a deep valley at around 200 bp occurs in the RM38-*r* seed coats (Figure 9), which suggests the absence of RNA-Seq reads that align to that region. The region of non-alignment occurs because of the "C"-nt deletion at 222 bp in the RM38-*r* allele. The transcripts from the *r* allele of the brown-seeded line (RM38-*r*) with the "C"-nt deletion in Exon2 also reach full length. If the processing of the *r* allele transcripts occurs at the same exon-intron splice junctions, the translation product of this defective *r* allele will be terminated prematurely eliminating the activation (KPRPR[S/T]F), repression (LNL[E/D]L), and bHLH (DLx_2 Rx_3Lx_6Lx_3R) binding domains (Figure 8), rendering the transcription factor non-functional resulting in brown seed coats.

The MYB Factor Upregulates *ANS* Genes in the Anthocyanin Biosynthesis Pathway Late in Seed Coat Development

Although many of the genes in the anthocyanin pathway examined showed differences in expression between the RM30-*R** black revertant and the RM38-*r* brown isoline, the most significant differences detected were for the *ANS2* and *ANS3* genes at the 400–500 mg seed developmental stage (Table 2, Figure 6A). In another study in which the transcript profiles of seed coats from a standard black and brown-seeded near-isogenic pair were compared at different stages of seed development, significant differences in *ANS2/ANS3* and *UGT78K1* expression were also seen between the two isolines with higher levels in the black seed coats [5]. For *ANS2/ANS3,* the significant differential expression ($P<0.01$) manifested early in seed development, (75–100 mg), while the *UGT78K1* differences were first detected at 200–300 mg weight range and reaching maximum at 300–400 mg seed size. Gillman et al, (2011) [4] measured *ANS* and *UF3GT* expression via qRT-PCR in seed coats of a brown- (PI 567115 B) and black- (PI 84970) seeded lines with similar results. As in those instances, our results show that the *UFGT* genes are expressed at very low level at early stages of seed development, while *ANS* expression is very high at those early stages and tapers down as seed development advances. This decrease is slowed down or reversed in the black-seeded phenotypes. The higher level of *ANS* gene expression at early stages of seed development can be explained by an additional role of the ANS enzyme in directing the synthesis towards the flavanol and proanthocyanidin branches of the flavonoid pathway (Figure 7). In contrast, the UFGT enzyme function is limited to the last step in the synthesis of anthocyanins (Figure 7).

Whether the MYB transcription factor encoded at the *R* locus regulates all or one of the genes in the anthocyanin pathway is not clear from the available data. A search in MEME Version 4.9.1 [33] for MYB transcription factors binding motifs in the 2-kb upstream sequence of each of seven genes from the anthocyanin pathway found two motifs in all seven genes. One of those motifs is a MYB binding site (JASPAR_CORE_2014) for Arabidopsis R2R3 MYBs (AtMYB84, AtMYB15, AtMYB77) of subgroups other than number five that contain the activation motif KPRPR[S/T]F.

The structural genes of the flavonoid metabolic pathways are regulated for the most part, at the transcriptional level and, with some exceptions, the majority of the regulatory genes have been found to be members of a large family of MYB transcription factors. Most of the MYB genes were predicted to function as transcriptional activators of the flavonoid biosynthetic pathway but in a few instances transcriptional repression guided by MYB transcription factor has been demonstrated [21]. In maize, *C1-I* was found to be a dominant inhibitor allele of the transcription activator *C1-myb*-like gene [34,35]. A different *C1* homologue of

Table 4. Bismark bisulfite-seq methylation results for the TgmR* CACTA transposon insertion and upstream/downstream regions of Glyma09g36983 in three soybean lines: RM30-*R** black, RM55-r^m striped and UC44-*R* black seed coat phenotypes.

DNA reference	Upstream (35,513 bp Glyma09g36983)			*TgmR** (13,024 bp in Glyma09g36983)			Downstream (25,417 bp Glyma09g36983)		
Soybean lines sequenced	**RM30**	**RM55**	**UC44**	**RM30**	**RM55**	**UC44**	**RM30**	**RM55**	**UC44**
	(black, *R**)	(mutable, r^m)	(black, *R*)	(black, *R**)	(mutable, r^m)	(black, *R*)	(black, *R**)	(mutable, r^m)	(black, *R*)
No of alignments w/unique best hits	2,678	2,874	3,257	13,134	15,168	15,670	1,895	2,145	2,483
Total methylated C's in CpG context	512	600	678	43,284	45,140	50,347	392	464	650
Total methylated C's in CHG context	144	203	203	50,016	49,405	53,631	85	89	129
Total methylated C's in CHH context	395	256	252	18,596	11,048	10,957	204	134	201
Total unmethylated C's in CpG context	5,649	5,951	6,890	6,943	13,044	9,108	4,265	4,890	5,562
Total unmethylated C's in CHG context	8,510	9,134	10,130	17,341	28,873	27,425	6,129	6,986	7,960
Total unmethylated C's in CHH context	60,635	65,005	73,402	301,746	355,884	369,037	42,634	47,710	55,145
C methylated in CpG context	8.3%	9.2%	9.0%	86.2%	77.6%	84.7%	8.4%	9.6%	10.5%
C methylated in CHG context	1.7%	2.2%	2.0%	74.3%	63.1%	66.2%	1.4%	1.9%	1.6%
C methylated in CHH context	0.6%	0.4%	0.3%	5.8%	3.0%	2.9%	0.5%	0.4%	0.4%

maize, *Zm38*, was found to be structurally similar to two other R2R3 MYB genes, *AmMYB308* and *AtMYB4*, that are transcriptional repressors of *4Cl* (b-coumaroyl 4-CoAligase) and the *C4H* (cinnamate 4-hydroxylase) genes in *Antirrhinum* and *Arabidopsis*, respectively [36,22]. Another TF with the motif shown to be involved in repression of transcription in *Arabidopsis* (*AtMYB4*) is the *FaMYB1* gene expressed during fruit maturation in strawberry [21]. *AmMYB308*, like *C1-I*, competes out and inhibits the action of their respective transcriptional activator alleles. In contrast, *AtMYB4* and *FaMYB1* are direct repressors of their respective target genes and contain sequence motifs that recognize specific promoter repressor domains. Although little is known about repression mechanisms in plants, in both those instances, AtMYB4 and FaMYB1 C-termini possess a peptide motif, pdLNL$^{D}/_{E}$L-xi$^{G}/_{S}$, shown to be required for suppression activity [22,21,20]. The soybean *R* locus MYB factor has the core portion of such motif, **LLDL**, at its C-terminus which may suffice to function as a direct repressor of late genes in the anthocyanin biosynthetic pathway. Likewise, the *R* gene possesses four of the amino acids in the motif, **KPRPR**[S/T]F, present in the subgroup five (AtMYB113) involved in anthocyanin regulation by activation [10]. The differential expression of the *R* MYB gene and *ANS* genes supports its role as an inducer of anthocyanidin synthase in the black soybean seed coats.

Our results show that the Intron2 with the *TgmR** insertion is spliced out of the *R** allele transcripts correctly allowing a small amount of the MYB to be translated in sufficient levels to induce the transcription of *ANS2/ANS3* and *UFGT* whose enzyme products will complete the last two steps of anthocyanin synthesis.

The RNAseq expression results also demonstrate that the activation of the anthocyanin pathway in RM30-*R** is due to the Glyma09g36983 with the *TgmR** transposon and not to either of the two other closely related MYB transcription factor genes, Glyma09g36966 and Glyma09g37010, since these two genes appear not to be expressed in the seed coats of either the RM30-*R** line with the black seed and the RM38-*r* line with the brown seed phenotypes. This eliminates the likelihood that the *TgmR** insertion occurred in any of those two related genes that could in turn affect the expression of the Glyma09g36983 gene in a similar fashion as the *C1* homologue of maize, *Zm38*, that was found to be a transcriptional repressor of *C1* [37].

Hypomethylation is Associated with the Variegated RM55-r^m Mutable Allele Compared to the Stable RM30-*R** Black Revertant

The soybean line with the mutable r^m allele (RM55-r^m) and striped seed phenotype showed instability with the frequent generation of revertant seed genotypes some being all black which in subsequent generations would switch back to variegated at a low frequency [6]. To find out whether methylation contributes to the variegated or stable phenotype, we compared the BS-sequences from the striped RM55-r^m and the stable revertant isoline RM30-*R**. As shown in Table 4, the level of methylated C residues in the CHH context for *TgmR** in the RM30-*R** stable line was double the percentage of that found in RM55-r^m. The more prevalent methylated CpG and CHG regions were also higher in comparison to RM55-r^m. More interestingly, distribution of the hypomethylated regions in the element were concentrated near the subterminal repeats and a region of the 3′ end of ORF2 (Figure 10).

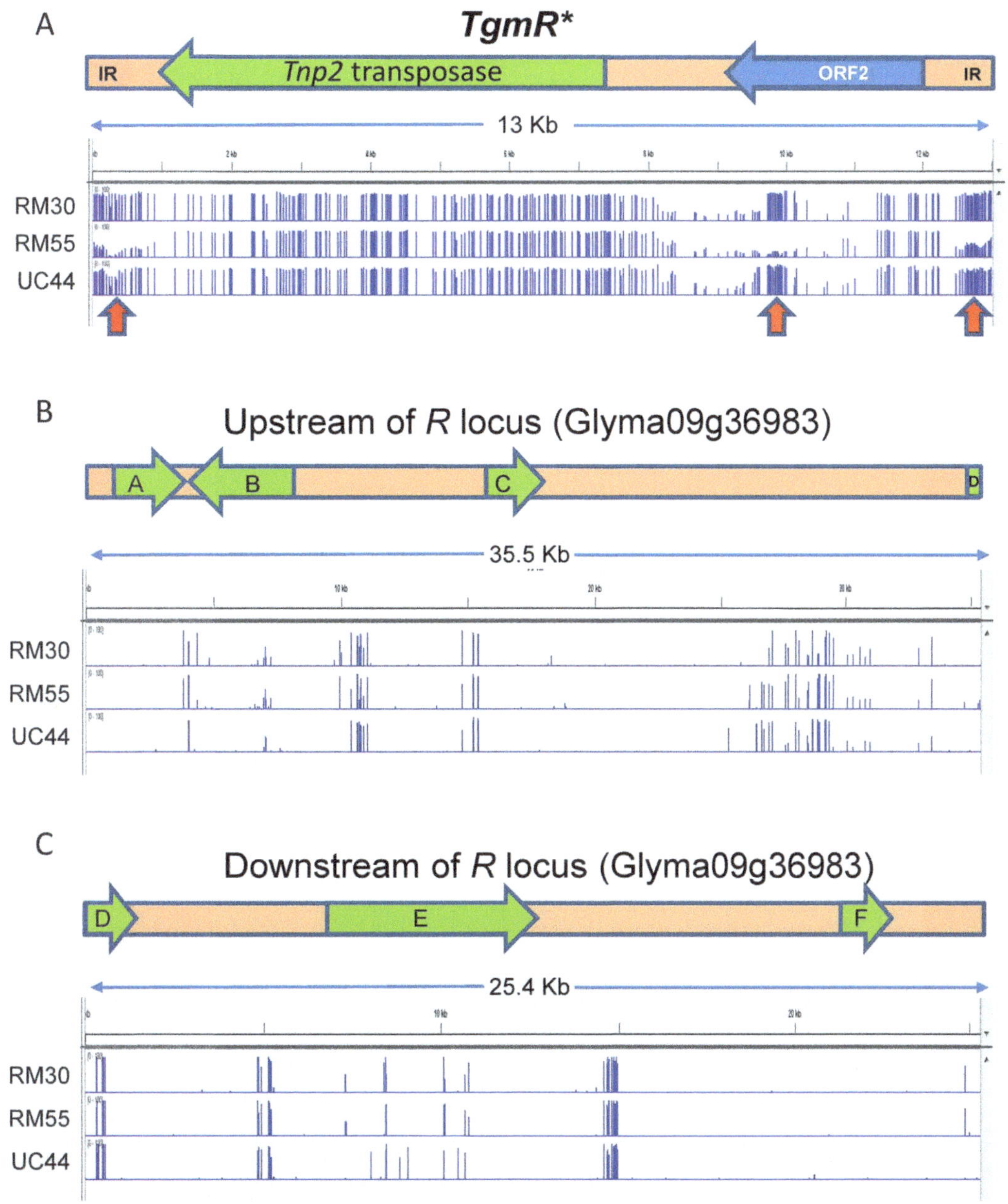

Figure 10. Distribution of Methylation in the *TgmR Transposon (A) and in the Upstream (B) and Downstream Regions (C) of the *R* locus in the Black-seeded RM30-*R** and the Mutable RM55-r^m Isolines, and of the UC44 line with a Standard *R* Allele.** All graphs were created by the sequence alignment tool of the Integrative Genomics Viewer (IGV) browser. (A) *TgmR** methylation regions. The 13 bp CACTA inverted repeat ends and subterminal repeats are marked IR. Lower levels of methylated sequences (marked by red arrows) are found aligning to the subterminal repeats of the *TgmR** element and the 3′-end of the predicted ORF2 transcript in the striped mutable RM55-r^m line. The black-seeded RM30-*R** line containing the *TgmR** insertion and the black-seeded UC44-*R*, line which lacks the *TgmR** insertion, present nearly identical patterns of methylation at those three differentiated areas of the *TgmR** element. (B) Upstream regions and (C) downstream regions of the *TgmR** insertion site in the *R* gene. A few methylation differences appear between the three lines in the upstream and downstream regions, but the most distinct are those for UC44-*R* (Williams self-black seed) which may represent varietal differences. Gene represented by green arrows are: A (Glyma09g36941); B (Glyma09g36950); C (Glyma09g36966); D (Glyma09g36983); E (Glyma09g37000) and F (Glyma09g37010). The upstream and downstream schematics are not drawn to scale.

The RM30-*R** and RM55-r^m Alleles resemble Maize *Spm*-Suppressible Alleles that are Responsive to the Methylation States of the Element

Involvement of methylation causing the variegated phenotype is consistent with the observation that there appear to be no structural changes in the MYB gene when the element excises from Intron2 since the 929-bp amplicon from the RM55-r^m DNA aligned with the Glyma09g36983 gene sequence in Phytozome (Williams 82 which has a standard *R* allele) with no rearrangement of Exon2 or the Exon2-Intron2 splice site junction. This suggests that when the element excises by transposition, a fully functional MYB protein is produced leading to the black sectors on the brown background in the RM55-r^m line.

The brown background in the RM55-r^m seed coats being the result of the 14 kb amplicon 13 kb *TgmR** (ie the 14kb amplicon) is harder to explain especially in light of the fact that we have shown that the revertant RM30-*R** line containing the 13 kb *TgmR** element in its intron has black seed coats. This would suggest that the phenotype of the allele (black or brown) is determined by the level of methylation of the *TgmR** element. When the element residing in Intron2 is less methylated as in the RM55-r^m, the sectors are brown implying the MYB gene is not expressed. When *TgmR** is more methylated, as in the RM30-*R** allele, the seed coats are black and the MYB gene is expressed as we have shown experimentally. We propose that the mechanism for this phenenomon is similar to that shown for the *Spm*-suppressible alleles of maize in that the phenotype of the affected allele can depend upon where the transposon sits in the gene and whether an active element is present (reviewed in [12,13]). The CACTA ends of *Spm* elements as well as the Activator (Ac) elements sometimes mimic splicing sites and are recognized as intron ends [12,38]. In some cases as shown in maize, binding of transposase from an active element will interfere with intron processing if the element is near the Exon-Intron junction leading to the gene not being processed correctly. In the absence of transposase, the introns are processed and the gene is expressed, although sometimes at a lower level. In maize, it has been shown that the autonomous *Spm* element itself can switch between an active state (less methylated with transposase produced) and an inactive state (more methylated and transposase not produced) based on the level of its methylation [11]. Thus, in the hypomethylated RM55-r^m line, transposase from the *TgmR** element in the MYB gene or other locations could bind to the element ends and interfere with intron processing leading to the brown background color. On the other hand, in the stable black line RM30-*R**, the opposite effect occurs as the element is more methylated and transposase is lacking or cannot bind the element termini and interfere with intron processing. Thus, the element is spliced out with the intron and the MYB gene is expressed at a low level leading to activation of the ANS genes and production of the anthocyanin pigments. There is one phenotypic difference between the standard *R* allele and the *R** allele in the RM30 line in that the hilum region, which is the point of attachment of the seed to the pod, is less pigmented than in the standard *R* allele and retain a brownish cast [6]. A genetic difference between the two alleles is that the r^m is normally recessive to the standard *R* allele producing progeny with black seed, but it appears to be dominant to the *R** allele with variegated seed predominating and an increased rate of somatic and germinal mutability results from crossing RM30-*R** and RM55-r^m [39]. The activation *in trans* of the *TgmR** element within the RM30-*R** allele by the transposase from the hypomethylated r^m allele is consistent with its dominance over the *R** allele.

Summary

In conclusion, our results confirm the soybean *R* locus to be Glyma09g36983 and demonstrate that the transcription of the *R* gene in the black seed coats of the RM30-*R** revertant line, may be hindered by the processing of the very large Intron2 containing the *TgmR** (13kb) insertion, but it does not render the processed transcript non-functional. Instead, the translated MYB protein drives an increase in transcript accumulation of the *ANS2/3*, *UFGT2* and *AOMT* genes whose products function in the late steps of anthocyanin synthesis, particularly in production of the cyanidin-3-glycoside that accounts for the black pigmentation of the seed coats. In contrast, the observed higher level of transcripts found for the defective *r* allele in RM38-*r* seed coats may translate into truncated peptides that are non-functional and do not activate the transcription of the late anthocyanin pathway genes resulting in brown seeds which reflects accumulation of only proanthocyanidins. The *TgmR** found in the striped RM55-r^m and black-seeded RM30-*R** isolines is a transposon of a second CACTA subfamily that encodes a full length *Tnp2* transposase gene that may drive the observed instability of the isolines with the r^m and *R** alleles. The level of instability of the RM55-r^m and RM30-*R** isolines is influenced by the total level of C methylation and also methylated C in the CHH context, judging by the bisulfite sequencing results of three genomes. The *TgmR** element in the variegated RM55-r^m line was hypomethylated especially at the subterminal repeat ends compared to the RM30-*R** element. Stress and other environmental factors may in turn affect the methylation level of the germinal and somatic tissues and consequently lead to the variegated phenotype and affect the stability of the mutable (r^m) and revertant (*R**) alleles by a mechanism similar to that elucidated for CACTA elements in maize including the *Spm*-suppressible alleles first described by Barbara McClintock (reviewed in [12,13]). This explains the opposite phenotypes resulting from the influence of the *TgmR** transposase on intron processing of the MYB protein when the element is less methylated in the RM55-r^m line (brown coat sectors) versus the more methylated element in the RM30-*R** revertant line with black seed coats.

Materials and Methods

Plant Material and Genotypes

The *Glycine max* cultivars and isolines used for this study, including their genotypes and phenotypes, are described in Table 1. All lines are homozygous for the indicated alleles at the *R* locus, and only one of the alleles at each locus is shown for brevity in the table and text. Plants were grown in the greenhouse and shoot tips (meristems surrounded by the primordial leaves) and seed coats dissected from seeds at varying stages of seed development were frozen in liquid nitrogen, freeze dried (VirTis *Benchtop* K) and stored at −20°C. The seed coats used for this study were those of seeds with fresh weight of the whole seed ranging from 100–500 mg. An additional, more advanced developmental stage used was from seeds with 300–400 mg fresh weight that had entered the dehydration process (Figure 4).

DNA Extractions and Long Range (LA) PCR Reactions

Genomic DNA was isolated from freeze dried shoot tips dissected from two weeks old soybean plants using a microprep method [40] with minor modifications [3].

Based on the Glyma09g36983 genomic sequence from Phytozome, multiple DNA oligo-primer pairs were designed to be use in PCR reactions that could amplify the Glyma09g36983 gene in all three soybean lines (RM30, RM38 and RM55) with the mutable and stable *R* loci. The DNA oligo-primers were synthesized at IDT (Integrated DNA Technologies) and the name and sequence of one forward and two reverse oligo-primers that successfully amplified a large portion of the *R* alleles in all the lines are:

FP: R6990FP1 (GAGTTGAAGGAATTGAGTTATATACG-TACACCTGAAC) (37 bases)

RP: R6990RP1 (CATCGTATAAAACCTTTATGCTGTC-CATGTC) (31 bases)

RP: R6990RPB (CAACGACAACAGTCATAATGACGGT-GATGATAACAG) (36 bases).

The PCR reaction conditions used to successfully amplify the transposon insertion at the left border of Intron2 were the following: Initial denaturation step at 94°C for 2 min followed by 30 cycles of 94°C for 30 sec denaturation, 1 min annealing at 68°C and 9 min extension at 68°C. This was followed by a 10 min extension at 72°C. The reagents for the reactions were those provided with the *TaKaRa LA Taq* Hot Start Version kit No. RR042A (TaKaRa Bio Inc, Dalian, CO). DNA template was 1 µg in 50 µl reaction. The amplification fragments were separated in a 0.8% preparative SeaPlaque Agarose (Lonza Rockland, ME) gel in 1x TA buffer. The large ~14 kb and smaller 929-nt amplified DNA fragments were extracted and cleaned from the gel pieces using The Zymoclean Large Fragment DNA Recovery Kit (#D4045) (Zymo Research Co) according to their protocol. The cleaned larger ~14 kb DNA fragments were sequenced via the 454 method while the smaller 929-nt fragments were sequenced using the Sanger method at the W. M. Keck Center for Comparative and Functional Genomics at the University of Illinois at Urbana-Champaign.

Sequencing on the Roche GS-FLX+ System

Two ~14 kb PCR fragments amplified with two oligomer primers (R6990FP1 and R6990RPB, sequences shown above) from the mutable RM55-r^m and revertant RM30-R^* soybean lines were submitted to the W. M. Keck Center for Comparative and Functional Genomics at the University of Illinois at Urbana-Champaign. A shotgun genomic DNA library was prepared from each PCR sample using the Roche GS Rapid Library Prep Kit following the Roche Rapid Library Preparation Manual instructions (Roche Applied Sciences, Indianapolis, IN), with the exception that the sheared PCR products were size selected on agarose gel from 600–1000nt after sonication. The final libraries were quantitated using a Qubit fluorometer (Invitrogen, CA) and average fragment sizes were determined by analyzing 1 µl of the library on the Bioanalyzer (Agilent, CA) using a High-Sensitivity DNA LabChip (shotgun library).

Libraries were pooled evenly and diluted to 1×10^6 molecules/µl for sequencing. Emulsion-based clonal amplification and sequencing on the 454 Genome Sequencer FLX+ system (400 flow cycles) were performed according to the manufacturer's instructions (454 Life Sciences, Branford, CT). The pool was sequenced on two 1/16th lanes of a 70×75 PicoTiter Plate with the Roche XL+ sequencing kit (454 Life Sciences) with software version 2.8, flow pattern B. Amplicon signal processing and base calling were performed using the bundled 454 Data Analysis Software version 2.8.

The DNAs from the two PCR fragments were barcoded (RM30: #11; RM55: #12) and pooled in the two 1/16 lanes for sequencing. A total of 36,359 sequence reads with an average length of 507 nt were obtained from the two 1/16 lanes after quality trimming. The sequence reads were assembled into contigs using the 454 Sequencing System Software V.2.6.

Construction of Shotgun Genomic DNA Libraries and Sequencing on the Illumina HiSeq2000

Shotgun genomic DNA libraries were constructed using the TruSeq DNA Sample prep kit (San Diego, CA). Briefly, 1 µg of genomic DNA from shoot tips was nebulized at 32 psi for 1 minute. After nebulization, DNA was blunt-ended, 3′-end A-tailed and ligated to indexed adaptors. The adaptor-ligated genomic DNA was amplified by PCR to selectively enrich for those fragments that have adapters on both ends. The libraries were loaded onto 25 Ex-Gels (Life Technologies, CA) and the fraction 600–800 bp was excised from the gel. The final libraries were quantitated by qPCR on an ABI 7900. Final amplified libraries are also run on Agilent bioanalyzer DNA 7500 LabChips (Agilent, Santa Clara, CA) to determine the average fragment size and to confirm the presence of DNA with the expected size range.

The libraries were multiplexed and loaded onto 8-lane flowcells for cluster formation and sequenced on an Illumina HiSeq2000. The DNAs in these libraries were sequenced from both ends of the molecules to a total read length of 100 nt from each end using a TruSeq SBS sequencing kit version 3. The run generated raw basecall files (.bcl) which were converted into demultiplexed compressed fastq files using Casava 1.8 (Illumina, CA). The total numbers of read counts of approximately 200 million at each end are shown in Table S1.

RNA Extraction, High-throughput RNA Sequencing and Alignment to *G. max* Gene Models

Total RNA was isolated from seed coats of seeds at five stages of seed development using a phenol-chlorophorm and lithium chloride precipitation method [41] that was modified to prevent RNA adhesion to procyanidins (polyphenols) [17,18].

The purified RNA samples were sequenced at the W. M. Keck Center for Comparative and Functional Genomics at the University of Illinois at Urbana-Champaign. Starting with 1 µg total RNA per sample, RNAseq libraries were constructed with the TruSeq RNA Sample Preparation Kit (Illumina San Diego, CA). These libraries were multiplexed and loaded onto 8-lane flowcells for cluster formation and sequenced on an Illumina HiSeq2000. One of the lanes was loaded with a PhiX Control library that provides a balanced genome for calculation of matrix, phasing and prephasing, which are essential for accurate base-calling. The libraries were sequenced from one end (single-reads) of the molecules to a total read length of 100-nt.

The sequencing run generated.bcl files that were converted into demultiplexed compressed fastq files using Casava 1.8.2 (Illumina, San Diego, CA). A secondary pipeline decompressed the fastq files, generated plots with quality scores using FastX Tool Kit, removed perfect matches to reads that contain only adaptor and generated a report with the number of reads per sample/library.

The resulting sequence reads were aligned to the 78,773 Glyma1 cDNA soybean gene models determined by the Soybean Genome Project, Department of Energy, Joint Genome Institute (Schmutz et al, 2010), using the alignment program Bowtie [42]. Bowtie parameters allowed up to three mismatches and up to 25 alignments per read; reads aligned more than 25 times were discarded, with none of their alignments added to the final count. The total read counts, generally from 30 to 50 million per sample at shown in Table S1.

Bisulfite (BS) Sequencing

The shotgun DNA libraries were prepared at the W. M. Keck Center for Comparative and Functional Genomics at the University of Illinois at Urbana-Champaign with the Library Construction kit from Kapa Biosystems with one modification: after adaptor ligation, libraries were treated with the EZ DNA Methylation-Lightning kit (Zymo Research). Bisulfite-treated libraries were amplified with the Kapa HiFi Uracil+ DNA Polymerase. The libraries were quantitated by qPCR and each library was sequenced on one lane for 101 cycles from each end of the fragments on a HiSeq2000 using a TruSeq SBS sequencing kit version 3. The average DNA fragment size was 500 bp (ranging from 350 bp to 700 bp) and the sequence reads were 100-nt in length. Fastq files were generated with the software Casava 1.8.2 (Illumina). The total number of read counts of approximately 209, 226, and 234 million for the three samples are shown in Table S1.

The sequence of adaptors used to make the libraries were: (AGATCGGAAGAGCACACGTCTGAACTCCAGTCACNNN-NNNATCTCGTATGCCGTCTTCTGCTTG (NNNNNN = 6 nt index)) for adaptor sequence in read 1, and (AGATCGGAAGAG-CGTCGTGTAGGGAAAGAGTGTAGATCTCGGTGGTCG-CCGTATCAT) for adaptor sequence in read 2.

The Integrative genomics viewer (IGV) browser was used to visualize the methylated C-nt in alignments to the genomic sequences of the *R** locus and surrounding upstream and downstream sequences (http://www.broadinstitute.org/software/igv/home) [31].

Sequencing Data Values, Clustering of Sequences and Gene Model Annotations

The RNAseq results are given in reads per kilobase of gene model per million mapped reads (RPKMs) [43]. The total number of mapped reads, allowing up to 3 mismatches, takes into account reads that aligned up to 25 different times, including multiple segment matches within the same gene model using Bowtie v1 [42]. Annotations of gene model were obtained from the Soybean Genome Project in the Phytozome database version 9.0 [44] with Gmax_109 genome or Gmax_189 and other previously characterized *G. max* cDNAs that have been entered in GenBank.

Phytozome Glyma Models Corresponding to Various Anthocyanin Pathway Genes, GenBank Accession Numbers and Information Sources

The choices of Glyma models representing the different genes of the anthocyanin pathways and listed in Table 2 were based on previous gene characterization studies. Here we provide the Glyma models, GenBank accession numbers and references for each one of the genes whose expression we analyzed in the soybean lines used for the current study. The *I* locus (Glyma01g43880.1) *chalcone synthase 7* (*CHS7*) Ac.N.: M98871 [45] and (Glyma11g01350.1) *CHS8* Ac.No.: AY237728; Glyma20g38560.1 *chalcone isomerase 1A* (*CHI1A*) Ac.No.: AY595413 [46] and Glyma20g38580.1 (*CHI2*) Ac.No.: AY595415 [46]; *T* locus (Glyma06g21920.1) *flavonoid 3'hydroxylase* (*F3'H*) Ac.No.: EU190438 [3]; *W1* locus (Glyma13g04210.1) *flavonoid 3'-5'-hydroxylase* (*F3'5'H*) Ac.No.: EF174665 [47]; *Wp* locus (Glyma02g05450.1) *flavanone 3-hydroxylase* (*F3H*) Ac.No.: AY669325, [47]; *W3* locus (Glyma14g07940.1) *dihydroflavonol-4-reductase 1* (*DFR1*) Ac.No.: AF167556; *W4* locus (Glyma17g37060.1) *DFR2* Ac.No.: EF187612 [9]; Glyma01g42350.1 *anthocyanidin synthase 2* (*ANS2*) Ac.No.: AY382829.1 and Glyma11g03010.1 (*ANS3*) Ac.No.: AY382830; Glyma08g07130, *UDP-glucose:flavonoid 3-O-glucosyltransferase* (*UF3GT*) (*UGT78K2*) Ac.No.: HM591298.1 [5] and Glyma07g30180 (*UGT78K1*) Ac.No.: GU434274.1 [48]. The latter, *UGT78K1*, was found to be expressed in the cotyledons in addition to the seed coats [4]. Glyma08g06630.1, *anthocyanidin reductase* (*ANR1*) Ac.No.: NM_001254984.1 and Glyma08g06640.1 (*ANR2*) (LOC100817081) Ac.No.: NM_001256143.1 [49]. The *ANR* genes function in diverting the anthocyanidins towards the synthesis of proanthocyanidins (tannins) which provide the color to the brown seed coats (Figure 1). Lastly, Glyma05g36210.1, *anthocyanin O-methyltransferase* (*AOMT*) Ac.No.: NM_001255526.1 [5] which methylates the anthocyanins.

Supporting Information

Figure S1 Distribution of RNA sequence reads expressed in seed coats of 400 mg seed in the RM38 line with the defective *r*-allele when aligned to (A) the wild type *R*-gene transcript sequence and (B) the mutant *r*-allele transcript lacking the "C"-nt at position 222.

Figure S2 Distribution of RNA-Seq Reads from Seed Coats of the RM30-*R** and RM38-*r* when Aligned to the Genomic Sequence of Glyma09g36983 with the *TgmR** Insertion in Intron2.

File S1 Glyma09g36983 DNA Sequence and Map Locations of Oligonucleotide Primers.

File S2 *TgmR** DNA Sequence with Highlighted ORFs and Terminal End Repeats.

File S3 *TgmR** Terminal Direct Repeats.

Table S1 Summary of Sequence Reads from 454, Genomic DNA, Methylation and RNA-Seq Sequenced Libraries Made from RM30-*R**, RM55-r^m, and RM38-*r* Soybean Lines.

Acknowledgments

We thank Alvaro Hernandez (Director of the DNA Sequencing W.M. Keck Center at the University of Illinois at Urbana-Champaign), Chris Wright (Assistant Director of DNA Sequencing) and Leslie D. Benson (Laboratory Supervisor) for coordinating and directing the Illumina genomic, RNA and Bisulfite sequencing, the 454 and Sanger fragment sequencing respectively. We are grateful to undergraduate assistants Cameron Lowe, Drew Metz and Achira Kulasekara for their assistance with the analysis of the large sequence data sets.

Author Contributions

Conceived and designed the experiments: GZ LV. Performed the experiments: GZ. Analyzed the data: GZ LV. Contributed reagents/materials/analysis tools: LV GZ. Wrote the paper: GZ LV.

References

1. Todd JJ, Vodkin LO (1996) Duplications that suppress and deletions that restore expression from a CHS multigene family. Plant Cell 8: 687–699.
2. Tuteja JH, Zabala G, Varala K, Hudson M, Vodkin LO (2009) Endogenous, tissue-specific short interfering RNAs silence the chalcone synthase gene family in *Glycine max* seed coats. (2009) Plant Cell 21: 3063–3077.
3. Zabala G, Vodkin L (2003) Cloning of the pleitropic *T* locus in soybean and two recessive alleles that differentially affect structure and expression of the encoded flavonoid 3′ hydroxylase. Genetics 163: 295–309.
4. Gillman JD, Tetlow A, Lee J-D, Shannon JG, Bilyeu K (2011) Loss-of-functions affecting a specific Glycine max R2R3 MYB transcription factor result in Brown hilum and brown seed coats. BMC Plant Biology 11: 155.
5. Kovinich N, Saleem A, Arnason JT, Miki B (2011) Combined analysis of transcriptome and metabolome data reveals extensive differences between black and brown nearly-isogenic soybean (*Glycine max*) seed coats enabling the identification of pigment isogenes. BMC Genomics 12: 381.
6. Chandlee J.M. and Vodkin L.O (1989) Unstable expression of a soybean gene during seed coat development. Theor Appl Genetics 77: 587–594.
7. Vodkin LO, Rhodes PR, Goldberg RB (1983) A lectin gene insertion has the structural features of a transposable element. Cell 34: 1023–1031.
8. Zabala G, Vodkin L (2008) A putative autonomous 20.5 kb-CACTA transposon insertion in an *F3'H* allele identifies a new CCACTA transposon subfamily in *Glycine max*. BMC Plant Biol 8: 124.
9. Xu M, Brar HK, Grosic S, Palmer RG, Bhattacharyya MK (2010) Excision of an active CACTA-like transposable element from DFR2 causes variegated flowers in soybean [Glycine max (L.) Merr.]. Genetics 184(1): 53–63.

10. Stracke R, Werber M, Weisshaar B (2001) The R2R3-MYB gene family in *Arabidopsis thaliana*. Curr Opin Plant Biol 4(5): 447–456.
11. Banks JA, Masson P, Fedoroff N (1988) Molecular mechanisms in the developmental regulation of the maize *Suppressor-mutator* transposable element. Genes & Development 2: 1364–1380.
12. Wessler SR (1988) Phenotypic diversity mediated by the maize transposable elements *Ac* and *Spm*. Science 242: 399–405.
13. Fedoroff NV (1989) About Maize Transposable Elements and Development. Cell 56: 181–191.
14. Rhodes PR, Vodkin LO (1985) Highly structured sequence homology between an insertion element and the gene in which it resides. PNAS 82: 493–497.
15. Schmutz J, Cannon SB, Schlueter J, Ma J, Mitros T, et al (2010) Genome sequence of the paleopolyploid soybean. Nature 463: 178–183.
16. Rhodes PR, Vodkin LO (1988) The *Tgm* family of transposable elements in soybean. Genetics 120: 597–604.
17. Wang CS, Vodkin LO (1994) Extraction of RNA from tissues containing high levels of procyanidins that bind RNA. Plant Mol Biol Reporter 12: 132–145.
18. Wang CS, Todd J, Vodkin LO (1994) Chalcone synthase mRNA and activity are reduced in yellow seed coats with dominant *I* alleles. Plant Physiol 105: 739–748.
19. Lin-Wang K, Bolitho K, Grafton K, Kortstee A, Karunairetnam S, et al. (2010) An R2R3 MYB transcription factor associated with regulation of the anthocyanin biosynthetic pathway in Rosaceae. BMC Plant Biol 10: 50.
20. Kranz HD, Denekamp M, Greco R, Jin H, Levya A, et al. (1998) Towards functional characterization of the members of the R2R3-MYB gene family from *Arabidopsis thaliana*. Plant J 16: 263–276.
21. Aharoni A, Ric-De-Vos CH, Wein M, Sun Z, Greco R, et al. (2001) The strawberry *FaMYB1* transcription factor suppresses anthocyanin and flavonol accumulation in transgenic tobacco. Plant J 28(3): 319–332.
22. Jin H, Cominelli E, Bailey P, Parr A, Mehrtens F, et al. (2000) Transcriptional repression by AtMYB4 controls production of UV-protecting sunscreens in *Arabidopsis*. The EMBO J 19(22): 6150–6161.
23. Zimmermann IM, Heim MA, Weisshaar B, Uhrig JF (2004) Comprehensive identification of *Arabidopsis thaliana* MYB transcription factors interacting with R/B-like BHLH proteins. Plant J 40(1): 22–34.
24. Winkel-Shirley B (2001) Flavonoid biosynthesis. A colorful model for genetics, biochemistry, cell biology, and biotechnology. Plant Physiol 126: 485–493.
25. Quattrocchio F, Wing J, van der Woude K, Souer E, de Vetten N, et al. (1999) Molecular analysis of the anthocyanin2 gene of petunia and its role in the evolution of flower color. Plant cell 11: 1433–1444.
26. Nesi N, Debeaujon I, Jond C, Pelletier G, Caboche M, et al. (2000) The TT8 gene encodes a basic helix-loop-helix domain protein required for expression of DFR and BAN genes in arabidopsis siliques. Plant Cell 12: 1863–1878.
27. Nesi N, Jond C, Debeaujon I, Caboche M, Lepiniec L (2001) The Arabidopsis TT2 gene encodes an R2R3 MYB domain protein that acts as a key determinant for proanthocyanidin accumulation in developing seed. Plant Cell 13: 2099–2114.
28. Jofuku KD, Schipper RD, Goldberg RB (1989) A frameshift mutation prevents Kunitz trypsin inhibitor mRNA accumulation in soybean embryos. Plant Cell 1: 427–435.
29. Cokus SJ, Feng S, Zhang X, Chen Z, Merriman B, et al. (2008) Shotgun bisulphite sequencing of the *Arabidopsis* genome reveals DNA methylation patterning. Nature 452: 215–219.
30. Krueger F, Andrews SR (2011) Bismark: a flexible aligner and methylation caller for Bisulfite-Seq applications. Bioinformatics 27: 1571–1572.
31. Robinson JT, Thorvaldsdóttir H, Winckler W, Guttman M, Lander ES, et al. (2011) Integrative genomics viewer. Nature biotechnology 29(1): 24–26.
32. Zabala G, Vodkin L (2005) The *wp* mutation of *Glycine max* carries a gene-fragment-rich transposon of the CACTA superfamily. Plant Cell 17: 2619–2632.
33. Bailey TL, Elkan C (1994) Fitting a mixture model by expectation maximization to discover motifs in biopolymers. Proceedings of the Second International Conference on Intelligent Systems for Molecular Biology. AAAI Press, Menlo Park, California: 28–36. (http://meme.nbcr.net/meme/cgi-bin/meme.cgi).
34. Paz-Ares J, Ghosal D, Saedler H (1990) Molecular analysis of the *C1-I* allele from *Zea mays*: A dominant mutant of the regulatory *C1* locus. EMBO J 9: 315–321.
35. Goff SA, Cone KC, From ME (1991) Identification of functional domains in the maize transcriptional activator C1: comparison of wild-type and dominant inhibitor proteins. Genes and Development 5: 298–309.
36. Tamagnone L, Merida A, Parr A, Mackay S, Culianez-Macia FA, et al. (1998) The AmMYB308 and AmMYB330 transcription factors from Anthirrhinum regulate phenylpropanoid and lignin biosynthesis in transgenic tobacco. Plant Cell 10: 135–154.
37. Franken P, Schrell S, Peterson PA, Saedler H, Wienand U (1994) Molecular analysis of protein domain function encoded by the myb-homologous maize genes C1, Zm 1 and Zm 38. Plant J 6: 21–30.
38. Menssen A, Hohmann S, Martin W, Schnable PS, Peterson PA, et al. (1990) The En/Spm transposable element of *Zea mays* contain splice sites at the termini generating a novel intron from a dSpm element in A2 gene. The EMBO J 9: 3051–3057.
39. Wang CS (1993) Genetics and expression of the anthocyanin biosynthetic pathway genes in soybean. Ph.D. thesis, University of Illinois.
40. Dellaporta SL (1993) Plant DNA miniprep version 2.1–2.3. In the Maize Handbook, M. Freeling and V. Walbot, eds (New York: Springer-Verlag): 522–525.
41. McCarty D (1986) A simple method for extraction of RNA from from maize tissue. Maize Genet Coop Newsl 60: 61.
42. Langmead B, Trapnell C, Pop M, Salzberg SL (2009) Ultrafast and memory efficient alignment of short DNA sequences to the human genome. Genome Biol 10: R25.
43. Mortazavi A, William BA, McCue K, Schaeffer L, Wold B (2008) Mapping and quantifying mammalian transcriptomes by RNA-Seq. Nat Methods 5: 621–628.
44. Goodstein DM, Shu S, Howson R, Neupane R, Hayes RD, et al (2012) Phytozome: a comparative platform for green plant genomics. Nucleic Acids Res. 40: D1178–D1186.
45. Akada S, Kung SD, Dube SK (1993) Nucleotide sequence and putative regulatory elements of a nodule-development-specific member of the soybean (Glycine max) chalcone synthase multigene family, Gmchs 7. Plant Physiol 102(1): 321–323.
46. Ralston L, Subramanian S, Matsuno M, Yu O (2005) Partial reconstruction of flavonoid and isoflavonoid biosynthesis in yeast using soybean type I and type II chalcone isomerases. Plant Physiol 137 (4): 1375–1388.
47. Zabala G, Vodkin L (2007) A rearrangement resulting in small tandem repeats in the F3′5′H gene of white flower genotypes is associated with the soybean *W1* locus. Crop Science 47: S113–S124.
48. Kovinich N, Saleem A, Arnason JT, Miki B (2010) Functional characterization of a UDP-glucose:flavonoid 3-O-glucosyltransferase from the seed coat of black soybean. Phytochemistry 71(11–12): 1253–1263.
49. Kovinich N, Saleem A, Arnason JT, Miki B. (2012). Identification of two anthocyanidin reductase genes and three red-brown soybean accessions with reduced anthocyanidin reductase 1 mRNA, activity, and seed coat proanthocyanidin amounts. JOURNAL J. Agric Food Chem 60 (2): 574–584.

16

The Symbiotic Biofilm of *Sinorhizobium fredii* SMH12, Necessary for Successful Colonization and Symbiosis of *Glycine max* cv Osumi, Is Regulated by Quorum Sensing Systems and Inducing Flavonoids via NodD1

Francisco Pérez-Montaño[1]*, Irene Jiménez-Guerrero[1], Pablo Del Cerro[1], Irene Baena-Ropero[2], Francisco Javier López-Baena[1], Francisco Javier Ollero[1], Ramón Bellogín[1], Javier Lloret[2], Rosario Espuny[1]

1 Departamento de Microbiología, Facultad de Biología, Universidad de Sevilla, Sevilla, Spain, 2 Departamento de Biología, Facultad de Ciencias, Universidad Autónoma de Madrid, Madrid, Spain

Abstract

Bacterial surface components, especially exopolysaccharides, in combination with bacterial Quorum Sensing signals are crucial for the formation of biofilms in most species studied so far. Biofilm formation allows soil bacteria to colonize their surrounding habitat and survive common environmental stresses such as desiccation and nutrient limitation. This mode of life is often essential for survival in bacteria of the genera *Mesorhizobium*, *Sinorhizobium*, *Bradyrhizobium*, and *Rhizobium*. The role of biofilm formation in symbiosis has been investigated in detail for *Sinorhizobium meliloti* and *Bradyrhizobium japonicum*. However, for *S. fredii* this process has not been studied. In this work we have demonstrated that biofilm formation is crucial for an optimal root colonization and symbiosis between *S. fredii* SMH12 and *Glycine max* cv Osumi. In this bacterium, *nod*-gene inducing flavonoids and the NodD1 protein are required for the transition of the biofilm structure from monolayer to microcolony. Quorum Sensing systems are also required for the full development of both types of biofilms. In fact, both the *nodD1* mutant and the lactonase strain (the lactonase enzyme prevents AHL accumulation) are defective in soybean root colonization. The impairment of the lactonase strain in its colonization ability leads to a decrease in the symbiotic parameters. Interestingly, NodD1 together with flavonoids activates certain quorum sensing systems implicit in the development of the symbiotic biofilm. Thus, *S. fredii* SMH12 by means of a unique key molecule, the flavonoid, efficiently forms biofilm, colonizes the legume roots and activates the synthesis of Nod factors, required for successfully symbiosis.

Editor: Roy M. RoopII, East Carolina University School of Medicine, United States of America

Funding: This work was supported by the following grants: AGL2012-38831 from the Spanish Ministerio de Economía y Competitividad, http://www.idi.mineco.gob.es/portal/site/MICINN/ P10-AGR-5821 and P11-CVI-7050 from the Junta de Andalucía, Consejería de Innovación, Ciencia y Empresas. http://www.juntadeandalucia.es/organismos/economiainnovacioncienciayempleo.html. Dr. Pérez-Montaño's work was supported by an FPU fellowship from the Spanish Ministerio de Ciencia y Tecnología and a contract from the University of Seville. http://www.idi.mineco.gob.es/portal/site/MICINN/ http://www.us.es/. The funders had no role in study design, data collection and analysis, decision to publish, or preparation of the manuscript.

Competing Interests: The authors have declared that no competing interests exist.

* Email: fperezm@us.es

Introduction

Rhizobia are bacteria that interact with the roots of leguminous plants to develop symbiotic nodules when environmental nitrogen is limited. In these symbiotic structures atmospheric nitrogen is reduced to ammonium making it available to the plant. This process requires the exchange of molecular signals between both members of the symbiosis. Legume roots exude flavonoids which induce, via a mechanism involving the NodD family of transcription activators, the biosynthesis and secretion of strain-specific lipo-chito-oligosaccharides, also known as nodulation factors (Nod factors), which trigger the initiation of nodule organogenesis [1]. Plant flavonoids, besides inducing Nod factor production, attract the bacteria to the legume root [2]; induce type III secretion machinery via NodD1 by activation of the transcriptional regulator TtsI [3,4]; and activate the rhizobial Quorum Sensing (QS) systems [5].

QS is defined as a regulation mode of bacterial gene expression in response to changes in their population density, which is mediated by small diffusible signal molecules called autoinducers (AI) [6,7]. QS-regulated genes are involved in adaptive changes in the physiology of the bacterial population, which modify their behaviour when a certain cell density is reached. QS modulates a broad variety of phenotypes, such as biofilm formation, toxin and exopolysaccharide production, virulence, plasmid transfer and motility, which usually are essential for successful establishment of a symbiotic or a pathogenic relationship with the eukaryotic hosts [8,9,10,11,12]. In plant-associated bacteria QS coordinates the expression of genes involved in virulence, colonization and symbiosis [13].

In all species studied so far, bacterial surface components (flagella, lipopolysaccharides and especially exopolysaccharides) in combination with the presence of bacterial QS signals are crucial for the formation of biofilms [12]. Biofilms are defined as bacterial communities surrounded by a self-produced polymeric matrix, and attached to an inert or a biotic surface [14]. In the course of biofilm formation, the initial reversible attachment to the surface is followed by an irreversible attachment and multiplication of the bacteria forming microcolonies that develop mature communities with a three-dimensional structure, in some cases permeated by channels, which act as the biofilm circulatory system. All these processes are coordinated by bacterial QS systems [14,15]. Many rhizobia have been described as forming microcolonies or biofilms when they colonize legume roots. These biofilms are mainly composed of water and bacterial cells. However, the three-dimensional structure of the biofilm is due to an extracellular matrix, which is formed by exopolysaccharides (EPS) [16]. In *S. meliloti*, the biofilm formation is affected and/or regulated by nutritional and environmental conditions [17,18]; exopolysaccharides and flagella [19]; ExoR with the ExoS–ChvI two-component system [20]; Nod factors synthesized by *nod* genes [19]; and regulation of exopolysaccharide biosynthesis by means of QS systems [21]. Development of biofilms by rhizobia is crucial to overcoming environmental stresses and in certain species the biofilm formation is clearly an important feature of symbiotic ability [12,19,22,23,24]. *S. fredii* SMH12, a wide host-range rhizobium that nodulates soybean and dozens of other legumes, produces at least three AHLs, *N*-octanoyl homoserine lactone (C8-HSL), 3-oxo *N*-octanoyl homoserine lactone (3-oxo-C8-HSL) and *N*-tetradecanoyl homoserine lactone (C14-HSL). In the presence of the flavonoid genistein, an activator of its nodulation genes, the overall production of AHLs is enhanced, being detected the C14-HSL only in bacterial cultures supplemented with inducing flavonoids [5]. SMH12 possesses at least one gene, *traI*, which is responsiblefor the synthesis of the 3-oxo-C8-HSL [5]; and at least two non-identified *luxI*-type gene involved in the synthesis of the C8-HSL and the C14-HSL. Furthermore, in this bacterium the AHLs and the QS systems are involved in the biofilm formation on the abiotic surface [25].

In this work, the role and regulation of biofilm formation in *S. fredii* SMH12 have been studied during symbiosis with *Glycine max* cv. Osumi. For this purpose, the wild-type strain, a *nodD1* mutant and a lactonase overproducing strain of SMH12 were constructed. The lactonase enzyme hydrolyzes the ester bond of the homoserine lactone ring of acylated homoserine lactones preventing these signalling molecules from binding to their target transcriptional regulators. Both *nodD1* and lactonase strains showed an impaired ability for colonization of *Glycine max* roots with respect to the wild-strain, probably due to an abnormal symbiotic biofilm formation which determines a less effective symbiosis. Interestingly, QS-biofilm formation and effective nodulation are connected through flavonoids, since these molecules initiate the molecular dialogue between bacteria and plants and allow the symbiotic biofilm formation. In summary, this report unequivocally demonstrates that the development of biofilm is crucial for successful root colonization and optimal symbiosis in *S. fredii* SMH12.

Materials and Methods

Strains and media

Bacterial strains and plasmids used in this work are listed in Table 1. *S. fredii* SMH12 and derivative strains were grown at 28°C in tryptone-yeast extract (TY) medium [26], yeast extract mannitol (YM) medium [27] with a lower mannitol concentration (3 g l^{-1}) and low-phosphate minimal glutamate mannitol (MGM) medium [21], supplemented when necessary with genistein 3.7 μM as inducing flavonoid, umbelliferone 6.2 μM as non-inducing flavonoid, or with commercial AHLs [25]. *Agrobacterium tumefaciens* NT1 (pZLR4) was grown at 28°C in YM with carbenicillin (100 μg ml^{-1}) and gentamicin (30 μg ml^{-1}). *Escherichia coli* strains were cultured in Luria-Bertani (LB) medium (28) at 37°C. When required, the media were supplemented with the appropriate antibiotics as described by Lamrabet et al. [29]. Commercial AHLs were dissolved in methanol and used at different concentrations. Flavonoids and AHLs were purchased from Fluka (Sigma-Aldrich, USA).

Plasmids were transferred from *E. coli* to SMH12 by conjugation as described by Simon [30] using plasmid pRK2013 as helper. Recombinant DNA techniques were performed according to the general protocols of Sambrook et al. [28]. For hybridization, DNA was blotted to Hybond-N nylon membranes (Amersham, UK), and the DigDNA method of Roche was employed according to the manufacturer's instructions. PCR amplifications were performed as previously described [31]. Using this methodology the plasmid pMUS534 was employed for the homogenotization of the mutated version of the *nodD1* gene in *S. fredii* SMH12, generating the mutant strain SVQ648. The double recombination event was confirmed by Southern blotting (data not shown).

For the growth curves of the different strains, bacteria were grown in 5 ml of YM medium to early stationary phase and then diluted to OD_{600} values around 0.03 in fresh YM medium with or without flavonoids. Growth was monitored by measuring the OD_{600} for at least 46 h. Each experiment was performed two days, eight replicates each day.

RNA isolation, cDNA synthesis and Quantitative RT-PCR

Bacterial RNA was extracted from bacterial cultures from the microtiter plates in the same conditions employed for the biofilm assays and described below. For the RNA extraction the PowerBiofilm RNA Isolation Kit was employed following the manufacturer's instructions (MO BIO, USA). Three independent RNA extractions were performed.

To quantify the expression of the *S. fredii* SMH12 *traI* gene using quantitative RT-PCR, primers rt-traI-F and rt-traI-R described by Pérez-Montaño et al. [5] were used. Expression was calculated relative to bacteria grown without flavonoid. The *S. fredii* SMH12 RNA *16S* gene was used as internal control to normalize gene expression. RNA *16S* primers used are described in the same work [5]. The expression data shown are the mean (± standard deviation of the mean) for three biological replicates. The fold change in the target gene, normalized to RNA *16S*, and relative to gene expression in the culture without flavonoids was calculated. PCR was conducted on the LightCycler 480 (Roche, Switzerland) with the following conditions: 95°C, 10 min; 95°C, 30 s; 50°C, 30 s; 72°C, 20 s; forty cycles, followed by the melting curve profile from 60 to 95°C to verify the specificity of the reaction. The threshold cycles (Ct) were determined with the iCycler software and the individual values for each sample were generated by averaging three technical replicates that varied less than 0.5 per cycle.

Table 1. Resistance phenotypes: Gm^R, Km^R, Nx^R, Ap^R and Tc^R, gentamicin, kanamycin, nalidixic acid, ampicillin and tetracycline, respectively.

Strain or plasmid	Relevant properties	Reference
Agrobacterium tumefaciens NT1 (pZRL4)	*A.tumefaciens* without pTiC58; with pZRL4, which carries the fusion *traG::lacZ* and the gene *traR*, Gm^R	[13]
Escherichia coli DH5α	*SupE44*, *ΔlacU169*, 5hsdR17, *recA1*, *endA1*, *gyrA96*, *thi-1*, *relA1*, Nx^R	[25]
Sinorhizobium fredii SMH12	Wild-type strain, Ap^R	[40]
SMH12 (pME6000)	SMH12 with the plasmid pME6000, Tc^R	This work
SMH12 (pME6863)	SMH12 with the plasmid pME6863, which carries the lactonase gene, Tc^R	This work
SMH12 (pMP2463)	SMH12 with the plasmid pMP2463, which carries the gene of the GFP, Gm^R	This work
SMH12 (pMP6000) (pMP2463)	SMH12 with the plasmids pME6000 and pMP2463, which carry the empty plasmid and the green fluorescent protein, respectively, Tc^R Gm^R	This work
SMH12 (pMP6863) (pMP2463)	SMH12 with the plasmids pME6863 and pMP2463, which carry the gene that encode the lactonase and the green fluorescent protein, respectively, Tc^R Gm^R	This work
SVQ648	SMH12 *nodD1::lacZ*-Gm^R	This work
SVQ648 (pMP2463)	SVQ648 with the plasmid pMP2463, which carries the gene of the GFP, Gm^R	This work
pBBR1MCS-5	Broad-host-range cloning vector, Gm^R	[41]
pK18mobsacB	Cloning vector (suicide in rhizobia), Km^R	[42]
pME6000	Broad-host-range cloning vector, Tc^R	[43]
pME6863	pME6000::*aiiA*, plasmid carrying the lactonase gene, Tc^R	[44]
pMP2463	pBBR-MCS-5::*egfp-1*, plasmid carrying the green fluorescent protein gene, Gm^R	[45]
pMUS534	Plasmid pK18mob derivative containing *nodD1::lacZ*-Gm^R	[46]
pRK2013	Helper plasmid, Km^R	[47]

Confocal laser scanning microscopy

The different events of biofilm formation were visualized by confocal laser scanning microscopy using a method described by Russo et al. [32]. Bacterial cultures were placed in 8 well chambered cover glass slides containing a borosilicate glass base 1 μm thick (Thermo Fischer Scientific Inc., USA) for 4 days without shaking. To avoid desiccation, the chambers were incubated in a humid sterilized petri dish. Confocal microscope image capture was carried out with Leica TCS SP2 (Leica Microsystems, Germany). *In silico* 3D reconstruction analysis was executed using the computer program ImageJ (Java, USA).

Biofilm formation assay

The biofilm formation assay was based on the method described by O'Toole and Kolter [33] with modifications [34]. Cultures were grown in 5 ml of low-phosphate MGM medium, diluted to

Table 2. Plant responses to inoculation of *Glycine max* cv. Osumi with *S. fredii* SMH12 and derivatives.

Inoculant	Number of nodules	Fresh mass of nodules (g)	Plant-top dry mass (g)
None	0±0*	0±0*	0.48±0.11*
SMH12	124.30±12.06	1.56±0.20	1.34±0.18
SVQ648	0±0*	0±0*	0.46±0.10*
SMH12 (pME6863)	76.20±16.02*	1.04±0.29	1.10±0.30
SMH12 (pME6000)	131.00±22.68	1.60±0.38	1.35±0.25

Data represent means ± sd of six soybean jars. Each jar contained two soybean plants. Determinations were made six weeks after inoculation. Mutant *nodD1* and the lactonase strain parameters were individually compared with the parental strain SMH12 parameters by using the Mann-Whitney non-parametric test. Values tagged by * are significantly different at the level $\alpha = 5\%$.

Table 3. *Glycine max* cv. Osumi root attachment.

Inoculant	CFU/cm of proximal root	CFU/cm of lateral root
None	0 ± 0*	0 ± 0*
SMH12	$1.68\pm0.18\times10^{5}$	$1.22\pm0.29\times10^{4}$
SVQ648	$1.10\pm0.42\times10^{5}$	$3.83\pm0.5\times10^{3}$*
SMH12 (pME6863)	$5.00\pm0.10\times10^{4}$*	$8.45\pm0.46\times10^{2}$*
SMH12 (pME6000)	$1.55\pm0.27\times10^{5}$	$1.20\pm0.39\times10^{4}$

Soybean plants were inoculated with the test strains, one centimetre of the proximal root and of the lateral roots of each plant were collected after 7 days, and the bacteria present were resuspended and plated. Data are the mean ± SD of at least three independent experiments performed in triplicate. Mutant *nodD1* and the lactonase strain attachment were individually compared with the parental strain SMH12 attachment by using the Mann-Whitney non-parametric test. Values tagged by * are significantly different at the level $\alpha = 5\%$.

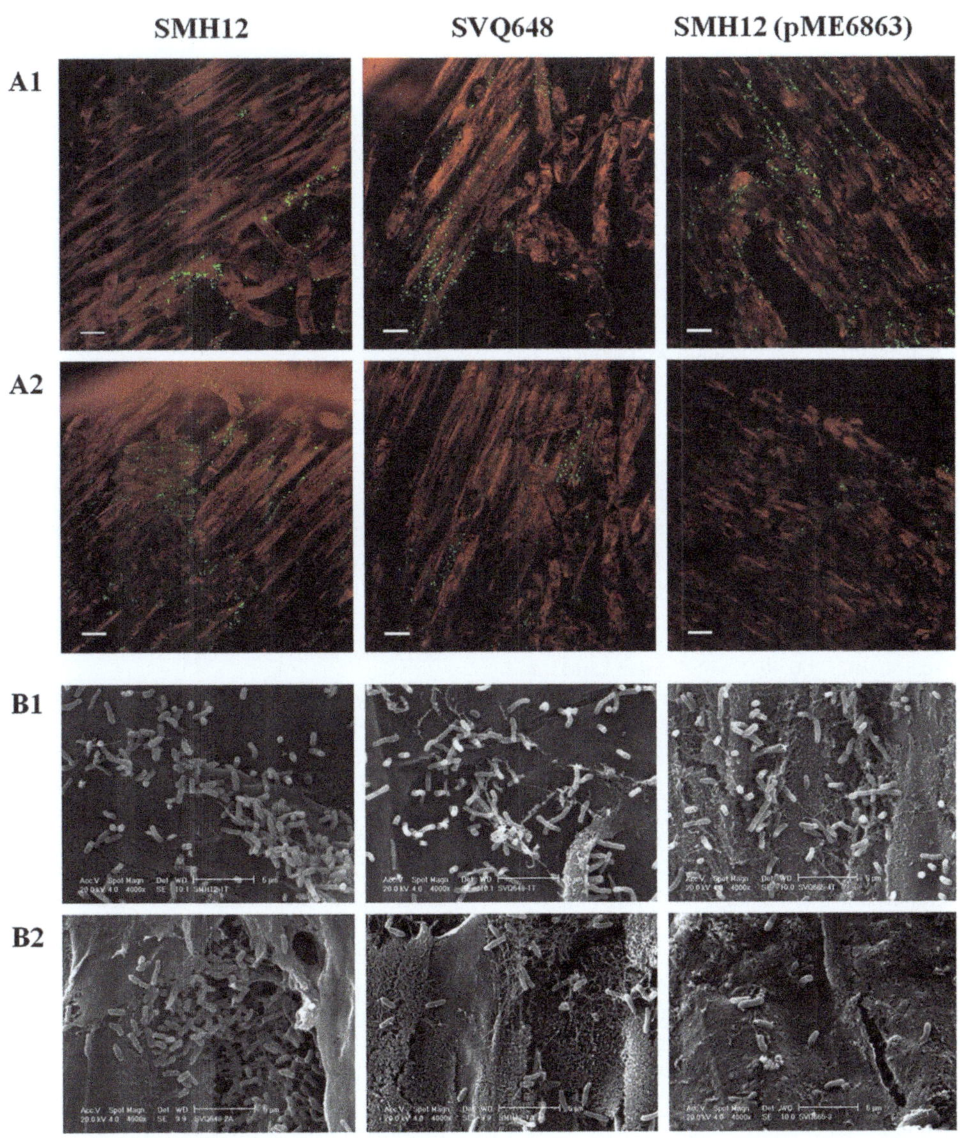

Figure 1. Visualization of the soybean root colonization. A. Epifluorescence microscopy analysis of the colonization of the soybean rhizosphere by gfp-tagged bacteria [SMH12, SVQ648, SMH12 (pME6863)]. Roots were visualized 7 days after inoculation. 1. Proximal root. 2. Lateral roots. Bar, 100 μm. B. Scanning microscopy analysis of the colonization of the soybean rhizosphere by SMH12, SVQ648 and SMH12 (pME6863). Roots were visualized 7 days after inoculation. 1. Proximal root. 2. Lateral roots. Bar, 5 μm. SMH12: wild-type, SVQ648: *nodD1* mutant, SMH12 (pME6863): lactonase strain.

an OD_{600} of 0.2, with or without flavonoids and AHLs, and inoculated with 100 µl aliquots and placed on polystyrene microtiter plates, U form (Deltalab S.L., Spain). The plates were inverted and incubated at 28°C for 7 days with gentle rocking. Cell growth was analyzed by measuring OD_{600} using a microtiter reader Synergy HT (Biotek, USA). The culture in each well was removed carefully; the wells were dried, washed three times with 0.9% NaCl and dried again. Biofilms in each well were stained with 100 µl of 0.1% crystal violet for 20 minutes, then washed with water three times and dried again. Finally, 100 µl of 96% ethanol were added to each well and the OD_{570} was measured. Every experiment was performed six times with eight replicates each time.

Quantification of overall AHLs

Supernatants of the rhizobial strains grown for 7 days in 96 wells on U shaped polystyrene microtiter plates (Deltalab S.L., Spain) were collected and sterilized by microfiltration. To determine the overall autoinducer production in each condition the method described by Pérez-Montaño et al. was used. [5]. Briefly, 25 µl, 2.5 µl or 0.25 µl of the supernatants were mixed with YM to obtain a final volume of 2.5 ml reaching supernatant concentrations of 1%, 0.1% and 0.01% (v/v). The mixtures were inoculated with approximately 10^7 cells ml^{-1} of the *A. tumefaciens* NT1 (pZLR4), incubated with shaking for 12 h at 28°C and assayed for β-galactosidase activity [35]. As controls, 125 µl of distilled water or 125 µl of *N*-(3-oxo-hexanoyl)-L-homoserine lactone (3-oxo-C6-HSL) 5.5 µM, were used. To obtain the standard curve with synthetic AHLs, 125 µl of 3-oxo-C8-HSL, C8-HSL or C14-HSL at different concentration were added. The experiments were repeated independently five times with 3 replicates.

Thin Layer Chromatography

For TLC analysis, cultures previously removed from each well of the microtiter plate were extracted with the same volume of dichloromethane, evaporated to dryness and analyzed by thin-layer chromatography as described by Pérez-Montaño et al. [25]. Briefly, 1 µl of each culture extract was loaded on TLC plates (HPTLC plates RP-18 $_{F254s}$ 1.13724 and 1.05559, Merck, Germany) using methanol:water (60:40 v/v) as eluent, dried and overlaid with a soft agar culture of the biosensor *A. tumefaciens* NT1 (pZLR4).

Plant assays

Nodulation assays on *Glycine max* (L.) Merrill cultivar Osumi were performed as described by de Lyra et al. [36]. Plants were inoculated with approximately 5×10^8 bacteria and were grown in Leonard jars with Fåhraeus nutrient solution [27] for 42 days with a 16 hour-photoperiod at 26°C in light and 18°C in the dark with 70% of humidity. Shoots were dried at 70°C for 48 h and weighed. Experiments were performed three times. For root colonization assays and microscopy, seeds were surface sterilized, germinated and inoculated with a modified method described in Pérez-Montaño et al. [25]. Briefly, *Glycine max* (L.) Merrill cultivar Osumi seeds were soaked for 30 seconds in 96% ethanol and for 8 minutes in commercial bleach. Then, seeds were washed repeatedly with sterilized distilled water, germinated and checked for sterility in LB medium. Each plant was inoculated with approximately 5×10^8 bacteria. Plants were grown under hydroponic controlled conditions in Fåhraeus solution for 7 days in the same conditions as those mentioned above.

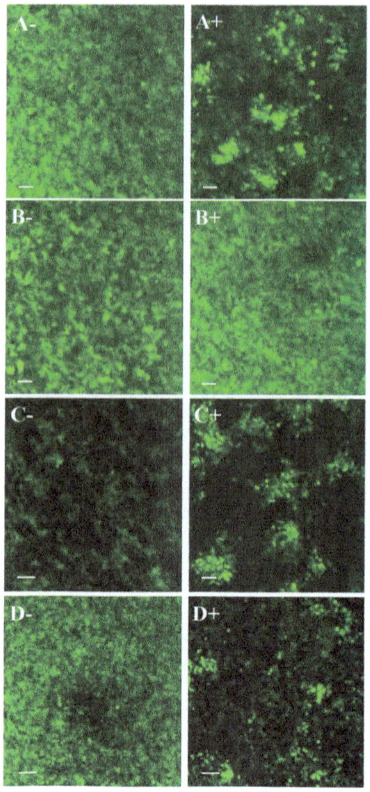

	Main fluorescence value
SMH12	100 ± 20.06
SMH12 + genistein (3.7 µM)	57.32 ± 5.76 *
SVQ648	80.16 ± 7.85
SVQ648 + genistein (3.7 µM)	105.86 ± 13.39
SMH12 (pME6863)	43.81 ± 2.51 *
SMH12 (pME6863) + genistein (3.7 µM)	51.37 ± 6.88 *
SMH12 (pME6000)	93.23 ± 17. 15
SMH12 (pME6000) + genistein (3.7 µM)	56.37 ± 3.72*

Figure 2. Biofilm structure of *S. fredii* SMH12 and derivatives on glass surfaces: reconstruction of the Z-stacks and measure of the surface coverage. Main fluorescence value of the wild-type strain was arbitrarily given a value of 100. Averages and standard deviations of five randomized optical fields per strain corresponding to two independent experiments are shown. The asterisks indicate a significant different at the level α = 5% with respect to wild-type strain by using the Mann-Whitney non-parametrical test. Left side corresponds to cultures without flavonoids. Right side corresponds to cultures with inducing flavonoid. A. SMH12. B. SVQ648. C. SMH12 (pME6863). D. SMH12 (pME6000). Bar, 20 µm. SMH12: wild-type, SVQ648: *nodD1* mutant, SMH12 (pME6863): lactonase strain. SMH12 (pME6000): carrying the empty plasmid.

Epifluorescence and scanning electron microscopy

Epifluorescence microscopy assays were carried out with *S. fredii* SMH12 and derivatives carrying GFP expressing plasmid pME2463. After growth, roots were excised (one centimetre of the proximal area of the main root and the last centimetre of a lateral root) and thoroughly washed with water to eliminate any loosely

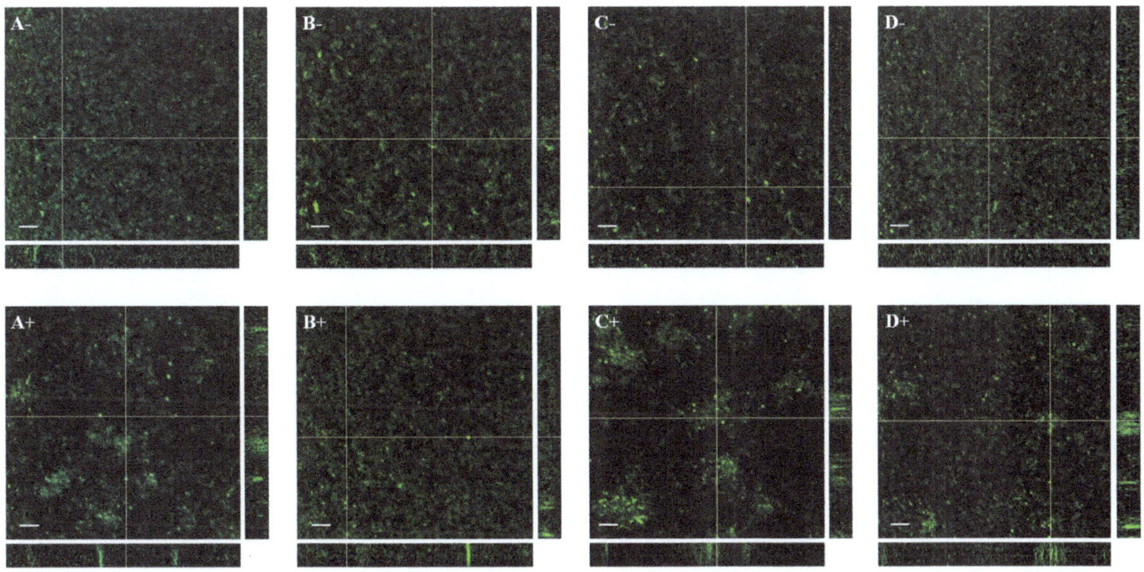

Figure 3. Biofilm structure of *S. fredii* SMH12 and derivatives on glass surfaces: reconstruction of the XY-axis, XZ-axis and YZ-axis. The top corresponds to cultures without flavonoids. The bottom corresponds to cultures with inducing flavonoid. A. SMH12. B. SVQ648. C. SMH12 (pME6863). D. SMH12 (pME6000). Bar, 20 μm. SMH12: wild-type, SVQ648: *nodD1* mutant, SMH12 (pME6863): lactonase strain. SMH12 (pME6000): carrying the empty plasmid.

attached bacteria. Afterwards, roots were soaked for 3 minutes in 0.5% crystal violet to avoid auto-fluorescence, following 3 washes with distilled water. Microscopy analysis was carried out using an Olympus BH2-FRCA microscope with 40x and 100x magnifications (Olympus, Japan). Images of GFP-labelled bacterial cells were obtained by using a filter set consisting of a 400 to 490 nm (BP490) bandpass exciter, a 505 nm dicroic filter and a 530 nm longpass emitter (EO530). In all cases, exposure length of red and green channels was 30 seconds.

For electron scanning microscopy, 7 day-soybean plants were collected and the roots were excised as described above. Sample preparation and visualization were done according to Barahona *et al.* [37].

Root attachment

Whole plants were carefully taken from the hydroponic solution and roots were excised and thoroughly washed with water to eliminate any loosely attached bacteria. Remaining attached bacteria were recovered from the root by vortexing (one centimetre of the proximal area of the main root or the last centimetre of a lateral root) for 2 min in a tube containing YM medium and plating the appropriate dilutions on YM medium plates. Experiments were performed three times with three plants per assay.

Table 4. β-galactosidase activity obtained using an adapted assay with *A. tumefaciens* NT1 (pZLR4) as bioreporter and grown in the presence of supernatants from biofilm cultures (1% v/v).

	Miller units	n (%)
Control (YM)	143.5±6.4	
Genistein (37 nM)	122.1±25.6	
Umbelliferone (62 nM)	131.2±7.2	
3-oxo-C6-HSL (5.5 μM)	824.7±21.2	
SMH12	717.6±23.6	100
SMH12+genistein (3.7 μM)	843.9±24.6	117*
SVQ648	702.7±28.4	97
SVQ648+genistein (3.7 μM)	745.8±39.2	104
SMH12 (pME6863)	150.2±53.8	21*
SMH12 (pME6863)+genistein (3.7 μM)	152.2±9.71	21*

Data are the mean ± SD of three independent experiments performed in triplicate.
n: percentage of induction of each supernatant with respect to SMH12 without flavonoids, defined as 100%.
Each β-galactosidase activity using biofilm supernatant was individually compared to that obtained in SMH12 without flavonoids by using the Mann-Whitney non-parametrical test. Numbers on the percentage of induction column followed by * are significantly different at the level $\alpha = 5\%$.

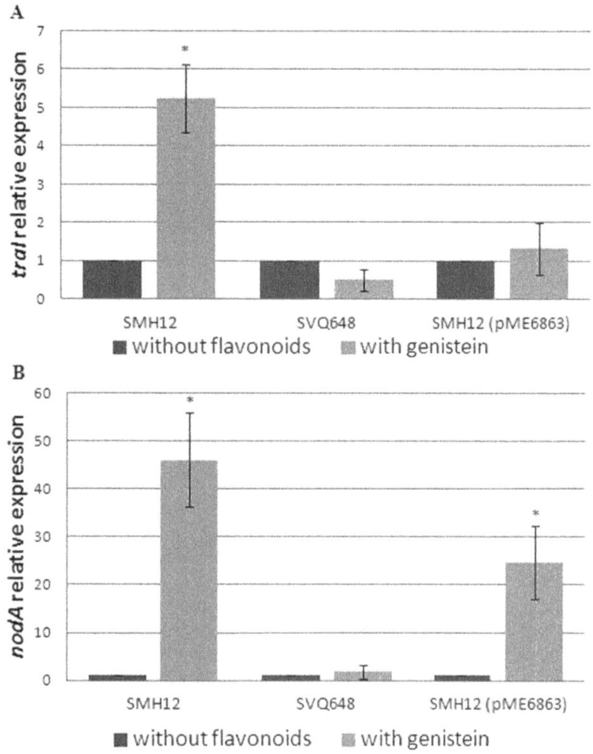

Figure 4. Quantitative RT-PCR analysis of the expression of *traI* and *nodA* from *S. fredii* SMH12 and derivatives from biofilm cultures. Expression data shown are the mean (± standard deviation of the mean) for three biological replicates. Expression was calculated relative to the expression without flavonoids of the wild-type strain by using the Mann-Whitney non-parametrical test. The asterisks indicate a significant different at the level $\alpha = 5\%$. White bars: biofilm cultures without flavonoids. Gray bars: biofilm cultures with genistein. A. *traI* relative expression. B. *nodA* relative expression. SMH12: wild-type, SVQ648: *nodD1* mutant, SMH12 (pME6863): lactonase strain.

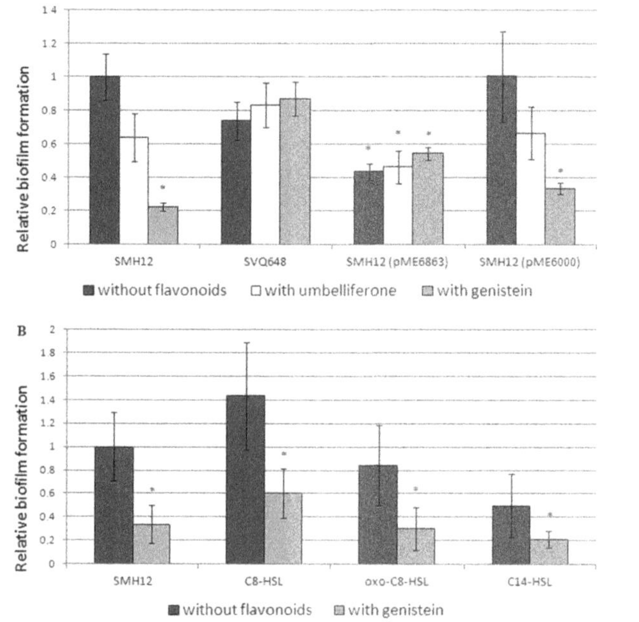

Figure 5. Adhesion of *S. fredii* SMH12 and derivatives on polystyrene surfaces. Biofilms were measured as the amount of crystal violet absorbed by the biofilm formed on multi-well plates and determined by absorbance at 570 nm after de-staining with ethanol (see methods). Absorbance of the wild-type strain was arbitrarily given a value of 1. Averages and standard deviations of eight replicas per strain corresponding to five independent experiments are shown. The asterisks indicate a significant different at the level $\alpha = 5\%$ with respect to wild-type strain by using the Mann-Whitney non-parametrical test. A. Dark gray bars correspond to experiments performed without flavonoids, white to experiments with umbelliferone and light grey bars to experiments with genistein. B. Dark gray bars correspond to experiments performed without flavonoids and light grey bars to experiments with genistein. 3-oxo-C8-HSL and C8-HSL are used at 5.5 μM. C14-HSL is used at 55 μM. SMH12: wild-type, SVQ648: *nodD1* mutant, SMH12 (pME6863): lactonase strain. SMH12 (pME6000): carrying the empty plasmid.

Results

Nodulation Test

In a previous work [5] we described that certain plant flavonoids, besides inducing synthesis of Nod factors, increase the overall production of QS signals in *S. fredii* SMH12. In many bacteria, these systems regulate a broad variety of phenotypes which are important for the successful establishment of a relationship with the eukaryotic hosts. For this reason, a nodulation test was performed to elucidate if SMH12 QS systems are regulating any important phenotype during the symbiosis with *Glycine max*. The symbiotic properties of the wild-type strain, SVQ648 (SMH12 *nodD1*::LacZ::Gm^R), hereafter *nodD1* mutant, SMH12 (pME6863), hereafter lactonase strain (this enzyme prevents AHL accumulation), and SMH12 (pME6000), which carries the empty plasmid, were determined in plant infection tests with soybean cultivar Osumi which are effectively nodulated by the wild-type strain *S. fredii* SMH12.

As expected, the *nodD1* mutant did not induce nodule development due to its inability to produce Nod factors, which are responsible for the formation of these structures in the soybean roots. Consequently, the plant top dry mass was significantly lower (a 60%, where * is the statistical significance of the differences observed using the Mann–Whitney non-parametric test) in plants inoculated with the *nodD1* mutant than in those inoculated with the parental strain SMH12. In the cases of the plants inoculated

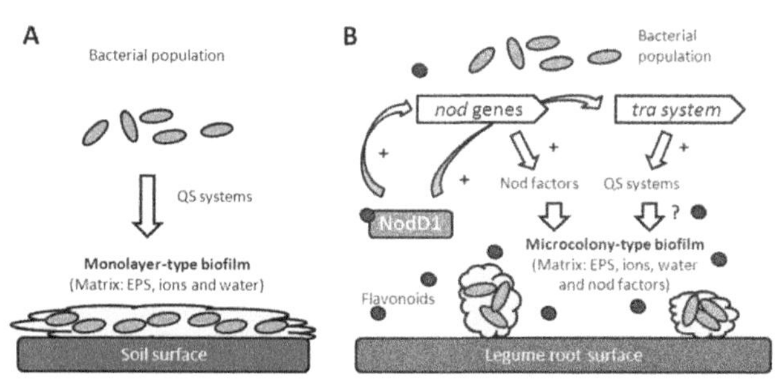

Figure 6. Model of biofilm formation in *S. fredii* SMH12. A. In soil. B. In rizosphere. *S. fredii* SMH12 forms monolayer-type biofilm when colonize soil surfaces. When the bacterium colonizes the legume root (in presence of inducing flavonoids), it forms microcolony-type biofilm, which is necessary for a successful root colonization and symbiosis between rhizobia and legume.

with the lactonase strain, the top dry mass, number of nodules and fresh mass of nodules formed were lower than in the wild-type strain (20%, 40% and 35% respectively). Only in the case of the number of nodules were these differences statistically significant (where * is the statistical significance of the differences observed using the Mann–Whitney non-parametric test) (Table 2). To confirm that differences in this strain are not due to the presence of the plasmid, the symbiotic parameters of SMH12 carrying the empty plasmid (pME6000) were studied. No changes were observed with respect to the wild-type strain. In summary, these results indicate that QS systems could regulate some important phenotypes during the symbiosis with *Glycine max*.

Plant root colonization

As commented in the introduction, in plant-associated bacteria one of the processes controlled by bacterial QS systems is the root colonization. For this reason three independent assays to study root colonization were carried out: root attachment quantification, observation studies of root by epifluorescence microscopy and electronic barrier microscopy. In these studies, proximal and lateral roots were analyzed separately to test differences in bacterial colonization in the whole root.

Root attachment assays showed a statistically significant reduction in the bacterial number per cm on the proximal area of the main root in the lactonase strain, the detected bacteria being around 10-fold less than in the other two strains. In lateral roots, colonization differences (statistically significant) were even more pronounced, these differences being 100-fold less than in the wild-type strain. A slight reduction in the number of bacteria per cm on the lateral root was also observed in the *nodD1* mutant (10-fold less than SMH12). No changes in root attachment were observed in plants inoculated with the SMH12 strain carrying the empty plasmid with respect to those inoculated with the wild-type strain (Table 3). No differences in the bacterial growth-curves were detected among the four strains (data not shown).

Epifluorescence microscopy and electronic barrier microscopy assays (Fig. 1) showed that in lateral roots of plants inoculated with *nodD1* mutant and lactonase strain, the soybean root surface presented lower bacterial density with respect to the wild-type strain. Interestingly, the wild-type strain seems to be also distributed more clustered than the other two strains along the main root and especially in the lateral roots, which could indicate different biofilm structures in these strains. All these results suggest that the *nodD1* mutant and especially the lactonase strain are defective in root colonization.

Influence of inducing flavonoids on biofilm structure

When rhizobia colonize the legume root surface microcolonies or biofilms are formed. Could the observed colonization differences be a consequence of abnormal biofilm formation in the lactonase strain and in the *nodD1* mutant? To study the role of the biofilm formation during symbiosis, the biofilm structures both in the presence/absence of nod-gen inducing flavonoid (genistein) were observed using confocal microscopy experiments. Biofilm images of the wild-type, the *nodD1* mutant, the lactonase and the control strain carrying the empty plasmid, all labelled with GFP, were obtained (Fig. 2 and Fig. 3). Results showed that after 4 days of growth without flavonoid, the formed biofilm consisted of a monolayer in all the studied strains being the surface coverage statistically lower (about 50%) in the lactonase strain biofilm (Fig. 2c). In the presence of genistein, the formed biofilm changed from monolayer to microcolony type in SMH12, being the coverage statistically lower (almost 45%) (Fig. 2a). In the case of the *nodD1* mutant, no change was observed in the biofilm three-dimensional structure or surface coverage with genistein (Fig. 2b). Moreover, the lactonase strain showed a less surface-coverage (43%) without genistein with respect to the wild-type, but interestingly, in the presence of genistein the biofilm development underwent the transition to microcolony but no reduction in the coverage was obtained with respect to the lactonase strain without flavonoid (Fig. 2c). The control strain carrying the empty plasmid showed the same phenotype as the wild-type strain (Fig. 2d). These results suggest that, in *S. fredii* SMH12, the nodulation gene inducing flavonoids via NodD1 protein provoke a decrease in the bacterial attachment to an abiotic surface, since the monolayer biofilm structure (covering the whole surface) developed in the absence of flavonoid changed to microcolony-type (covering only some parts of the surface). Furthermore, these data indicate that QS signals must to be involved at least in the full formation of the monolayer-type biofilm on the glass surface.

Influence of inducing flavonoids on the QS systems

Our earlier results demonstrated that *nod* gene inducing flavonoids increase the overall production of QS signals in *S. fredii* SMH12 [5]. To investigate if the activation of the QS systems takes place during the biofilm formation a quantification of the overall AI production was carried out in each condition and strain during the biofilm formation experiments on microtiter plates using the *A. tumefaciens* NT1 (pZRL4) biosensor. Firstly, the optimal concentration of bacterial supernatant required to obtain the wider differences between each condition without reaching saturation were determined adding different concentrations of bacterial supernatants (1%, 0.1% and 0.01%). The optimal concentration for these assays was 1% of the total volume of the biosensor culture (data not shown). Under these experimental conditions, results showed that the presence of genistein provoked a statistically significant higher production of AIs only in SMH12 (17% more), indicating that *nod* gene inducing flavonoids enhanced the QS signals accumulation via NodD1. Furthermore, β-galactosidase activity only in the presence of these flavonoids was similar to that obtained in the negative control (Table 4). As expected, a strong decrease in AI accumulation was detected in the supernatants from the lactonase strain, with or without genistein (Table 4). A supplementary TLC with extracts of the supernatants of biofilm experiments in microtiter plates showed a complete degradation of all AHLs in the lactonase strain. In the SMH12 strain carrying the empty plasmid no changes in the AHL profile were observed with respect to the wild-type strain (data not shown).

Finally, to ascertain if the increase in AHL production with flavonoids in biofilm culture assays was correlated with an increase of gene transcription, a quantitative real time RT-PCR assays was carried out. Results showed that the expression of *traI*, an AHL-synthesis gene from SMH12 [5], significantly increased (5-fold) in the presence of genistein compared to the control without flavonoids. No changes in the *traI* expression were observed in the presence of flavonoid in the case of the *nodD1* mutant or in the lactonase strain, suggesting that the induction of the transcription takes place via NodD1 and that QS systems are necessary for the gene expression enhancement probably due to the typical positive feedback of the QS systems at high cellular density (Fig. 4a). As control, the expression of *nodA*, a gene positively regulated via NodD1 and flavonoids, was measured. As expected, a statistically significant increase was observed in *nodA* gene expression (46-fold and 24-fold, respectively) in both SMH12 and lactonase strains in the presence of genistein, but not in the *nodD1* mutant. Interestingly, the *nodA* gene expression in the lactonase strain in the presence of genistein was significantly lower than in the wild-type (Fig. 4b), which could explain both the reduced soybean nodulation and the less rizosphere attachment capacities of the lactonase strain, connecting the Nod factor production with these capacities (Table 2, Table 3, Fig. 1).

All these observations suggest that SMH12 *nod* gene inducing flavonoids enhance the transcription of certain AHL synthesis genes and the overall AHL production in biofilm experiments via NodD1 when the bacteria have reached the necessary cell density threshold.

Role of the QS signals on biofilm formation

Finally, for a more exhaustive analysis of the role of each QS signal in the biofilm formation, bacterial adhesion to polystyrene surface (biofilm formation experiments) was studied after 7 days of growing in low-phosphate MGM medium (Fig. 5). Firstly, it was confirmed that the glass coverage results obtained in confocal microscopy experiments are correlated with the bacterial adhesion values obtained in biofilm formation in polystyrene microtiter plate experiments. Thus, we could use this experimental approach to study the different biofilm structures. As shown in Figure 5, in the presence of genistein, SMH12 showed a significant reduction in the adhesion value to the abiotic surface (80%, where * is the statistical significance of the differences observed using the Mann–Whitney non-parametric test). Only a slight reduction was obtained using umbelliferone, a non-inducing flavonoid [5]. In the case of *nodD1* mutant, the adhesion values with or without flavonoids were similar to those obtained in the case of SMH12 without flavonoid. However, in the lactonase strain a lower adhesion was observed to polystyrene surfaces (50%) in all cases with respect to the wild-type strain without flavonoids, indicating that QS signals must be implied in the full differentiation of both types of biofilm on polystyrene surface since neither the high values without flavonoids nor the low values with inducing flavonoids are reached. To confirm that the differences observed in this strain were not due to the presence of the plasmid, the adhesion values were studied using SMH12 carrying the empty plasmid (pME6000). The experiment confirmed that this strain behaved like the wild-type strain (Fig. 5a). The absorbance at 600 nm after bacterial growth was similar in all strains and conditions (data not shown). Thus, the adhesion values to polystyrene surface in the wild type strain have a correlation with the values of surface coverage obtained on a glass surface with confocal microscopy experiments (Fig. 2).

Once this method was validated as a reporter of the formation of the different biofilm types (monolayer with high values of coverage/adhesion and microcolony with low values of coverage/adhesion), the influence of QS signals in the biofilm formation of SMH12 with or without inducing flavonoids was studied. For this purpose, the cognate AHLs of SMH12, 3-oxo-C8-HSL, C8-HSL and C14-HSL [5], were added to wild-type bacterial cultures at physiological concentrations (data not shown) in biofilm formation experiments on microtiter plates (Fig. 5b). A slight increase or decrease in the adhesion to the polystyrene surface was observed when the medium was supplemented with C8-HSL or C14-HSL, respectively. However, these differences were not statistically significant (Fig. 5b). The absorbance at 600 nm after bacterial growth was similar in all bacteria and conditions (data not shown).

Discussion

During the first stages of root colonization, leguminous plants exudate molecules which chemotactically attract rhizobia. Once in the rhizosphere, the rhizobial population colonizes the root surface, forming a bacterial community surrounded by a matrix produced by its own bacteria, the biofilm. In the family *Rhizobiaceae*, the bacterial intrinsic factors that are required for the biofilm formation are mainly EPS and bacterial QS systems as well as in the case of *S. meliloti* the Nod factor production [12]. A fundamental question is whether the process of biofilm formation significantly affects nodulation in the symbiosis *S. fredii* SMH12-*G. max* cv. Osumi. To answer this question, the significance of QS systems and the *nod* gene inducing flavonoids on the biofilm formation, the colonization and the symbiosis with soybean were studied by means of both a *nodD1* mutant and a lactonase strain of *S. fredii* SMH12.

As expected, nodulation tests with soybean showed that the symbiotic phenotype of the *nodD1* mutant is similar to that obtained in non-inoculated plants, since this bacterium is unable to detect via NodD1 the flavonoids exuded by plants. Consequently the Nod factor production is blocked. Interestingly, results indicate that SMH12 QS systems could regulate some important process during the symbiosis, since the lactonase strain showed reductions in both the plant top dry mass and the fresh mass of nodules formed with respect to the wild-type strain but only in the number of nodules the reduction was statistically significant (Table 2). As described in the introduction, QS regulates a broad variety of phenotypes, including the biofilm formation. In the review of Rinaudi and Giordano [12] the mechanisms involved in both the rhizobial biofilm formation and the attachment to the plant roots are summarized. Taking into consideration the previous reports, they concluded that the biofilm lifestyle allows rhizobia to survive under unfavourable conditions and in certain species the biofilm formation is clearly an important feature of symbiotic ability [19,22,23,24]. However, the role and regulation of biofilm formation during the symbiosis *S. fredii* SMH12-soybean has not been reported. Our results suggest that there is a correlation between QS systems, biofilm formation, root colonization and symbiosis. Firstly, three independent experiments of bacterial root colonization showed a significant reduction in bacterial root colonization in the *nodD1* mutant and especially in the lactonase strain. These differences were more visible in the lateral root colonization (Table 3). Interestingly, in addition to the differences in the bacterial number, these two strains showed more of a spread surface distribution on roots compared to the wild-type, in which the bacteria appeared clustered (Fig. 1). These observations would indicate that the *nodD1* mutant and specially the lactonase strain do not colonize the root surface optimally due to an alteration in their biofilm formation ability. The impaired biofilm formation in the case of the lactonase strain would explain its less nodulation capacity in the soybean test (Table 2).

To corroborate this hypothesis, the biofilm structure was analyzed by confocal microscopy. Two types of biofilms were observed, the monolayer-type in the absence of flavonoid (bacterial population cover the entire surface homogeneously), and the microcolony-type in the presence of genistein (bacterial population is clustered) (Fig. 2a and Fig. 3a). These observations would indicate that on the surface of the legume root, with a high concentration of flavonoids, the bacterial biofilm could undergo a transition from monolayer to microcolony. Perhaps this biofilm structure allows the bacterial colonization of specific root areas (those with high flavonoid exudation), which would be the optimal for the symbiosis initiation (i.e. root hairs).

Furthermore, analysing these results we infer that QS systems is involved at least in the formation of the monolayer type, since the lactonase strain developed incomplete biofilm phenotypes without inducing flavonoids (Fig. 2c and Fig. 3c). In fact, most rhizobial biofilms investigated so far are regulated directly or indirectly by QS systems [12]. Interestingly, the transition to microcolony-type biofilm requires Nod factor production since this biofilm was not developed in the presence of genistein in the *nodD1* mutant

(Fig. 2b and Fig. 3b). Fujishige *et al.* [19] reported that *nodD1ABC* of *S. meliloti*, involved in the synthesis of Nod factors, are necessary for the three-dimensional architecture of the biofilm and, in fact, these molecules are present in the biofilm matrix. The mutation of any of these genes generates a monolayer-type biofilm. However, in contrast with our results, the presence of inducing flavonoids is not necessary for the development of three dimensional structures in the biofilm of *S. meliloti* [19].

Interestingly, our findings also suggest that flavonoids, beyond inducing the synthesis of Nod factors (which integrate the symbiotic biofilm matrix), are enhancing the QS systems in biofilm formation experiments (Table 4 and Fig. 4a). In the presence of inducing flavonoids, an increase in the *traI* expression and in the overall AI production was detected. These effects were not observed in the lactonase strain due to the positive feed-back regulation of the QS systems which only occurs when the threshold in cellular density is reached [6] (Fig. 4a and Table 4). Furthermore, it is clear that the enhancement of the *traI* gene expression takes place via NodD1, because in the *nodD1* mutant this over expression was not observed. He et al. [38] studied the QS systems of *Rhizobium* sp. NGR234, a *Sinorhizobium fredii* related bacterium, but they only unequivocally identified the *tra* system. This QS system is homologous to the one responsible for the plasmid Ti transfer in *A. tumefaciens*. This transference occurs only in the presence of opines, compounds that are produced by plants. Control of the Ti plasmid transference is modulated by TraM, a small protein that binds to and inactivates TraR, which senses the AI concentration. The induction in the presence of opines allows the synthesis of TraR at levels that overcome the inhibitory activity of TraM, activating the QS systems and the plasmid transfer [39]. In NGR234, the over expression of TraR activates not only the *tra* system but also other QS systems of the bacteria, since an increase in the overall AI production was detected [34]. However, so far, no natural compounds exuded by plants (similar to opines for *A. tumefaciens* QS systems) have been identified as inductor molecules for NGR234. Interestingly, results obtained in this paper and those reported in our previous work [5] indicate that the *nod* gene inducing flavonoids could be these inducing compounds for *S. fredii* SMH12.

In summary, experiments show that SMH12 QS systems must be involved at least in the full differentiation of monolayer-type biofilm and, moreover, in the presence of inducing flavonoids, these systems increase the AHL production and the biofilm undergoes a transition to microcolony-type (Fig. 4a, Table 4 and Fig. 2). Logically, this expression increase in the presence of flavonoids could be related to the developement of the microcolony-type biofilm, which would only occur in the rhizosphere. This fact would allow, through a unique key molecule, the efficient colonization of the legume root and the activation of the synthesis of Nod factors, both required for successful symbiosis. Interestingly, the study of the role of QS signals on biofilm formation experiments (Fig. 5) showed no statistically differences in adhesion values with or without flavonoid after addition of the wild-type cognate AHLs (Fig. 5b). However, in the case of the lactonase strain, the decrease in the adhesion values in the presence of inducing flavonoids did not reach the values obtained in the wild-type strain. This result suggests that, despite the transition to microcolony-type biofilm is regulated directly by NodD1 protein and inducing flavonoids (Fig. 2), the QS signals could also be involved in the full differentiation of this biofilm on the polystyrene surface (Fig. 5a).

Thus, taking into consideration all the results, we propose the following model (Fig. 6): *S. fredii* SMH12 forms a QS-regulated monolayer-type biofilm (whose matrix would be mainly composed of EPS, water and ions) when bacterium colonizes soil. In the rhizosphere, the exudation of inducing flavonoids, besides inducing the synthesis of Nod factors which are necessary for the microcolony-type biofilm, potentiate via NodD1 the *tra* QS system, which leads to an overproduction of the QS signals. This accumulation and the synthesis of Nod factors are required for the full development of the microcolony-type biofilm, whose matrix should be composed of EPS, water, ions and Nod factors. The complete formation of this symbiotic biofilm is necessary for a successful root colonization and symbiosis between rhizobia and legume.

Despite demonstrating in this work that the rhizobial biofilm formation is important for colonization and symbiosis of *S. fredii* SMH12 with the soybean, there are still many aspects that must to be studied in the future to further clarify this process.

Acknowledgments

We would like to thank the Laboratorio de Microscopía de Barrido y Análisis por Energía Dispersiva de Rayos X from the Universidad Autónoma de Madrid for electronic barrier microscopy assays. Thanks to Servicio General de Biología of the CITIUS from the University of Seville for allowing us to use their laboratory equipment. Thanks to Diane Haun for English style supervision. Special thanks to Miguel Cámara, for his comments on an earlier version of the manuscript.

Author Contributions

Conceived and designed the experiments: FPM RB JL RE. Performed the experiments: FPM IJG PDC IBR. Analyzed the data: FPM RB JL RE. Contributed reagents/materials/analysis tools: FJO JL. Contributed to the writing of the manuscript: FPM FJL RE.

References

1. Murray JD. (2011) Invasion by invitation: rhizobial infection in legumes. Mol Plant-Microbe Interact 24: 631–639.
2. Somers E, Vanderleyden J, Srinivasan M. (2004). Rhizosphere bacterial signalling: a love parade beneath our feet. Crit Rev Microbiol 30: 205–240.
3. Krause A, Doerfel A, Göttfert M. (2002) Mutational and Transcriptional Analysis of the Type III Secretion System of *Bradyrhizobium japonicum*. Mol Plant Microbe Interact 12: 1228–1235.
4. López-Baena FJ, Vinardell JM, Pérez-Montano F, Crespo-Rivas JC, Bellogín RA, et al. (2008) Regulation and symbiotic significance of nodulation outer proteins secretion in *Sinorhizobium fredii* HH103. Microbiology 154: 1825–1836.
5. Pérez-Montaño F, Guasch-Vidal B, González-Barroso S, López-Baena FJ, Cubo T, et al. (2011) Nodulation-gene-inducing flavonoids increase overall production of autoinducers and expression of *N*-acyl homoserine lactone synthesis genes in rhizobia. Res Microbiol 162: 715–723.
6. Fuqua WC, Winans SC, Greenberg EP. (1994) Quorum sensing in bacteria: the LuxR-LuxI family of cell density-responsive transcriptional regulators. J Bacteriol 176: 269–275.
7. Miller MB, Bassler BL. (2001) Quorum sensing in bacteria. Annu Rev Microbiol 55: 165–199.
8. Marketon MM, Glenn SA, Eberhard A, Gonzalez JE. (2003) Quorum sensing controls exopolysaccharide production in *Sinorhizobium meliloti*. J Bacteriol 185: 25–331.
9. Ohtani K, Hayashi H, Shimizu T. (2002) The *luxS* gene is involved in cell-cell signalling for toxin production in *Clostridium perfringens*. Mol Microbiol 44: 171–179.
10. Quiñones B, Dulla G, Lindow SE. (2005) Quorum sensing regulates exopolysaccharide production, motility, and virulence in *Pseudomonas syringae*. Mol Plant-Microbe Interact 18: 682–693.
11. Rice SA, Koh KS, Queck SY, Labbate M, Lam KW, et al. (2005) Biofilm formation and sloughing in *Serratia marcescens* are controlled by quorum sensing and nutrient cues. J Bacteriol 187: 3477–3485.
12. Rinaudi LV, Giordano W. (2010) An integrated view of biofilm formation in rhizobia. FEMS Microbiol Lett 304: 1–11.
13. Cha C, Gao P, Chen YC, Shaw PD, Farrand SK. (1998) Production of acyl-homoserine lactone quorum-sensing signals by gram-negative plant-associated bacteria. Mol Plant-Microbe Interact 11: 1119–1129.

14. Costerton JW, Lewandowski Z, Caldwell DE, Korber DR, Lappin-Scott HM. (1995) Microbial biofilm. Annu Rev Microbiol 49: 711–745.
15. Stanley NR, Lazazzera BA. (2004) Environmental signals and regulatory pathways that influence biofilm formation. Mol Microbiol 52: 917–924.
16. Sutherland IW. (2001) Biofilm exopolysaccharides: a strong and sticky framework. Microbiology 147: 3–9.
17. Rinaudi L, Fujishige NA, Hirsch AM, Banchio E, Zorreguieta A, et al. (2006) Effects of nutritional and environmental conditions on *Sinorhizobium meliloti* biofilm formation. Res Microbiol 157: 867–875.
18. Fujishige NA, Kapadia NN, De Hoff PL, Hirsch AM. (2006) Investigations of *Rhizobium* biofilm formation. FEMS Microbiol Ecol 56: 195–206.
19. Fujishige NA, Lum MR, De Hoff PL, Whitelegge JP, Faull KF, et al. (2008) Rhizobium common nod genes are required for biofilm formation. Mol Microbiol 67: 504–515.
20. Wells DH, Chen EJ, Fisher RF, Long SR. (2007) ExoR is genetically coupled to the ExoS-ChvI two-component system and located in the periplasm of *Sinorhizobium meliloti*. Mol Microbiol 64: 647–664.
21. Rinaudi LV, González JE. (2009) The low-molecular-weight fraction of exopolysaccharide II from *Sinorhizobium meliloti* is a crucial determinant of biofilm formation. J Bacteriol 191: 7216–7224.
22. González JE, Marketon MM (2003) Quorum sensing in nitrogen-fixing rhizobia. Microbiol Mol Biol Rev 67: 574–592.
23. Loh JT, Yuen-Tsai JP, Stacey MG, Lohar D, Welborn A, et al. (2001) Population density-dependent regulation of the Bradyrhizobium japonicum nodulation genes. Mol Microbiol 42: 37–46.
24. Jitacksorn S1, Sadowsky MJ (2008) Nodulation gene regulation and quorum sensing control density-dependent suppression and restriction of nodulation in the *Bradyrhizobium japonicum*-soybean symbiosis. Appl Environ Microbiol 74: 3749–3756.
25. Pérez-Montaño F, Jiménez-Guerrero I, Sánchez-Matamoros RC, López-Baena FJ, Ollero FJ, et al. (2013) Rice and bean AHL-mimic quorum-sensing signals specifically interfere with the capacity to form biofilms by plant-associated bacteria. Res Microbiol 164: 749–760.
26. Beringer JE. (1974) R factor transfer in *Rhizobium leguminosarum*. J Gen Microbiol 84: 188–198.
27. Vincent JM (1970) The modified Fahraeus slide technique. In *A manual for the practical study of root nodule bacteria*, 144–145. Edited by J. M. Vincent. Oxford, UK: Blackwell Scientific Publications.
28. Sambrook J, Fritsch EF, Maniatis T(1989) Molecular cloning: a laboratory manual, 2nd edn. Cold Spring Harbor NY: Cold Spring Harbor Laboratory.
29. Lamrabet Y, Bellogín RA, Cubo T, Espuny R, Gil A, et al. (1999) Mutation in GDP-fucose synthesis genes of *Sinorhizobium fredii* alters Nod factors and significantly decreases competitiveness to nodulate soybeans. Mol Plant-Microbe Interact 12: 207–217.
30. Simon R. (1984) High frequency mobilization of gram-negative bacterial replicons by the in vitro constructed Tn5-Mob transposon. Mol Gen Genet 196: 413–420.
31. López-Baena FJ, Monreal JA, Pérez-Montano F, Guasch-Vidal B, Bellogín RA, et al. (2009) The absence of Nops secretion in *Sinorhizobium fredii* HH103 increases *GmPR1* expression in Williams soybean. Mol Plant-Microbe Interact 22: 1445–1454.
32. Russo DM, Williams A, Edwards A, Posadas DM, Finnie C, et al. (2006) Proteins exported via the PrsD-PrsE type I secretion system and the acidic exopolysaccharide are involved in biofilm formation by *Rhizobium leguminosarum*. J Bacteriol 188: 4474–4486.
33. O'Toole GA, Kolter R. (1998) Initiation of biofilm formation in *Pseudomonas fluorescens* WCS365 proceeds via multiple, convergent signalling pathways: a genetic analysis. Mol Microbiol 28: 449–461.
34. Mueller K, González JE. (2011) Complex regulation of symbiotic functions is coordinated by MucR and quorum sensing in *Sinorhizobium meliloti*. J Bacteriol 193: 485–496.
35. Miller JH. (1972) Experiments in molecular genetics. Cold Spring Harbor Laboratory Press. Cold Spring Harbor, New York (USA).
36. de Lyra, MCCP, López-Baena FJ, Madinabeitia N, Vinardell JM, Espuny MR, et al. (2006) Inactivation of the *Sinorhizobium fredii* HH103 *rhcJ* gene abolishes nodulation outer proteins (Nops) secretion and decreases the symbiotic capacity with soybean. Int Microbiol 9: 125–133.
37. Barahona E, Navazo A, Yousef-Coronado F, Aguirre de Cárcer D, Martínez-Granero F, et al. (2010) Efficient rhizosphere colonization by *Pseudomonas fluorescens* f113 mutants unable to form biofilms on abiotic surfaces. Environ Microbiol 12: 3185–3195.
38. He X, Chang W, Pierce DL, Seib LO, Wagner J, et al. (2003) Quorum sensing in *Rhizobium* sp. strain NGR234 regulates conjugal transfer (tra) gene expression and influences growth rate. J Bacteriol 185: 809–822.
39. Piper KR, Farrand SK. (2000) Quorum sensing but not autoinduction of Ti plasmid conjugal transfer requires control by the opine regulon and the antiactivator TraM. J Bacteriol 182: 1080–1088.
40. Rodríguez-Navarro DN, Bellogín R, Camacho M, Daza A, Medina C, et al. (2003) Field assessment and genetic stability of *Sinorhizobium fredii* strain SMH12 for commercial soybean inoculants. Eur J Agron 19: 299–309.
41. Kovach ME, Elzer PH, Hill DS, Robertson GT, Farris MA, et al. (1995) Four new derivatives of the broad-host-range cloning vector pBBR1MCS, carrying different antibiotic-resistance cassettes. Gene 166: 175–176.
42. Schäfer A, Tauch A, Jager W, Kalinowski J, Thierbach G, et al. (1994) Small mobilizable multi-purpose cloning vectors derived from the *Escherichia coli* plasmids pK18 and pK19: selection of defined deletions in the chromosome of *Corynebacterium glutamicum*. Gene 145: 69–73.
43. Maurhofer M, Reimmann C, Schmidli-Sacherer P, Heeb S, Haas D, et al. (1998) Salicylic acid biosynthetic genes expressed in *Pseudomonas fluorescens* strain P3 improve the induction of systemic resistance in tobacco against tobacco necrosis virus. Phytopathology 88: 678–684.
44. Reimmann C, Ginet N, Michel L, Keel C, Michaux P, et al. (2002) Genetically programmed autoinducer destruction reduces virulence gene expression and swarming motility in *Pseudomonas aeruginosa* PAO1. Microbiology 148: 923–932.
45. Stuurman N, Pacios-Bras C, Schlaman HR, Wijfjes AH, Bloemberg G, et al. (2000) Use of green fluorescent protein color variants expressed on stable broad-host-range vectors to visualize rhizobia interacting with plants. Mol Plant Microbe-Interact 13: 1163–1169.
46. Vinardell JM, Ollero FJ, Hidalgo A, López-Baena FJ, Medina C, et al. (2004) NolR regulates diverse symbiotic signals of *Sinorhizobium fredii* HH103. Mol Plant Microbe-Interact 17: 676.
47. Figurski DH, Helinski DR. (1979) Replication of an origin-containing derivative of plasmid RK2 dependent on a plasmid function provided in trans. Proc Natl Acad Sci USA 76: 1648–1652.

Molecular Characterization of a Highly-Active Thermophilic β-Glucosidase from *Neosartorya fischeri* P1 and Its Application in the Hydrolysis of Soybean Isoflavone Glycosides

Xinzhuo Yang[1]ᵜ, Rui Ma[1,2]ᵜ, Pengjun Shi[1], Huoqing Huang[1], Yingguo Bai[1], Yaru Wang[1], Peilong Yang[1], Yunliu Fan[2], Bin Yao[1]*

1 Key Laboratory for Feed Biotechnology of the Ministry of Agriculture, Feed Research Institute, Chinese Academy of Agricultural Sciences, Beijing, China, **2** Biotechnology Research Institute, Chinese Academy of Agricultural Sciences, Beijing, China

Abstract

Isoflavone occurs abundantly in leguminous seeds in the form of glycoside and aglycone. However, isoflavone glycoside has anti-nutritional effect and only the free type is beneficial to human health. In the present study we identified a β-glucosidase from thermophilic *Neosartorya fischeri* P1, termed NfBGL1, capable of efficiently converting isoflavone glycosides into free isoflavones. The gene, belonging to glycoside hydrolase family 3, was successfully overexpressed in *Pichia pastoris* at high cell density in a 3.7-l fermentor. Purified recombinant NfBGL1 had higher specific activity (2189±1.7 U/mg) and temperature optimum (80°C) than other fungal counterparts when using *p*-nitrophenyl β-D-glucopyranoside as the substrate. It retained stable at temperatures up to 70°C and over a broad pH range of 3.0−10.0. NfBGL1 had broad substrate specificity including glucosidase, cellobiase, xylanase and glucanase activities, and displayed preference for hydrolysis of β-1,2 glycosidic bond rather than β-1,3, β-1,4, β-1,6 bonds. The enzyme showed high bioconversion ability for major soybean isoflavone glycosides (daidin, gensitin and glycitin) into free forms. These properties make NfBGL1 potential for the wide use in the food, feed, pharmacy and biofuel industries.

Editor: Spencer J. Williams, University of Melbourne, Australia

Funding: This work was supported by the National High Technology Research and Development Program of China (2012AA022208), the National Science Foundation for Distinguished Young Scholars of China (31225026), the National Science and Technology Support Program (2013BAD10B01-2), and the China Modern Agriculture Research System (CARS-42). The funders had no role in study design, data collection and analysis, decision to publish, or preparation of the manuscript.

Competing Interests: The authors have declared that no competing interest exists.

* Email: binyao@caas.cn

ᵜ These authors contributed equally to this work.

Introduction

Plant isoflavones are important secondary metabolites that are mainly distributed in leguminous plants with high levels [1]. It can occur in glycoside and free types. Most of isoflavones exist in natural materials as glycoside, which represent one of the anti-nutritional factors that reduce the nutrient utilization and/or food intake of plants or plant products as human foods or animal feeds [2], while their free form plays the major beneficial roles to prevent certain cancers, osteoporosis and cardiovascular disease [3–5]. Therefore, how to convert isoflavone glycosides to free isoflavones is the key for extensive utilization of isoflavone in the food and feed industries.

β-Glucosidases (EC3.2.1.21) catalyzes the hydrolysis of β-glucosidic linkages of various oligosaccharides and glycosides to release glucose and debranched oligosaccharides. According to the CAZy database (http://www.cazy.org) [6], these enzymes are grouped into 6 of 133 glycoside hydrolase (GH) families, i.e. 1, 3, 5, 9, 30 and 116. β-Glucosidase plays a vital role in the biological conversion of cellulose by degrading the cellooligosaccharide products released by endoglucanase and cellobiohydrolase to form glucose and a debranched oligosaccharide. To avoid product inhibition, β-glucosidase with high tolerance for glucose is heated in these fields [7]. Recently, β-glucosidase has attracted much interest due to its important roles in the hydrolysis of isofalvone glycosides. Those from *Dalbergia* sp. [8], *Paecilomyces thermophila* [9], *Dictyoglomus turgidum* [10], *Sulfolobus solfataricus* [11], and *Rhizopus* spp. [12] have shown ability to degrade isofalvone glycosides. However, these β-glucosidases have low enzymatic activities and low commercial values. What's more, their conversion efficiencies of glucosidic isoflavones were further reduced due to their sensitivity to glucose inhibition. Thus a highly efficient and glucose-tolerate β-glucosidase with isofalvone glycoside hydrolysis ability is mostly urgently required.

Thermophilic/thermostable enzymes are favorable for hydrolysis of isoflavone glycosides since high temperatures enhance the mass transfer rate, reduce the substrate viscosity, and reduce the risk of contamination [13]. The present study aims to obtain a new thermophilic β-glucosidase for efficient hydrolysis of isoflavone glycosides and other saccharides. A family 3 β-glucosidase, termed NfBGL1, was identified in the thermophilic fungal strain *Neosartorya fischeri* P1 and overexpressed in *Pichia pastoris*. This enzyme was characterized to be thermophilic and highly active and have wide substrate specificity and high soybean isoflavone glycoside-hydrolysis ability. Its supplementation significantly enhanced the conversion of isoflavone glycosides to aglycone types.

Methods

2.1. Materials

Soybean flour extract was prepared following the method of previous report with some modifications [14]. The soybean flour was defatted in three volumes of *n*-hexane for 30 min with agitation at room temperature, centrifuged at 13,000 *g* for 20 min, and air-dried to a fine powder.

The substrates cellobiose, carboxymethyl cellulose sodium (CMC-Na), Avicel, laminiarin, lichenin, amygdalin, daidzin, genistin, glycitin, daidzein, genistein, glycitein, birchwood xylan, beechwood xylan, sophorose, gentiobiose, salicin, lactose, *p*-nitrophenyl β-D-glucopyranoside (*p*NPG) and other *p*-nitrophenyl glycosides, and Novozyme 188 (cellobiase from *Aspergillus niger*) were purchased from Sigma-Aldrich (St. Louis, MO). Other substrates including barley β-glucan and xyloglucan were obtained from Megazyme (Wicklow, Ireland). T4 DNA ligase and restriction endonucleases were purchased from Promega (Madison, WI).

The Fungal DNA Isolation Kit and RNeasy Plant Mini Kit were obtained from Omega Bio-tek (Norcross, GA) and QIAGEN (Hilden, Germany), respectively. The DNA Purification Kit, LA Taq DNA polymerase and restriction endonucleases were purchased from TaKaRa (Otsu, Japan). All other chemicals were of analytical grade and commercially available.

2.2. Strains

N. fischeri P1 CGMCC 3.15369 [15] was grown in a β-glucosidase inducing medium containing 10 g/l wheat bran, 10 g/l CMC-Na, 5 g/l $(NH_4)_2SO_4$, 1 g/l KH_2PO_4, 0.5 g/l $MgSO_4 \cdot 7H_2O$, 0.2 g/l $CaCl_2$, and 10 mg/l $FeSO_4 \cdot 7H_2O$ at 45°C and pH 5.0 for 4 days. The total β-glucosidase activity in *N. fischeri* culture supernatants was 11.9±0.2 U/ml.

Escherichia coli Trans1-T1 (TransGen, Beijing, China) was cultivated in Luria-Bertani (LB) medium with 100 μg/ml ampicillin at 37°C for gene cloning and sequencing. *P. pastoris* GS115 (Invitrogen, Carlsbad, CA) cultivated in yeast peptone dextrose (YPD) medium at 30°C was used for heterologous protein expression. The plasmids pEASY-T3 (TransGen) and pPIC9 (Invitrogen) were used as cloning and expression vectors, respectively.

2.3. Cloning and sequence analysis of the β-glucosidase gene (*Nfbgl1*)

Mycelia of strain P1 were collected after 4 days' growth in β-glucosidase producing medium as described above. Total DNA and RNA were extracted and purified according to the manufacturer's instructions, respectively. The cDNAs were synthesized *in vitro* using the ReverTra Ace-α- Kit (TOYOBO, Osaka, Japan) with the total RNA as the template. According to the genome sequence of a hypothetical β-glucosidase from *N. fischeri* NRRL 181 (XP_001261562), two specific primer sets (Nfbgl1-F: ATGCAGAACCTCTTCCTTTCTCTCC and Nfbgl1-R: CTTTCAGGGTCCGTCGCGGATGGTGA; and

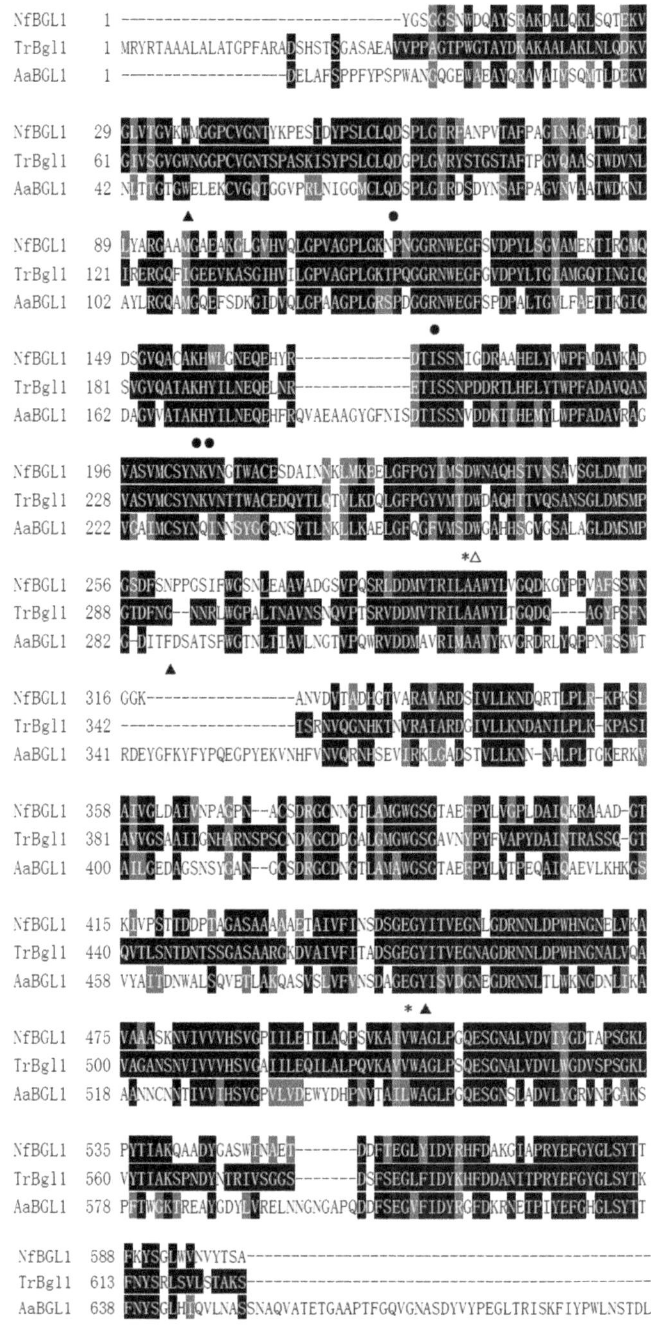

Figure 1. Multiple alignment of the deduced amino acid sequence of mature NfBGL1 (XP_001261562) with other fungal counterparts from *T. reesei* (TrBgl1, 1713235A), and *A. aculeatus* (AaBGL1, 4IIB). Identical and similar amino acids are indicated by solid black and gray boxes, respectively. The putative catalytic residues, D235 and E447, are indicated with asterisks. The residues probably related to subsite −1 (W236) and +1 (W36, N261, and Y449) are indicated with white and black triangles, respectively. The residues probably forming hydrogen bonds with substrate, D60, R124, K147 and H148, are indicated with black dots.

Nfbgl1-expF: GGGGAATTCTACGGTTCTGGTGGCAGCAACTGGG and Nfbgl1-expR: GGGGCGGCCGCTCACCATCCGCGACGGACCCTGAAAG, restriction sites underlined) were designed to amplify the full-length β-glucosidase gene, *Nfbgl1*, and its cDNA without the signal peptide coding

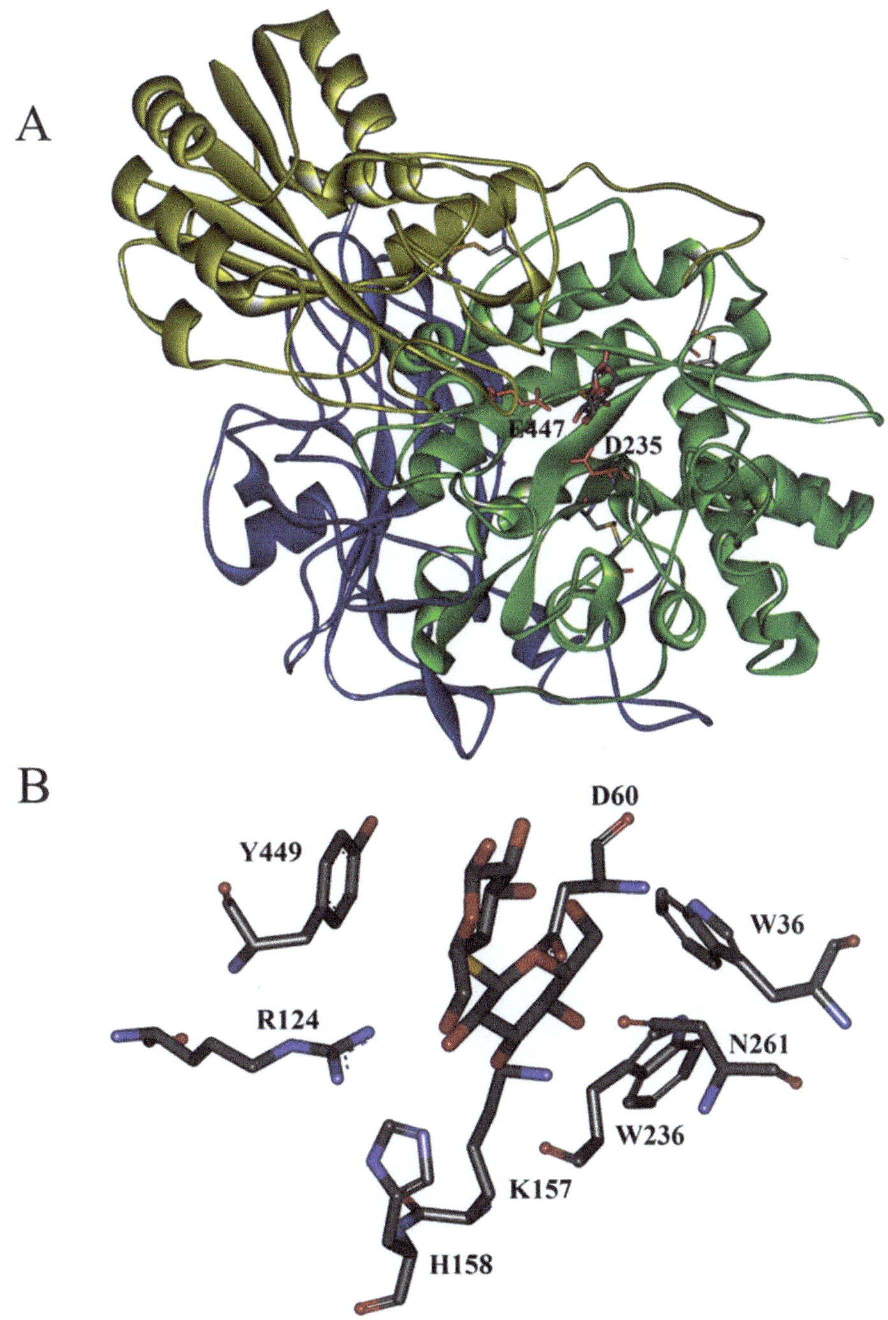

Figure 2. The homology-modeled NfBGL1 with AaBGL1 from *A. aculeatus* (4IIB, 44.1% identity) as the template. (A) Putative structure of NfBGL1. The catalytic residues D235 and E447 are indicated. (B) Putative interactions between the key residues of NfBGL1 and substrate cellobiose.

sequence. The PCR products with appropriate size were ligated into a pGEM-T3 Easy vector (TransGen) for sequencing.

The nucleotide sequence was analyzed using Vector NTI Advance 10.0 software (Invitrogen). BLASTn and BLASTp programs (http://www.ncbi.nlm.nih.gov/BLAST/) and AlignX from Vector NTI were used to analyze the nucleotide and deduced amino acid sequences, respectively. The online software FGENESH (http://linux1.softberry.com/berry.phtml) was used to predict the transcription initiation sites, introns and exons. NetNGlyc 1.0 Server (http://www.cbs.dtu.dk/services/NetNGlyc/) was used to predict the potential *N*-glycosylation sites. Multiple sequence alignment of deduced NfBGL1 and other fungal counterparts were conducted using the ClustalW program (www.genome.jp/tools/clustalw/●). The three-dimensional structure was predicted using the SWISS-MODEL (http://swissmodel.expasy.org/) with the GH3 β-glucosidase AaBGL1 from *Aspergillus aculeatus* (PDB: 4IIB; 44.1% identity) [16] as the template.

2.4. Heterologous expression and purification of NfBGL1

Heterologous expression of recombinant NfBGL1 was conducted following the *Pichia* Expression Kit (Invitrogen). The cDNA fragment of mature NfBGL1 without the signal peptide-coding sequence was digested with *Eco*RI and *Not*I, and then cloned into

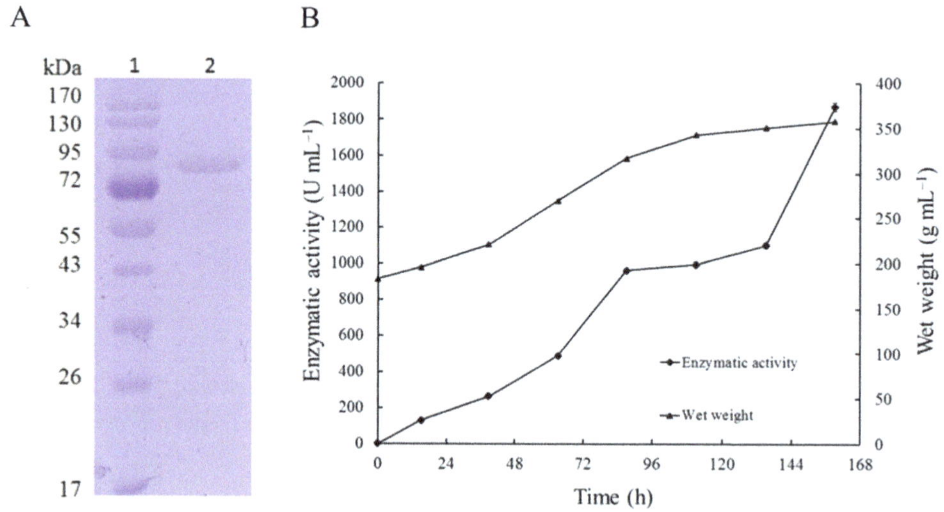

Figure 3. Purification and high-cell-density fermentation of recombinant NfBGL1. (A) SDS-PAGE analysis of purified recombinant NfBGL1. Lanes: 1, the molecular mass standards; 2, the purified recombinant of NfBGL1. (B) Time course of NfBGL1 production in a 3.7-l fermenter. Each value in the panel represents the means ± S.D. ($n = 3$).

the corresponding sites of vector pPIC9. The recombinant plasmid pPIC9-*Nfbgl1* was then linearized with *Bgl*II and transformed into *P. pastoris* GS115 competent cells by electroporation (Bio-Rad, Hercules, CA). The positive transformants were selected based on their enzymatic activities in shake tubes. The transformant that exhibited the highest β-glucosidase activity in culture supernatant was used for large-scale fermentation in 1-l Erlenmayer flasks. The cell-free culture supernatants were collected and loaded onto a HiTrap Q Sepharose XL FPLC column (GE Healthcare, Uppsala, Sweden) equilibrated with buffer A (20 mM McIlvaine buffer, pH 7.0). Proteins were eluted with NaCl gradients (0–1.0 M) in the same buffer at a flow rate of 3 ml/min. The fractions containing enzyme activity were collected and evaluated by sodium dodecyl sulfate-polyacrylamide gel electrophoresis (SDS-PAGE). The protein concentration was determined by the Bradford assay with bovine serine albumin as the standard.

2.5. High-cell-density fermentation in a 3.7-l fermentor

The positive transformant with the highest β-glucosidase activity in shake flask culture was selected for high-cell-density fermentation in a 3.7-l stirred-tank fermentor (Bioengineering KLF 2000, Wald, Switzerland) using a three-step fermentation strategy as described [17]. The inoculum ratio was 10% (v/v). The temperature and initial pH was set to be 30°C and 5.0. During the whole fermentation process, samples were periodically withdrawn for analysis of wet weights and enzyme activities.

2.6. Enzyme activity assays

The β-glucosidase activity was assayed using *p*NPG as the substrate. The reaction system consisted of 400 μl of 1.25 mM *p*NPG in McIlvaine buffer (200 mM Na_2HPO_4, 100 mM citric acid, pH 5.0) and 100 μl of appropriately diluted enzyme solution. The mixture was incubated at 80°C for 10 min, and then terminated by addition of 1.5 ml of 1 M Na_2CO_3. After cooling down to room temperature, the absorbance was measured at 405 nm. One unit of enzyme activity was defined as the amount of enzyme that released 1 μmol of *p*-nitrophenol per min. Each reaction was performed in triplicate, and the data were shown as mean ± S.D. ($n = 3$). The enzymatic activities towards other *p*-nitrophenyl glycosides were assayed as described above with different substrates.

Cellulase/glucanase/xylanase activity was determined according to the 3,5-dinitrosalicylic acid (DNS) assay [18]. The standard reaction system containing 100 μl of appropriately diluted enzyme and 900 μl of 1% (w/v) substrate (CMC-Na, Avicel, xyloglucan, barley β-glucan, beechwood xylan or birchwood xylan) in McIlvaine buffer (pH 5.0) was incubated at 80°C for 10 min, followed by addition of 1.5 ml of DNS reagent. After 5-min boiling and cooling down to room temperature, the absorbance was measured at 540 nm. One unit of cellulase/glucanase/xylanase activity was defined as the amount of enzyme that released 1 μmol of reducing sugar per min under the assay conditions.

The enzymatic activities on 10 mM of cellobiose, salicin, lactose, daidzin, glycitin, and genistin were estimated by measuring the amount of released glucose using GOD-POD method with a commercial kit (Biosino, Beijing, China). The absorbance was measured at 505 nm. One unit of enzyme activity was defined as the amount of enzyme that released 1 μmol of glucose per minute.

2.7. Biochemical characterization

The optimal pH values were determined at 80°C in the following solutions, 100 mM glycine-HCl for pH 2.0–3.0, McIlvaine buffer for pH 3.0–8.0, 100 mM Tris-HCl for pH 8.0–9.0, and 100 mM glycine-NaOH for pH 9.0–12.0. The pH stability was determined by measuring the residual enzyme activity under the standard conditions (pH 5.0, 80°C, 10 min) after pre-incubation of the enzyme in buffers of pH 2.0–12.0 at 37°C for 1 h without the substrate.

The optimal temperature was examined at optimal pH by measuring the enzyme activity over the temperature range of 20 to 90°C. The thermostability assay was carried out by incubating the enzyme at optimal pH and at 70°C or 80°C without substrate for 2, 5, 10, 20, 30, or 60 min, and then measuring the residual enzyme activities under the assay conditions.

To investigate the effects of different metal ions and chemical reagents on the activity of purified recombinant NfBGL1, the reaction system containing 5 mM of each Na^+, K^+, Li^+, Ag^+, Cu^{2+}, Ni^{2+}, Mn^{2+}, Ca^{2+}, Pb^{2+}, Co^{2+}, Zn^{2+}, Mg^{2+}, Fe^{3+}, Cr^{3+}, SDS,

Table 1. Property comparison of NfBGL1 from *N. fischeri* P1 and its fungal counterparts. [a]

Microbial source	Optimum activity		Specific activity (U/mg)	K_m (mM)	V_{max} (μmol/min/mg)	*Ki* for glucose (mM)	References
	pH	Temperature (°C)					
N. fischeri P1	5.0	80	2189	0.51	2172	13.4	This work
Aspergillus terreus NRRL265	5.0	60	-	2.5	-	13.6	[24]
P. thermophila	6.2	75	97.2	0.27	780	73	[25]
Penicillium funiculosum NCL1	4.0–5.0	60	1796	0.057	1920	1.5	[26]
Trichoderma citrinoviride	5.5	50	167	0.27	200	-	[27]
Stachybotrys sp.	6.0	50	-	1.85	211	-	[28]
Fomitopsis palustris	4.5	70	191	0.117	-	0.35	[29]
Chaetomium thermophilum	5.5	65	-	0.76	-	-	[30]
Talaromyces emersonii	4.0	71.5	483	0.13	512	0.254	[31]

[a] The substrate used for enzyme characterization is *p*NPG.

EDTA, and β-mercaptoethanol was subject to enzyme activity assay under the standard conditions and compared to the blank control without any additives.

2.8. Substrate specificity and kinetic parameters

The substrate specificities of NfBGL1 were determined under standard assay conditions (pH 5.0 and 80°C for 10 min) in the presence of 1 mM of *p*-nitrophenyl derivatives and oligosaccharides, or 10 mg/ml of polysaccharides and isoflavone glycosides.

The K_m and V_{max} values of NfBGL1 were determined in McIlvaine buffer (pH 5.0) containing 0.01–2 mM *p*NPG at 80°C for 5 min. The experiments were carried out three times, and each experiment included three replicates. The data were calculated using the Lineweaver-Burk method with a non-linear regression computer program GraFit (Version 7.0, Erithacus Software, Horley, UK).

2.9. Tolerance to glucose inhibition

The inhibitory effect of glucose on NfBGL1 activity was determined by fitting to the Dixon plot [19,20]. Aliquots of enzyme solution (5 μl) were added to 120 μl of 0.01–2.0 M glucose solution and incubated at room temperature for 1 h. The mixtures were added with 125 μl of 3 or 4 mM *p*NPG and 250 μl of McIlvaine buffer (pH 5.0) containing the same amount of glucose. The residual enzyme activity was measured at 80°C, pH 5.0 for 10 min. The K_i value was calculated by plotting the reciprocal of the reaction velocity and inhibitor concentration at each *p*NPG concentration.

2.10. HPLC analysis of the hydrolysis products of soybean four extract

To compare the hydrolysis efficiencies of NfBGL1 and Novozyme 188 towards isoflavone glycosides, 50 μl of 10% (w/v) soybean flour was incubated with 200 μl of each enzyme solution (0.05 U) in McIlvaine buffer (pH 5.0) in a thermostatically controlled incubator at 50°C. The reactions were terminated by ice water cooling. After centrifugation at 13,000 *g*, 4°C for 10 min, 20 μl of each supernatant was subject to HPLC analysis by using Waters HP1100 HPLC system (Milford, MA) equipped with a C18 column (5 μm, 4.6×250 mm). The solvent was composed of acetonitrile and 100 mM phosphate buffer (pH 5.0) at the ratio of 70:30 (v/v) and run at a flow rate of 0.8 ml/min. The chromatograms were detected at 260 nm.

Results and Discussion

3.1. Sequence analysis of *Nfbgl1*

The cDNA of *Nfbgl1* from *N. fischeri* P1 consists of 2220 base pairs and encodes a polypeptide of 739 amino acids with a putative signal peptide of 17 residues at the N-terminus. The molecular mass and *p*I value were estimated to be 78.8 kDa and 5.54, respectively. Two putative *N*-glycosylation sites (Asn207 and Asn381) were identified in deduced NfBGL1.

The deduced amino acid sequence of mature NfBGL1 showed the highest identity of 100% with the hypothetical β-glucosidase from *N. fischeri* NRRL 181 (XP_001261562), and of 61% identity with the *Trichoderma reesei* β-glucosidase (1713235A) with the experimentally evidenced activity [21]. Multiple sequence alignment and structure analysis with other β-glucosidases revealed that NfBGL1 is a typical β-glucosidase of GH3 (Fig. 1). Different from the two-domain barley β-glucosidase ExoI from *Hordeum vulgare* [22] and four-domain yeast β-glucosidase BglI from *Kluyveromyces marxianus* [23], the homology-based model of mature NfBGL1 with AaBGL1 as the template [16] was predicted to have

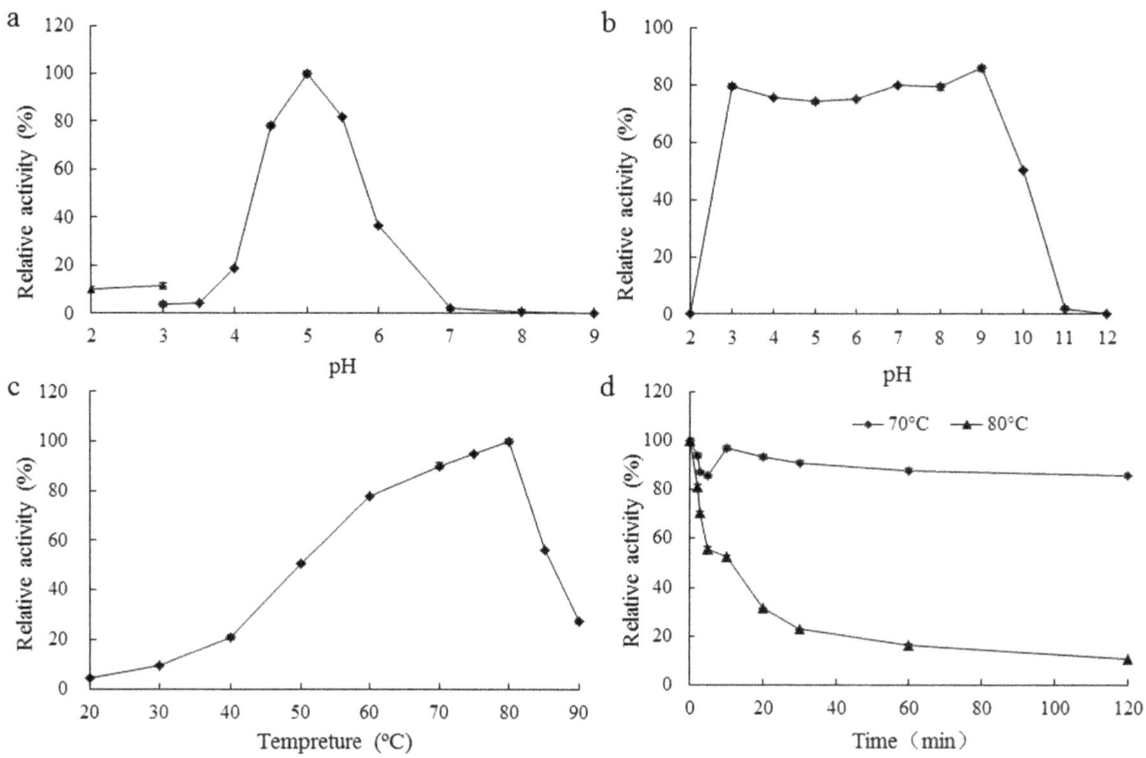

Figure 4. Characterization of purified recombinant NfBGL1. (A) Effect of pH on β-glucosidase activity. The enzyme assay was performed at 80°C for 10 min. (B) pH stability. The enzyme was pre-incubated without substrate at 37°C for 60 min, and then subjected to residual activity assay under standard conditions (pH 5.0, 80°C, 10 min). (C) Effect of temperature on β-glucosidase activity determined at pH 5.0 for 10 min. (D) Thermostability. The residual enzyme activities were measured under standard conditions after pre-incubation of the enzyme without substrate in McIlvaine buffer (pH 5.0) for various periods. Each value in the panel represents the means ± S.D. ($n = 3$).

three domains (Fig. 2A): the TIM barrel-like domain (Y1−S313), the α/β sandwich domain (T324−G526), and the C-terminal FnIII domain (Y585−W722). The putative catalytic residues, D235 in the barrel domain and E447 in the α/β sandwich domain, are located at the interface of the first two domains. Three disulfide bonds may form between C41–C57, C201–C212, and C374–C379 to stabilize the protein structure. Residues D60, R124, K157, and H158 are supposed to form hydrogen bonds with the sugar hydroxyls of substrate. Consistent with the structural features of AaBGL1, the subsite −1 of modeled NfBGL1 has a strict stereochemcial requirement, forming an aromatic stacking interaction with sugar ring by W236 (Fig. 2B)

Table 2. Substrate specificities of NfBGL1.

Substrate	Specific activity (U/mg)[a]	Substrate	Specific activity (U/mg)
*p*NPG	2189±1.7	Cellobiose	203.8±0.9
p-Nitrophenyl β-D-cellobioside	601.0±0.9	Lichenin	93.3±1.2
p-Nitrophenyl β-D-xylopyranoside	6.6±0.3	Daidzin	90.1±0.7
p-Nitrophenyl β-D-galactoside	–	Glycitin	75.2±0.6
p-Nitrophenyl α-L-arabinofuranoside	–	Genistin	45.6±0.3
p-Nitrophenyl α-L-arabinopyranoside	–	Beechwood xylan	6.4±0.5
Sophorose	896.1±0.9	Birchwood xylan	6.2±0.7
Laminiarin	675.4±1.1	Lactose	3.5±0.3
Gentiobiose	602.4±0.8	Xyloglucan	–
Amygdalin	510.8±0.7	Avicel	–
Barley β-glucan	269.4±1.3	CMC-Na	–
Salicin	246.6±0.6		

[a] Values are means ± S.D. ($n = 3$); –, no activity detected.

Table 3. Conversion of soybean isoflavone glycosides into free isoflavones by NfBGL1 and Novozyme 188. [a]

Enzyme	Isoflavone glycosides (mg/g)			Free isoflavonones (mg/g)		
	Daidzin	Genistin	Glycitin	Daidzein	Genistein	Glycitein
Control	146.8±0.4	68.7±0.3	54.0±0.2	100.8±0.7	220.7±1.1	21.7±0.4
NfBGL1	0	25.9±0.4	0	222.4±0.9	598.2±0.6	73.4±1.0
Novozyme 188	0	32.7±0.6	16.9±0.1	253.6±0.8	746.5±1.0	70.3±1.1

[a] The reaction system without enzyme addition was treated as the control; the data are shown as mean ± S.D. ($n = 3$).

the equivalent of W281 of AaBGL1 [16]. The amino acid residues related to subsite +1 are less conserved, one (N261 of modeled NfBGL1) of the three residues (W36, N261 and Y449) is different. The diverse structures forming subsite +1 may account for the large substrate specificity ranges of GH3 β-glucosidases.

3.2. Overexpression, purification and high-cell-density production of NfBGL1

The cDNA fragment of *Nfbgl1* without the signal peptide-coding sequence was amplified by PCR and cloned into the pPIC9 vector for heterologous expression. The β-glucosidase activity of recombinant NfBGL1 was determined to be 33.5±0.4 U/ml (pH 5.0, 50°C, 10 min) after 2-day growth in 1-l Erlenmeyer flasks at 30°C. The culture supernatant was purified to electrophoretic homogeneity by one-step anion exchange chromatography. A single band of approximately 80 kDa migrated in SDS-PAGE (Fig. 3A), which is essentially identical to the calculated molecular weight of NfBGL1 (78.8 kDa). The result indicates that no *N*-glycosyaltion occurred in recombinant NfBGL1 during heterologous expression in *P. pastoris*. The specific activity of purified recombinant NfBGL1 against *p*NPG was 2189±1.7 U/mg, quite higher than that of all known fungal β-glucosidases (Table 1).

The transformant that showed the highest β-glucosidase activity in shaker flask cultures was subject to high-cell-density fermentation in a 3.7-l fermentor. The cell wet weight kept increasing, but no β-glucosidase activity was detected in the culture supernantant before the induction phase. Upon methanol induction, the β-glucosidase activity in the supernatant increased to a maximum of 1873±1.5 U/ml at 159 h, which was 56 fold of that of shake flask cultures (Fig. 3B). This significant activity increase might be ascribed to the higher dissolved oxygen levels and more efficient mass transfer in aerated system [32]. The high yield, easy purification and highest specific activity make NfBGL1 favorable for large-scale industrial production.

3.3. Enzymatic properties of purified recombinant NfBGL1

Generally fungal β-glucosidases have an acidic pH optimum, ranging from 4.0 to 6.5, and are usually stable over a wide pH range (Table 1). NfBGL1 from *N. fischeri* P1 showed similar properties. It had a pH optimum of 5.0, and remained >80% of maximum activity at pH 4.5–5.5 (Fig. 4A). In comparison with other fungal counterparts, NfBGL1 was stable over a broader pH range, retaining >80% of the activity after incubation at pH 3.0 to 9.0, 37°C for 1 h (Fig. 4B). NfBGL1 represents a typical thermophilic/thermostable β-glucosidase. Of all characterized fungal β-glucosidases, that from *P. thermophila* [25] has the highest temperature optimum of 75°C, which is still 5°C lower than that of NfBGL1 (Fig. 4C). Moreover, NfBGL1 showed distinct adaptability and stability to high temperatures. It remained >50% of the activity at 50–85°C and 27% activity even at 90°C, and was thermostable at 70°C for 120 min (Fig. 4D). When increased the incubation temperature to 80°C, the enzyme retained 50% of the activity for 10 min and 11% activity for 120 min. NfBGL1 was resistant to most chemicals including Li^{+}, Pb^{2+}, Mn^{2+}, Mg^{2+}, Zn^{2+}, Cr^{3+}, Ni^{+}, Cu^{2+}, SDS, and EDTA, but was strongly inhibited by Ag^{+}, Co^{3+}, Fe^{3+}, Ca^{2+} and β-mercaptoethanol (data not shown). Considering its acidic, thermophilic, wide pH/temperature range-stable and chemical-resistant advantages in biomass conversion, handling, post-processing and transportation [30], NfBGL1 has potentials for wide application in the biofuels, detergent, textile, and other industries.

3.4. Substrate specificity

NfBGL1 exhibited broad substrate specificity including β-glucosidase, cellobiase, glucanase, xylanase and isoflavone glycoside hydrolysis activities (Table 2). Besides *p*NPG, it was active towards the β-linked synthetic substrate *p*-nitrophenyl β-D-cellobioside and *p*-nitrophenyl β-D-xylopyranoside. Of all tested saccharides, sophorose (β-1,2-linked) was the most preferred substrate, followed by laminiarin (β-1,3, β-1,6-linked), salicin (β-linked), cellobiose (β-1,4-linked), lichenan (β-1,3, β-1,4-linked), and lactose (β-1,4-linked). Although most β-glucosidases have higher activities toward β-1,4-linked saccharides, NfBGL1 as well as the β-glucosidases from *Trichoderma koningii* [33] and a compost microbial metagenome [34] were found to be highly efficient in hydrolysis of β-1,2, 1,3, or 1,6-linkages. Such β-glucosidases may be involved in more biological activities than cellulose saccharification. Besides β-glucosidase and cellobiose activities, NfBGL1 displayed weak xylanase activities against beechwood xylan and birchwood xylan. No activity was detected against β-1,4-linked CMC-Na and Avicel. This broad substrate specificity suggests that NfBGL1 may have loose stereochemcial specificity at subsite +1 as indicated by the structure analysis described above. Thus NfBGL1 can be used in various biotechnological fields for diverse saccharide degradation.

3.5. Kinetic analysis and tolerance to glucose inhibition

With *p*NPG as the substrate, the K_m, V_{max} and k_{cat} values for NfBGL1 was 0.51 mM, 2172 μmol/min/mg, and 2853/s, respectively. The K_m value of NfBGL1 fell within the range of fungal β-glucosidase K_m values, but its V_{max} value was quite high (Table 1). Catalytic efficiency (k_{cat}/K_m) is the true kinetic parameter to evaluate enzyme catalytic performance. The k_{cat}/K_m value of NfBGL1 was 5594/mM/s, higher than that of β-glucosidases from *P. thermophila* (42.8/mM/s) [25], *Stachybotrys* sp. (494.6/mM/s) [28], and *C. thermophilum* (193.4/mM/s) [30].

Glucose accumulation during enzymatic hydrolysis can significantly lower the cellulose hydrolysis rate by blocking the active site or preventing the release of hydrolysis products [35]. Thus high tolerance towards glucose accumulation is a requirement of fungal β-glucosidases. To our best knowledge, the GH3 β-glucosidases from *Candida peltata* [36] and *Aspergillus oryzae* [37] exhibited the strongest glucose tolerance with the *K*i values of 1.4 and 1.36 M, respectively. Compared with these two fungal counterparts, NfBGL1 showed less resistance to glucose inhibition with a *K*i of 13.4 mM, but the glucose tolerance was much higher than the β-glucosidases from *T. reesei*, *F. palustris*, and *P. funiculosum* NCL1 (Table 1).

3.6. Hydrolysis of soybean isoflavone glycosides by NfBGL1

β-Glucosidases with hydrolysis ability to degrade soybean isoflavone glycosides have application potential in food and feed industry. Based on the HPLC analysis, the soybean flour extract contained both isoflavone glucosides (270 mg/g; daidzin, genistin and glycitin) and isoflavone aglycones (343 mg/g; daidzein, genistein and glycitein) (Table 3). After 5-min incubation at 50°C, the contents of daidzin, genistin and glycitin were remarkably decreased, which suggested that most soybean isoflavone glucosides (>90%) were converted to free isoflavones by NfBGL1. The conversion efficiency of daidzin and glycitin reached up to 100%, and 90% for the isoflavone glucosides tested. Under the same conditions, the commercial β-glucosidase Cellobiase (Novozyme 188) was a little less efficient in isoflavone glycoside hydrolysis, converting 82% of glycosides to aglycones. In addition, more free isoflavonones were detected in the soybean flour extract after treatment with Cellobiase than with NfBGL1 (1120 mg/g v.s. 920 mg/g). It might be ascribed to the presence of other unidentified soybean isoflavone glycosides like malonyl or acetyl.

Conclusion

In the present study, a highly-active, thermophilic, soybean isoflavone glycoside-degrading β-glucosidase of GH3 was identified in *N. fischeri* P1. The enzyme had an optimum temperature and specific activity higher than that of all known fungal counterparts, was stable over a broader pH and temperature ranges, and had strong resistance to most chemicals tested. Its substrate specificity was remarkably broad, including glucosidase, cellobiase, xylanase and glucanase activities. Its high efficiency to convert isoflavone glycosides of different soybean products into aglycons was much attractive. These characteristics make it a preeminent enzyme candidate in the food, feed, pharmacy, and biofuel industries.

Author Contributions

Conceived and designed the experiments: RM BY. Performed the experiments: XY RM PS. Analyzed the data: XY RM HH PY. Contributed reagents/materials/analysis tools: YB YW. Contributed to the writing of the manuscript: XY RM YF.

References

1. Franke AA, Custer LJ, Cerna CM, Narala KK (1994) Quantitation of phytoestrogens in legumes by HPLC. J Agric Food Chem 42: 1905–1913.
2. Soetan KO, Oyewole OE (2009) The need for adequate processsing to reduce the anti-nutritional factors in plants used as human foods and animal feed: A review. Afr J Food Sci 3: 223–232.
3. Goodman-Gruen D, Kritz-Silverstein D (2001) Usual dietary isoflavone intake is associated with cardiovascular disease risk factors in postmenopausal women. J Nutr 131: 1202–1206.
4. Ravindranath MH, Muthugounder S, Presser N, Viswanathan S (2004) Anticancer therapeutic potential of soy isoflavone, genistein. In: Cooper EL, Yamaguchi N, editors. Complementary and Alternative Approaches to Biomedicine. 121–165.
5. Kuo LC, Cheng WY, Wu RY, Huang CJ, Lee KT (2006) Hydrolysis of black soybean isoflavone glycosides by *Bacillus subtilis* natto. Appl Microbiol Biot 73: 314–320.
6. Cantarel BL, Coutinho PM, Rancurel C, Bernard T, Lombard V, et al. (2009) The Carbohydrate-Active EnZymes database (CAZy): an expert resource for glycogenomics. Nucleic Acids Res 37 (Suppl 1): D233−D238.
7. Pei J, Pang Q, Zhao L, Fan S, Shi H (2012) *Thermoanaerobacterium thermosaccharolyticum* β-glucosidase: a glucose-tolerant enzyme with high specific activity for cellobiose. Biotechnol Biofuels 5: 31.
8. Chuankhayan P, Rimlumduan T, Svasti J, Cairns JRK (2007) Hydrolysis of soybean isoflavonoid glycosides by *Dalbergia* β-glucosidases. J Agric Food Chem 55: 2407–2412.
9. Yang S, Wang L, Yan Q, Jiang Z, Li L (2009) Hydrolysis of soybean isoflavone glycosides by a thermostable β-glucosidase from *Paecilomyces thermophila*. Food Chem 115: 1247–1252.
10. Kim YS, Yeom SJ, Oh DK (2011) Characterization of a GH3 family β-glucosidase from *Dictyoglomus turgidum* and its application to the hydrolysis of isoflavone glycosides in spent coffee grounds. J Agric Food Chem 59: 11812–11818.
11. Kim BN, Yeom SJ, Kim YS, Oh DK (2012) Characterization of a β-glucosidase from *Sulfolobus solfataricus* for isoflavone glycosides. Biotechnol Lett 34: 125–129.
12. Cheng KC, Wu JY, Lin JT, Liu WH (2013) Isoflavone conversion of black soybean under solid state fermentation by *Rhizopus* spp. 2013. Eur Food Res Technol 236: 1107–1113.
13. Badieyan S, Bevan DR, Zhang C (2012) Study and design of stability in GH5 cellulases. Biotechnol Bioeng 109: 31–44.
14. Xue Y, Yu J, Song X (2009) Hydrolysis of soy isoflavone glycosides by recombinant β-glucosidase from hyperthermophile *Thermotoga maritima*. J Ind Microbiol Biot 36: 1401–1408.
15. Wang H, Shi P, Luo H, Huang H, Yang P, et al. (2014) A thermophilic α-galactosidase from *Neosartorya fischeri* P1 with high specific activity, broad substrate specificity and significant hydrolysis ability of soymilk. Bioresour Technol 153: 361–364.
16. Suzuki K, Sumitani JI, Nam YW, Nishimaki T, Tani S, et al. (2013) Crystal structures of glycoside hydrolase family 3 β-glucosidase 1 from *Aspergillus aculeatus*. Biochem J 452: 211–221.
17. Luo H, Wang Y, Wang H, Yang J, Yang Y, et al. (2009) A novel highly acidic β-mannanase from the acidophilic fungus *Bispora* sp. MEY-1: gene cloning and overexpression in *Pichia pastoris*. Appl Microbiol Biot 82: 453–461.
18. Miller GL (1959). Use of dinitrosalicylic acid reagent for determination of reducing sugar. Anal Chem 31: 426–428.
19. Dixon M (1953). The determination of enzyme inhibitor constants. Biochem J 55: 170−171.
20. Uchima CA, Tokuda G, Watanabe H, Kitamoto K, Arioka M (2012) Heterologous expression in *Pichia pastoris* and characterization of an endogenous thermostable and high-glucose-tolerant β-glucosidase from the termite *Nasutitermes takasagoensis*. Appl Environ Microb 78: 4288–4293.
21. Barnett CC, Berka RM, Fowler T (1991) Cloning and amplification of the gene encoding an extracellular β-glucosidase from *Trichoderma reesei*: evidence for improved rates of saccharification of cellulosic substrates. Nature Biot 9: 562–567.
22. Varghese JN, Hrmova M, Fincher GB (1999) Three-dimentional structure of a barley β-D-glucan exhohydrolase, a family 3 glycosyl hydrolase. Structure 7: 179–190.
23. Yoshida E, Hidaka M, Fushinobu S, Koyanagi T, Minami H, et al. (2010) Role of a PA14 domain in determining substrate specificity of a glycoside hydrolase family 3 β-glucosidase from *Kluyveromyces marxianus*. Biochem J 431: 39–49.
24. Elshafei AM, Hassan MM, Morsi NM, Elghonamy DH (2014) Purification and some kinetic properties of β-glucosidase from *Aspergillus terreus* NRRL 265. Afr J Biotechnol 10: 19556–19569.
25. Yang S, Jiang Z, Yan Q, Zhu H (2007) Characterization of a thermostable extracellular β-glucosidase with activities of exoglucanase and transglycosylation from *Paecilomyces thermophila*. J Agric Food Chem 56: 602–608.
26. Ramani G, Meera B, Vanitha C, Rao M, Gunasekaran P (2012) Production, purification, and characterization of a β-glucosidase of *Penicillium funiculosum* NCL1. Appl Biochem Biot 167: 959–972.
27. Chandra M, Kalra A, Sangwan NS, Sangwan RS (2013) Biochemical and proteomic characterization of a novel extracellular β-glucosidase from *Trichoderma citrinoviride*. Mol Biotechnol 53: 289–299.
28. Saibi W, Amouri B, Gargouri A (2007) Purification and biochemical characterization of a transglucosilating β-glucosidase of *Stachybotrys* strain. Appl Microbiol Biot 77: 293–300.
29. Yoon JJ, Kim KY, Cha CJ (2008) Purification and characterization of thermostable β-glucosidase from the brown-rot basidiomycete *Fomitopsis palustris* grown on microcrystalline cellulose. J Microbiol 46: 51–55.

30. Venturi LL, de Lourdes Polizeli M, Terenzi HF, dos Prazeres Melo Furriel R, Jorge JA (2002) Extracellular β-D-glucosidase from *Chaetomium thermophilum* var. *coprophilum*: production, purification and some biochemical properties. J Basic Microb 42: 55–66.
31. Murray P, Aro N, Collins C, Grassick A, Penttilä M, et al. (2004) Expression in *Trichoderma reesei* and characterisation of a thermostable family 3 β-glucosidase from the moderately thermophilic fungus *Talaromyces emersonii*. Protein Express Purif 38: 248–257.
32. Chisti Y (2007) Biodiesel from microalgae. Biotechnol Adv 25: 294–306.
33. Wood TM, Mccrae SI (1982) Purification and some properties of the extracellular β-D-glucosidase of the cellulolytic fungus *Trichoderma koningii*. J Gen Microbiol 128: 2973–2982.
34. Uchiyama T, Miyazaki K, Yaoi K (2013) Characterization of a novel β-glucosidase from a compost microbial metagenome with strong transglycosylation activity. J Biol Chem 288: 18325–18334.
35. Andrić P, Meyer AS, Jensen PA, Dam-Johansen K (2010) Reactor design for minimizing product inhibition during enzymatic lignocellulose hydrolysis: I. Significance and mechanism of cellobiose and glucose inhibition on cellulolytic enzymes. Biotechnol Adv 28: 308–324.
36. Saha BC, Bothast RJ (1996) Production, purification, and characterization of a highly glucose-tolerant novel β-glucosidase from *Candida peltata*. Appl Environ Microb 62: 3165–3170.
37. Riou C, Salmon JM, Vallier MJ, Günata Z, Barre P (1998) Purification, characterization, and substrate specificity of a novel highly glucose-tolerant β-glucosidase from *Aspergillus oryzae*. Appl Environ Microb 64: 3607–3614.

18

Carbon Sequestration Efficiency of Organic Amendments in a Long-Term Experiment on a Vertisol in Huang-Huai-Hai Plain, China

Keke Hua, Daozhong Wang*, Xisheng Guo*, Zibin Guo

Soil and Fertilizer Research Institute, Anhui Academy of Agricultural Sciences, Hefei, China

Abstract

Soil organic carbon (SOC) sequestration is important for improving soil fertility of cropland and for the mitigation of greenhouse gas emissions to the atmosphere. The efficiency of SOC sequestration depends on the quantity and quality of the organic matter, soil type, and climate. Little is known about the SOC sequestration efficiency of organic amendments in Vertisols. Thus, we conducted the research based on 29 years (1982–2011) of long-term fertilization experiment with a no fertilizer control and five fertilization regimes: CK (control, no fertilizer), NPK (mineral NPK fertilizers alone), NPK+1/2W (mineral NPK fertilizers combined with half the amount of wheat straw), NPK+W (mineral NPK fertilizers combined with full the amount of wheat straw), NPK+PM (mineral NPK fertilizers combined with pig manure) and NPK+CM (mineral NPK fertilizers combined cattle manure). Total mean annual C inputs were 0.45, 1.55, 2.66, 3.71, 4.68 and 6.56 ton/ha/yr for CK, NPK, NPKW1/2, NPKW, NPKPM and NPKCM, respectively. Mean SOC sequestration rate was 0.20 ton/ha/yr in the NPK treatment, and 0.39, 0.50, 0.51 and 0.97 ton/ha/yr in the NPKW1/2, NPKW, NPKPM, and NPKCM treatments, respectively. A linear relationship was observed between annual C input and SOC sequestration rate (SOCsequestration rate = 0.16 Cinput −0.10, R = 0.95, P<0.01), suggesting a C sequestration efficiency of 16%. The Vertisol required an annual C input of 0.63 ton/ha/yr to maintain the initial SOC level. Moreover, the C sequestration efficiencies of wheat straw, pig manure and cattle manure were 17%, 11% and 17%, respectively. The results indicate that the Vertisol has a large potential to sequester SOC with a high efficiency, and applying cattle manure or wheat straw is a recommendable SOC sequestration practice in Vertisols.

Editor: Xiujun Wang, University of Maryland, United States of America

Funding: This work was supported by the Natural Science Foundation of China (Grant No. 41401331), Special Fund for Agro-scientific Research in the Public Interest of China (Grant No. 201203030), the President Distinguished Youth Innovation Fund of Anhui Academic of Agricultural Science (Grants No. 14B1008 and 14B1007), and by the Anhui Province Science and Technology Program (Grant No. 1206c0805033). The funders had no role in study design, data collection and analysis, decision to publish, or preparation of the manuscript.

Competing Interests: The authors have declared that no competing interests exist.

* Email: wdzhong-3@163.com (DZW); 1078681598@qq.com (XSG)

Introduction

Soil organic carbon (SOC) sequestration contributes to the mitigation of greenhouse gas emissions and to the improvement of soil fertility [1]. Net SOC sequestration is the balance of organic C inputs into the soil (via crop residues, organic amendments in compost, animal manure, etc.) and organic C decomposition by soil microbes. SOC sequestration efficiency is commonly expressed by the relationship between annual C input and SOC accumulation rate, which is an indicator of soil C sequestration ability [2]. Therefore, information about the C sequestration efficiency is useful for seeking high efficiency management strategies of enhancing the SOC stock and soil fertility.

C sequestration efficiency is regulated by climate, the quantity, quality of added organic materials, and soil inherent properties [3–5]. Within a climate zone, soil inherent properties (i.e. initial SOC content, soil texture, clay type and aggregates) and cultivation practices have effects on SOC sequestration efficiency. A negative linear relationship has been reported between C sequestration efficiency and initial SOC content [6,7], mainly because SOC tends to increase faster if initial SOC content is far from its saturation level. Feller and Beare [8] reported a positive relationship between SOC sequestration and soil clays. Because clay type (i.e. 2:1 smectite clay minerals versus 1:1 allophanic clay minerals) has influence on the stabilization of SOC [9]. Elliott and Coleman [10] indicated that soil aggregates may physically protect SOC against decomposition by soil microbes. Recently, several studies have examined the SOC sequestration efficiency of different soil types based on long-term fertilization experiment [11–15]. However, no information is available about SOC sequestration efficiency and C sequestration efficiencies of organic amendments in Vertisols [16], which occupy approximately 4.0 million km^2 in Huang-Huai-Hai Plain of China. Vertisols are characterized by a low SOC content (less than 0.6% in the topsoil) a high clay content (more than 35%) and shrinking and swelling properties which contribute to self-mulching [17].

The purpose of this study was to evaluate the SOC sequestration efficiency under long-term fertilization practices in Vertisols, with a winter wheat–soybean double cropping system. Specific objectives were (1) to assess C input and SOC dynamics under various long-term fertilization practices (2) to estimate the SOC sequestration efficiency of different organic amendments.

Materials and Methods

Ethics Statement

The administration of the department of agricultural of Anhui Province gave permission for this research at the study site. We confirm that the field studies did not involve endangered or protected species.

Site description

The experiment is located at the Madian Agro-Ecological Station in Huang-Huai-Hai Plain, Eastern China (N33°13′, E116°37′). The areas has a sub-humid climate, with annual mean, maximum and minimum air temperature of 16.5°C, 36.5°C and −7.4°C, respectively (Fig. 1a). Annual precipitation ranged from about 457 to 1478 mm during the last 24 years (Fig. 1b), about 70% of which occurs from May to September. There are strong inter-annual and seasonal variations in precipitation. For example, a 90-day period of summer drought with a total precipitation of only 138 mm occurred in 1988. Precipitation in the soybean period ranged from 244 mm to 1049 mm with an average of 616 mm, which was much higher than that in the wheat period (259 mm).

The predominant Vertisols have developed in fluvial and lacustrine deposits. They are classified as Calcic Kastanozems, according to the soil classification system of the Food and Agriculture Organization (FAO). Soil pH ranges from 6.0 to 8.6 and SOC content from 5.8 to 7.5 g/kg in the area. The initial (1982) topsoil (0–20 cm) had a total SOC of 5.8±0.08 g/kg, total N 0.96±0.04 g/kg, total P 0.28±0.02 g/kg, bulk density 1.45 g/cm^3 and pH (1:2.5 w/v) 7.4, respectively. The topsoil under NPK treatment in 2013 had coarse sand (2 to 0.2 mm) 8.5±2.0 g/kg, fine sand (0.2 to 0.02 mm) 322.9±21.7 g/kg, silt (0.02 to 0.002 mm) 262.3±24.0 g/kg and clay (<0.002 mm) 406.3±24.0 g/kg, respectively.

Experimental design

The long-term field experiment was initiated in 1982. The experiment had six treatments (Table 1): no fertilizer (CK), mineral nitrogen-phosphorus-potassium fertilizers alone (NPK), mineral NPK fertilizers combined with 2.5 ton/ha/yr (dry base) of wheat straw (NPKW1/2), mineral NPK fertilizers combined with 5.0 ton/ha/yr of wheat straw (NPKW), mineral NPK fertilizers combined with 7.8 ton/ha/yr of pig manure (NPKPM), and mineral NPK fertilizers combined with 12.5 ton/ha/yr of cattle manure (NPKCM). The treatments were laid out in a randomized block design with four replications. The net plot size was 70 m^2 (14.9 m ×4.7 m). Each plot was isolated by 50 cm deep plates. Mineral N, P and K fertilization was urea, calcium superphosphate and potassium chloride, respectively. The amounts applied were similar to the amounts applied by local farmers, i.e., 180 kg N, 90 kg P_2O_5, and 135 kg K_2O/ha/yr. All fertilizers were applied as base fertilizer at the start of the wheat growing season. No fertilizers were applied to soybean growing season because of the high background value of soil N, P and K, which could supply sufficient available nutrients for the growth of soybean. Moreover, extra nitrogen was also obtained by nitrogen fixation by soybean itself. The mean chemical characteristics of pig manure were 360 g C/kg (dry base), 17 g N/kg and 8.9 g P/kg, and those of cattle manure were 370 g C/kg, 7.9 g N/kg and 4.2 g P/kg.

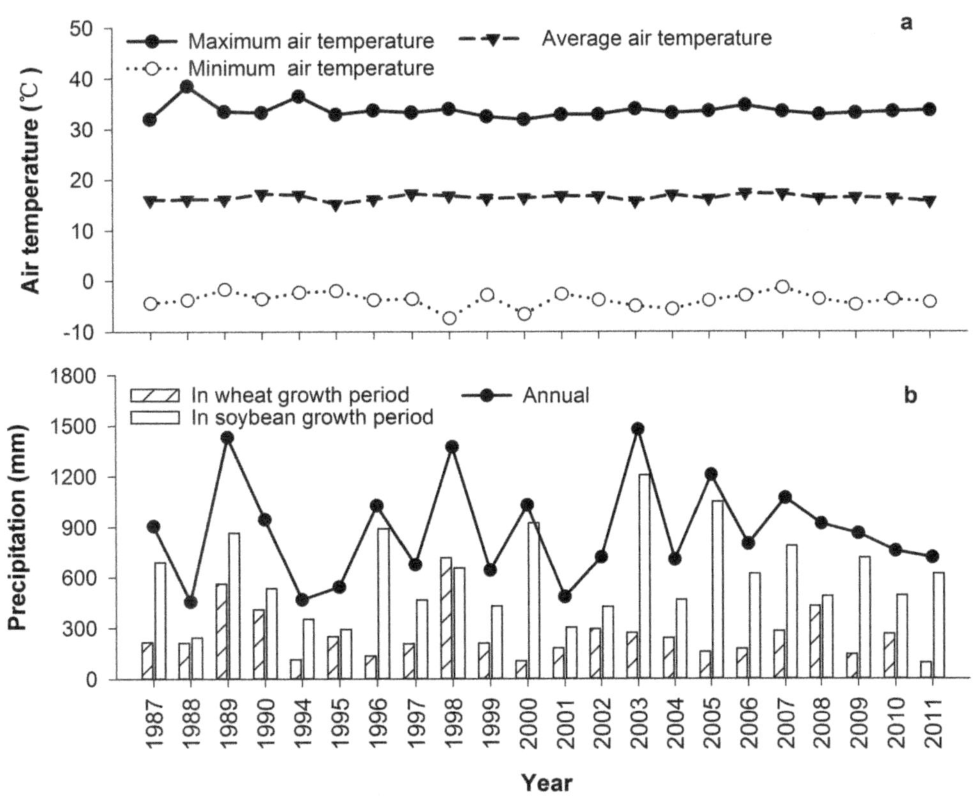

Figure 1. Air temperature and precipitation during the period 1987 to 2011 (no data for 1991 and 1993). (a) Maximum, minimum and average air temperatures. (b) Accumulated precipitation during the wheat growing period (from November to May), the soybean growing period (from June to October), and for the annual total.

Table 1. Mean application rates of mineral and organic fertilizers (kg/ha/yr) per treatment.

Treatment	N	P_2O_5	K_2O	Wheat straw	Pig manure	Cattle manure
CK	0	0	0	0	0	0
NPK	180	90	135	0	0	0
NPKW1/2	180	90	135	2500	0	0
NPKW	180	90	135	5000	0	0
NPKPM	180	90	135	0	7800	0
NPKCM	180	90	135	0	0	12500

Treatments were no fertilization (CK), mineral NPK fertilizers alone (NPK), mineral NPK fertilizers combined with 2.5 ton/ha/yr of wheat straw (NPKW1/2), mineral NPK fertilizers combined with 5.0 ton/ha/yr of wheat straw (NPKW), mineral NPK fertilizers combined with pig manure (NPKPM), and mineral NPK fertilizers combined with cattle manure (NPKCM).

A double cropping system of winter wheat-soybean is the common crop rotation in the region. At the Madian Agro-Ecological Station, plots were plowed to 20 cm depth after crop harvest. Winter wheat (*Triticum aestivum L.*), variety Yedan 13, was grown in rows from late October to May, and soybean (*Glycine max*), variety Zhonghuang 13, from June to September. Herbicides and pesticides were applied during the growth periods when it was needed. Weed residues in all the fields were manually removed after herbicides application. Hence, biomass of weeds left in the field was negligible. Wheat and soybean were harvested manually and all above-ground biomass, except for stubble, were removed from the experimental fields. Wheat grain yields were measured in 1983, 1985, and 1989–2011. Soybean grain yields were measured in 1983, 1985, 1989–1991 and 1998–2011. Grains were air dried, threshed and then weighted separately.

Soil sampling and analyses

Soil samples were collected from the top 20 cm each year after the soybean harvest (in October). Soil samples were randomly taken from three locations in each plot using a soil core sampler (inner diameter 7 cm). The samples were air dried and passed through an 8 mm sieve. Visible pieces of crop residues and roots were removed. Representative sub-samples were then passed through a 2 mm sieve and ground further to pass a 0.25 mm sieve for determination of SOC, total N, total P and other properties. The sieved soil was stored in glass jars until analysis. Soil bulk density was measured using the core method [18]. Soil pH was measured by the potentiometric method in a soil-water extract (1:2.5, w/v water) [18]. SOC content was determined following the vitriol acid-potassium dichromate oxidation method [19]. Total N was determined by using the method described by [19], and total P by Black [20]. Soil texture was measured following the gravitometer methods described by Lu [18]. Four replicates were carried out for each analysis.

Estimation of C inputs, and SOC sequestration rate and efficiency

C inputs into the topsoil (0–20 cm) included materials from crops residues (i.e., mainly roots and stubble), organic amendments (i.e. wheat straw, pig manure, and cattle manure). Annual amount of C inputs by roots from wheat and soybean were estimated at 30% [21] and 19% [22] of above-ground biomass, respectively. Annual C inputs by stubble from wheat and soybean were estimated at 15% of straw biomass. We used a straw to grain ratio of 1.2:1 and 1.6:1 to calculate wheat and soybean straw biomass, respectively [23]. The C contents of wheat and soybean were estimated at 400 g/kg and 453 g/kg (dry base), respectively [23]. The annual C input via crop residues was estimated from

$$C_{input} = ((Y_g + Y_s) \times R \times D_r + R_s \times Y_s) \times (1 - W) \times C/1000 \quad (1)$$

where Y_g, and Y_s were grain yield and straw yield (kg/ha); R was ratio of root biomass to above ground biomass (AGB); D_γ was ratio of root biomass in topsoil (0–20 cm) to root biomass in the soil profile, which were set at 0.753 for wheat [24] and at 0.984 for soybean [25]; R_s and W represented ratio of stubble to straw biomass and water content of air-dried grain (14% [23]); C was the organic C content of the crop (g/kg).

SOC stock ($SSOC$, ton/ha) in topsoil was calculated using the equation:

$$SSOC = SOC \times BD \times d/10 \quad (2)$$

where SOC was soil organic C content (g/kg), BD was the bulk density (g/cm^3), and d was the thickness of the soil layer (20 cm).

SOC sequestration rate (SSR, ton/ha/yr) was estimated for the topsoil by the following equation [4]:

$$SSR = (SSOC_t - SSOC_0)/t \quad (3)$$

where $SSOC_t$ and $SSOC_0$ were the stock of SOC at time t and in the initial year (1982); t was the duration of experiment. The SOC sequestration efficiency was derived from the slope of the linear regression between annual SOC sequestered and C input [14].

ΔSOC sequestration rates (ΔSSR, ton/ha/yr) via wheat straw, pig manure and cattle manure were estimated for topsoil by the following equation.

$$\Delta SSR = (SSR_{NPK+OM} - SSR_{NPK}) \quad (4)$$

where SSR_{NPK+OM} and SSR_{NPK} were the SOC sequestration rate in treatments with organic amendments combined with mineral NPK fertilizers (NPKW1/2, NPKW, NPKPM and NPKCM) and the treatment with mineral NPK fertilizers alone (NPK), respectively.

C sequestration efficiency of organic amendment (SEO) was calculated by the following equation.

$$SEO(\%) = \Delta SSR/CI \times 100\% \quad (5)$$

where CI was annual C input via organic amendment (ton/ha/yr).

Data analysis

All the statistical analyses were performed using SPSS 13.0 software package (SPSS, Inc., USA). Significant differences were analyzed using *LSD* test at significance level $P = 0.05$ or $P = 0.01$. Graphs were prepared with Sigma plot 10.0 software (Systat Software, Inc., Chicago, IL, USA).

Results

Grain yield and C inputs into the soil

Wheat and soybean yields are shown in Fig. 2. Average annual grain yields followed the order NPKCM> NPKPM> NPKW> NPKW1/2> NPK> CK. Grain yields tended to decrease under the CK treatment, but to increase under all fertilization treatments over time.

Similarly, the CK treatment had a low and decreasing crop C input, while the fertilization treatments had a high and increasing trend of C input via roots and stubble over time (Table 2). Average annual C input via roots and stubble was 2.4 times higher in the NPK treatment than in the CK treatment. Soil amendments with straw and manure slightly increased the C inputs via roots and stubble in the NPKW1/2, NPKW, NPKPM and NPKCW treatments, respectively, comparing to the NPK treatment. Similarly, CK had the lowest annual total C input t (0.45 ton/ha). Compared with CK, the average annual total C inputs in the NPK, NPKW1/2, NPKW, NPKPM and NPKCW treatments were increased by a factor of 2.4, 4.9, 7.2, 9.4, and 13.6, respectively.

SOC dynamics

Changes in SOC contents are shown in Fig. 3. As expected, CK had a low and slightly decreasing SOC content, whereas other treatments had an increasing SOC content. Changer over time in SOC content could be described by linear regression or linear plateau models. In CK, the changes in SOC content were best described by simple linear regression ($y = -0.03\,x+6.70$, $R = 0.59$, $P<0.01$), suggesting that SOC decreased by 3% per year. Changes in SOC contents of the NPK, NPKW1/2 and NPKW were also best described by simple linear regression, but here significant increases in SOC content were observed. Changes in SOC contents of the NPKPM and NPKCM were best described by linear plateau model. For NPKPM and NPKCW, SOC contents tended to increase rapidly in the first twenty years, but leveled off thereafter. However, a longer duration of soil C sequestration was observed in NPKPM (26 years) than in NPKCM (20 years).

C sequestration rate

Soil bulk density (BD) decreased in all fertilization treatments but not in the CK treatment (Table 3). The decrease was largest in the treatments receiving wheat straw and animal manures, which suggests that long-term application of organic amendments significantly improved soil physical conditions.

The mean SOC sequestration rate over the 29 years experimental period ranged from -0.08 ± 0.03 ton/ha/yr in CK to 0.97 ± 0.04 ton/ha/yr in NPKCM. Mineral fertilizer application (NPK) reversed the SOC decline in the CK treatment into a net SOC sequestration of 0.20 ± 0.02 ton/ha/yr. Mean SOC sequestration rates followed the order: NPKCM> NPKPM> NPKW> NPKW1/2> NPK. These findings are consistent with a large body of evidence indicating that long-term fertilizer, crop residues and manure applications increase the SOC content of arable soils [4,14]. Average ΔSOC sequestration rates (ΔSSR) of wheat straw at low input rate, wheat straw at high input rate, pig manure and cattle manure were 0.19 ± 0.03, 0.30 ± 0.04, 0.31 ± 0.03 and 0.77 ± 0.02 ton/ha, respectively. The result indicated that SOC sequestration rate changes were related to total C inputs.

Relationship between C input and SOC sequestration rate

There was a significantly positive linear correlation between C input and SOC sequestration rate (Fig. 4). The slope of the equation (0.16) indicates that on average 16% of the total C input into the soil was sequestered as SOC; the mean soil C sequestration efficiency was 16%. The results also indicate that 0.63 ton/ha/yr was needed to maintain SOC level constant at the initial level.

The mean relationships between C inputs and ΔSOC sequestration rates through organic amendments are also shown in Fig. 4. Interestingly, C sequestration efficiencies for wheat straw (c

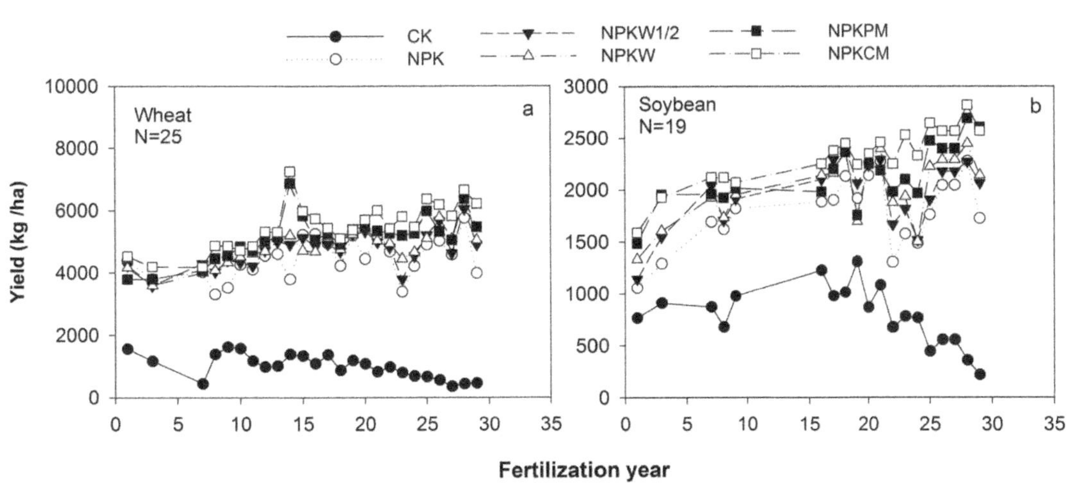

Figure 2. Average annual grain yields in six treatments during the period 1983 to 2011. (a) Wheat yields (b) Soybean yields. Treatments were no fertilization (CK), mineral NPK fertilizers alone (NPK), mineral NPK fertilizers combined with 2.5 ton/ha/yr of wheat straw (NPKW1/2), mineral NPK fertilizers combined with 5.0 ton/ha/yr of wheat straw (NPKW), mineral NPK fertilizers combined with 7.8 ton/ha/yr of pig manure (NPKPM), and mineral NPK fertilizers combined with 12.5 ton/ha/yr of cattle manure (NPKCM). Results are means (n = 4).

Table 2. Average annual crop C input (roots and stubble) and organic amendment C inputs for each treatment during the period 1987 to 2011.

Treatment	Annual crop C input (ton/ha/yr)				Organic amendment C input	Total
	Wheat	CV%	Soybean	CV%	(ton/ha/yr)	(ton/ha/yr)
CK	0.23c	38.3	0.22c	36.2	0	0.45
NPK	1.04b	14.3	0.51b	19.0	0	1.55
NPKW1/2	1.10b	11.8	0.56b	16.7	1.00	2.66
NPKW	1.13b	12.5	0.57b	16.1	2.00	3.71
NPKPM	1.20a	13.7	0.61a	14.0	2.87	4.68
NPKCM	1.28a	13.5	0.66a	12.3	4.62	6.56

Treatments were no fertilization (CK), mineral NPK fertilizers alone (NPK), mineral NPK fertilizers combined with 2.5 ton/ha/yr of wheat straw (NPKW1/2), mineral NPK fertilizers combined with 5.0 ton/ha/yr of wheat straw (NPKW), mineral NPK fertilizers combined with 7.8 ton/ha/yr of pig manure (NPKPM), and mineral NPK fertilizers combined with 12.5 ton/ha/yr of cattle manure (NPKCM). Data are present as means (n = 4). Those with the same letter are not significantly different ($P<0.05$ or 0.01).

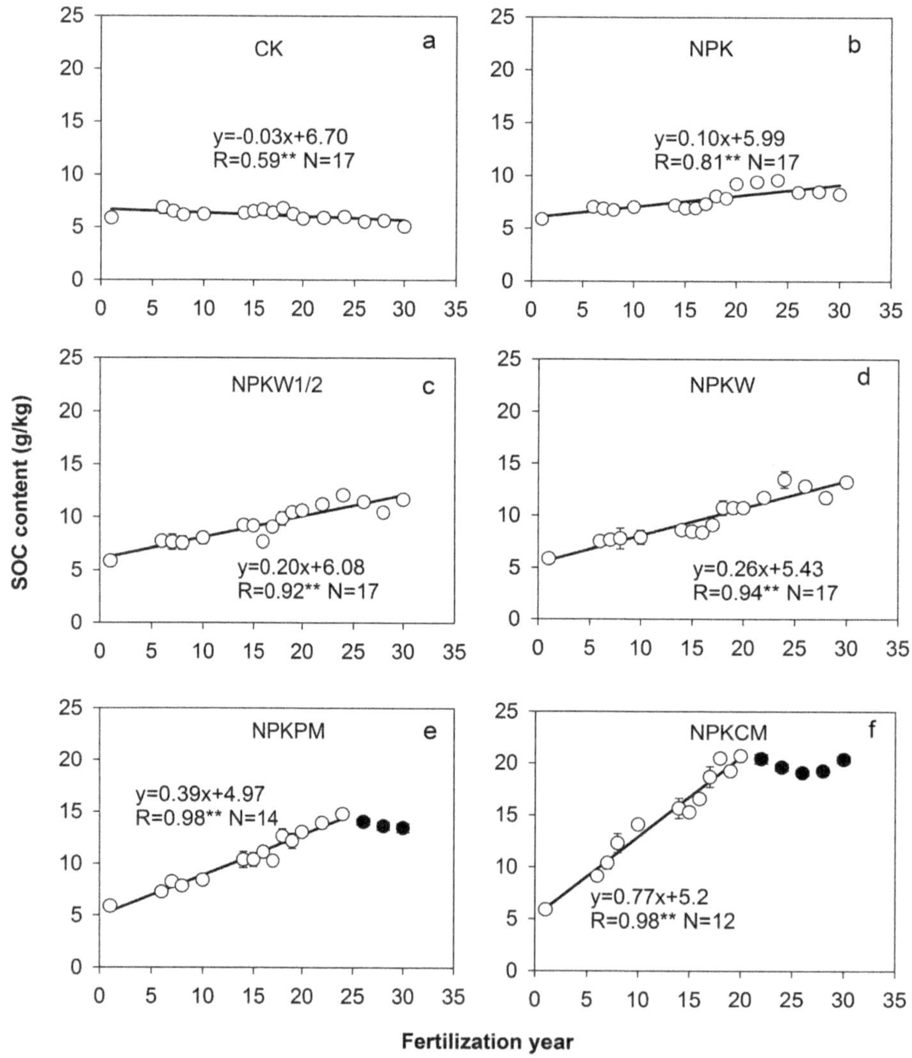

Figure 3. Changes in soil organic carbon (SOC) content over time in the six fertilization treatments during the period 1983 to 2011. CK: no fertilizer (a); NPK: mineral NPK fertilizers alone (b); NPKW1/2: mineral NPK fertilizers combined with 2.5 ton/ha/yr of wheat straw (c); NPKW: mineral NPK fertilizers combined with 5.0 ton/ha/yr of wheat straw (d); NPKPM: mineral NPK fertilizers combined with pig manure (e); NPKCM: mineral NPK fertilizers combined with cattle manure (f); ** indicates significant correlation at $P<0.01$, respectively. Error bars indicated for some years are standard errors (n = 4).

Table 3. Soil bulk density (BD), SOC contents, the estimated mean SOC sequestration rate (SSR) and mean ΔSOC sequestration rate (ΔSSR) during the 29-years period, for all treatments.

Treatment	1982			2011			SSR (ton/ha/yr)	ΔSSR (ton/ha/yr)
	BD (g/cm^3)	SOC (g/kg)	SSOC (ton/ha)	BD (g/cm^3)	SOC (g/kg)	SSOC (t/ha)		
CK	1.45	5.86	16.99	1.44±0.03a	5.08±0.17e	14.63±0.79e	−0.08±0.03e	-
NPK	1.45	5.86	16.99	1.38±0.01a	8.24±0.07d	22.74±0.47d	0.20±0.02d	-
NPKW1/2	1.45	5.86	16.99	1.21±0.02b	11.66±0.46c	28.22±1.58c	0.39±0.05c	0.19±0.03
NPKW	1.45	5.86	16.99	1.19±0.02b	13.23±0.55b	31.49±1.84b	0.50±0.06b	0.30±0.04
NPKPM	1.45	5.86	16.99	1.18±0.02b	13.46±0.42b	31.77±1.53b	0.51±0.05b	0.31±0.03
NPKCM	1.45	5.86	16.99	1.11±0.01c	20.36±0.32a	45.20±1.12a	0.97±0.04a	0.77±0.02

CK: no fertilizer; NPK: mineral fertilization; NPKW1/2: mineral fertilizer combined with 2.5 ton/ha/yr of wheat straw; NPKW: mineral fertilizer combined with 5.0 ton/ha/yr of wheat straw; NPKPM: mineral fertilizer combined with pig manure; NPKCM: mineral fertilizer combined with cattle manure. Data are presented as means (n = 4), and those designated with the same letter are not significantly different ($P<0.01$).

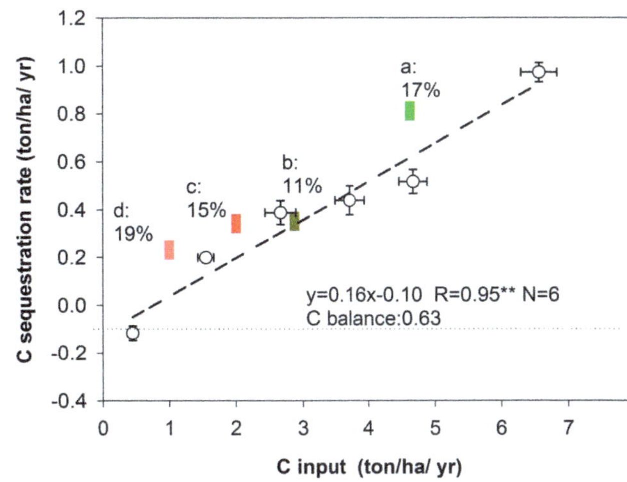

Figure 4. Relationship between C input and SOC sequestration rate. Solid line indicated linear regression curve between C inputs and SOC sequestration rates for the six treatments. a, b, c, and d represent C sequestration efficiencies for wheat straw at low input rate, wheat straw at high input rate, pig manure and cattle manure (calculated by equation (5)), respectively. ** indicated significant at $P<0.01$; error bars indicate standard errors (n = 4).

and d), pig manure (b) and cattle manure (a) differed. The C sequestration efficiencies for wheat straw was 17% (mean value), pig manure 11% and for cattle manure 17%. Hence, application cattle manure or wheat straw has a higher efficiency to enhance SOC.

Discussion

Soil C sequestration duration

Our results clearly show that trends in SOC depend on long-term fertilization regime. The changes of SOC content over time for the three treatments of NPK, NPK1/2W and NPKW are consistent with the observations reported by Zhou et al (2013) [26]. The time of soil C steady state occurring represents soil C sequestration duration [27]. West and Six [28] reported that the duration of C sequestration varies between ecosystems, climate regimes, and fertilization management (e.g., soil organic amendments inputs). West and Post [29] estimated that a change from conventional tillage to no-till results in an average 14% increase in soil C, during a mean experimental period of 15 years. Rui and Zhang [30] found that there was a negative correlation between soil C sequestration rates and duration of soil C sequestration. The C sequestration duration in the NPKW, NPKPM and NPKCM treatments reflect differences in the quantity and quality of C input. Our results indicate that the recommended mean C sequestration period of 20 years for National Greenhouse Gas according to the IPCC Guidelines [31] may not be suitable for all the organic amendments application.

Relation between C input and soil C sequestration

The relationship between C input into the soil and SOC sequestration may present in the SOC sequestration efficiency. Several studies showed that there was no general relationship between C input and SOC sequestration. Some soils might be C saturated [32–34]. Other studies demonstrated that SOC sequestration was linear or logarithmically related to the C input into soil [2,4,15]. In the present study, a clear linear relationship between annual C input and SOC sequestration rate was found, which was similar to findings in upland [4,35], but different from those findings in paddy soils [14,15,36]. The strong linear relationship indicated that the Vertisol had a large capacity to store SOC. Moreover, we did not find indications that the Vertisol had a clear, uniform saturation level for SOC sequestration. The results presented in Fig. 3b, c and d indicate that these maxima (equilibrium levels) have not been reached in 29 years. However, we observed that SOC sequestration rate leveled off after about 20 years in the NPKPM and NPKCM treatments at SOC levels of about 15 and 20 g/kg. The results suggest that SOC saturation level depends on C inputs of organic amendments.

Our results revealed that 0.63 ton/ha/yr (Fig. 4) was needed to maintain the initial SOC level, and that the unfertilized control treatment (CK) did not provide sufficient C input to maintain the initial SOC content. This finding is in agreement with the result from Yan et al. [15], who showed that the C input from unfertilized crops cannot sustain the native SOC content of upland soils. Interestingly, NPK fertilization did provide sufficient C input to contribute to a net SOC sequestration, indicating that the C input provided through the incorporation of only roots and stubble was sufficient to maintain and even enhance the C stock. This contrasts with result reported by Yang [37]. They reported that NPK fertilization was insufficient for maintaining the SOC level under conditions of traditional management with the removal of all aboveground crop biomass. These differences may relate to differences in crop yield levels and in SOC sequestration efficiency. Our study indicates C inputs from wheat and soybean stubbles and roots under conventional NPK fertilization seem sufficient to replenish the C loss through SOC decomposition in Vertisols.

C sequestration efficiency

The initial SOC content and soil texture (i.e. clay content) are important factors regulating SOC sequestration efficiency [9]. The Vertisol in our study had a low initial C content and a high clay content, which suggest that the SOC sequestration might be high. Indeed, we observed a relatively high efficiency irrespective of the type of C input (Fig. 4). Zhang [14] reported a C sequestration efficiency of 6.8% for alluvial soils in Zhengzhou, China, in the same climatic zone as our study. Stewart [38] argued that a greater SOC saturation deficit results in greater sequestration efficiency of added C. We estimated the soil C saturation deficit of the Vertisol at 80%, using the relationship between soil mineral (silt + clay) and C contents (g/kg soil) as proposed by Stewart [33]:

$$\textit{Csaturation deficit} = (1\textit{-initial SOC content/protective capacity}) \times 100\%$$

$$\textit{Protective capacity} = 0.21 \times (\textit{silt} + \textit{clay content}) + 14.76$$

Evidently, the high clay+silt content and the low initial SOC content contributed to the relatively high C saturation deficit. The protection of SOC by clay particles was well established [39,40]. Stewart [38] reported a larger enrichment of microbial derived carbohydrates in the clay fraction compared with that of the sand fraction. Hassink [41] reported that C associated with organo-mineral complexes in soil were chemically protected, and the protection increased with an increase in clay content. Six [9] found that soils with a high clay content had a low SOC decomposition rate, likely because SOC in clay soils was chemically stabilized and absorbed onto negatively charged clay minerals [42,43]. Therefore, the clay content has a positive effect on SOC sequestration efficiency [44].

Many factors have been suggested to affect the humification coefficient, including the type of input material, soil type, climate factors and soil nutrients status [45,46]. Maillard and Angers [47] reported that the quality of the organic amendment was the dominant driver determining C sequestration efficiencies of organic amendments, based on a meta-analysis of worldwide published reports. Likely, the quality of the organic amendment in our experiment was also the major reason explaining the differences in C sequestration efficiencies for wheat straw, pig manure and cattle manure. Especially lignin is considered to be one of the most chemical recalcitrant components [48]. In soils, lignins are synthesized from L-phenylalanine and cinnamic acids via various metabolic ways to form lignin precursors such as sinapyl and coniferylalcohols [49]. The lignin structure consists of aromatic rings with side chains and –OH and $-OCH_3$ groups linked by strong covalent bonds. Therefore, lignins are considered as stabilized component of SOC, influencing its pool-size and its turnover [50]. In our study, we observed that C sequestration efficiency of cattle manure was higher than that of pig manure, likely because the cattle manure was composted. Ghosh et al. [51] reported that the lignin content of cattle manure may reach 23.7%, which was significantly higher than that of pig manure. Therefore, application of composted cattle manure seems the preferred strategy for enhancing SOC sequestration in the Vertisol due to its high C sequestration efficiency.

There are a number of uncertainties in our study, which are quite common to many long-term field studies. We estimated the C inputs via roots and stubble from fixed percentages of aboveground biomass. Likely, the C input via roots and stubble depends on the level of fertilization, and varies from year to year due to variations in rainfall. Also, the C content of the pig and cattle manure was not measured each year, while there was quite a big annual variation in composition. Further, there are missing data, especially during the first half of the experiment, due to organizational matters. We obtained C sequestration efficiencies of wheat straw, pig manure and cattle manure based on the two main hypotheses. We assumed that the difference in C input via stubble and roots was negligible small between the treatment NPK and the treatments NPKW1/2, NPKW, NPKPM and NPKCM, and no significant priming effect caused by organic amendments application was existed. These uncertainties in primary data may also lead to uncertainties in the estimated mean SOC sequestration efficiency and C sequestration efficiencies of organic amendments, but do not undermine our main conclusions. Our future studies focus on the accurate quantification of C inputs via roots and stubble, effects of organic amendments application on the C inputs, and the accurate quantification of priming effect by organic amendments application using isotope technique in this ongoing field experiment. Also, SOC fractionation studies are undertaken to establish relationships between C input and SOC sequestration in different SOC pools (i.e. light and heavy C fractions).

Conclusions

The changes in SOC content over time were described by linear equations. However, the SOC sequestration rates clearly decreased in the NPKPM and NPKCM treatments after about 20 years. A significant linear relationship was observed between annual C input and SOC sequestration rate. The Vertisol in our study had a high SOC sequestration potential. Also, the overall mean SOC sequestration efficiency of the wide range of C inputs via roots, stubble, straw and manure was equally high, with an overall mean efficiency of 16% over the 29-years period. The C sequestration efficiencies of wheat straw and cattle manure are higher than pig manure, indicating that cattle manure or wheat straw application is a recommendable SOC sequestration practice in Vertisols.

Acknowledgments

We thank the staff of Madian Agro-Ecological Station and the Ministry of Agriculture, China. We sincerely thank Professor Oene Oenema from Environmental Sciences, Wageningen University, for the useful suggestions and comments. We are also greatly grateful for anonymous reviewers for their very constructive and valuable comments for our manuscript.

Author Contributions

Conceived and designed the experiments: DZW XSG. Performed the experiments: KKH DZW ZBG. Analyzed the data: KKH. Contributed reagents/materials/analysis tools: KKH DZW XSG ZBG. Wrote the paper: KKH.

References

1. Lal R (2004) Soil carbon sequestration impacts on global climate change and food security. Science 304: 1623–1627.
2. McLauchlan KK (2006) Effects of soil texture on soil carbon and nitrogen dynamics after cessation of agriculture, Geoderma 136: 289–299.
3. Tong C, Xiao H, Tang G, Wang H, Huang T, et al. (2009) Long-term fertilizer effects on organic carbon and total nitrogen and coupling relationships of C and N in paddy soils in subtropical China. Soil and Tillage Research 106: 8–14.
4. Zhang WJ, Xu MG, Wang XJ, Huang QH, Nie J, et al. (2012) Effects of organic amendments on soil carbon sequestration in paddy fields of subtropical China. Journal of Soils and Sediments 12: 457–470.
5. Freibauer A, Rounsevell MD, Smith P, Verhagen J (2004) Carbon sequestration in the agricultural soils of Europe. Geoderma 122: 1–23.
6. Li ZP, Lin XX, Che YP (2002) Analysis for the balance of organic carbon pools and their tendency in typical arable soils of Eastern China. Acta Pedologica Sinica 39: 351–360. (In Chinese).
7. Zhou P, Pag GX, Li LQ, Zhang XH (2009) SOC Enhancement in major types of paddy soils in a long-term agro-ecosystem experiment in South China. V. Relationship between carbon Input and soil carbon sequestration. Scientia Agricultura Sinica 42: 4260–4268. (In Chinese).
8. Feller C, Beare MH (1997) Physical control of soil organic matter dynamics in the tropics. Geoderma 79: 69–116.
9. Six J, Conant RT, Paul EA, Paustian K (2002) Stabilization mechanisms of soil organic matter: Implications for C-saturation of soils. Plant Soil 241: 155–176.
10. Elliott ET, Coleman DC (1988) Let the soil work for us. Ecological Bulletins 39: 23–32.
11. Gong W, Yan XY, Wang JY, Hu TX, Gong Y (2009) Long-term manuring and fertilization effects on soil organic carbon pools under a wheat-maize cropping system in North China Plain. Plant Soil 314: 67–76.
12. Zhang JB, Zhu TB, Cai ZC, Qin SW, Muller C (2012) Effects of long-term repeated mineral and organic fertilizer applications on soil nitrogen transformations. European Journal of Soil Science 63: 75–85.
13. Chander K, Goyal S, Mundra MC, Kapoor KK (1997) Organic matter, microbial biomass and enzyme activity of soils under different crop rotations in the tropics. Biology and Fertility of Soils 24: 306–310.
14. Zhang WJ, Wang XJ, Xu MG, Huang SM, Liu H, et al. (2010) Soil organic carbon dynamics under long-term fertilizations in arable land of northern China. Biogeosciences 7: 409–425.
15. Yan X, Zhou H, Zhu QH, Wang XF, Zhang YZ, et al. (2013) Carbon sequestration efficiency in paddy soil and upland soil under long-term fertilization in southern China. Soil and Tillage Research 130: 42–51.
16. Li DC, Zhang GL, Gong ZT (2011) On taxonomy of Shajiang Black Soils in China. Soils 43: 623–629. (In Chinese).
17. Guo ZB, Hua KK, Wang J, Guo XS, He CL, et al. (2014) Effects of different regimes of fertilization on soil organic matter under conventional tillage. Spanish Journal of Agricultural Research 2014 12: 801–808.
18. Lu RK (2000) Analytical methods of soil agricultural chemistry. China Agricultural Science and Technology Press, Beijing. (In Chinese).
19. Walkley A, Black IA (1934) An examination of the degtjareff method for determining soil organic matter and a proposed modification of the chromic acid titration method. Soil science 37: 29–38.
20. Black CA (1962) Methods of soil analysis. Part 2. Chemical and microbiological properties. Madison, Wisc: ASA.
21. Kuzyahov Y, Domenski G (2000) Carbon input by plants into the soil. Journal of Plant Nutrition and Soil science 163: 421–431.
22. IPCC (2006) IPCC guidelines for national greenhouse gas inventories. Prepared by the national greenhouse gas inventories programme. In: Eggleston HS, Buendia L, Miwa K, Ngara T and Tanabe K (Eds), Institute for Global Environmental Strategies. IPCC, Hayama, Japan.
23. NCATS (1994) Chinese organic fertilizer handbook. National Center for Agricultural Technology Service, Chinese Agricultural Publisher. (In Chinese).
24. Miao GY, Zhang YT, Jun Y, Hou YS, Pan XL (1989) A study on the development of root system in winter under inirrigated conditions in semi-arid Loss Plateau. Acta Agronomica Sina 15: 104–115. (In Chinese).
25. Lin WG, Wu JJ, Dong DJ, Zhong P, Wang JS, et al. (2012) Impact of different residue retention system on soybean root distribution in soil profile. Soybean Science 31: 584–588. (In Chinese).
26. Zhou ZC, Gan ZT, Shangguan ZP, Zhang FP (2013) Effects of long-term repeated mineral and organic fertilizer applications on soil organic carbon and total nitrogen in a semi-arid cropland. European Journal of Agronomy 45: 20–26.
27. Six J, Conant RT, Paul EA, Paustian K (2002) Stabilization mechanisms of soil organic matter: implications for C-saturation of soils. Plant Soil 241: 155–176.
28. West TO, Six J (2007) Considering the influence of sequestration duration and carbon saturation on estimates of soil carbon capacity. Climatic Change 80: 25–41.
29. West TO, Post WM (2002) Soil organic carbon sequestration rates by tillage and crop rotation: a global analysis. Soil Science Society of America Journal 66: 1930–1946.
30. Rui WY, Zhang WJ (2010) Effect size and duration of recommended management practices on carbon sequestration in paddy field in Yangtze Delta Plain of China: A meta-analysis. Agriculture, Ecosystems and Environment 135: 199–205.
31. Houghton JT, MeiraFilho LG, Lim B, Tréanton K, Mamaty I, et al. (eds) (1997) Revised 1996 IPCC Guidelines for National Greenhouse Gas Inventories, volumes 1–3. Hadley Centre Meteorological Office, United Kingdom.
32. Gulde S, Chung H, Amelung W, Chang C, Six J (2008) Soil carbon saturation controls labile and stable carbon pool dynamics. Soil Science Society of America Journal 72: 605–612.
33. Stewart CE, Paustian K, Conant RT, Plante AF, Six J (2009) Soil carbon saturation: Implications for measurable carbon pool dynamics in long-term incubations. Soil Biology and Biochemistry 41: 357–366.
34. Chung H, Kathie JN, Plantec A, Six J (2009) Evidence for carbon saturation in a highly structured and organic-matter-rich. Soil Science Society of America Journal 74: 130–138.
35. Kong AY, Six J, Bryant DC, Denison RF, Kessel CA (2005) The relationship between carbon input, aggregation, and soil organic carbon stabilization in sustainable cropping systems. Soil Science Society of America Journal 69: 1078–1085.
36. Cai ZC, Qin SW (2006) Dynamics of crop yields and soil organic carbon in a long-term fertilization experiment in the Huang-Huai-Hai Plain of China. Geoderma 136: 708–715.
37. Yang XM, Zhang XP, Fang HJ, Zhu P, Ren J, et al. (2003) Long-term effects of fertilization on soil organic carbon changes in continuous corn of northeast China: RothC model simulations. Environ Management 32: 459–465.
38. Stewart CE, Paustian K, Conant RT, Plante AF, Six J (2008b) Soil C saturation: evaluation and corroboration by long-term incubations. Soil Biology and Biochemistry 40: 1741–1750.
39. Feller C, Beare MH (1997) Physical control of soil organic matter dynamics in the tropics. Geoderma 79: 69–116.
40. Follett RF, Paul EA, Pruessner EG (2007) Soil carbon dynamics during a long-term incubation study involving ^{13}C and ^{14}C measurements. Soil Science 172: 189–208.
41. Hassink J (1997) The capacity of soils to preserve organic C and N by their association with clay and silt particles. Plant Soil 191: 77–87.
42. Guggenberger G, Christensen BT, Zech W (1994) Land-use effects on the composition of organic matter in particle-size separates of soil: 1. Lignin and carbohydrate signature. European Journal of Soil Science 45: 449–458.
43. Franzluebbers AJ, Haney RL, Hons FM, Zuberer DA (1996) Active fractions of organic matter in soils with different texture. Soil Biology and Biochemistry 28: 1367–1372.
44. Shi XZ, Wang HJ, Yu DS, Weindorf DC, Cheng XF, et al. (2009) Potential for soil carbon sequestration of eroded areas in subtropical China. Soil and Tillage Research 105: 322–327.
45. Lin XX, Wen QX (1987) Utilization of organic wastes in the Taihu region of Jiangsu Province of China. Resources and Conservation 13: 109–116.
46. Galantini JA, Rosell RA, Andriulo A, Miglierina A, Iglesias J (1992) Humification and nitrogen mineralization of crop residues in semi-arid Argentina. Science of the Total Environment 118: 263–270.
47. Maillard E, Angers DA (2014) Animal manure application and soil organic carbon stocks: a meta-analysis. Global Change Biology 20: 666–679.
48. Torres IF, Bastida F, Hernández T, Bombach P, Richnow HH, et al. (2014) The role of lignin and cellulose in the carbon-cycling of degraded soils under semiarid climate and their relation to microbial biomass. Soil Biology and Biochemistry 75: 152–160.
49. Higuchi T (1971) Formation and biological degradation of lignins. Advances in Enzymology and Related Areas of Molecular Biology 34: 207–283.

50. Thévenot M, Dignac MF, Rumpel C (2010) Fate of lignins in soils: a review. Soil Biology and Biochemistry 42: 1200–1211.
51. Ghosh PK, Ramesh P, Bandyopadhyay KK, Tripathi AK, Hati KM, et al. (2004) Comparative effectiveness of cattle manure, poultry manure, phosphor compost and fertilizer-NPK on three cropping systems in Vertisols of semi-arid tropics. I. Crop yields and system performance, Bioresource Technology 95: 77–83.

19

Genome Re-Sequencing of Semi-Wild Soybean Reveals a Complex *Soja* Population Structure and Deep Introgression

Jie Qiu[1]☯, Yu Wang[1]☯, Sanling Wu[1], Ying-Ying Wang[1], Chu-Yu Ye[1], Xuefei Bai[1], Zefeng Li[1], Chenghai Yan[1], Weidi Wang[1], Ziqiang Wang[1], Qingyao Shu[2], Jiahua Xie[3], Suk-Ha Lee[4], Longjiang Fan[1]*

1 Department of Agronomy & James D. Watson Institute of Genome Sciences, Zhejiang University, Hangzhou, China, **2** Institute of Nuclear Agricultural Science, Zhejiang University, Hangzhou, China, **3** Department of Pharmaceutical Sciences, North Carolina Central University, Durham, North Carolina, United States of America, **4** Department of Plant Science and Research Institute for Agriculture and Life Sciences, Seoul National University, Seoul, Korea

Abstract

Semi-wild soybean is a unique type of soybean that retains both wild and domesticated characteristics, which provides an important intermediate type for understanding the evolution of the subgenus *Soja* population in the *Glycine* genus. In this study, a semi-wild soybean line (Maliaodou) and a wild line (Lanxi 1) collected from the lower Yangtze regions were deeply sequenced while nine other semi-wild lines were sequenced to a 3-fold genome coverage. Sequence analysis revealed that (1) no independent phylogenetic branch covering all 10 semi-wild lines was observed in the *Soja* phylogenetic tree; (2) besides two distinct subpopulations of wild and cultivated soybean in the *Soja* population structure, all semi-wild lines were mixed with some wild lines into a subpopulation rather than an independent one or an intermediate transition type of soybean domestication; (3) high heterozygous rates (0.19–0.49) were observed in several semi-wild lines; and (4) over 100 putative selective regions were identified by selective sweep analysis, including those related to the development of seed size. Our results suggested a hybridization origin for the semi-wild soybean, which makes a complex *Soja* population structure.

Editor: Matthew E. Hudson, University of Illinois, United States of America

Funding: This work was supported by the National Basic Research Program of China (2011CB109306) (to L.F.), the Next Generation BioGreen 21 Program (PJ008117), Rural Development Administration, Republic of Korea (to S.-H.L.) and National Science Foundation of China (31301046) and China Postdoctoral Science Foundation (2013M531467) (to Y.W.). The funders had no role in study design, data collection and analysis, decision to publish, or preparation of the manuscript.

Competing Interests: The authors have declared that no competing interests exist.

* Email: fanlj@zju.edu.cn

☯ These authors contributed equally to this work.

Introduction

The genus *Glycine* has two subgenera: *Glycine* and *Soja*. The latter one consists of the cultivated soybean (*Glycine max*) and its progenitor wild soybean (*G. soja*). *G. max* is an important cash crop for dietary protein and oil world-wide. It is generally believed that *G. max* was domesticated from its annual wild relative *G. soja* in China around 5,000–6,000 years ago [1,2]. After domestication, *G. max* displays distinct differences in several traits from *G. soja*. For example, *G. soja* has much smaller seeds (<3.0 g per 100 seeds) and a darker seed coat, whereas *G. max* (generally >9.0 g per 100 seeds) has a yellow seed coat [3]. Besides the above two species with distinct morphological characters in the subgenus *Soja*, an intermediate type can be found in accessions of landrace soybean germplasm collections or wild line collections in China [4–6]. For example, an intermediate type known as *G. gracilis* has been described as a semi-wild soybean, which usually has an intermediate seed weight (>3.0 g per 100 seeds) with a dark seed coat [6,7]. In China, this form is also a popular type of soybean cultivated in the northeast and the Yangtze regions, because it usually has a very high seed germination rate and a short growth period as well as a robust adaptability to various environments [4,8]. For example, Maliaodou (refers to "beans for horse fodder" in Chinese) is a popular growing semi-wild soybean in the Yangtze regions. In the modern Chinese Soybean Breeds [9], Maliaodou (No. 495) was listed as a landrace in the Jinhua, Zhejiang province. It has a dark seed coat with a mud film, which in general is a typical phenotype for wild species (Fig. 1). However, it possesses an erected plant architecture with a main stem and large leaves, which are similar to cultivated soybean lines. In brief, the intermediate type, i.e. semi-wild soybean, provides an important population for the subgenus *Soja* and may be beneficial in shedding light on its origin and the process of soybean domestication.

As a new species, *G. gracilis* was first proposed by Skvortzow [10]. However, its origin has been a subject of intense debate. Some early studies denied its novelty as a new species and proposed that it be incorporated into *G. max* [11]. Two original

Figure 1. Phenotypes of cultivated, semi-wild and wild soybeans used in the present study. Seed weight (per 100 seeds) and plant architecture are shown.

views have been projected for *G. gracilis*: an intermediate evolutionary type between *G. soja* and *G. max* [12] and a hybridization origin from *G. max* and *G. soja* [13]. The latter hypothesis was supported by several studies based on the analyses of the frequency/distribution of alleles [14] and molecular markers [15–17]. A recent observation of inter-species gene flow, which captured the natural occurrence of introgression between cultivated and wild accessions, provided further phenotypic evidence to support the hybridization hypothesis [6,18,19]. However, the origin of this semi-wild type of soybean has not clearly been established.

Gene flow between *G. max* and the wild relative *G. soja* has been observed [7,19,20]. Despite significant phenotypic differences between the two species, no reproductive isolation has been found yet. Introgressions between wild and cultivated soybeans were revealed by RFLP [15] and SSR [3,6,17,19,21–23]. It has been proven that the introgression between wild and cultivated soybean is bidirectional rather than unidirectional, i.e. from wild to cultivated soybean populations [21,24,25], and also cultivated soybean into the wild population [18,19].

The *Soja* population has been investigated using diverse molecular markers such as SSR [3,6,17,19,21–23]. Based on 111 fragments from 102 soybean genes, the genetic bottleneck associated with artificial selection in soybean was first illustrated [26]. Later, SNPs based on genome-resequencing [25,27,28] or SNP chips [23,29] were identified in the wild and cultivated soybeans, and this provided the first investigation of population structure and the estimation of loci under domestication and genetic improvement from the whole soybean genome. With the low cost and high-throughput sequencing platforms, genome re-sequencing of a representative set of semi-wild soybean accessions is now possible, therefore providing us with a unique opportunity to investigate a more comprehensive subgenus *Soja* population structure and the origin of the semi-wild soybean at the genomic level.

In this study, we sequenced 10 semi-wild soybeans with a wild line and then analyzed our sequence data together with other available genomic data from wild and cultivated soybean lines. Our results demonstrate the hybridization origin of semi-wild soybean and a high rate of genetic introgression among the members of the subgenus *Soja*, which resulted in a mixed population structure including wild, semi-wild and cultivated soybean.

Materials and Methods

Plant materials

The semi-wild soybean Maliaodou and wild soybean (*G. soja*) Lanxi 1 were collected from the Jinhua basin in Zhejiang province, China (N29°04′, E119°38′). Other semi-wild soybeans were collected from wide geographical locations in China and kindly provided by Institute of Crop Sciences, Chinese Academy of Agricultural Sciences (CAAS) (Table S1). No specific permissions were required for the location, and the study did not involve endangered or protected species.

Genomic sequencing

Green leaves from a single plant of each accession were used for DNA extraction following previously described protocol [30]. Two libraries with 500 bp and 2 kb insertion sizes for Lanxi 1 and one library with a 500 bp insertion size for each of other lines were generated for Illumina Hiseq2000 sequencing platform. Paired-end (PE) reads with 100 bp were determined and a clean data set was collected from raw reads, which were pre-processed to remove adaptors to filter out low quality reads ($\geq$50% of its nucleotides with <Q20). E-corrections were performed with the program "Correction" to reduce the low frequent K-mer for better assembly [31].

Genome assembly

To get the draft genome sequences of Maliaodou and Lanxi1, different K-mer sizes ranging from 29-mer to 55-mer were tried to perform *de novo* assembly by SOAPdenovo v1.05 with clean paired-end reads [31]. The best assembly draft data (i.e., its contigs with the longest N50) was achieved at the 49-mer parameter for both Maliaodou and Lanxi1. Scaffold construction was performed based on the paired-end information of reads, and the gaps between the scaffolds were then closed by GapCloser v1.12 [31].

For the assembly of the Lanxi 1 chloroplast genome, the clean paired-end reads of Lanxi 1 that could be mapped to the known *G. max* chloroplast genome (NC_007942 [32]) by Bowtie2 v2.0.5 [33] were collected. The average mapping depth reached above 4000×, which is ~80 times of the whole genome mapping depth. Collected reads with low mapping depth (<250x) was filtered in our assembly effort. The remaining reads were used for *de novo* assembly using Velvet v1.2.07 [34] with 51-mer length. Considering the possibility that some reads from the chloroplast genome may be missed due to the used 250× threshold, we closed the gaps using all the clean reads by the software GapCloser v1.10 [35]. The overlap-based CAP3 [36] was utilized to merge redundant sequences for assembly refinement.

Genome annotation

Repeat regions in the assembled genome of Lanxi 1 were first identified using *de novo* methods implemented in RepeatScout v1.0.5 [37] and further masked by homology-based RepeatMasker v3.3.0 [38]. Genes were predicted using Augustus v2.5.5 [39] with Arabidopsis as a species parameter. Functions of the predicted genes were annotated by the BLASTP search against nr [40] and the annotated gene set of Williams 82 genome (ftp://ftp.jgi-psf.org/pub/compgen/phytozome/v8.0/Gmax_v1.0/annotation/Gmax_109_peptide.fa.gz) [41] with e-value<1e-5 as a threshold value.

SNP calling and structural variations

The clean reads of each soybean line were aligned to the Williams 82 reference genome (ftp://ftp.jgi-psf.org/pub/compgen/phytozome/v8.0/Gmax_v1.0/assembly/Gmax_109.fa.gz) using Bowtie2 v2.0.5 with default settings [33]. Consecutive steps were applied for SNPs detection. Samtools v0.1.18 [42] was used for SNP calling with the parameter –C = 50, which aims to reduce the effect of reads with excessive mismatches. In order to avoid paralogue inference, –q = 1 as threshold was used to filter reads that could align to multiple regions. Besides, to alleviate the false positive calling result due to relative low sequencing depth of some semi-wild soybean lines, we combined the variant calling information of all the 43 soybean accessions (31 by Lam *et al.* [27], one by Kim *et al.* [46] and 11 by this study), and selected SNPs present in at least two accessions or present in only one accession but with above 10 reads supported. The bam file produced from the mapping procedure was further analyzed for structural variations detection by BreakDancer v1.1 [43] with default parameters. Structural variations were displayed using Circos v0.62 [44].

Genetic diversity estimation

The average pairwise divergence (π, [45]) within a population was estimated for the wild, semi-wild and cultivated soybean populations. An in-house custom PERL script was applied for the estimation. Based on SNP calling results of each line, all variant sites across the whole genome were identified for each population and the number of nucleotide substitutions per site was estimated. The whole length of the reference genome was taken as a total number of nucleotides for π estimation.

Identification of cultivated and wild soybean-specific sequences

Nineteen wild lines (17 by Lam *et al.* [27] and two by Kim *et al.* [46] and this study) and 15 cultivars (14 by Lam *et al.* [27] and one Williams 82 by Schmutz *et al.* [41]) were used to identify the cultivated and wild soybean-specific sequences. Any sequences in the reference genome Williams 82 (ftp://ftp.jgi-psf.org/pub/compgen/phytozome/v8.0/Gmax_v1.0/assembly/Gmax_109.fa.gz), which could not be mapped by any reads from Lanxi 1 and other wild lines, were identified and defined as cultivated-specific sequences. For wild-specific sequences, the reads from the wild line Lanxi 1, which were not mapped into Williams 82, were collected and assembled by *De novo* into a 10 Mb set using Velvet v1.2.07 [34] with optimal K-mer size of 43 bp. The set was further mapped by the reads from the other 14 cultivars (Lam *et al.* [27]). Those sequences that could not be mapped by any reads from the 14 cultivated lines were considered as wild-specific sequences. Moreover, for identification of wild/cultivated common sequences, the genomic regions mapped by the all 14 cultivated lines but were not wild Lanxi 1 were considered as to be common sequences of cultivated soybeans, while the common-mapped regions in the *de novo* assembly of Lanxi 1 (as a wild reference genome) by the 18 wild lines, but not the Williams 82, were selected as common sequences of wild soybeans. In the process of identification specific or common regions, BWA v0.6.1 [54] was applied for mapping with default parameters. Mapping coverage and location information was obtained from the Samtools v0.1.18 using 'Samtools depth' parameter. The unmapped regions with at least 100 bp in length were extracted by in-house PERL scripts.

Phylogenetic tree and population structure

To collect a solid SNP dataset for constructing a *Soja* phylogenetic tree, the loci with a minimum coverage of three bases for each line across the genome were first selected, and a total number of 3,794,973 loci were found in all soybean lines. Heterozygous SNPs and the adjacent SNPs within 50 bp detected in each individual were further excluded for further genotyping. Finally, a total of 7,424 SNPs were used to construct the phylogenetic tree and population structure. The neighbor-joining tree was constructed by MEGA5.2 [47] with bootstrap support (1,000 replicates). The program STRUCTURE v2.3.4 [48] was applied for population structure; and the length of burn-in period was set to 100,000 with 100,000 MCMC reps afterwards. The number of genetic clusters was assigned using the Delta*K*-method described by Evanno *et al.* [49]. The number of clusters (*K*) was tested ranging from 2 to 9 with 8 replicates per *K*.

Selection analysis

To detect signals of recent selection, a method based on reduced pooled heterozygosity [50,51] was used. Considering the unbalanced sequencing data, in order to alleviate bias we preprocessed the sequences from each line by randomly choosing the fastq sequences to make all lines with 3~4× sequencing depth. Three subgroup pools (wild, semi-wild and cultivated) with pretreated sequence data from each line were created. By taking the genome of Williams 82 as a reference, variable sites in each pool were identified with the coverage ranging from 10 to 500. Reference and variant allele counts (n_{ref} and n_{alt}) at identified SNP positions from each pool were used to identify selections in 100 kb sliding windows with a step size of 50 kb. For a 100 kb window along the reference genome, the pooled heterozygous (H_p) was calculated by the formula: $H_p = 2\sum n_{ref} \sum n_{alt} / (\sum n_{ref} + \sum n_{alt})^2$, where $\sum n_{ref}$ and $\sum n_{alt}$ are the sums of n_{ref} and n_{alt} for all SNPs in the window. *Z* transformation ($ZH_p = (H_p - \mu H_p)/\sigma H_p$) was then applied to locate the putatively selected regions from the extreme tails by a threshold of 4 standard deviations as previously used [50,51].

GO enrichment analysis

The genes located in the selected regions were extracted from the soybean gene annotation file in Phytozome (ftp://ftp.jgi-psf.org/pub/compgen/phytozome/v8.0/Gmax_v1.0/annotation/Gmax_109_gene.gff3.gz), and the GO enrichment study conducted through AgriGO and '*Glycine max*' was set as the species setting (http://bioinfo.cau.edu.cn/agriGO/) [52]. The p-value and FDR criteria for the considered enrichment GO terms were <0.0001 and <0.05, respectively.

Experimental validation

To confirm the genomic variations in Maliaodou, 25 variations (10 SNPs and 15 indels) were selected for experimental validation. Primers were designed based on the flanking sequences of the variation sites using the on-line software Primer3 (http://frodo.wi.mit.edu/) (Table S2). The PCR reaction was carried out as described previously [53]. The PCR amplification products were checked by electrophoresis on 1.0% (w/v) agarose, then purified and sequenced directly or through clone sequencing using the

Sanger platform (ABI 3737XL) (Sunny, Shanghai). In the clone sequencing effort, at least five clones were sequenced for each sample.

Results

Genome re-sequencing

Most of the semi-wild soybeans possess an erected plant architecture with a main stem and/or large leaves, which are similar to cultivated soybean lines (Maliaodou is shown in Fig. 1). The ten semi-wild soybeans have an intermediate seed size (4.34–12.32 g), which is between the wild line (e.g. Lanxi 1, 1.22±0.04 g) and the cultivated line (e.g. Williams 82, 22.44±1.28 g) (Fig. 1; Table S1).

All semi-wild soybeans were sequenced into a ~3-fold genome coverage by high-throughput sequencing; Maliaodou was further sequenced into a 41-fold coverage (Table S3). In order to obtain a reference for the wild soybean genome for comparative analysis, a wild line Lanxi 1 was collected from the same location as Maliaodou and was sequenced into a 55-fold genome coverage by two libraries with different insertion sizes. After *de novo* assembly, a 929.9 Mb genome with contig/scaffold N50 sizes of 21.7/51.0 kb was obtained and used as a wild-type reference in the following studies. A total of 56,298 genes were further annotated, of which 48,240 were able to find their orthologs (hits) in the cultivated reference genome Williams 82 (Table S3). For the assembly of the Lanxi 1 chloroplast genome, we finally obtained complete chloroplast genome of Lanxi 1 with length of 152,199 bp, which harbors 139 protein coding genes. The assembly information is available in GenBank under the accession number KC779227.

Genetic diversity

Variation calling for 43 soybean accessions was carried out using the Williams 82 genome as a reference (Table S4ab). After SNP calling by Samtools v0.1.18 [42] and further steps to alleviate false positive calling by combining the variant calling information of all 43 soybean accessions, a total number of 7,704,637 SNPs were identified. Summary of shared (supported by at least two accessions) and unique variations (supported by at least 10 reads) of each accession is listed in Table S4a. In general, relatively lower variations were observed in the semi-wild soybeans than in the wild lines. The average variation of semi-wild lines had 910,373 SNPs and 38,258/32,907 insertion/deletions. Based on the available genomic data of 19 wild soybean genomes from previous studies [27,46] and the current study (Lanxi 1), the wild lines had an average of 1,628,253 SNPs and 110,181/94,292 insertion/deletions (Table S4b). For Maliaodou, 1,587,320 SNPs and 111,399/92,112 small insertions/deletions (indels) (<5 nt) were observed compared to the reference (Fig. 2). About 14,157 large regions (>1 kb) or 27.8 Mb cumulated sequences (about 2.8% of the reference genome) in the reference genome were not mapped by any reads of Maliaodou (Table S5a). Among 14,157 regions, 257 were over 10 kb in size (the largest one with 83.5 kb); they are shown in Fig. 2. The results were similar to those obtained by another approach based on information of paired-ends reads using BreakDancer [43], which detected 2,983 deletion events with 8.7 Mb cumulated sequences (score >60). Meanwhile, limited inversion (total number of 60) and translocation (160 and 419 intra- and inter-chromosomal) events were observed in the semi-wild (Table S5b). We also measured genetic divergence of wild, semi-wild and cultivated soybeans using the parameter π [45]. The π value was 1.416×10^{-3} for semi-wild, which is intermediate between the wild (2.173×10^{-3}) and cultivated soybeans (1.332×10^{-3}).

Unusual high heterozygous rates (0.19–0.49) were observed in several semi-wild soybeans (Table S4b). For example, a heterozygous rate of 0.49 was observed among the 1,587,320 SNPs in Maliaodou. Using the same pipeline, Lanxi 1 and most other wild lines were found to have lower heterozygous rates (<0.05) except for three lines with 0.12–0.15 (Table S4b). Similar higher heterozygous rates (>0.35) were also observed for indels in Maliaodou but not in any wild lines.

To confirm the variations, especially the unexpected high heterozygosity of Maliaodou, an independent SNP calling a different mapping algorithm implemented in BWA v0.6.1 [54] was carried out for Maliaodou, which is less sensitive but more specific than Bowtie2 v2.0.5 [33]. The results showed 0.50, 0.35, and 0.36 heterozygous rates for SNP, insertion and deletion, respectively (Table S4b). When 25 variations (10 SNPs and 15 indels) were selected for experimental validation by traditional Sanger sequencing, 21 (84%) were confirmed (Table S2). Three indels caused by repetitive sequences were not validated, perhaps due to sequencing errors [55]. The high heterozygous rates were also consistent with our fragmented assembly result of the Maliaodou genome (contig N50 size = 0.5 kb). Previous studies have indicated that the heterozygous rate is one of major disruptive factors in the algorithm for genome assembly [56].

Phylogenetic relationship and population structure

A phylogenetic relationship of the 30 re-sequencing wild and cultivated soybean genomes has been analyzed. One of the cultivated soybean accessions (C19) was excluded due to its high admixture [27]. Based on their phylogenetic structure, 10 semi-wild soybean lines from this study and two additional wild soybeans (Lanxi 1 in this study and a line from South Korea by Kim *et al.* [46]) were added into the tree (Fig. 3a). The semi-wild soybeans were not grouped into an independent branch but scattered into wild subgroups.

Seed size is a key target trait of soybean domestication as cultivated soybeans usually have a bigger seed size. According to the phylogenetic tree (Fig. 3a), species with a large seed size (cultivated lines in green) are generally separated from those with a smaller seed size (wild lines in blue). However, in the mixed branch with semi-wild soybeans (in red), their seed weight seems to not be consistent with their evolutionary relationship (Fig. S1). The result indicated that seed size, the most important trait under domestication selection, could hardly be used as a single factor to estimate soybean divergence, especially for those in the intermediate evolutionary stage. This is consistent with the fact that seed size is merely one of the target traits during soybean domestication and is significantly affected by the environmental conditions [57].

Bayesian clustering (STRUCTURE) of soybean accessions was further carried out. Using the method described by Evanno *et al.* [49], ΔK reached a peak when K was set to 3 (Table S6), indicating that $K = 3$ is optimal for the population structure. When we ordered the accessions based on their genetic background, they could be divided into three subgroups: a cultivated, wild, and mixed subgroup (including both wild and semi-wild soybeans) (Fig. 3b). These results are consistent with our phylogenetic results (Fig. 3a). Similarly, semi-wild soybeans do not have an independent subpopulation and are mainly grouped with wild soybeans.

Selective sweep analysis

Crops usually experience two stages (domestication with subsequently genetic improvement) during the evolutionary

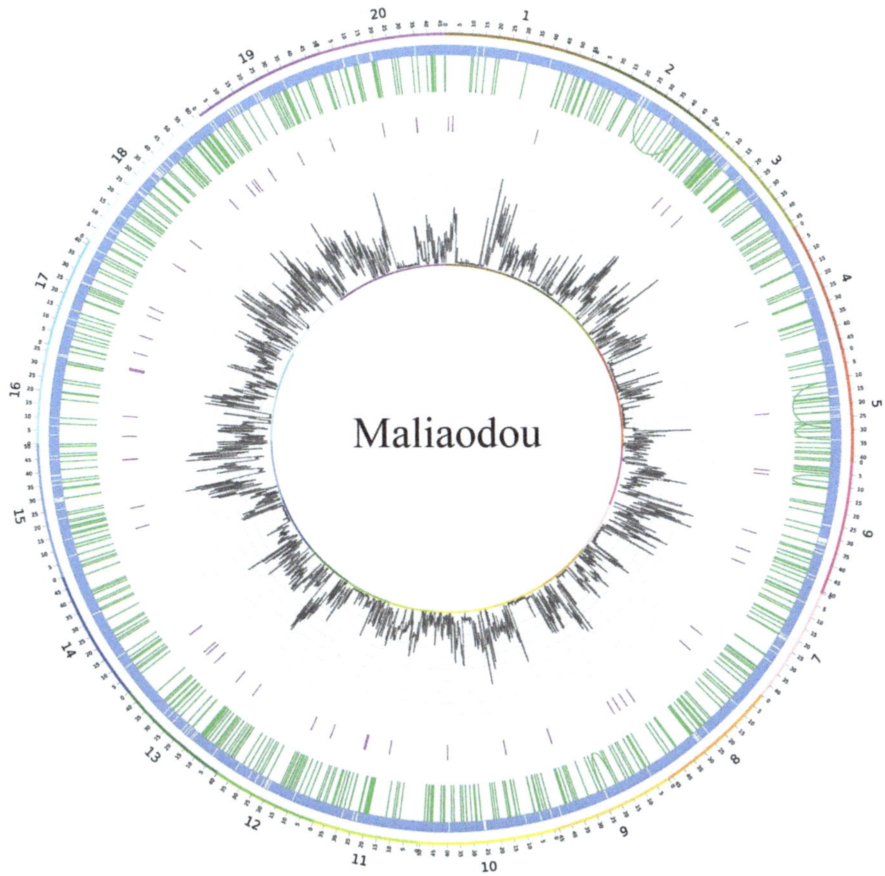

Figure 2. Genomic variations between the semi-wild genome Maliaodou and the reference genome Williams 82. Circles from outside: 20 chromosomes labeled with different colors; in the blue bars, each white vertical line presents a >10 kb un-mapped region in the reference genome; green and purple vertical lines present the intra-chromosomal translocations and inversion events, respectively. Distribution of the SNP density of the Maliaodou is labeled with grey lines.

process from wild-type, and semi-wild is a transitional point between these two stages [58]. To detect signals of artificial selection, we searched the wild, semi-wild, and cultivated soybean genome for selection regions with reduced pooled heterozygosity (H_p). Using autosomal 100 kb as the scanning window suggested by Axelsson *et al.* [51] and Chung *et al.* [28], several regions in cultivated and semi-wild soybean populations had extremely low heterozygosity rates (Fig. 4; Table S7a). By the threshold of at least four standard deviations away from the mean ($Z(H_p)<-4$), a total of 102 and 144 putatively selected regions could be detected in the semi-wild and cultivated populations, respectively. It is reasonable to believe that more selected loci could have been identified in the latter because cultivated population experience additional artificial selection in recent genetic improvements [26].

In the 144 selected loci, only 101 loci were detected in the cultivated soybeans, but not in the wild or semi-wild soybean were $Z(H_p)<-2$ was as a threshold, suggesting that they are the putative target loci under recent genetic improvements. The 101 loci include 282 protein-coding genes (Table S7bc), which could play a vital role in agronomy-related traits. We performed GO enrichment analysis on the 282 genes and identified 9 gene ontology terms with FDR (false discovery rate) <0.05 (Table S8). Interestingly, the GO categories were enriched in embryonic and embryo sac development as well as megagametogenesis, which are involved in the development of seed size. The variants within these genes may lay a genetic foundation to the seed size evolution.

Discussion

A mixed semi-wild group complicated population structure of the subgenus *Soja*

It is believed that the cultivated soybean (*G. max*) was domesticated from the wild *G. soja* [1,2]. Our results demonstrate a complex transitional stage from wild to cultivated soybean: a mixed population including a series of semi-wild types, which suggests no distinct or independent transitional period for soybean domestication. The complex evolutionary situation might be due to the hybridization events between domesticated and wild soybeans, which frequently occurred in natural field conditions (see next section for details). Many crops have had significant reproductive isolation from their wild progenitors after domestication, which made their phylogenetic relationship clearly (such as rice [59]). Our results are consistent with the observation by Wang *et al.* [6] who clearly revealed that there is no existence of a transitional intermediate ancestor between the wild and cultivated soybean. Furthermore, our results do not support the semi-wild soybean as an independent species (*G. gracilis*). Apparently, the mixture of semi-wild and wild accessions in the phylogenetic tree not as an independent sub-branch (Fig. 3a) strongly suggests that the semi-wild soybeans belong to the wild category and should be

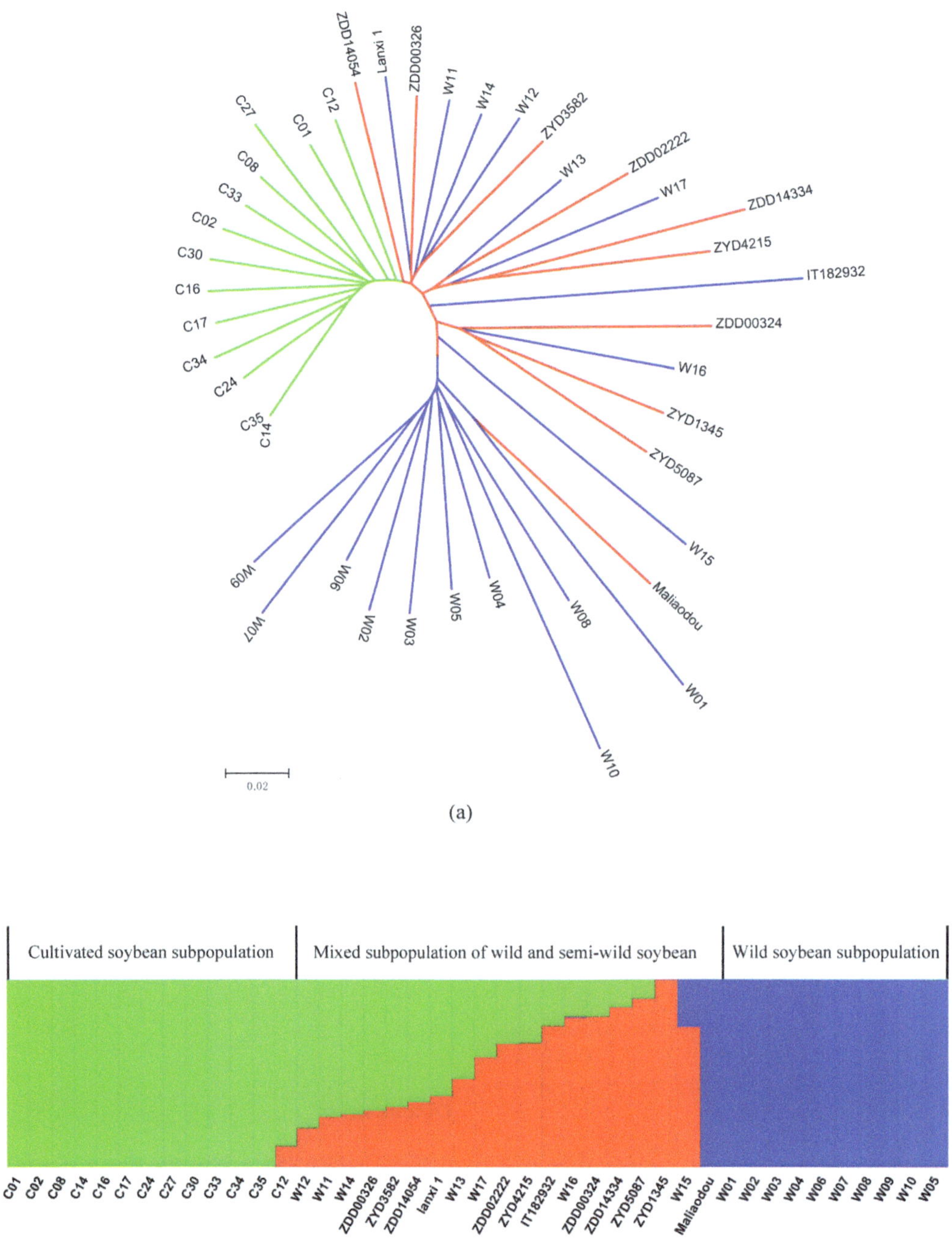

Figure 3. Phylogenetic tree and population structure of wild, semi-wild, and cultivated soybeans. (a) A neighbor-joining phylogenetic tree constructed using SNP data. Cultivated, semi-wild, and wild soybeans are labeled with green, red, and blue lines, respectively. (b) Bayesian clustering (STRUCTURE, $K = 3$) of soybean accessions were grouped based on their species. Cultivated lines were designated with 'C' as prefix while wild lines were with 'W' as prefix except IT182932 (Korean) and Lanxi 1 (this study). Semi-wild lines were titled with 'Z' as prefix except Maliaodou.

considered a variant or subclade of *G. soja*. However, in order to fully understand the *Soja* population structure, more semi-wild lines from wide geographical locations and their sequence data need to be sampled or generated in future.

The hybridization origin of the semi-wild soybean

The hybridization origin of the semi-wild soybean has been proposed [13,60]. However, direct evidence in support of this hypothesis is lacking, although hybridization events between cultivated and wild soybean have been observed in natural field conditions [6,18,19]. In this study, several genomic evidences favor the above hypothesis.

First, the semi-wild soybeans were not grouped into an independent branch in the *Soja* phylogenetic tree (Fig. 3a). Instead, they were scattered into the wild soybean subgroup. As mentioned above, our results failed to support an independent speciation and domestication of an intermediate evolutionary type between *G. soja* and *G. max* [12]. Second, the semi-wild soybean hosts both cultivated and wild soybean novel sequences. Based on the available genomic sequences of the wild lines including Lanxi 1

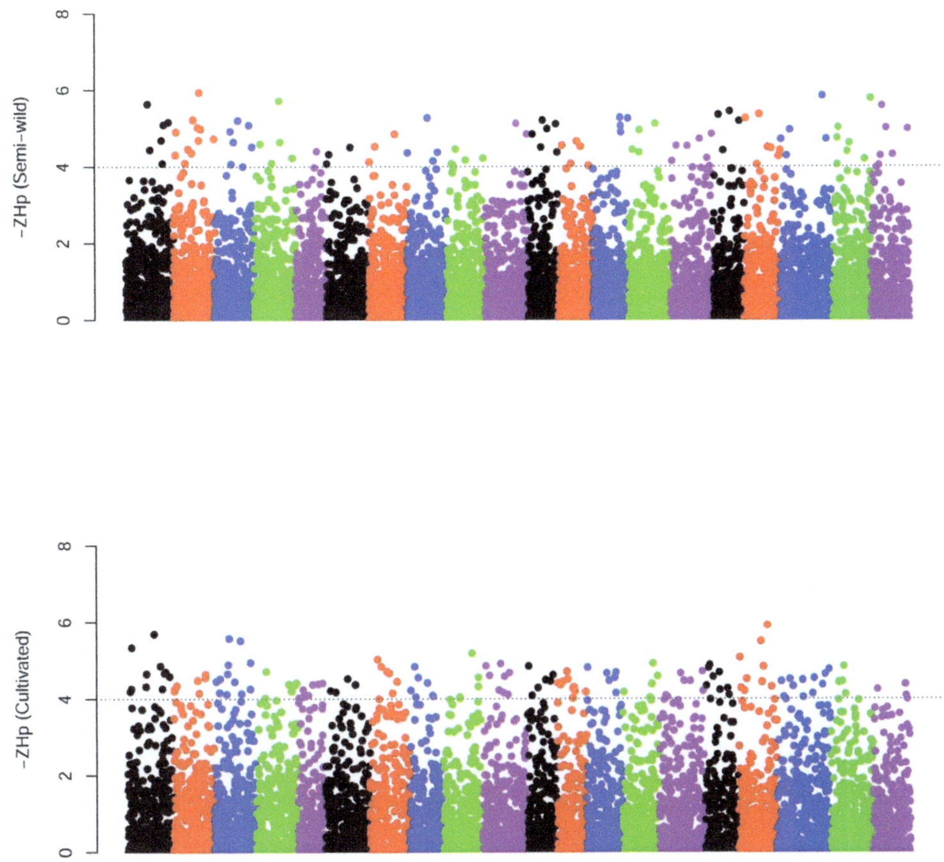

Figure 4. Summary of selective sweep analysis. Distribution of *Z*-transformed average pooled heterozygosity (H_p) in semi-wild (top) and cultivated soybeans (bottom), respectively. The negative end of the $Z(H_p)$ distribution plotted along soybean autosomes 1–20 are shown. A dashed horizontal line indicates the cut-off ($Z<-4$) used for extracting outliers.

and cultivars [27,46], we identified 3.07/0.25 Mb cultivated/wild soybean-specific sequences and 3.35/0.82 Mb cultivated/wild soybean-common sequences (for detailed definition see Methods section). Using the deeply sequenced Maliaodou as an example, more than 99% of 3.07 Mb cultivated-specific sequences were covered by the reads from the Maliaodou genome (Table S9). Meanwhile, most (>95%) of the 0.82 Mb wild-common sequences and part (>12%) of the 0.25 Mb wild-specific sequences were also covered by the Maliaodou genome. Again, these findings strongly support the above hypothesis that Maliaodou originated from the hybridization of the wild and domesticated soybean. Third, unexpectedly high heterozygous rates were observed in Maliaodou and three other semi-wild lines. Soybeans are predominantly self-pollinating and usually have low outcrossing rates of <3.0% in cultivated lines but relatively higher rates (2.4–13%) in wild lines [19,61]. This is the first time that the high heterozygous rate was observed and reported in the semi-wild soybean. Regarding the significantly higher heterozygous rate of the Maliaodou genome as compared with the 19 wild lines (*t*-test *P* value$<0.1\times10^{-3}$ for SNP and $P=0.4–0.9\times10^{-3}$ for indel), a reasonable explanation is that the semi-wild line may have a recent hybrid origin rather than an intermediate evolutionary type between *G. soja* and *G. max*.

In addition, the above hypothesis is also supported by chloroplast genome data. Xu *et al.* [17] investigated chloroplast DNA SSR in 326 wild and cultivated soybean accessions and identified 52 haplotypes. However, no cultivar-specific haplotypes were found and thus hybrid swarms between cultivated and wild forms were suggested. In this study, we collected green soybean leaves of Lanxi 1 for genome re-sequencing. This generated chloroplast genomic sequences and provided us with a wild soybean chloroplast genome (Accession number: KC779227) for comparison study. Using this and the known *G. max* chloroplast genome (NC_007942) as references, the wild and cultivated lines with available genomic data [27,46] were genotyped. No haplotype is shared by all cultivars, all wild lines or all semi-wild lines (i.e. within each group of material) used in this study (Table S10). Our results are similar to the findings by Xu and co-workers [17].

Taken together, we believe that introgressions from local wild soybeans to cultivated soybeans occurred frequently during the long domestication process of soybean, i.e. a kind of hybridization origin, which creates a complex genetic background for species in subgenus *Soja*. Several important crops are polyploidy and have originated from the hybridization of two or three ancient species, such as wheat, cotton and tobacco [62]. However, the hybridization origin of diploid crops from a domesticated line and its progenitor is rare, with only one case with genomic evidence coming from *indica* rice (*Oryza sativa*), which originated from a cross between the cultivated *japonica* rice and local wild rice [59]. For the semi-wild soybean, this should be another case of genomic evidence for the hybridization origin of diploid crops from cultivated and wild lines.

The extremely high heterozygous rates observed in Maliaodou and other semi-wild soybeans (Table S4b) suggest that their hybridization events occurred recently. This is in agreement with the fact that no record of "Maliaodou" could be found in the Chinese ancient literatures until last century (Hu & Tian 1993), although soybeans became one of the most important crops in China several thousand years ago. Wang *et al.* [18] also proposed a short creation history of semi-wild soybean based on their observation of some newly collected semi-wild lines. Additionally, the ancestry inference based on STRUCTURE result for Maliaodou indicated it is different from that of other semi-wild accessions. Based on the SNP calling result, we observed an unexpectedly high heterozygosity rate in the Maliaodou (~50%), which indicated that this semi-wild line might originate from a recent hybridization between cultivated and wild soybeans. Tracing the location in which the Maliaodou was collected, we found that many wild soybeans are also grown in the same field, and thus genetic introgressions from wild lines to Maliaodou probably frequently occurred.

Footprints of artificial selection in the soybean genome

Both natural and domestication selections target genes/loci controlling adaptive or agronomic traits and leave footprints of selection in the soybean genome. Several candidate domestication regions have been identified by recent genomic investigations of wild and cultivated soybeans [25,26,28]. Of the 282 genes located in the putative selective regions in this study (Table S7), at least 63 were also identified as putative selective genes by Li *et al.* [25] or Chung *et al.* [28] (Table S7). The gene coding for ABSCISIC ACID-INSENSITIVE 5-like protein 3 (Glyma13g03880) was detected by both of their studies as well as our study. It is one of the seed color related genes based on QTL mapping [63], and it suggests that this gene was targeted by the recent genetic improvement. Besides seed size, seed color is also a main target trait of domestication and subsequent improvement. In our study, a relatively strict criteria ($Z(H_p)<-4$) was applied, which may exclude some putative artificially selected genes. Moreover, as shown above, 14,157 large regions (>1 kb), or about 2.8% of the reference genome of cultivated Williams 82, were not mapped by any reads from the deep sequenced (~41×) semi-wild line Maliaodou (Table S5). These regions might be kept only in the cultivated soybean population during improvement. Taken together, the above-identified genes in the selective and unmapping regions in this study that may relate to the artificial selection and important agronomic traits provide some candidate targets for further functional investigation.

Supporting Information

Figure S1 A neighbor-joining phylogenetic tree of wild (blue), semi-wild (red) and cultivated soybeans (green) labelled with seed weight per 100 seeds.

Table S1 Detailed information of the Chinese accessions of subgenus *Soja* used in this study.

Table S2 Experimental validation results for the genomic variations and PCR primers used in this study.

Table S3 Global statistics of the wild soybean Lanxi 1 and semi-wild soybean genomes.

Table S4 Detailed information of genomic variations in the all soybean lines in this study relative to the reference genome Williams 82.

Table S5 Genomic structural variations between the semi-wild Maliaodou and domesticated reference genome Williams 82.

Table S6 Inference of best *K* for separating soybean subgroups using the delta *K* method.

Table S7 Position information for selective loci and genes in the selective loci in cultivated soybeans.

Table S8 GO enrichment analysis of the genes in selective regions of cultivated soybeans.

Table S9 Genomic coverage of the semi-wild soybean Maliaodou in the cultivated and wild-specific sequences of soybean.

Table S10 Chloroplast SNPs of cultivated, wild and semi-wild soybeans relative to the reference chloroplast genome (Accession no. NC_007942).

Acknowledgments

We thank Dr. Farooqahmed S Kittur (North Carolina Central University) for his critical reading of the manuscript.

Author Contributions

Conceived and designed the experiments: LF. Performed the experiments: YW. Analyzed the data: JQ YW SW CHY WW XB ZL CYY LF. Contributed reagents/materials/analysis tools: ZW. Wrote the paper: LF JQ. Contributed to the discussion: CYY ZW QS JX SHL.

References

1. Carter TE, Nelson RL, Sneller CH, Cui Z (2004) Soybean: improvement, production, and uses. Agronomy 16: 303–416.
2. Zhao TJ, Gai JY (2004) The origin and evolution of cultivated soybean (*Glycine max* (L.) Merr.) (in Chinese). Sci Agric Sin 37: 954–962.
3. Chen YW, Nelson RL (2004) Genetic variation and relationships among cultivated, wild, and semiwild soybean. Crop Sci 44: 316–325.
4. Li Y, Guan R, Liu Z, Ma Y, Wang L, et al. (2008) Genetic structure and diversity of cultivated soybean (*Glycine max* (L.) Merr.) landraces in China. Theor Appl Genet 117: 857–871.
5. Wang KJ, Li XH, Liu Y (2012) Fine-scale phylogenetic structure and major events in the history of the current wild soybean (*Glycine soja*) and taxonomic assignment of semi-wild type (*Glycine gracilis* Skvortz.) within the Chinese subgenus *Soja*. J Hered 103: 13–27.
6. Wang KJ, Li XH (2013) Genetic diversity and gene flow dynamics revealed in the rare mixed populations of wild soybean (*Glycine soja*) and semi-wild type (*Glycine gracilis*) in China. Genet Resour Crop Ev 60: 2303–2318.
7. Kuroda Y, Kaga A, Tomooka N, Vaughan D (2010) The origin and fate of morphological intermediates between wild and cultivated soybeans in their natural habitats in Japan. Mol Ecol 19: 2346–2360.
8. Wang KJ, Li FS (2000) General situation of wild soybean (*G. soja*) germplasm resources and its utilization of introgression into cultivated soybean in China. Rev Agric Sci Technol 2: 69–72.
9. Hu M, Tian P (1993) Chinese Soybean Breeds. Beijing: Agriculture Press.
10. Skvortzow BV (1927) The soybean–wild and cultivated in Eastern Asia. In: Proceedings of the Manchurian Research Society, Natural History Section Publication Series A, No. 22. Harbin, China.

11. Hermann FJ (1962) A revision of genus Glycine and its immediate allies. No. 1268. United States Department of Agriculture: Washington, DC.
12. Fukuda Y (1933) Cytogenetical studies on the wild and cultivated Manchurian soybeans. J Jpn Bot 6: 489–506.
13. Hymowitz T (1970) On the domestication of the soybean. Econ Bot 24: 408–421.
14. Broich SL, Palmer RG (1980) A cluster analysis of wild and domesticated soybean phenotypes. Euphytica 29: 23–32.
15. Abe J, Hasegawa A, Fukushi H, Mikami T, Ohara M, et al. (1999) Introgression between wild and cultivated soybeans of Japan revealed by RFLP analysis for chloroplast DNAs. Econ Bot 53: 285–291.
16. Close PS, Shoemaker RC, Keim P (1989) Distribution of restriction site polymorphism within the chloroplast genome of the genus *Glycine*, Subgenus *Soja*. Theor Appl Genet 77: 768–776.
17. Xu H, Abe J, Gai Y, Shimamoto Y (2002) Diversity of chloroplast DNA SSRs in wild and cultivated soybeans: evidence for multiple origins of cultivated soybean. Theor Appl Genet 105: 645–653.
18. Wang KJ, Li XH, Zhang JJ Chen H, Zhang ZL, et al. (2010) Natural introgression from cultivated soybean (*Glycine max*) into wild soybean (*Glycine soja*) with the implications for origin of populations of semi-wild type and for biosafety of wild species in China. Genet Res Crop Evol 57: 747–761.
19. Wang KJ, Li XH (2011) Interspecific gene flow and the origin of semi-wild soybean revealed by capturing the natural occurrence of introgression between wild and cultivated soybean populations. Plant Breed 130: 117–127.
20. Nakayama Y, Yamaguchi H (2002) Natural hybridization in wild soybean (*Glycine max* ssp. *soja*) by pollen flow from cultivated soybean (*Glycine max* ssp. *max*) in a designed population. Weed Biol Manag 2: 25–30.
21. Concibido VC, La Vallee B, McLaird P, Pineda N, Meyer J, et al. (2003) Introgression of a quantitative trait locus for yield from *Glycine soja* into commercial soybean cultivars. Theor Appl Genet 106: 575–582.
22. Wen Z, Ding Y, Zhao T, Gai J (2009) Genetic diversity and peculiarity of annual wild soybean (*G. soja* Sieb. et Zucc.) from various eco-regions in China. Theor Appl Genet 119: 371–381.
23. Li YH, Li W, Zhang C, Yang L, Chang RZ, et al. (2010) Genetic diversity in domesticated soybean (*Glycine max*) and its wild progenitor (*Glycine soja*) for simple sequence repeat and single-nucleotide polymorphism loci. New Phytol 188: 242–253.
24. Leroy AR, Fehr WR, Cianzio SR (1991) Introgression of genes for small seed size from *Glycine soja* into *G. max*. Crop Sci 31: 693–697.
25. Li YH, Zhao SC, Ma JX, Li D, Yan L, et al. (2013) Molecular footprints of domestication and improvement in soybean revealed by whole genome re-sequencing. BMC Genomics 14: 579.
26. Hyten DL, Song Q, Zhu Y, Choi IY, Nelson RL, et al. (2006) Impacts of genetic bottlenecks on soybean genome diversity. Proc Natl Acad Sci U S A 103: 16666–16671.
27. Lam HM, Xu X, Liu X, Chen W, Yang G, et al. (2010) Resequencing of 31 wild and cultivated soybean genomes identifies patterns of genetic diversity and selection. Nat Genet 42: 1053–1059.
28. Chung WH, Jeong N, Kim J, Lee WK, Lee YG, et al. (2013) Population Structure and Domestication Revealed by High-Depth Resequencing of Korean Cultivated and Wild Soybean Genomes. DNA Res.
29. Hao D, Cheng H, Yin Z, Cui S, Zhang D, et al. (2012) Identification of single nucleotide polymorphisms and haplotypes associated with yield and yield components in soybean (*Glycine max*) landraces across multiple environments. Theor Appl Genet 124: 447–458.
30. Sambrook J, Russell DW (2001) Molecular cloning: a laboratory manual. 3rd edn. New York: Cold Spring Harbor Laboratory Press.
31. Li R, Zhu H, Ruan J, Qian W, Fang X, et al. (2010) *De novo* assembly of human genomes with massively parallel short read sequencing. Genome Res 20: 265–272.
32. Saski C, Lee SB, Daniell H, Wood TC, Tomkins J, et al. (2005) Complete chloroplast genome sequence of *Glycine max* and comparative analyses with other legume genomes. Plant Mol Biol 59: 309–322.
33. Langmead B, Salzberg SL (2012) Fast gapped-read alignment with Bowtie 2. Nat Methods 9: 357–359.
34. Zerbino DR, Birney E (2008) Velvet: algorithms for *De novo* short read assembly using de Bruijn graphs. Genome Res 18: 821–829.
35. Boetzer M, Pirovano W (2012) Toward almost closed genomes with GapFiller. Genome Biol 13: R56.
36. Huang X, Madan A (1999) CAP3: A DNA sequence assembly program. Genome Res 9: 868–877.
37. Price AL, Jones NC, Pevzner PA (2005) *De novo* identification of repeat families in large genomes. Bioinformatics 21 Suppl 1: i351–358.
38. Smit AF (1996) The origin of interspersed repeats in the human genome. Curr Opin Genet Dev 6: 743–748.
39. Stanke M, Tzvetkova A, Morgenstern B (2006) AUGUSTUS at EGASP: using EST, protein and genomic alignments for improved gene prediction in the human genome. Genome Biol 7 Suppl 1: S11 11–18.
40. Coordinators NR (2013) Database resources of the National Center for Biotechnology Information. Nucleic Acids Res 41: D8–D20.
41. Schmutz J, Cannon SB, Schlueter J, Ma J, Mitros T, et al. (2010) Genome sequence of the palaeopolyploid soybean. Nature 463: 178–183.
42. Li H, Handsaker B, Wysoker A, Fennell T, Ruan J, et al. (2009) The Sequence Alignment/Map format and SAMtools. Bioinformatics 25: 2078–2079.
43. Chen K, Wallis JW, McLellan MD, Larson DE, Kalicki JM, et al. (2009) BreakDancer: an algorithm for high-resolution mapping of genomic structural variation. Nat Methods 6: 677–681.
44. Krzywinski M, Schein J, Birol I, Connors J, Gascoyne R, et al. (2009) Circos: an information aesthetic for comparative genomics. Genome Res 19: 1639–1645.
45. Nei M, Miller JC (1990) A simple method for estimating average number of nucleotide substitutions within and between populations from restriction data. Genetics 125: 873–879.
46. Kim MY, Lee S, Van K, Kim TH, Jeong SC, et al. (2010) Whole-genome sequencing and intensive analysis of the undomesticated soybean (*Glycine soja* Sieb. and Zucc.) genome. Proc Natl Acad Sci U S A 107: 22032–22037.
47. Tamura K, Peterson D, Peterson N, Stecher G, Nei M, et al. (2011) MEGA5: molecular evolutionary genetics analysis using maximum likelihood, evolutionary distance, and maximum parsimony methods. Mol Biol Evol 28: 2731–2739.
48. Pritchard JK, Stephens M, Donnelly P (2000) Inference of population structure using multilocus genotype data. Genetics 155: 945–959.
49. Evanno G, Regnaut S, Goudet J (2005) Detecting the number of clusters of individuals using the software STRUCTURE: a simulation study. Mol Ecol 14: 2611–2620.
50. Rubin CJ, Zody MC, Eriksson J, Meadows JR, Sherwood E, et al. (2010) Whole-genome resequencing reveals loci under selection during chicken domestication. Nature 464: 587–591.
51. Axelsson E, Ratnakumar A, Arendt ML, Maqbool K, Webster MT, et al. (2013) The genomic signature of dog domestication reveals adaptation to a starch-rich diet. Nature 495: 360–364.
52. Du Z, Zhou X, Ling Y, Zhang Z, Su Z (2010) agriGO: a GO analysis toolkit for the agricultural community. Nucleic Acids Res 38: W64–70.
53. Mullis K, Faloona F, Scharf S, Saiki R, Horn G, Erlich H (1992) Specific enzymatic amplification of DNA *in vitro*: the polymerase chain reaction. Biotechnol Ser, 24, 17–27.
54. Li H, Durbin R (2009) Fast and accurate short read alignment with Burrows-Wheeler transform. Bioinformatics 25: 1754–1760.
55. Nakamura K, Oshima T, Morimoto T, Ikeda S, Yoshikawa H, et al. (2011) Sequence-specific error profile of Illumina sequencers. Nucleic Acids Res 39: e90.
56. Kelley DR, Salzberg SL (2010) Detection and correction of false segmental duplications caused by genome mis-assembly. Genome Biol 11: R28.
57. Kumar V, Rani A, Solanki S, Hussain SM (2006) Influence of growing environment on the biochemical composition and physical characteristics of soybean seed. J Food Compos Anal 19: 188–195.
58. Yamasaki M, Tenaillon MI, Bi IV, Schroeder SG, Sanchez-Villeda H, et al. (2005) A large-scale screen for artificial selection in maize identifies candidate agronomic loci for domestication and crop improvement. Plant Cell 17: 2859–2872.
59. Huang X, Kurata N, Wei X, Wang ZX, Wang A, et al. (2012) A map of rice genome variation reveals the origin of cultivated rice. Nature 490: 497–501.
60. Hirota T, Sayama T, Yamasaki M, Sasama H, Sugimoto T, et al. (2012) Diversity and population structure of black soybean landraces originating from Tanba and neighboring regions. Breed Sci 61: 593–601.
61. Fujita R, Ohara M, Okazaki K, Shimamoto Y (1997) The extent of natural cross-pollination in wild soybean (*Glycine soja*). J Hered 88: 124–128.
62. Mallet J (2007) Hybrid speciation. Nature 446: 279–283.
63. Yang K, Jeong N, Moon JK, Lee YH, Lee SH, et al. (2010) Genetic analysis of genes controlling natural variation of seed coat and flower colors in soybean. J Hered 101: 757–768.

PERMISSIONS

All chapters in this book were first published in PLOS ONE, by The Public Library of Science; hereby published with permission under the Creative Commons Attribution License or equivalent. Every chapter published in this book has been scrutinized by our experts. Their significance has been extensively debated. The topics covered herein carry significant findings which will fuel the growth of the discipline. They may even be implemented as practical applications or may be referred to as a beginning point for another development.

The contributors of this book come from diverse backgrounds, making this book a truly international effort. This book will bring forth new frontiers with its revolutionizing research information and detailed analysis of the nascent developments around the world.

We would like to thank all the contributing authors for lending their expertise to make the book truly unique. They have played a crucial role in the development of this book. Without their invaluable contributions this book wouldn't have been possible. They have made vital efforts to compile up to date information on the varied aspects of this subject to make this book a valuable addition to the collection of many professionals and students.

This book was conceptualized with the vision of imparting up-to-date information and advanced data in this field. To ensure the same, a matchless editorial board was set up. Every individual on the board went through rigorous rounds of assessment to prove their worth. After which they invested a large part of their time researching and compiling the most relevant data for our readers.

The editorial board has been involved in producing this book since its inception. They have spent rigorous hours researching and exploring the diverse topics which have resulted in the successful publishing of this book. They have passed on their knowledge of decades through this book. To expedite this challenging task, the publisher supported the team at every step. A small team of assistant editors was also appointed to further simplify the editing procedure and attain best results for the readers.

Apart from the editorial board, the designing team has also invested a significant amount of their time in understanding the subject and creating the most relevant covers. They scrutinized every image to scout for the most suitable representation of the subject and create an appropriate cover for the book.

The publishing team has been an ardent support to the editorial, designing and production team. Their endless efforts to recruit the best for this project, has resulted in the accomplishment of this book. They are a veteran in the field of academics and their pool of knowledge is as vast as their experience in printing. Their expertise and guidance has proved useful at every step. Their uncompromising quality standards have made this book an exceptional effort. Their encouragement from time to time has been an inspiration for everyone.

The publisher and the editorial board hope that this book will prove to be a valuable piece of knowledge for researchers, students, practitioners and scholars across the globe.

LIST OF CONTRIBUTORS

Xia Yan
Key Laboratory of Inland River Ecohydrology, Cold and Arid Regions Environmental and Engineering Research Institute, Chinese Academy of Sciences, Lanzhou, P. R. China
Key Laboratory of Stress Physiology and Ecology in Cold and Arid Regions of Gansu Province, Cold and Arid Regions Environmental and Engineering Research Institute, Chinese Academy of Sciences, Lanzhou, P. R. China

Xicun Dong
Department of Radiobiology, Institute of Modern Physics, Chinese Academy of Sciences, Lanzhou, P. R. China

Wen Zhang, Hengxia Yin and Xiao-Fei Ma
Key Laboratory of Stress Physiology and Ecology in Cold and Arid Regions of Gansu Province, Cold and Arid Regions Environmental and Engineering Research Institute, Chinese Academy of Sciences, Lanzhou, P. R. China

Honglang Xiao
Key Laboratory of Inland River Ecohydrology, Cold and Arid Regions Environmental and Engineering Research Institute, Chinese Academy of Sciences, Lanzhou, P. R. China

Peng Chen
Faculty of Plant Science and Technology, Huazhong Agricultural University, Wuhan, P. R. China

Bingjun Jiang, Tianfu Han, Youfei Gao, Yanlei Yue, Liming Ma, Shi Sun, Cunxiang Wu and Wensheng Hou
The National Key Facility for Crop Gene Resources and Genetic Improvement and MOA Key Lab of Soybean Biology (Beijing), Institute of Crop Sciences, The Chinese Academy of Agricultural Sciences, Beijing, China

Haiyang Nan and Sijia Lu
The Key Laboratory of Soybean Molecular Design Breeding, Northeast Institute of Geography and Agroecology, Chinese Academy of Sciences, Harbin, China
University of Chinese Academy of Sciences, Beijing, China

Lili Tang, Dong Cao, Jialin Wang, Xiaohui Yuan, Fanjiang Kong and Baohui Liu
The Key Laboratory of Soybean Molecular Design Breeding, Northeast Institute of Geography and Agroecology, Chinese Academy of Sciences, Harbin, China

Guo-Hong Yu, Lin-Lin Jiang, Meng-Meng Liu, Shu-Guang Shan and Xian-Guo Cheng
Key Lab. of Plant Nutrition and Fertilizer, Ministry of Agriculture, Institute of Agricultural Resources and Regional Planning, Chinese Academy of Agricultural Sciences, Beijing, China

Xue-Feng Ma
Key Lab. of Plant Nutrition and Fertilizer, Ministry of Agriculture, Institute of Agricultural Resources and Regional Planning, Chinese Academy of Agricultural Sciences, Beijing, China
Institute of Agro-Products Processing Science and Technology, Chinese Academy of Agricultural Sciences, Beijing, China

Zhao-Shi Xu
Institute of Crop Science, Chinese Academy of Agricultural Sciences (CAAS)/National Key Facility for Crop Gene Resources and Genetic Improvement, Key Laboratory of Biology and Genetic Improvement of Triticeae Crops, Ministry of Agriculture, Beijing, China

Atanu Mukherjee and Rattan Lal
Carbon Management and Sequestration Center, School of Environment and Natural Resources, The Ohio State University, Columbus, Ohio, United States of America

Satoko Ohfuji and Wakaba Fukushima
Department of Public Health, Osaka City University Faculty of Medicine, Osaka, Japan

Kenji Watanabe and Hirokazu Yamagami
Department of Gastroenterology, Osaka City University Faculty of Medicine, Osaka, Japan

Satoshi Sasaki
Department of Social and Preventive Epidemiology, School of Public Health, The University of Tokyo, Tokyo, Japan

Masakazu Nagahori and Mamoru Watanabe
Department of Gastroenterology and Hepatology, Tokyo Medical and Dental University, Tokyo, Japan

Yoshio Hirota for the Japanese Case-Control Study Group for Ulcerative Colitiś
Department of Public Health, Osaka City University Faculty of Medicine, Osaka, Japan
Clinical Epidemiology Research Center, Medical Co. LTA, Fukuoka, Japan

P. Dilip Venugopal, Peter L. Coffey, Galen P. Dively and William O. Lamp
Department of Entomology, University of Maryland, College Park, Maryland, United States of America

Marcelo Hermes-Lima
Department of Cell Biology, University of Brasília, Campus Universitário Darcy Ribeiro, Brasília, Distrito Federal, Brazil

Renata Velozo Timbó
Department of Cell Biology, University of Brasília, Campus Universita´rio Darcy Ribeiro, Brasília, Distrito Federal, Brazil
Department of Biological Control, Embrapa Genetic Resources and Biotechnology, Brasília, Distrito Federal, Brazil

Luciano Paulino Silva, Angela Mehta, Maria Carolina Blassioli Moraes and Débora Pires Paula
Department of Biological Control, Embrapa Genetic Resources and Biotechnology, Brasília, Distrito Federal, Brazil

Ling Wang, Shuai Yang, Lanlan Han, Dong Fan and Kuijun Zhao
College of Agriculture, Northeast Agricultural University, Harbin, China

Xiaobo Wang, Haowei Zhang, Yali Gao and Wenming Zhang
School of Agronomy, Anhui Agricultural University, Hefei, China

Genlou Sun
School of Agronomy, Anhui Agricultural University, Hefei, China
Biology Department, Saint Mary's University, Halifax, NS, Canada

Lijuan Qiu
National Key Facility for Gene Resources and Genetic Improvement/Key Laboratory of Crop Germplasm Utilization, Ministry of Agriculture, Institute of Crop Science, Chinese Academy of Agricultural Science, Beijing, China

Sujan Mamidi, Rian K. Lee and Phillip E. McClean
Genomics and Bioinformatics Program, North Dakota State University, Fargo, North Dakota, United States of America
Department of Plant Sciences, North Dakota State University, Fargo, North Dakota, United States of America

Jay R. Goos
Department of Soil Science, North Dakota State University, Fargo, North Dakota, United States of America

Hungyen Chen and Hirohisa Kishino
Graduate School of Agricultural and Life Sciences, The University of Tokyo, Tokyo, Japan

Junko Yamagishi
Institute for Sustainable Agro-ecosystem Services, The University of Tokyo, Tokyo, Japan

Konstantinos A. Aliferis and Suha Jabaji
Department of Plant Science, Macdonald Campus of McGill University, Sainte-Anne-de-Bellevue, Quebec, Canada

Denis Faubert
Institut de Recherches Cliniques de Montréal, Montréal, Quebec, Canada

Xiaolian Zhang, Ning Wang, Pei Chen, Mengmeng Gao, Juge Liu, Yufeng Wang, Tuanjie Zhao, Yan Li and Junyi Gai
National Key Laboratory of Crop Genetics and Germplasm Enhancement, National Center for Soybean Improvement, Key Laboratory for Biology and Genetic Improvement of Soybean (General, Ministry of Agriculture), Nanjing Agricultural University, Nanjing, Jiangsu, China

Gracia Zabala and Lila O. Vodkin
Department of Crop Sciences, University of Illinois, Urbana, Illinois, United States of America

Francisco Pérez-Montaño, Irene Jiménez-Guerrero, Pablo Del Cerro, Francisco, Javier López-Baena, Francisco Javier Ollero, Ramón Bellogín, Rosario Espuny
Departamento de Microbiología, Facultad de Biología, Universidad de Sevilla, Sevilla, Spain

Javier Lloret and Irene Baena-Ropero
Departamento de Biología, Facultad de Ciencias, Universidad Autónoma de Madrid, Madrid, Spain

Xinzhuo Yang, Pengjun Shi, Huoqing Huang, Yingguo Bai, Yaru Wang, Peilong Yang and Bin Yao
Key Laboratory for Feed Biotechnology of the Ministry of Agriculture, Feed Research Institute, Chinese Academy of Agricultural Sciences, Beijing, China

Rui Ma
Key Laboratory for Feed Biotechnology of the Ministry of Agriculture, Feed Research Institute, Chinese Academy of Agricultural Sciences, Beijing, China

Yunliu Fan
Biotechnology Research Institute, Chinese Academy of Agricultural Sciences, Beijing, China

Keke Hua, Daozhong Wang, Xisheng Guo and Zibin Guo
Soil and Fertilizer Research Institute, Anhui Academy of Agricultural Sciences, Hefei, China

Jie Qiu, Yu Wang, Sanling Wu, Ying-Ying Wang, Chu-Yu Ye, Xuefei Bai, Zefeng Li, Chenghai Yan, Weidi Wang, Ziqiang Wang and Longjiang Fan
Department of Agronomy & James D. Watson Institute of Genome Sciences, Zhejiang University, Hangzhou, China

Qingyao Shu
Institute of Nuclear Agricultural Science, Zhejiang University, Hangzhou, China

Jiahua Xie
Department of Pharmaceutical Sciences, North Carolina Central University, Durham, North Carolina, United States of America

Suk-Ha Lee
Department of Plant Science and Research Institute for Agriculture and Life Sciences, Seoul National University, Seoul, Korea

Index